W9-DEW-818

e Elements

			III	IV	V	VI	VII	O
								2 | 2 He 4.0026
			2 3 | 5 B 10.811	2 4 | 6 C 12.01115	2 5 | 7 N 14.0067	2 6 | 8 O 15.9994	2 7 | 9 F 18.9984	2 8 | 10 Ne 20.183
			2 8 3 | 13 Al 26.9815	2 8 4 | 14 Si 28.086	2 8 5 | 15 P 30.9738	2 8 6 | 16 S 32.064	2 8 7 | 17 Cl 35.453	2 8 8 | 18 Ar 39.948
8 li .71	2 8 18 1 | 29 Cu 63.546	2 8 18 2 | 30 Zn 65.38	2 8 18 3 | 31 Ga 69.72	2 8 18 4 | 32 Ge 72.59	2 8 18 5 | 33 As 74.9216	2 8 18 6 | 34 Se 78.96	2 8 18 7 | 35 Br 79.904	2 8 18 8 | 36 Kr 83.80
6 d 6.4	2 8 18 18 1 | 47 Ag 107.868	2 8 18 18 2 | 48 Cd 112.40	2 8 18 18 3 | 49 In 114.82	2 8 18 18 4 | 50 Sn 118.69	2 8 18 18 5 | 51 Sb 121.75	2 8 18 18 6 | 52 Te 127.60	2 8 18 18 7 | 53 I 126.9044	2 8 18 18 8 | 54 Xe 131.30
8 't .09	2 8 18 32 18 1 | 79 Au 196.967	2 8 18 32 18 2 | 80 Hg 200.59	2 8 18 32 18 3 | 81 Tl 204.37	2 8 18 32 18 4 | 82 Pb 207.19	2 8 18 32 18 5 | 83 Bi 208.980	2 8 18 32 18 6 | 84 Po (210)	2 8 18 32 18 7 | 85 At (210)	2 8 18 32 18 8 | 86 Rn (222)

3 u .96	2 8 18 25 9 2 | 64 Gd 157.25	2 8 18 27 8 2 | 65 Tb 158.924	2 8 18 28 8 2 | 66 Dy 162.50	2 8 18 29 8 2 | 67 Ho 164.930	2 8 18 30 8 2 | 68 Er 167.26	2 8 18 31 8 2 | 69 Tm 168.934	2 8 18 32 8 2 | 70 Yb 173.04	2 8 18 32 9 2 | 71 Lu 174.97

5 m 43)	2 8 18 32 25 9 2 | 96 Cm (247)	2 8 18 32 26 9 2 | 97 Bk (247)	2 8 18 32 27 9 2 | 98 Cf (249)	2 8 18 32 28 9 2 | 99 Es (254)	2 8 18 32 29 9 2 | 100 Fm (253)	2 8 18 32 30 9 2 | 101 Md (256)	2 8 18 32 31 9 2 | 102 No (254?)	2 8 18 32 32 9 2 | 103 Lw† (257)

Atomic weights are based on carbon-12;
values in parentheses are for the most stable or the most familiar isotope.
† Symbol is unofficial.

Organic and Biological Chemistry

Organic and Biological Chemistry

John R. Holum

PROFESSOR OF CHEMISTRY, AUGSBURG COLLEGE

JOHN WILEY & SONS

New York • Chichester • Brisbane • Toronto

547
HO

Copyright © 1978, by John Wiley & Sons, Inc.

All rights reserved. Published simultaneously in Canada.

Reproduction or translation of any part of this work beyond that permitted by Sections 107 or 108 of the 1976 United States Copyright Act without the permission of the copyright owner is unlawful. Requests for permission or further information should be addressed to the Permissions Department, John Wiley & Sons, Inc.

Library of Congress Cataloging in Publication Data:
Holum, John R
Organic and biological chemistry.

Consists of chapters 11-30 of the author's Fundamentals of general, organic, and biological chemistry.
Includes bibliographies and index.
1. Chemistry, Organic. 2. Biological chemistry.
I. Title.
QD251.2.H63 1978 547 78-634
ISBN 0-471-40872-7

Printed in the United States of America

10 9 8 7 6 5 4

Gift 8-26-83

Preface

This book is the last 20 chapters of another textbook, *Fundamentals of General, Organic and Biological Chemistry* (which we may call *Fundamentals*). *Fundamentals* is designed for students at the freshman or sophomore college level who are preparing for careers in the health sciences, home economics, or physical education, and who need one year of college chemistry geared to their needs. At many institutions, however, students of biology or allied health sciences begin their year of chemistry with one term in a section of a general chemistry course. Then, they move into a different course that provides a one-term study of selected topics of organic and biological chemistry. This book is intended for such a course.

Where organic and biological chemistry are taught in one term, a premium is placed both on the careful selection of topics and on the level at which they are presented. Since the applications of greatest relevance for those entering allied health careers are in biological chemistry, the organic chemistry is here limited to what is needed for biochemistry (or to the background required for the study of the later topics of organic chemistry). Functional groups that rarely occur among living systems—for instance, the nitro group, aliphatic halides, or triple bonds—either are omitted or are drastically played down. Reagents that cannot operate in living systems and have no counterparts are given little if any treatment, for example, metallic magnesium (Grignard reaction), aluminum chloride (Friedal-Crafts reaction), and sodium alkoxides (Williamson reaction).

The functional groups that must be included are:

alkenes	mercaptans	aldehydes	acetals and ketals
alcohols	disulfides	ketones	carboxylic acids and salts
phenols	amines and	hemiacetals and	esters
ethers	amine salts	hemiketals	anhydrides
			amides

These groups are all that can be reasonably expected to be of interest in the limited time given to the study of biological chemistry.

The reagents that must be studied are those that can occur in living systems or their close counterparts. Although sulfuric acid destroys protoplasm (if sufficiently concentrated), its inclusion here is defended on the grounds that it is simply an acid catalyst. Strong bases can similarly be defended. Even metal hydrides and molecular hydrogen may be studied

Copy 1

because they are hydride donors and have counterparts among biological systems. The most important inorganic reagents thus fall into these general categories:

1. Bases or acids that catalyze other changes.
2. Water, whether under acid or base or enzyme catalysis.
3. Reducing or hydrogenating agents.
4. Oxidizing or dehydrogenating agents.

The types of reactions that must be introduced among simple organic substances are those that do small operations on large molecules—those that take out water, hydrogen, or carbon dioxide, for example; or reactions that do large operations on small molecules—the aldol or Claisen types (and their reverses). In addition, there are other reactions wholly between organic reactants, which include those that result in ethers, acetals, esters, and amides.

Optical isomerism has its own chapter. The D and L families have been chosen to represent optical families among carbohydrates and amino acids because they are still commonly used in biochemistry. The R/S system is described in an appendix, however, and it can be included in the course or left for future reference as time permits.

Since plastics are of growing importance in medicine, a separate chapter on them is provided. Except for two or three basic terms—monomer, polymer, and polymerization—this chapter may be left out or assigned as reading, put into the "future reference" category, or studied in full. The polymers selected for inclusion are those with present or possibly future medical applications.

The chapters on biochemistry are organized in a traditional way. They begin with a chapter on carbohydrates that opens with an overview of the material discussed in the rest of the book. The chapter about lipids includes a study of biological membranes and active transport. In the chapter on enzymes, considerable attention is given to the several ways by which enzymes and the reactions they catalyze are controlled. The clonal selection theory is also included in this chapter as part of a brief introduction to immunochemistry.

Those who may work in respiratory care centers or emergency rooms will profit from the detailed study of the chemistry of respiration, an area of increasingly vital importance.

The chapters on metabolism provide enough material for a good introduction to this topic, particularly as it applies to human health and disease. Because good nutrition is important and because students may not have the opportunity to study it elsewhere in a way that utilizes a background in biochemistry, the final chapter of this book is a survey of this topic.

Each chapter closes with a brief summary that is a topic-by-topic review of the main units of the chapter. A number of in-chapter exercises are included, and their answers are provided at the end of the book.

This text is written with the assumption that the students in the course will have studied at least one term of freshman college chemistry. Since some students, however, will have completed such a course several months or even one or two years earlier, this book includes a special appendix that reviews the basic principles of general chemistry. For some students it will be sufficient; for others, the appendix will enable them to identify the topics they must review more deeply.

The Student Study Guide

A softcover *Study Guide* for students has been prepared that contains additional examples, sample examination questions, answers to end-of-chapter exercises (many with extra explanations), and complete glossaries of each chapter. These have also been collected and re-alphabetized to make a complete glossary for the text. This *Guide* is the same one used to accompany *Fundamentals* (and should be ordered as *Study Guide for Fundamentals of General, Organic and Biological Chemistry).* Those needing a review of general chemistry will find its first third helpful.

The Laboratory

The laboratory manual that accompanies this book is the same as the one for *Fundamentals.* It is entitled *Laboratory Manual for Fundamentals of General, Organic, and Biological Chemistry* and has enough experiments for a laboratory to the one-term organic and biological chemistry course. The organic experiments include several that involve molecular models, others on physical properties, and many that study chemical reactions. Not included, however, are experiments specifically designed to teach the special techniques of the organic chemist—for example, distillation, crystallization, and extraction. The experiments in biochemistry likewise emphasize tests of properties and substances as well as molecular models. Professor Ruth Denison was the chief architect of this manual, although several experiments also appear in the laboratory manual I wrote for *Elements of General and Biological Chemistry,* a briefer book for allied health students. Professor Denison not only clarified several of my experiments and made a number of significant modifications but also contributed many excellent experiments. Some of these experiments involve "unknowns," and others involve the identification and analysis of foods, vitamin C, and analgesic drugs (by thin layer chromatography). A teachers' manual accompanies the laboratory manual. It includes answers to questions posed in the latter as well as all the usual aids in managing the laboratory and its stockroom.

Acknowledgments

Dr. Oscar A. Anderson, president of Augsburg College, and Dr. Earl R. Alton, chairman of its chemistry department, have been unflagging in their support of my writing efforts. I am deeply grateful to them for their personal encouragement as well as for providing a climate in which this writing could take place. Dr. Margaret Etter, now of the 3M Company, taught my courses one year during the preparation of this book. She is an outstanding teacher, and several of her insights concerning course content and topic presentations were incorporated in the writing of the book.

The following individuals reviewed the entire manuscript or substantial parts of it. I was pleased to have their suggestions and adopted most of them as well as accepted all of their corrections.

Ruth C. Denison
Western Connecticut State College

Floyd W. Kelly
Casper College

Arne Langsjoen
Gustavus Adolphus College

D. A. McQuarrie
Indiana University

Nancy Sandelin Paisley
Montclair State College

Michael A. Peterson
North Seattle Community College

William Scovell
Bowling Green State University

Cornelius Steelink
University of Arizona

To my typist, Doris Berg, goes my appreciation for careful and diligent work. At John Wiley, John Balbalis, Kathy Bendo, and my editor, Gary Carlson, all contributed not only great skill and imagination but have been delightful people to know.

Finally to my wife, Mary, and our children, Liz, Ann, and Kathryn, a special thanks for lifetimes of support.

JOHN R. HOLUM

Augsburg College
Minneapolis, Minnesota

Contents

chapter 5

chapter 6

Organic and Biological Chemistry

chapter 1

Introduction to Organic Chemistry

Each model has 10 hydrogens (light balls) and four carbons (dark balls), but they are joined differently together. We see how that matters in this chapter. (Photo by K. T. Bendo.)

1.1 Wöhler's Experiment

The term **organic** arose from an early belief that compounds of carbon could come only from living organisms. This belief began to crumble in 1828. Friedrich Wöhler, a professor at the University of Göttingen (Germany), discovered that urea, an organic compound, could be made simply by heating a substance regarded as a mineral, ammonium cyanate:

$$\underset{\text{Ammonium cyanate}}{NH_4NCO} \xrightarrow{\text{heat}} \underset{\text{Urea}}{NH_2-\overset{\overset{\displaystyle O}{\|}}{C}-NH_2}$$

According to the theory accepted before 1828—the **vital force theory**—organic compounds could be prepared only with the catalytic help of a mysterious "vital force." People believed that this force resided only in living organisms or in chemicals made by living organisms and that inorganic matter, minerals, lacked this vital force. The theory was firmly believed for one simple, most compelling reason; all attempts to synthesize organic substances from inorganic matter had failed. Even with Wöhler's experiment on record it was many years before most of the scientific world was convinced. After that, however, organic chemistry surged forward, buttressed by accurate methods to determine molecular formulas and, after August Kekulé (1829–1869), useful theories of valence and structure. Literally thousands of organic compounds were synthesized from inorganic sources, many of them substances not hitherto observed to occur naturally. Today millions of organic compounds have been made in the continuing search for better drugs, fabrics, dyes, and plastics, and for ways to make vitamins and many other substances we need for health. Wöhler's experiment was simple but its implications were far reaching.

1.2 Organic Compared with Inorganic Compounds

Although the definitions of inorganic and organic compounds originated in a theory no longer used in science, these classes are retained because their members differ in several important ways: in elementary composition, in the chemical bonds present, and in some characteristic physical properties. Both classes obey all the laws of chemistry and physics.

1.3 Differences in Elementary Composition

All organic compounds contain carbon, and most inorganic compounds do not. Carbon is central because it forms the backbones and skeletons of organic molecules. In addition to carbon, all but a small handful of organic compounds also contain hydrogen.[1]

1.4 Differences in Bonds

Ionic bonds are more prevalent among inorganic compounds than organic. Thus most inorganic compounds consist of aggregations of oppositely charged ions. Organic compounds are largely collections of neutral, although usually polar, molecules. Within each molecule, covalent bonds hold the nuclei together. Between molecules, electrical forces of attraction between polar sites hold the molecules more or less tightly to each other.

1.5 Differences in Physical Properties

Because oppositely charged ions hold each other by unusually strong forces of attraction, ionic compounds have both high melting points and high boiling points. Most ionic inorganic compounds melt well above 350 °C. In contrast, most organic compounds usually melt below 350 °C because they consist of collections of molecules, not ions. Some are even gases at room temperature. Others consist of larger, more polar molecules and are liquids. Still others are solids.

The fact that most inorganic compounds are ionic, not molecular, accounts for their ability to conduct electricity, either dissolved in water or as molten material. With some exceptions organic compounds cannot provide ions, at least in sufficient concentrations, and therefore do not conduct electricity either as molten fluids or in solutions.

Relatively few organic compounds are soluble in water, whereas many inorganic compounds dissolve in water. Water can solvate ions or small polar molecules. Substances whose molecules are only slightly polar or are very large are not easily solvated, however, and they cannot dissolve in water. Most organic substances are in this category.

The differences between ionic and molecular compounds are summarized in Table 1.1.

1.6 The Uniqueness of Carbon

Alone among the elements, carbon consists of atoms that can bond to each other successively many times while at the same time forming equally strong bonds to atoms of other nonmetallic elements such as hydrogen, oxygen, sulfur, nitrogen, and the halogens. One carbon can bond covalently to another, and this in turn with another, and that with still another theoretically an unlimited number of times. In polyethylene plastic hundreds of carbon atoms are covalently joined in succession while also holding enough hydrogen atoms to fill out four bonds from each carbon. Figure 1.1 illustrates a **carbon**

[1] Recall that this is a manner of speaking. Organic substances contain *nuclei* of carbon and *nuclei* of hydrogen and *nuclei* of a few other elements plus electrons, too. Intact atoms are not present as if they were so many jelly beans in a jar.

TABLE 1.1
Contrasting Ionic and Molecular Substances

A Common Characteristic

In the pure state at room temperature only very rarely will substances of either category conduct electricity. (Metals are omitted from both families; carbon, a nonmetal, does conduct electricity, but it would be classified as covalent.)

	IONIC SUBSTANCES	MOLECULAR SUBSTANCES
Most Significant Differences	If soluble in water, the solution will conduct electricity	If soluble in water, the solution will not conduct electricity (with only a few exceptions)
	If insoluble in water but fusible, the molten mass will conduct electricity	If insoluble in water but fusible, the melt will not conduct electricity
Other Common Differences	Percent of members of this class that are somewhat soluble in water is relatively high	Small percent soluble in water
	Virtually no members dissolve in solvents such as carbon tetrachloride, gasoline, benzene, ether, alcohol, etc. (organic solvents)	Members are generally much more soluble in the organic solvents
	Rarely does the element carbon furnish nuclei for these compounds (exceptions: bicarbonates, carbonates, cyanides)	Very common to find carbon nuclei in these compounds (This class includes virtually all organic compounds)
	At room temperature, all are solids	Includes all the gases, all the liquids, and innumerable solids (room temperature being assumed)
	Virtually none will burn	Nearly all will burn
	Melting points are usually above 350 °C and are commonly much higher	Melting points are usually well below 350 °C, but some char and decompose before they melt
Examples	Salts (e.g., sodium chloride, barium sulfate)	Water, alcohols, sugars, fats and oils, lacquers, perfumes, most drugs and dyes
	Metal oxides (sodium oxide, iron oxides)	
	Carbonates and bicarbonates	

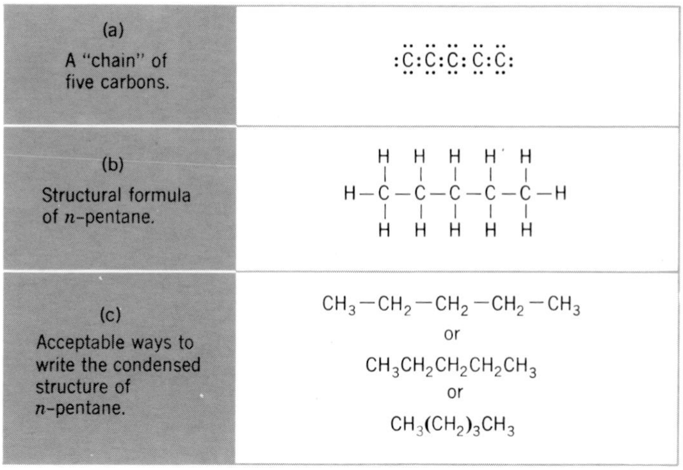

<table>
<tr><td>(a)
A "chain" of five carbons.</td><td>:C̈:C̈:C̈:C̈:C̈:</td></tr>
<tr><td>(b)
Structural formula of n-pentane.</td><td>H H H H H
| | | | |
H—C—C—C—C—C—H
| | | | |
H H H H H</td></tr>
<tr><td>(c)
Acceptable ways to write the condensed structure of n-pentane.</td><td>$CH_3-CH_2-CH_2-CH_2-CH_3$
or
$CH_3CH_2CH_2CH_2CH_3$
or
$CH_3(CH_2)_3CH_3$</td></tr>
</table>

Figure 1.1 Pentane, a substance whose molecules have a five carbon chain.

chain of five. Carbons on the interior of the chain have two bonds each to other carbons and two bonds each to hydrogens. (Recall that a straight line represents one shared pair of electrons—one covalent bond.)

1.7 Structural Formulas or Structures

The symbol shown in Figure 1.1*b* is called a **structural formula,** or simply, a **structure.** The molecular formula of that substance, *n*-pentane, is C_5H_{12}, but there will be almost no occasion when a molecular formula conveys enough information. To study and understand the properties of organic compounds, we have to know what atoms are joined to which and by what bonds, details displayed only by a structure.

1.8 Condensed Structures

A full structure, as in Figure 1.1*b*, is quite stylized. As we shall see in the next section, it falsely depicts the geometry of the molecule, both its bond angles and its three-dimensional character. Its virtue is that it is easy to set in type or write. Since the full structure is stylized anyway, we may simplify writing, typing, and taking notes even more by reducing or condensing full structures. Part *c* of Figure 1.1 illustrates three ways that a full structure may be condensed. The principal rules for condensing full structures are summarized in Figure 1.2, and illustrated with a typical branched chain compound, one with a carbon-containing group stuck somewhere on a chain other than on an end. In using these rules, we take advantage of the fact that carbon always has four bonds and hydrogen always has one. We always include vertical bonds as seen in the bottom of Figure 1.2, but we may leave horizontal bonds, if they are single bonds, understood or not as we please. Multiple bonds are always shown. Occasionally, two or three identical groups

```
   H
   |
H—C—          condenses to CH3—
   |
   H

   H
   |
 —C—          condenses to —CH2—
   |
   H

   H                          |
   |          condenses to —CH—
 —C—
   |

           H
           |
   H   H—C—H  H  H  H                                CH3
   |       |    |  |  |                               |
E.g. H—C———C———C—C—C—H     condenses to CH3—CH—CH2—CH2—CH3
   |       |    |  |  |
   H       H    H  H  H
                                                       CH3
                                                       |
                           Equally acceptable is:  CH3CHCH2CH2CH3
                                                       or
                                                  (CH3)2CHCH2CH2CH3
```

Figure 1.2 Rules for condensing structural formulas. The bottom box illustrates how these rules work with a typical branched-chain compound. If a group is written above the main chain, the vertical bond to it is always written.

attached to the same carbon at a chain's end may be put in parentheses, as illustrated in Figure 1.2. We shall seldom use that practice, however.

Exercise 1.1 Condense each full structure.

```
     H  H               H  H  H
     |  |               |  |  |
(a) H—C—C—H       (b) H—C—C—C—H
     |  |               |  |  |
     H  H               H  |  H
                          H—C—H
                            |
                            H

           H           H           H
           |           |           |
     H  H—C—H     H—C—H     H—C—H  H
     |     |           |           |     |
(c) H—C———C———————C———————C———C—H
     |     |           |           |     |
     H  H—C—H          H           H     H
           |
           H
```

Exercise 1.2 Write the full structure of each.

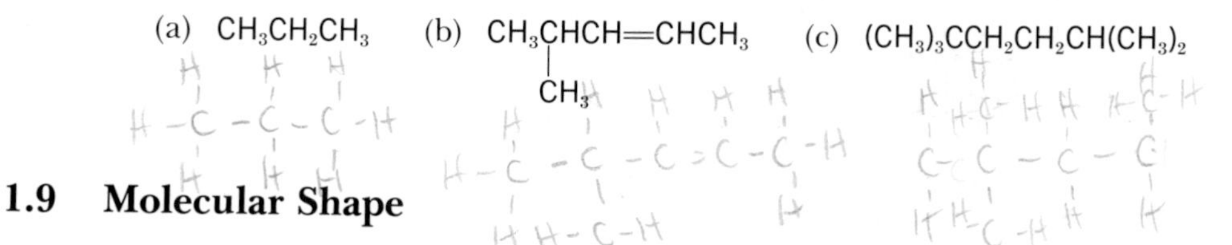

1.9 Molecular Shape

Life at the molecular level depends as much on the shapes of molecules as on their structures, particularly at the places where genes, enzymes, antibodies, and poisons work. Fortunately, some simple relations exist between structure and shape, and to a considerable extent the former determines the latter.

When four single bonds come from the same carbon as in methane, shown in Figure 1.3, they point to the corners of a regular tetrahedron with the carbon in the exact center. (A tetrahedron is a solid with four, identical triangular sides.) A carbon with four single bonds is therefore called a **tetrahedral carbon,** and the angle between any two covalent bonds at such a carbon is 109.5°.

1.10 Molecular Orbitals at a Tetrahedral Carbon

In atomic orbital notation the electronic configuration of carbon, atomic number 6, is $1s^2 2s^2 2p_x^{\,1} 2p_y^{\,1}$. In the outside level, the second level, the $2s$ orbital is filled, the $2p_z$ orbital is empty and one electron is in each of the $2p_x$ and $2p_y$ orbitals. If carbon used these atomic orbitals to overlap with $1s$ orbitals of hydrogen, methane's molecules could in no way exist in the form of regular tetrahedra. The bond angles could not all be identical, 109.5°, because bond angles are shaped primarily by angles taken by the orbitals of the central atom. Yet the fact remains that the bond angles in methane are 109.5°. Another fact, here introduced in our study for the first time, is that all four hydrogens in methane are chemically equivalent. One is no different in response to some attacking chemical than any other.

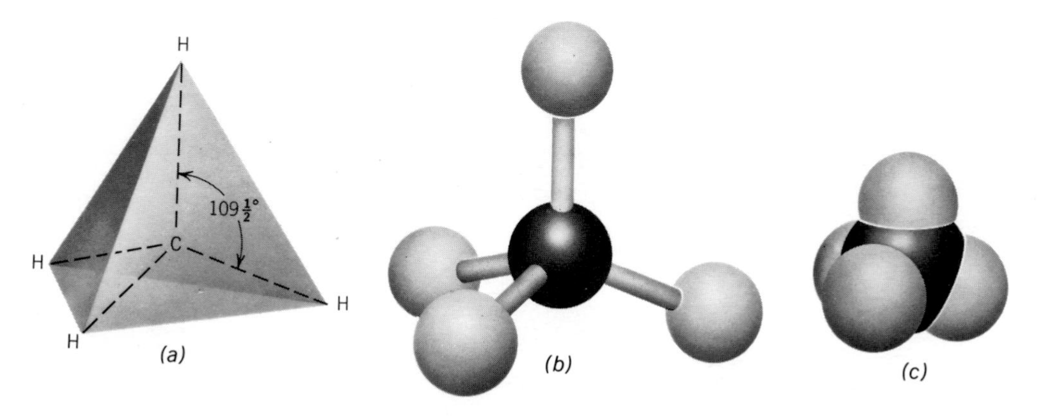

Figure 1.3 The tetrahedral carbon. (*a*) The dashed lines are the axes of the molecular orbitals. (*b*) A "ball-and-stick" model of the methane molecule. (*c*) A scale model of methane that shows the relative volumes of space occupied by the "electron clouds" in the molecule.

A straight application of the rules for writing electronic configurations from atomic numbers leads with carbon to a result that is apparently untrue, at least insofar as carbon exists within molecules instead of as atoms. The carbon in methane is tetrahedral. A tetrahedral carbon, as contrasted to atomic carbon, must have four equivalent atomic orbitals having axes that point to the corners of a regular tetrahedron if the shape, geometry, and the equivalence of the hydrogens of methane are to be explained. When a carbon atom is in a molecule such as methane, its four atomic orbitals at level 2—one *s*-orbital and three *p*-orbitals—reorganize their spaces and shapes into four new atomic orbitals. The reason for doing this, as calculations have supported, is that a more stable system of bonds can then form, and nature always takes up as stable an arrangement as possible if paths are open. The four new orbitals are called hybrid orbitals, and each is named an ***sp*³ orbital** after the "parent" orbitals, one *s*-orbital and three *p*-orbitals.

In the living world a hybrid is a cross, a genetic cross between different parents. In the world of orbitals, the sp^3 hybrid orbital is also a cross, one between its "parents," one part of *s*-orbital and three parts of *p*-orbital. A genetic cross resembles its parents, of course, and the sp^3 orbital has some character of the *s*-orbital (it includes the nucleus) and quite a lot of the character of the *p*-orbital (it has two lobes, but one is much larger than the other). The tetrahedral array is the only way these lobes can fit and be bunched around a common nucleus.

Figure 1.4 shows how we visualize the mixing of the orbitals—they undergo **hybridization**—and how we place the four electrons of carbon's level-2 into the four new sp^3 orbitals. Figure 1.5 shows a cross section of one sp^3 orbital and a perspective of the array of four such orbitals around a central point, the carbon nucleus. Figure 1.6 shows how these four sp^3 orbitals overlap with 1*s*-orbitals from four hydrogen atoms to form the four equivalent molecular orbitals in methane. Each molecular orbital in methane is a region of space lying around the axis of the bond and having the general symmetry of a cylinder. If you travel directly outward from some point on the bond axis a fixed distance, you encounter the same electron density in any direction. Any bond with this cylindrical symmetry is called a **sigma bond,** symbolized σ-bond.

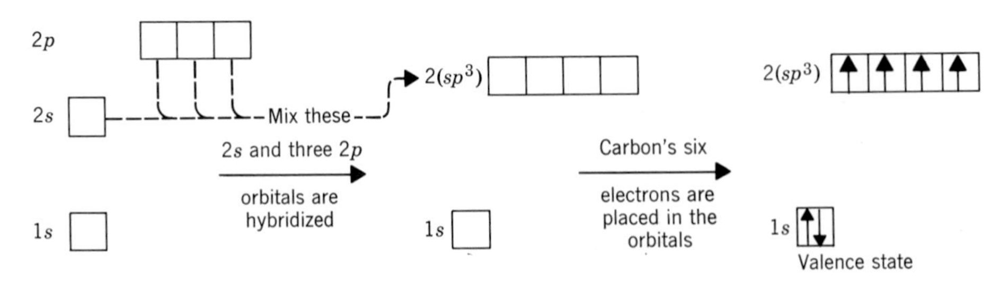

Figure 1.4 Making hybrid atomic orbitals. (*a*) The atomic orbitals of carbon as predicted by the rules for constructing electronic configurations. (*b*) The 1*s* orbital does not change, but the 2*s* and the three 2*p* orbitals mix to form four new hybrid atomic orbitals. (*c*) Carbon's six electrons are inserted into available orbitals. Now we have four equivalent one-electron orbitals that can form four equivalent bonds to hydrogen.

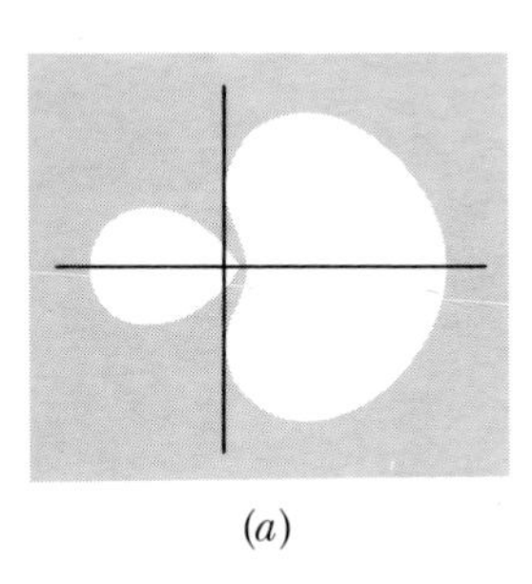

(*a*)

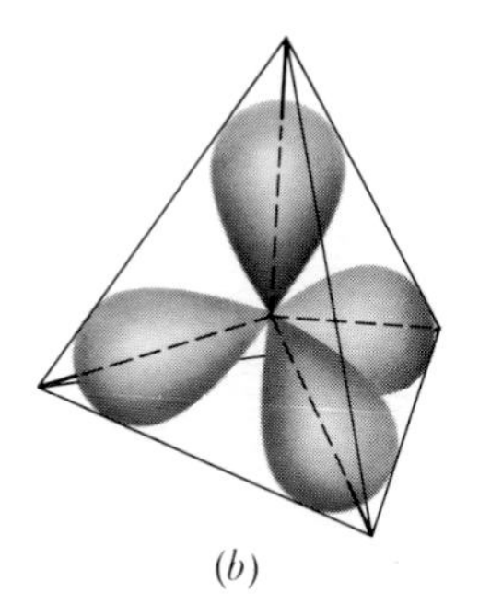

(*b*)

Figure 1.5 The sp^3 hybrid orbitals of carbon. (*a*) This cross-sectional view shows how the orbital includes the nucleus and has two lobes. (*b*) The larger lobes of four sp^3 hybrid orbitals of carbon take up positions that minimize contact between them. The result is that their axes point to the corners of a regular tetrahedron.

1.11 Free Rotation at Single Bonds

The molecular orbital model of a molecule of ethane, C_2H_6, is shown in Figure 1.7. Each carbon is tetrahedral, and the two are bonded by the shared electrons in the molecular orbital created by the overlap of two sp^3 orbitals. One of the principles in the theory of covalent bonds, the **principle of maximum overlap,** is that the more the overlap between the atomic orbitals approaches its maximum, the stronger the final bond. This principle implies that any action that reduces overlap weakens a bond and costs energy. Moreover, an action that does not weaken overlap is allowed because it costs no energy. One such action at a single bond is **free rotation.**

As seen in Figure 1.7, the two CH_3-groups joined by the single bond are free to rotate with respect to each other because that action cannot reduce the orbital overlap, at least not to any significant degree. The important implication of free rotation is that organic molecules have freedom to flex, twist, gyrate, and undulate. They are not frozen or welded by their bonds into some rigid state. We cannot imagine life without flexible molecules. Muscle contraction, to cite one example, involves the coiling and uncoiling of parts of ultra-long molecules of proteins.

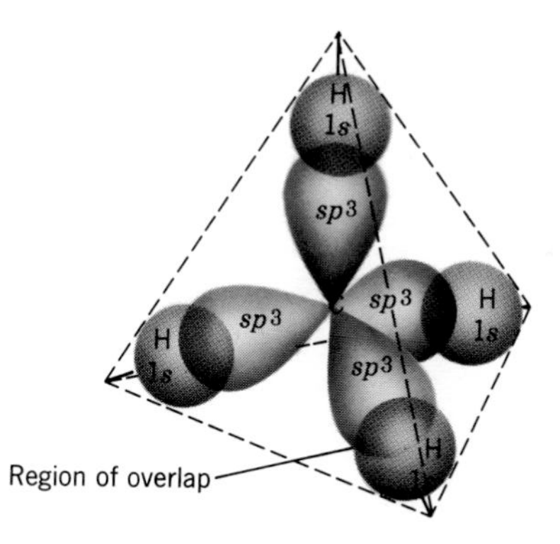

Figure 1.6 Bonds in methane. Each shared pair of electrons is in a molecular orbital formed by the overlap of an sp^3 orbital of carbon and the 1*s* orbital of hydrogen. Since overlap can occur best only where atomic orbitals are, the bonds point to the corners of a regular tetrahedron outlined by the dashed lines.

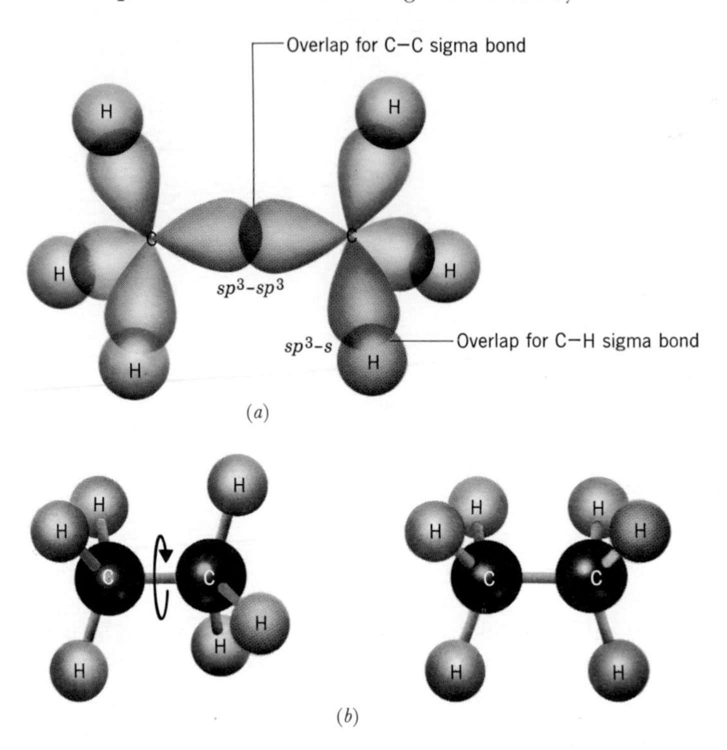

Figure 1.7 The ethane molecule. (*a*) Molecular orbitals in ethane. (*b*) Free rotation about the carbon-carbon single bond does not affect the sp^3-sp^3 overlap and therefore is allowed.

The pentane molecule, seen in Figure 1.8 without its carbon-hydrogen bonds, is a simple illustration of how a molecule can twist and flex. Only a few of the innumerable possible contortions are shown. A sample of pentane contains billions and billions of molecules, and some are in each of the conformations shown and in countless others not shown. Moreover, each molecule twists constantly as it suffers random collisions. When we measure physical and chemical properties of pentane, we get the average effects that all of the twisting, colliding molecules have on whatever instrument or chemical we use to measure some property. No one molecular conformation is unique. Hence we should not read into a structural formula too much about the conformation of the molecule.

1.12 Isomerism

Consider the molecular formula, C_2H_6O. In how many ways within the rules of valence can the nuclei of two carbons, six hydrogens, and one oxygen be arranged? Trial and error will show just two, given as ball-and-stick models in Figure 1.9. Substances exist consisting of molecules of each kind—ethyl alcohol and dimethyl ether. Some of the properties of these two compounds are summarized in Table 1.2.

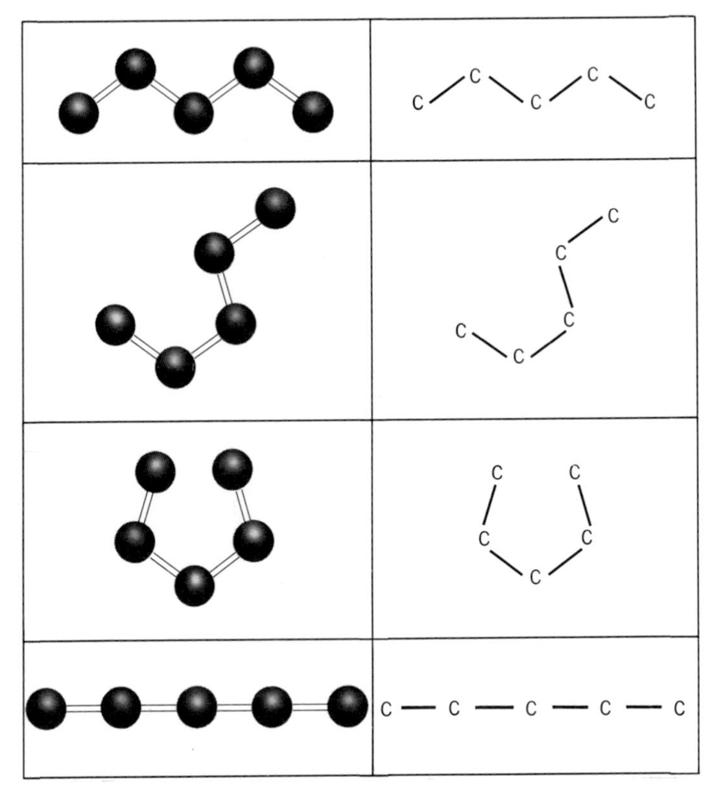

Figure 1.8 Free rotation about single bonds in the pentane molecule makes possible innumerable conformations, a few of which are shown here. (Only the carbon skeletons are given.)

Compounds with identical molecular formulas but different structures are called **isomers** of each other (Greek *isos,* equal; *meros,* parts). The phenomenon of the existence of isomers is called **isomerism.** The structural differences between isomers must be such that simple rotation about single bonds will not convert one into the other. Isomerism is one of the reasons for

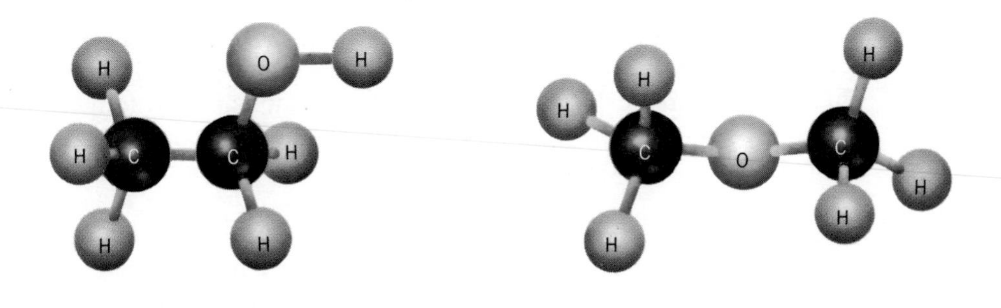

Figure 1.9 Isomers of C_2H_6O, ethyl alcohol and dimethyl ether. Ball and stick models of molecules of these isomers show how their molecules differ in atom-to-atom sequences. Although free rotation is possible about bonds in these molecules, it cannot change one into the other.

TABLE 1.2
Properties of Two Isomers. Ethyl Alcohol and Dimethyl Ether

Property	Ethyl Alcohol	Dimethyl Ether
Molecular formula	C_2H_6O	C_2H_6O
Boiling point	78.5 °C	−24 °C
Melting point	−117 °C	−138.5 °C
Density	0.789 g/ml	2 g/liter
Solubility in water	Completely soluble in all proportions	Slightly soluble
Action of metallic sodium	Vigorous reaction, hydrogen evolved	No reaction
Structural formula	H H | | H—C—C—O—H | | H H	H H | | H—C—O—C—H | | H H
Condensed structural formula	CH_3CH_2OH	CH_3OCH_3

the existence of so many organic compounds. We have studied only one type in this section, structural isomerism. The molecular formula is the only completely shared feature of isomers. Otherwise they are truly different compounds with different structures, different molecular shapes, and different physical and chemical properties.

1.13 How Organic Chemistry Is Organized for Study

The several million organic compounds can be classified into relatively few groups on the basis of similarities in structure and, therefore, similarities in chemical properties. Just as zoologists classify animals into families having structural likenesses, so chemists classify organic compounds according to structural similarities. A few such families of compounds are shown in Table 1.3.

In the next few chapters, we shall be doing something very similar to what you once did a long time ago when you learned how to read maps. You learned the meaning of a few simple map signs, and with that knowledge you became able to understand an enormous number and variety of complicated maps. Our study of simple organic compounds and their reactions will be like learning map signs. We shall then use them to understand some of the properties of the more complicated molecules found in living systems.

TABLE 1.3
Some Important Classes of Organic Compounds

Class	Characteristic Structural Features of Molecules	Example
Hydrocarbons	Contain only carbon and hydrogen; may have carbon chains or carbon rings. Subclasses according to presence of multiple bonds	
	Alkanes: all single bonds	CH_3CH_3, ethane
	Alkenes: at least one double bond	$CH_3CH{=}CH_2$, propylene
	Alkynes: at least one triple bond	$HC{\equiv}CH$, acetylene
	Aromatic: at least one benzene-like ring system	(benzene ring) benzene
Alcohols	At least one —OH joined to a tetrahedral carbon	CH_3CH_2OH, ethyl alcohol
Aldehydes	$-\overset{O}{\overset{\Vert}{C}}-H$ or aldehyde group	$CH_3\overset{O}{\overset{\Vert}{C}}H$, acetaldehyde
Ketones	$-\underset{\vert}{\overset{\vert}{C}}-\overset{O}{\overset{\Vert}{C}}-\underset{\vert}{\overset{\vert}{C}}-$ or keto group	$CH_3\overset{O}{\overset{\Vert}{C}}CH_3$, acetone
Carboxylic acids	$-\overset{O}{\overset{\Vert}{C}}-OH$ or carboxyl group	$CH_3\overset{O}{\overset{\Vert}{C}}OH$, acetic acid
Amines	$-NH_2$, or $-NH-$ or $-\overset{\vert}{N}-$	CH_3NH_2, methylamine CH_3NHCH_3, dimethylamine

Brief Summary

Organic and Inorganic Compounds. Most organic compounds are molecular, and the majority of inorganic compounds are ionic. Molecular and ionic compounds differ in electrical conductivity, in composition, in types of bonds, and in several physical properties. The most important difference is in the forces of attraction between particles—between oppositely charged ions or between neutral molecules whose polarities may vary from zero to the equivalents of opposite, full charges.

Structural Organic Chemistry. The ability of carbon atoms to join to each other many times in succession and in chains and branches while strongly holding atoms of other nonmetals makes possible the existence of several million carbon compounds. In forming single bonds, carbon does not use the orbitals we write for its atomic state, but rather hybrid, sp^3 orbitals whose axes point to the corners of a regular tetrahedron. The resulting single bonds permit groups to rotate with respect to each other. Most organic compounds are members of a set of compounds called isomers that share the same molecular formula but are put together with different atom-to-atom sequences and geometries.

Selected Reference

Article

1. J. B. Lambert. "The Shapes of Organic Molecules." *Scientific American*, January 1970, page 58. The 1969 Nobel Prize in chemistry was shared by Odd Hassel (Norway) and D. H. R. Barton (England) for work on the concept of conformation in chemistry. This article discusses that work.

Questions and Exercises

1. Using the simple covalences of nonmetals—carbon, 4; hydrogen, 1; nitrogen, 3; oxygen or sulfur, 2; and any of the halogens, 1—by trial and error write full structures of each molecular formula. In some you will have to use double bonds and in a few, triple bonds.

   (a) CH_5N (b) CH_2Br_2 (c) $CHCl_3$ (d) C_2H_6
   (e) CH_2O_2 (f) CH_2O (g) NH_3O (h) C_2H_2
   (i) N_2H_4 (j) HCN (k) C_2H_3N (l) CH_4O

2. Examine each structure carefully, concentrating on atom-to-atom sequences.

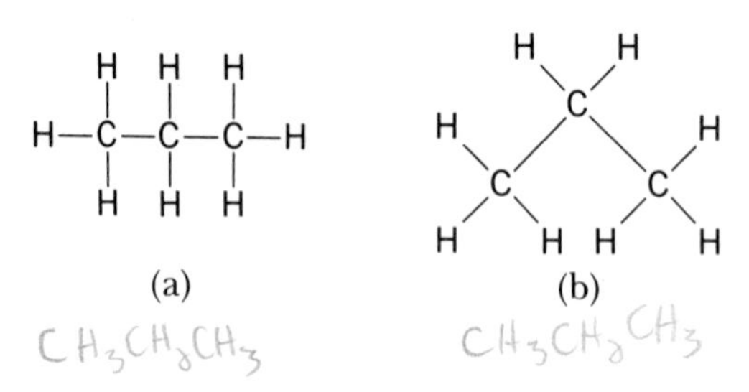

(a) (b)

   (1) What is the molecular formula of each?
   (2) How many groups does each have consisting of one carbon holding three hydrogens plus one two-carbon–five hydrogen groups?
   (3) How many groups does each have consisting of one carbon holding two hydrogens plus two one-carbon–three-hydrogen groups?
   (4) If there is any difference between the two in fundamental atom-atom sequences, describe what that difference is.

(5) Are the two structures of identical compounds, or are they isomers, or do they differ in some other way?

(6) If you have access to a ball-and-stick molecular model kit, make models of each and compare them. In any case, write a condensed structure of each arranging all carbons as much as possible on one line.

3. Examine the two following condensed structures.

$H—O—CH_2CH_2CH_3$ (a) $CH_3CH_2CH_2—O—H$ (b)

(1) What is the molecular formula of each?

(2) Are the two isomers or identical? If they are isomers describe one structural difference. One the O is connected to the C, the other the H

4. Decide whether members of each pair are identical, are isomers, or are unrelated.

(a) CH_3 | CH_3 and $CH_3—CH_3$ Identical

(b)  $CH_3 \diagdown CH_2 \diagdown CH_3$ and $H_3C \diagup CH_2 \diagdown CH_3$ Ident

(c) $CH_3CH_2—OH$ and $CH_3CH_2CH_2—OH$ unrelated

(d) 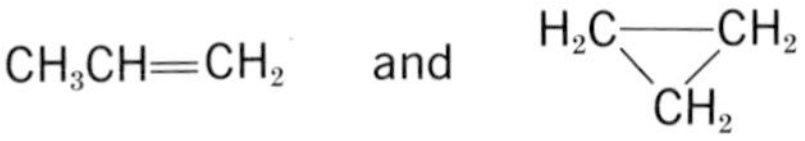 $CH_3CH{=}CH_2$ and $H_2C—CH_2$ (ring, with CH_2 below) unrelated

(e) $H—\overset{O}{\overset{\|}{C}}—CH_3$ and $CH_3—\overset{O}{\overset{\|}{C}}—H$ Isomers

(f) $CH_3\underset{}{\overset{CH_3}{\overset{|}{C}}}HCH_3$ and $CH_3\overset{CH_3}{\overset{|}{C}}H$ with CH_3 below Isomers

(g) $CH_3CH_2—NH_2$ and $CH_3—\underset{H}{\underset{|}{N}}—CH_3$ unrelated

(h) $CH_3CH_2\overset{O}{\overset{\|}{C}}—O—H$ and $H—O—\overset{O}{\overset{\|}{C}}CH_2CH_3$ isomers

(i) $H—\overset{O}{\overset{\|}{C}}—O—CH_2CH_3$ and $CH_3CH_2—\overset{O}{\overset{\|}{C}}—O—H$ isomers

(j) $H—\overset{O}{\overset{\|}{C}}—O—CH_2CH_2OH$ and $HOCH_2CH_2—\overset{O}{\overset{\|}{C}}—OH$ isomers

(k) $CH_3\overset{O}{\overset{\|}{C}}CH_2CH_3$ and $CH_3CH_2\overset{O}{\overset{\|}{C}}CH_3$ isomers

(l)
```
CH3—CH—CH3
    |
    CH2—CH2  CH3   CH3
         |    |   /
         CH2—C—CH
              |   \
              CH3  CH3
```
and
```
     CH3                       CH3 CH3
     |                          |   |
CH3—CH—CH2—CH2—CH2—C—CH—CH3
                                |
                               CH3
```

(m) $CH_3—NH—\overset{\overset{\large O}{||}}{C}—CH_3$ and $CH_3CH_2\overset{\overset{\large O}{||}}{C}NH_2$

(n) H—O—O—H and H—O—H

5. Each compound described below is either ionic or molecular. State which it most likely is and give one reason.
   (a) The compound is a colorless gas at room temperature.
   (b) This compound dissolves in water. When hydrochloric acid is added, the solution fizzes and an odorless, colorless gas is released. The gas can extinguish a burning match.
   (c) This compound melts at 345 °C and burns in air.
   (d) This compound melts at 675 °C and becomes white when heated.
   (e) This compound is a liquid that will not dissolve in water.

6. The compound C_5H_{12} exists in a few isomeric forms. How many are there and what are their condensed structural formulas?

7. Suppose that the oxygen atom (atomic number 8) in a molecule of water uses hybrid atomic orbitals of the sp^3 type. Fill in the boxes below to show how the electronic configurations of atomic oxygen in (a) the conventional form, and (b) the hybridized form would compare. (The analogy to the situation of the carbon atom in methane is exact.) Disregarding the possibility that initially formed bond angles might be distorted in the final molecule, what bond angle must one predict for the water molecule if configuration (a) is used? If configuration (b) is used?

$$1s^2 2s\square 2p_x\square 2p_y\square 2p_z\square \qquad 1s^2 2(sp^3)\square 2(sp^3)\square 2(sp^3)\square 2(sp^3)\square$$

(a) (b)

The actual bond angle in water is 104.5°. Which model, (a) or (b) is in closer agreement with this fact? Another fact about water is that the unshared pairs of electrons—there are two such pairs—are chemically equivalent. Which model, (a) or (b), is in closer agreement with that fact?

8. The nitrogen atom in the ammonia molecule, $:NH_3$, is believed to make available sp^3 hybrid orbitals for overlapping with hydrogen atoms. What bond angle is predicted for the H—N—H network in ammonia? (The actual bond angle in ammonia is 107°.)

chapter 2

Saturated Hydrocarbons. Alkanes and Cycloalkanes

Molecules with carbon-hydrogen bonds store solar energy from eons ago in pools of underground petroleum. Scarcely any geological feature has for long thwarted our quest for this fossil fuel. Seen here is an above-ground section of the trans Alaska pipeline as it zigzags south from the North Slope toward the Brooks Mountains, top. (Photo courtesy Alyeska Pipeline Service Company.)

2.1 Families of Hydrocarbons

Hydrocarbons are substances whose molecules contain only carbon and hydrogen.[1] These molecules are held together by covalent bonds, not ionic bonds, and in some hydrocarbons carbon-carbon double or triple bonds exist. The families of hydrocarbons, defined in Figure 2.1, are characterized by the presence or absence of such bonds. Nature's richest sources of these substances are petroleum, coal, and natural gas, the three fossil fuels.

2.2 Saturated and Unsaturated Organic Compounds

We use the word "saturated" with both substances and solutions, and it carries a similar meaning in both. A "saturated solution" is one that is "full"; it cannot hold any more solute in the dissolved state. A **saturated compound** is one whose molecules are "full"; their atoms are holding as many other atoms or groups as the rules of valence permit. The **alkanes,** for example, are saturated hydrocarbons. Each carbon in an alkane holds its maximum of four atoms or groups. The alkenes are **unsaturated compounds.** While each carbon at a double bond does display its normal covalence of four, it holds in fact only three groups, one of them by a double bond. Alkenes, alkynes, and aromatic hydrocarbons are all unsaturated hydrocarbons. In fact, nearly all organic substances found in a living system are unsaturated compounds. Carbon-oxygen double bonds are particularly common.

Just as we may add more solute to an unsaturated solution—a physical change—we may add more atoms to an unsaturated molecule—a chemical change—to make all atoms saturated. That is why the unsaturated compounds have a richer chemistry than the saturated.

The Alkanes

2.3 Structural Features

The kinds of bonds in molecules of ethane, studied in the previous chapter, are in molecules of all alkanes. A carbon may be joined to as many as four other carbons. A compound with molecules whose carbon atoms follow one another in the way links in a chain are joined is called a **straight-chain compound.** If somewhere in the chain something is attached as an appendage, the system is called **branch-chained.**

Neither with real chains nor with molecules does "straight chain" imply a fixed, rigid geometry. Flexing, coiling, and kinking of molecular chains is allowed; relatively free rotation exists.

The structures and important physical properties of the 10 smallest straight-chain alkanes are given in Table 2.1.

[1] See footnote 1, Chapter 1.

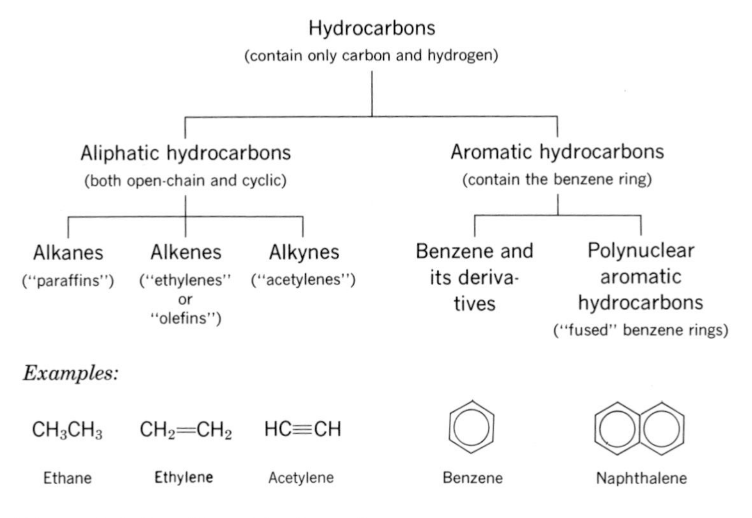

Figure 2.1 Families of hydrocarbons

The series in Table 2.1 is said to be **homologous series** because members differ from each other in a consistent, regular way. Here each member differs from the one just before or after it by one CH_2 unit. In chemical terminology butane, for example, is "the next higher **homolog** of propane."

TABLE 2.1
Straight-Chain Alkanes

Name	Number of Carbon Atoms	Molecular Formula[a]	Condensed Structural Formula	Bp (°C at atmospheric pressure)	Mp (°C)	Density (in g/ml at 20 °C)
Methane	1	CH_4	CH_4	−161.5		
Ethane	2	C_2H_6	CH_3CH_3	−88.6		
Propane	3	C_3H_8	$CH_3CH_2CH_3$	−42.1		
Butane	4	C_4H_{10}	$CH_3CH_2CH_2CH_3$	−0.5	−138.4	
Pentane	5	C_5H_{12}	$CH_3CH_2CH_2CH_2CH_3$	36.1	−129.7	0.626
Hexane	6	C_6H_{14}	$CH_3CH_2CH_2CH_2CH_2CH_3$	68.7	−95.3	0.659
Heptane	7	C_7H_{16}	$CH_3CH_2CH_2CH_2CH_2CH_2CH_3$	98.4	−90.6	0.684
Octane	8	C_8H_{18}	$CH_3CH_2CH_2CH_2CH_2CH_2CH_2CH_3$	125.7	−56.8	0.703
Nonane	9	C_9H_{20}	$CH_3CH_2CH_2CH_2CH_2CH_2CH_2CH_2CH_3$	150.8	−53.5	0.718
Decane	10	$C_{10}H_{22}$	$CH_3CH_2CH_2CH_2CH_2CH_2CH_2CH_2CH_2CH_3$	174.1	−29.7	0.730

[a] The molecular formulas of the open-chain alkanes fit the general formula C_nH_{2n+2}, where n is the number of carbons in the molecule.

2.4 Isomerism among the Alkanes

Among the alkanes, from butane and on to higher homologs, isomerism is possible. Several examples are listed in Table 2.2. There are two isomers of formula C_4H_{10}, three of formula C_5H_{12}, and five of formula C_6H_{14}. As the homologous series is ascended, the number of possible isomers approaches astronomical figures. There are 75 possible isomers of decane, $C_{10}H_{22}$, and an estimated 6.25×10^{13} possible isomers of $C_{40}H_{82}$. Not all possible isomers have actually been prepared in the pure state and studied, for no useful purpose would be served and time would not permit.

TABLE 2.2
Properties of Isomeric Alkanes

Family	Common Name (except where noted)	Structure	Bp (°C at atmospheric pressure)	Mp (°C)	Density (in g/ml)
Butane isomers	*n*-Butane	$CH_3CH_2CH_2CH_3$	−0.5	−138.4	0.622 (−20 °C)
	Isobutane	CH_3CHCH_3 | CH_3	−11.7	−159.6	0.604 (−20 °C)
Pentane isomers	*n*-Pentane	$CH_3CH_2CH_2CH_2CH_3$	36.1	−129.7	0.626 (20 °C)
	Isopentane	$CH_3CHCH_2CH_3$ | CH_3	27.9	−159.9	0.620
	Neopentane	CH_3 | CH_3CCH_3 | CH_3	9.5	−16.6	0.591
Hexane isomers	*n*-Hexane	$CH_3CH_2CH_2CH_2CH_2CH_3$	68.7	−95.3	0.659
	3-Methylpentane (no common name)	CH_3 | $CH_3CH_2CHCH_2CH_3$	63.3		0.664
	Isohexane	CH_3 | $CH_3CHCH_2CH_2CH_3$	60.3	−153.7	0.653
	2,3-Dimethylbutane (no common name)	CH_3 | $CH_3CHCHCH_3$ | CH_3	58.0	−128.5	0.662
	Neohexane	CH_3 | $CH_3CCH_2CH_3$ | CH_3	49.7	−99.9	0.649

Octane isomers, C_8H_{18}—total of 18
Decane isomers, $C_{10}H_{22}$—total of 75
Eicosane isomers, $C_{20}H_{42}$—total of 366,319
Tetracontane isomers, $C_{40}H_{82}$—estimated total of 6.25×10^{13} isomers

2.5 Nomenclature of Alkanes and Alkyl Groups

The earliest known organic compounds were named after their source. For example, formic acid (Latin *formica,* ants) can be made by grinding ants with water and distilling the result. Hundreds of compounds were named after their sources, but the system becomes impossibly difficult to extend to all compounds. These common names, however, are still used, and the beginning student is faced with the necessity of learning some of them. For more complicated structures there are rules for formal or systematic nomenclature.

2.6 Common Names of Alkanes

The straight-chain isomers are designated the **normal alkanes,** and their common names include *n*- for normal. Examples are *n*-butane, *n*-octane, but not *n*-ethane because ethane has no isomer. In Table 2.1 butane and all subsequent names would be common if *n*- were placed before each. As they stand, they are formal names, to be described later. Except where noted, the names in Table 2.2 are also common names. The names of all the normal alkanes through *n*-decane, as well as the names of the isomers through the four-carbon series, must be learned. The total carbon content of the alkane should be associated with the prefix portion of its name. Thus "eth-" is a word part for a two-carbon unit; "but-" signifies four carbons whether in *n*-butane or isobutane. The "-ane" ending is characteristic of all alkanes.

2.7 Formal Names. Geneva System. IUPAC Rules for Alkanes

Representatives from chemical societies all over the world, meeting as the International Union of Pure and Applied Chemistry (IUPAC), with sessions held usually in Geneva, have recommended adoption of the following rules for naming alkanes.

1. The general name for saturated hydrocarbons is *alkane.*

2. The names of the straight-chain members of the alkanes are those listed in Table 2.1. (The designation *n*- is not included. Names going beyond the 10-carbon alkanes are, of course, available, but we shall not need them.)

3. For branched-chain alkanes, base the name on the alkane that corresponds to the longest continuous chain of carbons in the molecule. For example, in the compound

```
              CH2—CH2
              |    |
CH2—CH—CH2   CH3
|    |
CH3  CH3
```

that when straightened out is

```
        CH3
        |
CH3CH2CHCH2CH2CH2CH3
```

3-Methylheptane

the longest continuous chain totals seven carbons. The last part of the complete name for this compound will therefore be heptane.

4. To locate branches, assign a number to each carbon of the longest continuous chain. Begin at whichever end of the chain will result in the smaller number for the carbon holding the first branch. In our example,

$$\begin{array}{c} \quad\quad\; CH_3 \\ \quad\quad\; | \\ CH_3CH_2CHCH_2CH_2CH_2CH_3 \\ 1\;\;\;2\;\;\;3\;\;4\;\;\;5\;\;\;6\;\;\;7 \end{array}$$

if this chain had been numbered from right to left, the carbon holding the branch would have the number 5.

5. If a side chain or branch consists only of carbon and hydrogen linked with single bonds, it is called an **alkyl group;** "alkyl" comes from changing the "-ane" ending of "alkane" to "-yl." This change is the key to making up names for alkyl groups, and Table 2.3 lists the names and structures for the most common ones. Their structures must be learned. The table shows four butyl groups, two related to *n*-butane and two to isobutane. All are butyl groups because each has four carbons. To distinguish them from each other, additional word parts are appended. The words "secondary" and "tertiary" (abbreviated *sec*- and *tert*- or simply *t*-) denote the condition of the carbon in the group having the unused bond. In the *sec*-butyl group this carbon is directly attached to two other carbons and is therefore classified as a **secondary carbon.** In the *t*-butyl group this carbon has direct bonds to three other carbons and is classified as a **tertiary carbon.** When a carbon is attached directly to but one other carbon, it is classified as a **primary carbon.** These classifications of carbons should be learned.

The prefix "iso-" has a special meaning. It can be used to name any group with the following feature:

$$\begin{array}{l} H_3C \\ \quad\quad \diagdown \\ \quad\quad\;\; CH{-}(CH_2)_n{-} \qquad (n = 0, 1, 2, 3, \text{etc.}) \\ \quad\quad \diagup \\ H_3C \end{array}$$

n = 0 Isopropyl n = 2 Isopentyl
= 1 Isobutyl = 3 Isohexyl

Note that the total carbon content of each group is also disclosed in the name. The specific groups given here are official names in the IUPAC system.

6. Whenever two identical groups are attached at the same place, numbers are supplied for each group. If there are two identical groups, we add the prefix "di-"; if three, "tri-"; etc. For example,

$$\begin{array}{l} \quad\;\; CH_3 \\ \quad\;\; | \\ CH_3CCH_2CH_2CH_2CH_3 \\ \quad\;\; | \\ \quad\;\; CH_3 \end{array}$$

Correct name: 2,2-Dimethylhexane
Incorrect: 2-Dimethylhexane
Incorrect: 2,2-Methylhexane

7. Whenever two or more different groups are affixed to a chain, two

TABLE 2.3
Alkyl Groups—Names and Structures

Parent Alkane	Structure of Parent Alkane	Structure of Alkyl Group from Alkane	Name of Alkyl Group
Methane	CH_4	$CH_3—$	Methyl
Ethane	CH_3CH_3	$CH_3CH_2—$	Ethyl
Propane	$CH_3CH_2CH_3$	$CH_3CH_2CH_2—$	*n*-Propyl[a]
		H_3C ╲ $CH—$ ╱ H_3C	Isopropyl
n-Butane	$CH_3CH_2CH_2CH_3$	$CH_3CH_2CH_2CH_2—$	*n*-Butyl[b]
		CH_3 │ $CH_3CH_2CH—$	Secondary butyl (sec-butyl)
Isobutane	CH_3 │ CH_3CHCH_3	CH_3 │ $CH_3CHCH_2—$	Isobutyl
		CH_3 │ $CH_3C—$ │ CH_3	Tertiary butyl (*t*-butyl)
Any normal alkane:	If the "free valence" extends from the *end* of the unbranched chain, change the "-ane" ending of the alkane to "-yl"; e.g., $CH_3CH_2CH_2CH_2CH_2CH_2CH_2—$ is *n*-heptyl		
Any alkane in general:	R—H	R—	Alkyl

[a] The IUPAC name is simply "propyl."
[b] The IUPAC name is "butyl."

ways are acceptable for organizing all the name parts into the final name. The last word part is always the name of the alkane corresponding to the longest chain.

a. The word parts can be ordered by increasing complexity of side chains (e.g., in order of increasing carbon content): methyl, ethyl, propyl, isopropyl, butyl, isobutyl, *sec*-butyl, *t*-butyl, etc.
b. They can be listed in simple alphabetical order. (This corresponds to the indexing system of *Chemical Abstracts,* a publication of the American Chemical Society.)

In this text we shall not be particular about the matter of order. For the following compound both names given are acceptable.

```
    CH3
    |
CH3CHCH2CHCH2CH2CH3
         |
         CHCH3
         |
         CH3
```

2-Methyl-4-isopropylheptane
or 4-Isopropyl-2-methylheptane

Note carefully the use of hyphens and commas in organizing the parts of names. Hyphens always separate numbers from word parts, and commas always separate numbers. The intent is to make the final name one word.

8. The formal names for several of the more important nonalkyl substituents are as follows:

-F	Fluoro	$-NO_2$	Nitro
-Cl	Chloro	$-NH_2$	Amino[2]
-Br	Bromo	-OH	Hydroxy[2]

Several examples of compounds correctly named according to these rules follow. Some common ways in which incorrect names are often devised are also shown. As an exercise, describe how each incorrect name violates one or more of the rules.

CH_3
|
$CH_3—C—CH_3$
|
CH_2
|
CH_3

2,2-Dimethylbutane
Not 2-Methyl-2-ethylpropane

$H_3C \quad CH_3$
$\diagdown \; \diagup$
$CH_3—CH_2 \quad CH$
| |
$CH_2—CH—CH_3$

2,3-Dimethylhexane
Not 2-Isopropylpentane

$CH_3 \quad CH_3$
| |
$CH_3—C——CH—CH_2—CH_3$
|
CH_3

2,2,3-Trimethylpentane
Not 2,3-Trimethylpentane
Not 2-*t*-Butylbutane

$CH_3—CH_2—CH—Cl$
|
Cl

1,1-Dichloropropane
Not 3,3-Dichloropropane
Not 3,3-Chloropropane
Not 1-Dichloropropane
Not 1,1-Chloropropane

CH_3
|
$CH_3—CH$
|
CH_3

2-Methylpropane
Not 1,1-Dimethylethane
Not Isobutane, which is its common name

CH_3
|
$CH_3CH_2CH_2CHCH_2CHCH_3$
|
$CH_3—C—CH_3$
|
CH_3

2-Methyl-4-*t*-butylheptane
Not 4-*t*-Butyl-6-methylheptane
But 4-*t*-Butyl-2-methylheptane is acceptable

[2] These two are used only in special circumstances. As we shall see in later chapters, amino compounds are usually named as amines and hydroxy compounds as alcohols, with special IUPAC rules.

Exercise 2.1 Write the structures (condensed) of each of the following compounds:

(a) 1-Bromo-2-nitropentane
(b) 2,2,3,3,4,4-Hexamethyl-5-isopropyloctane
(c) 2,2-Diiodo-3-methyl-4-isopropyl-5-*sec*-butyl-6-*t*-butylnonane
(d) 1-Chloro-1-bromo-2-methylpropane
(e) 5,5-Di-*sec*-butyldecane

Exercise 2.2 Write IUPAC names for each of the following compounds:

```
(a) CH3—CH2
          \
           CH—CH3
          /
    CH2—CH2
    |
    CH3

                     CH3
                     |
(b)       H3C CH3—C—CH3
             \       |
              CH—CH—CH2—CH2—CH3
             /
       CH3—CH
           |
           CH3

           CH3   CH3        CH3
           |     |          |
(c) CH3—CH2—CH—CH—CH—CH2—CH—CH3
                  |
                  CH2—CH2—CH2—CH3

                        I
                        |
(d) Cl—CH2—CH—CH2—Br

                    CH3—CH—CH3
                        |
(e) CH3—CH2—CH2—CH—CH—CH2—CH2—CH3
                |
           CH3—C—CH3
                |
                CH3

                     CH3
                     |
(f) NO2—CH2—C—CH2—NO2
                     |
                     CH3

                  CH2—Cl
                  |
(g) CH3—CH2—CH—CH—CH3
            |
            CH3

         CH3                CH2—CH3
         |                  |
(h) CH3—CH—CH2—CH2—CH—CH—CH3
                            |
                            CH3

(i) H3C
       \
        CH—CH2—CH2
       /         |
    H3C          CH3

(j) CH3—CH2—CH2        CH2—CH3
            |          |
            CH2—CH—CH—CH2CH2CH3
                |
                CH3
```

Exercise 2.3 Examine the structure of part (c) of Exercise 2.2. Underline each primary carbon; draw an arrow pointing toward each secondary carbon and circle each tertiary carbon.

2.8 Physical Properties of Hydrocarbons—Saturated and Unsaturated

We expect the molecules of a substance to be polar whenever they contain polar bonds whose polarities do not cancel. We expect a covalent bond to be polar whenever it joins atoms of different electronegativities. Hydrogen and carbon differ only slightly in relative electronegativities. Carbon-hydrogen bonds, therefore, are almost nonpolar. Carbon-carbon bonds, of course are nonpolar, also. The result is that molecules of any of the families of hydrocarbons—saturated or otherwise—are only weakly polar at best. That is how we understand certain trends in some of the physical properties of hydrocarbons, especially in boiling points and solubilities.

Molecules of methane (formula weight 16) and water (formula weight 18) have nearly the same masses. Yet methane, being nonpolar, boils at −161.5 °C. The higher the alkane is in the homologous series of normal alkanes (Table 2.1), the higher its boiling point, melting point, and density. All hydrocarbons are insoluble in water, a very polar solvent, and soluble in the relatively nonpolar solvents that include some common hydrocarbons or mixtures of hydrocarbons (benzene, gasoline, ligroin, and petroleum ether) and halogen derivatives (carbon tetrachloride, chloroform, and dichloromethane). The rule of thumb is that "likes dissolve likes," where "likes" refers to polarities. Because molecules of a polar solvent "stick" to themselves or other polar molecules, they cannot let in nonpolar molecules and dissolve them.

What we are seeking here are general correlations between structure and property. Just as we have now learned to associate hydrocarbons with low water solubility, we shall in the future be able to predict that any molecule of any family substantially hydrocarbonlike will also have this property.

2.9 Chemical Properties

Alkane chemistry is quite simple. Very few chemicals react with alkanes. That is why they are called *paraffins* or paraffinic hydrocarbons, after the Latin *parum affinus,* meaning "little affinity." To illustrate, at room temperature alkanes are not attacked by water, by strong, concentrated acids (e.g., sulfuric acid, hydrochloric acid), strong, concentrated alkalies (e.g., sodium hydroxide), active metals (e.g., sodium), most strong oxidizing agents (e.g., potassium permanganate, sodium dichromate), or any of the reducing agents. These facts explain why alkanelike portions (alkyl groups) of the molecules of other families of compounds are called **nonfunctional groups.** When molecules that do have functional groups also have alkanelike portions, those parts of them generally ride through chemical events unchanged. The compound 1-pentanol, an alcohol, illustrates what we mean. See Figure 2.2.

The alkanes do have some very important chemical properties, of course. They burn, for example. (Virtually all organic compounds do.) They can be "cracked" to alkenes and hydrogen. Alkanes are also attacked by fluorine, chlorine, bromine, and hot nitric acid. We shall look briefly here at combustion and chlorination.

2.10 Combustion

We burn certain alkanes or their mixtures to obtain energy. Natural gas is mostly methane. Gasoline, diesel fuel, and heating oil are all mixtures of hydrocarbons.

If enough oxygen is available, complete combustion occurs. Complete

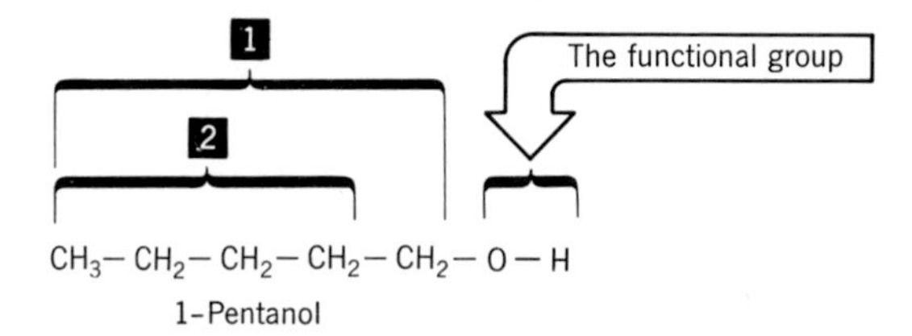

Figure 2.2 Functional and nonfunctional groups. Usually, the entire alkanelike group [1] is called the nonfunctional group, but often the part of it close to the functional group participates in the reactions of the latter. Only the segment marked [2] is left alone, for example, when this alcohol is oxidized.

combustion of any hydrocarbon or any oxygen derivative of a hydrocarbon (e.g., ethers, alcohols, esters, etc.) produces solely carbon dioxide, water, and energy. To illustrate, using propane:

$$CH_3CH_2CH_3 + 5O_2 \longrightarrow 3CO_2 + 4H_2O + 531 \text{ kcal of energy/mole of propane}$$

Exercise 2.4 Write balanced equations for the complete combustion of (a) pentane, and (b) acetylene (H—C≡C—H).

2.11 Chlorination

Chlorine, Cl_2, will not attack an alkane unless energy is provided—heat or ultraviolet light. When the energy is provided, the alkane molecule will have one (or more) of its hydrogens pulled off and replaced by an atom of chlorine, as we see in the case of methane and chlorine:

$$CH_4 + Cl_2 \xrightarrow[\text{ultraviolet light (e.g., sunlight)}]{\text{Heat or}} \underset{\substack{\text{Methyl chloride} \\ \text{(bp } -24\ ^\circ\text{C)}}}{CH_3Cl} + HCl$$

Methyl chloride also has replaceable hydrogens, and it can undergo chlorination to give methylene chloride, a solvent. Methylene chloride can also be chlorinated to chloroform, another solvent sometimes used as an anesthetic but one that poses danger to the liver. The chlorination of chloroform gives still another solvent, carbon tetrachloride, which is also dangerous to the liver. (Chlorinated solvents should always be handled carefully in well-ventilated areas only. They readily are absorbed through the skin.)

$$\underset{\substack{\text{Methane} \\ \text{(bp } -162\ ^\circ\text{C)}}}{CH_4} \xrightarrow[HCl]{Cl_2} \underset{\substack{\text{Methyl chloride} \\ \text{(bp } -24\ ^\circ\text{C)}}}{CH_3Cl} \xrightarrow[HCl]{Cl_2} \underset{\substack{\text{Methylene chloride} \\ \text{(bp 40 } ^\circ\text{C)}}}{CH_2Cl_2} \xrightarrow[HCl]{Cl_2} \underset{\substack{\text{Chloroform} \\ \text{(bp 61 } ^\circ\text{C)}}}{CHCl_3} \xrightarrow[HCl]{Cl_2} \underset{\substack{\text{Carbon tetrachloride} \\ \text{(bp 77 } ^\circ\text{C)}}}{CCl_4}$$

When 1 mole of methane is mixed with just 1 mole of chlorine, the final mixture includes some of each of these compounds. The reaction, therefore, is "messy" and unlike most reactions of inorganic, ionic compounds. Those of the latter usually go rapidly to completion and produce a desired product quantitatively. Organic reactions are seldom clean; the conversion of one reactant to one product is seldom 100%; and the desired compound must be separated from one or more by-products.

Bromine reacts with methane similarly, but iodine is unreactive. Fluorine reacts explosively with most organic compounds at room temperature stripping hydrogens from their molecules and producing polyfluoro compounds and hydrogen fluoride, among others.

Higher homologs of methane also react with chlorine (or bromine). For example,

$$CH_3CH_3 + Cl_2 \xrightarrow{\text{heat}} CH_3CH_2{-}Cl + HCl + \text{higher chlorinated products}$$

Ethane (bp −89 °C); Chlorine; Ethyl chloride (bp 13 °C)

$$CH_3CH_2CH_3 + Cl_2 \longrightarrow CH_3CH_2CH_2{-}Cl + CH_3\underset{\displaystyle Cl}{\underset{|}{C}}HCH_3 + HCl$$

Propane (bp −42 °C); *n*-Propyl chloride (45%) (bp 47 °C); Isopropyl chloride (55%) (bp 36 °C)

The reaction with propane produces a mixture more complicated than that shown; di- and polychlorinated compounds form also. Moreover, the equation is not balanced because not all products can be shown. Two monochloropropanes, however, are among the products; the percents given apply to that portion of the mixture only. These two compounds are isomers. The chain in each is a straight chain, but the substituent, the chlorine atom, is affixed at each of the two possible sites on that chain.

Exercise 2.5 (a) How many monochloro derivatives of *n*-butane are possible? Give both the common and the IUPAC names. (Be concerned only about mono-chloro compounds.) (b) How many monochloro derivatives of isobutane are possible? Name them according to the common and the IUPAC systems.

2.12 Cycloalkanes

A carbon skeleton can take the form of a closed ring as well as of an open chain. The simplest cyclic compounds are called the **cycloalkanes** or cycloparaffins. The following examples are illustrative:

Cyclopropane	Cyclobutane	Cyclopentane	Cyclohexane
(bp −33 °C)	(bp 13 °C)	(bp 49 °C)	(bp 81 °C)

Rings of higher carbon content (C_7 to over C_{30}) are also known. Since the molecules in these compounds have only single bonds, they generally exhibit the same types of physical and chemical properties as the open-chain alkanes. Cyclopropane and cyclobutane are exceptions. Their molecules are **strained,** because their carbon-carbon bond angles (60 and 90°, respectively) are too different from the normal tetrahedral angle (109° 28′). Cyclopropane reacts with bromine, for example, to form 1,3-dibromopropane.

$$\text{Cyclopropane (bp } -33\ ^\circ\text{C)} + Br_2 \longrightarrow BrCH_2CH_2CH_2Br\ \text{1,3-Dibromopropane (bp 167 °C)}$$

The strained ring opens up in the reaction. Cyclopentane, however, as well as all higher cycloalkanes react with bromine without opening the rings.

Cyclic five-membered and six-membered rings are very common in molecules of living things. We shall find them among all carbohydrates, proteins, and nucleic acids, and most vitamins and hormones. In many molecular rings one or more carbons have been replaced by some other atom, for example, oxygen or nitrogen. Many ring systems contain one or more double bonds.

2.13 Nonplanar Rings

The rings in saturated ring compounds having six ring atoms or more cannot all lie in the same geometric plane. Their rings cannot be flat and still allow each carbon to have its tetrahedral angles. Because the saturated six-membered ring is so common in carbohydrates, we include here some details about these rings in general.

The cyclohexane ring is not flat as in a normal hexagon. It is twisted out of the plane, and two conformations called chair forms having normal bond angles are shown in Figure 2.3. Important structural features of cyclohexane are discussed in the legend for this figure.

2.14 Interconversion of Chair Forms

In cyclic systems rotations about the bonds of the ring itself are severely limited. In cyclohexane a small amount of movement is possible; one chair form can interconvert via a boat form into the other equivalent but not iden-

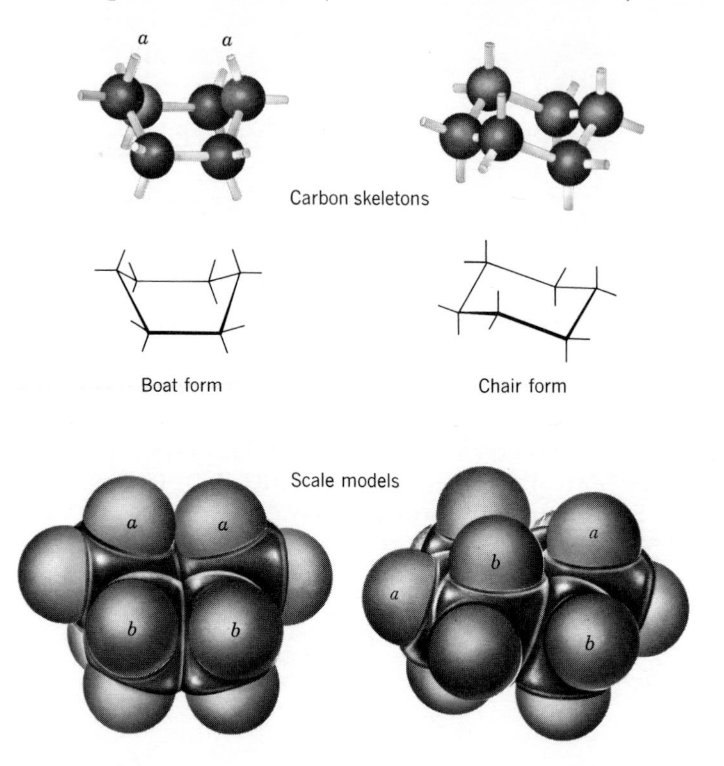

Figure 2.3 Boat and chair conformations of the cyclohexane ring. Carbon's bond angles in both are normal (109° 28′), but the chair form is still more stable. In the boat form hydrogen atoms from opposite corners (marked *a*) are brought close together and electron "clouds" about them repel each other. Adjacent hydrogen atoms (marked *b*) are also closer in the boat than in the chair form. Since these repulsions are minimized in either of the chair forms, they are the normal forms for cyclohexane and similar six-membered rings.

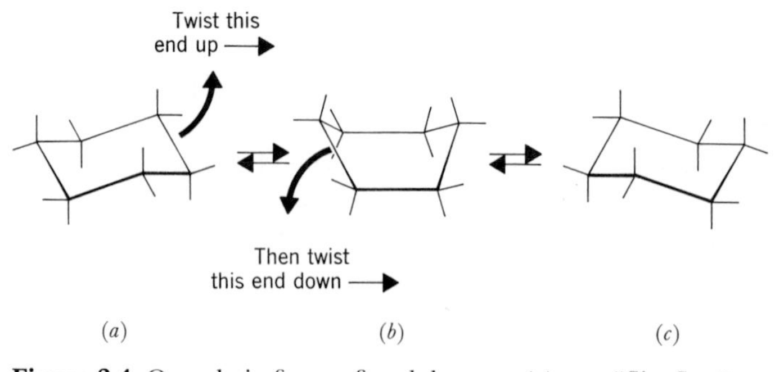

Figure 2.4 One chair form of cyclohexane (*a*) can "flip-flop" to the other chair form (*c*) and back again. The boat form (*b*) is an intermediate stage in this dynamic equilibrium. Interconversion from one chair to the other occurs rapidly enough at ordinary temperatures so that we may symbolize a six-membered ring as a flat hexagon (Figure 2.5) in most situations.

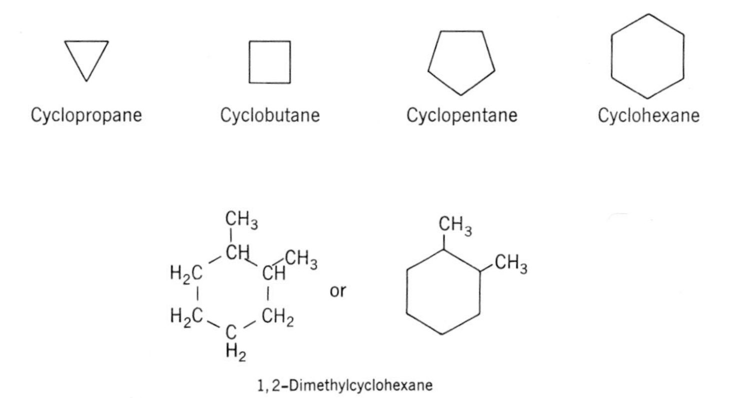

Figure 2.5 Symbolism and nomenclature in cycloalkane systems. At each corner in the geometric symbols a carbon is understood to be present. The lines from corner to corner are carbon-carbon bonds. Unless otherwise indicated, hydrogens are assumed to be present at the corner carbons in sufficient number to fill out carbon's tetravalence. When substituents occupy corners, the ring positions are numbered in a direction and from a beginning point that together yield the set of smallest numbers possible. Alkyl groups are usually given smaller numbers than halogens.

tical chair form, as illustrated in Figure 2.4. This ready interconvertibility makes it possible to use simple flat hexagons to represent cyclohexane, as though it were a planar molecule, and chemists usually represent cycloalkane systems with simple geometric forms such as those shown in Figure 2.5. Their symbolism is described in the legend and should be learned.

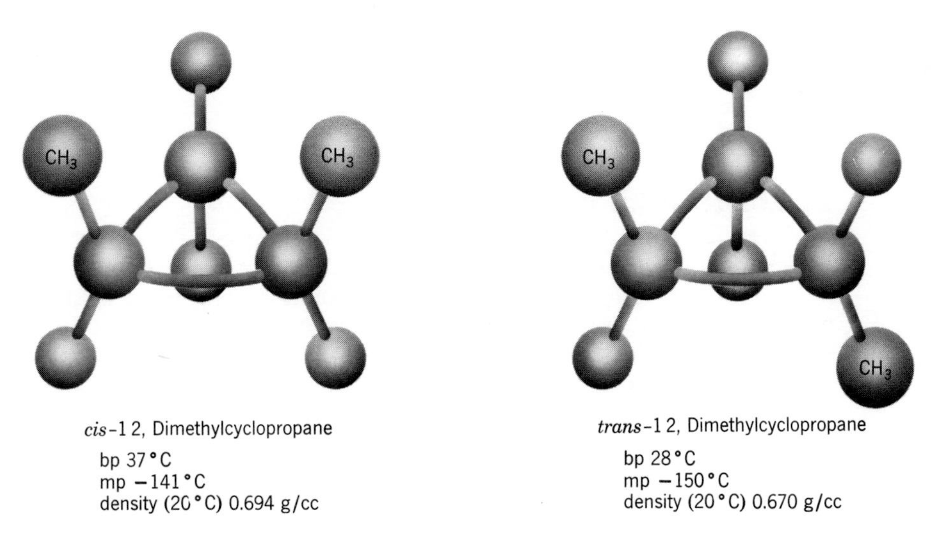

Figure 2.6 Geometrical isomerism. The differences between the two 1,2-dimethylcyclopropanes mean differences in properties.

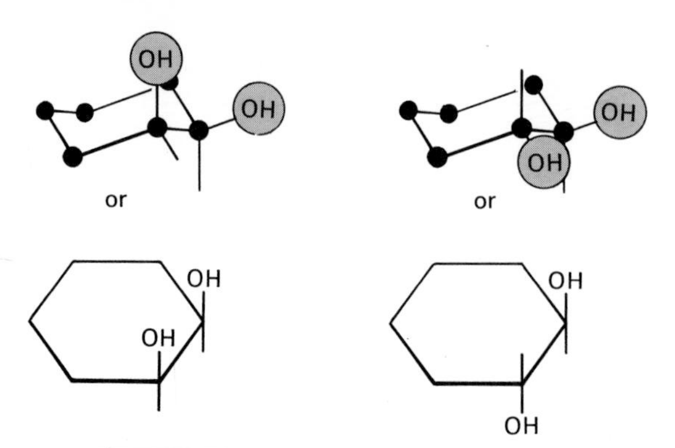

Figure 2.7 Two geometric isomers of the cyclohexanediols. Chair forms shown at the top may in most uses be replaced by the simpler forms given at the bottom where *cis* and *trans* relations are much more obviously apparent.

2.15 Geometrical Isomerism. *Cis-Trans* Isomerism

Two different 1,2-dimethylcyclopropanes (Figure 2.6) are known. The lack of free rotation about single bonds in ring systems makes their existence possible.

This kind of isomerism is called **geometric isomerism.** The difference between these isomers is not where substituents are located but instead how they are oriented. Because two different compounds cannot have the same name, chemists call the isomer in which the substituents project in generally opposite directions the ***trans* isomer.** In the ***cis* isomer** the substituents generally protrude in the same direction. Another example of *cis-trans* isomerism is given in Figure 2.7.

Brief Summary

Saturated Hydrocarbons. When the molecules of a substance consist only of carbon and hydrogen and contain only single bonds not only is the substance virtually nonpolar, but it is also unreactive, chemically, at room temperature with water, common acids and bases, or common oxidizing and reducing agents. They do burn. Hydrocarbons in general, whether saturated or not or cyclic or not, are relatively nonpolar and insoluble in water but soluble in common nonpolar organic solvents. ("Likes dissolve likes.")

Nomenclature. Rules of nomenclature insure that each compound will have a unique name and each name can correspond to one structural formula. If two structures positioned differently on paper turn out to have identical names, they must be identical compounds.

Geometrical Isomerism. These isomers differ only in relative orientations in space and not in fundamental atom-atom sequences. Unlike structural isomers, geometric isomers have identical names except that *cis* or *trans* becomes a prefix to specify a particular geometric orientation.

Cyclic Compounds. The fact that the molecules of a substance contain rings of atoms has, with few exceptions, little bearing on general physical or chemical properties. For example, cycloalkanes (cyclopropane and cyclobutane occasionally being exceptions) are just like open-chain relatives in being chemically unreactive to most reagents and nonpolar. Ring compounds can exhibit geometric isomerism that open-chain alkanes cannot. We use simple geometric forms to represent rings. Even though rings from five members and higher are not flat, we may treat them as if they were for most purposes.

Questions and Exercises

Where you are asked to write structures, provide condensed structures.

1. Write the name and structure of the compound that is the next higher homolog of hexane. heptane $CH_3CH_2CH_2CH_2CH_2CH_2CH_3$
2. Write the structures and the IUPAC names of all of the isomers of C_7H_{16}.
3. Write the structure and the IUPAC name for isononane.
4. Write structures of all the isomeric monochloro derivatives of 2,2,3-trimethylpentane.
5. If the common name of chloromethane is methyl chloride and if the common name for chloroethane is ethyl chloride, what are the common names for the following?

(a) $CH_3C(CH_3)_2—Cl$ (central C bearing CH₃ above and CH₃ below)

(b) CH_3CHCH_3 with Cl on the central C

(c) $Cl—CH(CH_3)_2$

(d) CH_3CHCH_2Cl with CH_3 on the central C

(e) $CH_3—CH_2—CH_2—Cl$

(f) $CH_3CHCH_2CH_3$ with Cl on C-2

6. Which would be the more soluble in water, (a) or (b)? Explain.

$HOCH_2CH_2OH$ (a) $CH_3CH_2CH_2CH_3$ (b)

7. Which of these can exist in two forms as *cis*- and *trans*-isomers?

(a) 1,1-Dimethylcyclopropane
(b) 1,2-Dimethylcyclobutane
(c) 2,3-Dimethylbutane

8. Explain why the cyclohexane ring is not flat.
9. Explain why the chair form of a six-membered ring is favored over the boat form.

chapter 3

Unsaturated Hydrocarbons. Alkenes and Aromatics

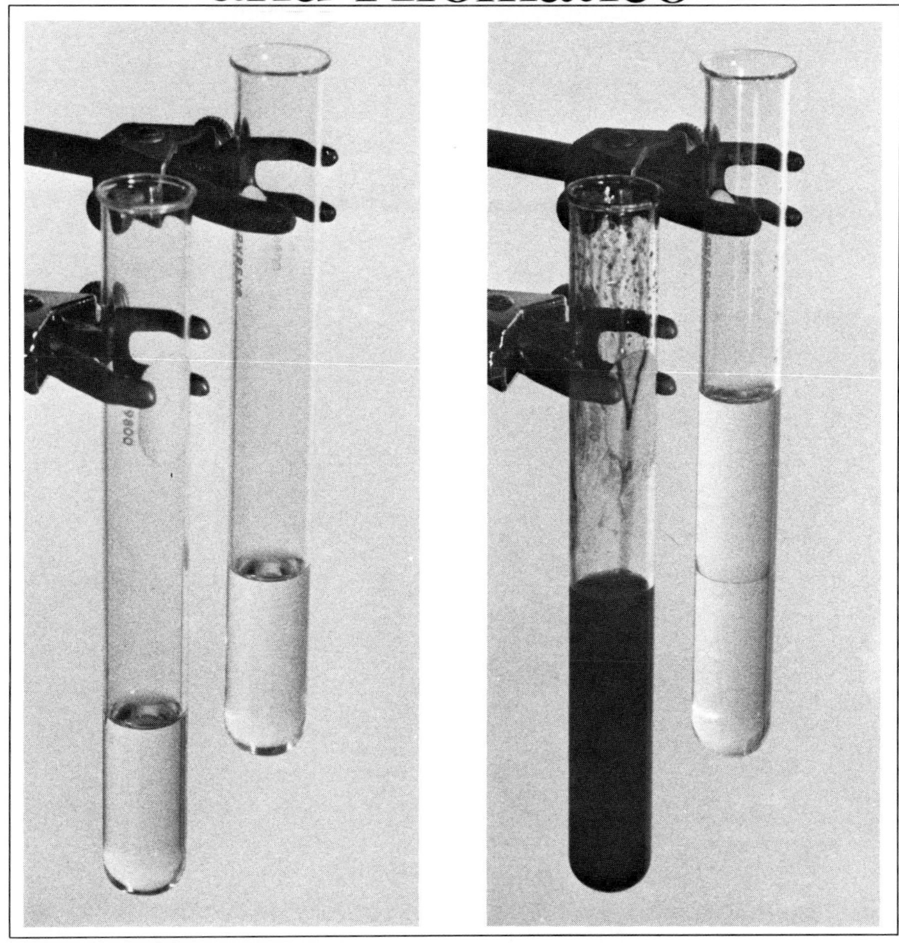

What a difference a double bond makes! On the left, two colorless liquids, both hydrocarbons but one an alkene and the other an alkane. On the right, after adding concentrated sulfuric acid, the alkene has turned into a tarry, black mess while the alkane floats untouched atop the acid. (Photos by K. T. Bendo)

3.1 Principal Types of Unsaturated Hydrocarbons

The three principal types of unsaturation involving carbon-carbon bonds are the alkene double bond, the alkyne triple bond, and the bonding network in aromatic compounds such as benzene. (See Figure 2.1.) The carbon-carbon double bond is sometimes called an "-ene" function, and it occurs widely in nature from substances in petroleum to fats and oils to vitamins and hormones. Simple examples of alkenes are given in Table 3.1. Some compounds have molecules with two double bonds and are called dienes. Trienes, tetraenes, and polyenes with, respectively, three, four, and many double bonds are also known.

The carbon-carbon triple bond is called an "-yne" function, as in ethyne ("acetylene," in common nomenclature). This function is seldom found among biochemically important substances, but certain drugs have them, such as some fertility control drugs.

The special network of bonds in aromatic compounds requires more than just a paragraph to define and illustrate, and we shall defer this system to later in the chapter. We add only that all compounds with a benzenelike ring are called **aromatic compounds** regardless of what else is present. All compounds without that structural feature are called **aliphatic compounds** (from the Greek for "fatlike").

TABLE 3.1
Properties of Some 1-Alkenes

Name (IUPAC)	Structure	Bp (°C)	Mp (°C)	Density (in g/ml at 10 °C)
Ethylene[a]	$CH_2{=}CH_2$	−104	−169	—
Propene	$CH_2{=}CHCH_3$	−48	−185	—
1-Butene	$CH_2{=}CHCH_2CH_3$	−6	−185	—
1-Pentene	$CH_2{=}CHCH_2CH_2CH_3$	+30	−165	0.641
1-Hexene	$CH_2{=}CHCH_2CH_2CH_2CH_3$	64	−140	0.673
1-Heptene	$CH_2{=}CHCH_2CH_2CH_2CH_2CH_3$	94	−119	0.697
1-Octene	$CH_2{=}CHCH_2CH_2CH_2CH_2CH_2CH_3$	121	−102	0.715
1-Nonene	$CH_2{=}CHCH_2CH_2CH_2CH_2CH_2CH_2CH_3$	147	−81	0.729
1-Decene	$CH_2{=}CHCH_2CH_2CH_2CH_2CH_2CH_2CH_2CH_3$	171	−66	0.741
Cyclopentene	(ring structure)	44	−135	0.722
Cyclohexene	(ring structure)	83	−104	0.811

[a] See footnote 1.

Alkenes

The Carbon-Carbon Double Bond

3.2 ***sp*^{2} Hybridization**

When carbon is attached to four other atoms or groups, as in molecules of the alkanes, it utilizes sp^3 hybrid orbitals. At a double bond, however, each carbon is attached to only three other groups. Carbon cannot use sp^3 orbitals here. It needs different orbitals. As illustrated in Figure 3.1, we start with carbon's atomic orbitals at the first and second energy levels. Then we "mix" or "hybridize" the 2*s* and just two of the 2*p* orbitals. We leave the third *p*-orbital at level-2 unchanged, for the moment. Because we "mix" one *s* and two *p* orbitals, we call the new hybrid orbitals ***sp*^{2} orbitals.** The shapes and distributions of these orbitals about a carbon atom are shown in Figure 3.2. Let us now see how these orbitals interact to give the network of bonds in ethylene.

3.3 **The Pi Bond**

Figure 3.3 illustrates two carbon atoms whose atomic orbitals have undergone sp^2 hybridization. To form the C—H bonds, each atom has used two sp^2 orbitals to accept overlap with 1*s* orbitals of two hydrogens. Next, there is the sp^2-sp^2 overlap between the two carbons to give one of the bonds of the double bond, a sigma bond. Like any ordinary single, covalent bond, we should expect free rotation about it. At one "stop" in such a rotation, the two (so far) unused p_z orbitals will be lined up side by side as seen in Figure 3.4. They can and do overlap, side by side, and this locks the system from further free rotation. The side-by-side overlap gives us a new molecular orbital with two electrons, the second bond of the double bond, but it does not have cylindrical symmetry as a sigma (σ) bond. Viewed end-on it has the cross section of a figure eight, and any molecular orbital with this symmetry is called a pi

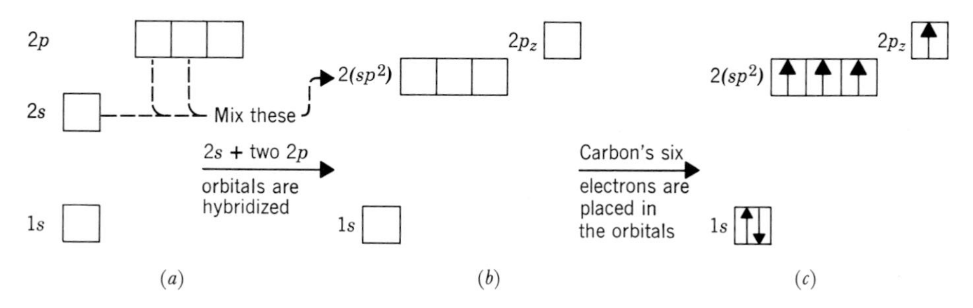

Figure 3.1 Origin of sp^2 hybrid orbitals of carbon. (*a*) The boxes represent atomic orbitals of carbon at levels 1 and 2. For the moment they are empty. (*b*) The orbitals when a carbon atom becomes sp^2 hybridized as the 2*s* orbital "mixes" with two of the 2*p* orbitals. The places for putting electrons have not changed in number but in energy level and geometry. The boxes are still empty; we do not hybridize electrons—we hybridize orbitals. (*c*) The valence state of the sp^2 hybridized carbon is $1s^2 2(sp^2)2(sp^2)2(sp^2)2p_z$.

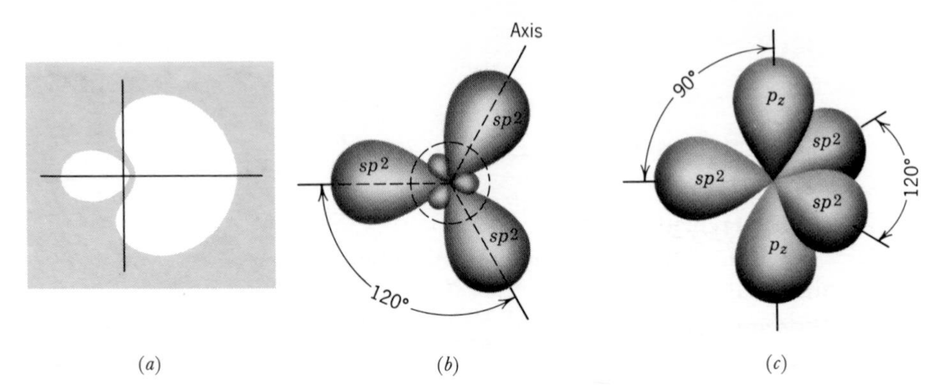

Figure 3.2 The sp^2 hybrid orbitals of carbon in ethylene. (*a*) Cross-sectional slice showing two lobes (*p*-character) and one including the center of the nucleus (*s*-character). (*b*) Three sp^2 orbitals of carbon all with axes in the same geometric plane; top view. (*c*) Perspective view of carbon in the sp^2-hybridized state showing also the one unhybridized p_z orbital.

bond, symbolized, **π-bond.** The double bond, therefore, consists of one sigma bond and one pi bond, and free rotation about the bond is normally impossible.

3.4 Importance of the Pi Bond

The electrons in the pi bond are called the pi electrons, and they are not held as close to the carbon skeleton as the sigma electrons. The pi electrons, therefore, are capable of reacting with electron-seeking species (e.g., protons). As we shall see, the chemistry of alkenes is much richer than that of alkanes.

3.5 Isomerism among the Alkenes

Two types of isomerism are possible in the alkene family. One is positional isomerism; the double bond may be located between different pairs of carbons, for example, 1-butene and 2-butene. The other is geometrical or ***cis-trans*** **isomerism.** Because of the normal lack of free rotation, the groups attached to the double bond may have *cis-trans* relations. For example, four alkenes of the formula C_4H_8 are known. Their structures are

$CH_2{=}CHCH_2CH_3$	$(H_3C)(H)C{=}C(CH_3)(H)$	$(H_3C)(H)C{=}C(H)(CH_3)$	$(H_3C)_2C{=}CH_2$
1-Butene	*cis*-2-Butene	*trans*-2-Butene	2-Methyl-1-propene
(α-Butylene)	(*cis*-β-Butylene)	(*trans*-β-Butylene)	(Isobutylene)
bp −6.3 °C	bp 3.7 °C	bp 0.9 °C	bp −6.9 °C

Differences in structure, however subtle, do mean differences in observable properties, as seen in the boiling points of the isomeric butenes. The

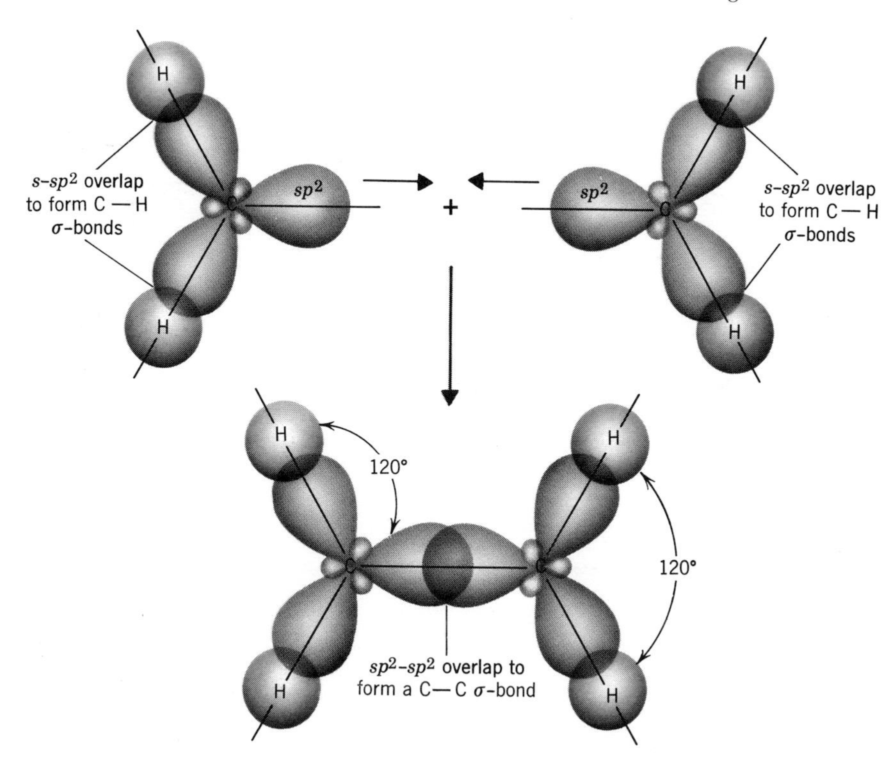

Figure 3.3 Sigma bond network in ethylene. (The p_z orbital of each carbon is not shown.)

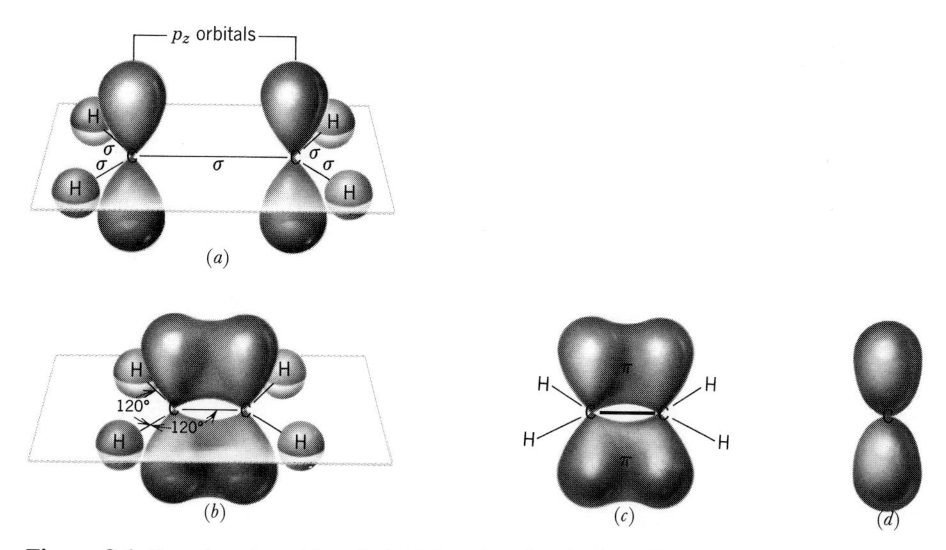

Figure 3.4 Forming the pi bond. (*a*) The situation before the p_z orbitals overlap; all the sigma bonds have formed (Figure 3.3). (*b*) The p_z orbitals achieve overlap, which they can do only in the orientation shown in which all nuclei at, or adjacent to, the double bond are in the same geometric plane. (*c*) One sigma bond (heavy line) and one pi bond (with two sausage-shaped lobes or halves) comprise the carbon-carbon double bond. (*d*) Looking down the axis of the bond, the pi bond in cross section looks like this.

word *cis* means that groups are on the same side of the double bond; *trans* means they are on opposite sides.

Exercise 3.1 Identify each of the following compounds by name.

(a) H_3C and H on the upper C; $C{=}C$ (vertical); H and CH_3 on the lower C

(b) $CH_3CH_2CH{=}CH_2$

(c) $CH_2{=}C{-}CH_3$ with CH_3 on the central C

(d) H, H_3C on the left C; $C{=}C$; H, CH_3 on the right C (H above, H_3C below; H above, CH_3 below)

Exercise 3.2 Examine each of the following structures; write trial structures for geometric isomers; and determine whether or not *cis-trans* isomerism is possible.

(a) CH_3CH_2, H on the left C; $C{=}C$; H, H on the right C

(b) Cl, H on the left C; $C{=}C$; Cl, H on the right C

(c) Cl, Cl on the left C; $C{=}C$; H, H on the right C

(d) Cl, H on the left C; $C{=}C$; Br, H on the right C

(e) Cl, Br on the left C; $C{=}C$; H, H on the right C

3.6 Writing Structures of Alkenes

While there are some chemical properties of *cis* and *trans* isomers that are remarkably different, most are not. Unless *cis-trans* distinctions are important, when we write structures of alkenes we may write them in a simple manner. For example, 2-butene is usually well enough represented just by $CH_3CH{=}CHCH_3$, which would mean either the *cis* or the *trans* isomer or a mixture of both. What is important is that, unlike single bonds, we never leave a double bond "understood."

Nomenclature of Alkenes

3.7 Common Names

The common names of alkenes have the same ending, "-ylene." An alkene with the carbon skeleton of the three-carbon alkane, propane, is called "propylene." The "prop-" prefix indicates a total of three carbons; the "-ylene" suffix places it in the ethylene or alkene family. The various butylenes are differentiated arbitrarily by the prefixes α, β, and "iso-." In the homologous series the common system loses most of its practicality above the butylenes, and the IUPAC system is used.

3.8 IUPAC Names

The IUPAC rules for naming the alkenes are as follows:

1. The characteristic name ending is "-ene."

2. The longest continuous chain that contains the double bond is selected. This chain is named by selecting the alkane with the identical chain length and changing the suffix in its name from "-ane" to "-ene."

3. The chain is numbered to give the first carbon of the double bond the lowest possible number (e.g., $CH_3CH_2CH{=}CH_2$ is 1-butene, not 1,2-butene and not 3-butene). In numbering the main chain, precedence is given to the double bond, not the locations of substituents.

4. The locations of the groups attached to the main chain are identified by numbers. The word parts are then assembled in a manner analogous to that in alkane nomenclature.

```
                                                 CH3
                                                 |
     CH3          CH3                   CH3    CHCH2CH3
     |            |                     |      |
CH3CH2CHCH2CH=CCH3                 CH3CHCH2C=CH2
2,5-Dimethyl-2-heptene           2-Isobutyl-3-methyl-1-pentene
                               Not 2-sec-Butyl-4-methyl-1-pentene
```

$CH_2{=}CH_2$	$CH_3CH{=}CH_2$	$CH_3CH{=}CHCH_2CH_3$
Ethylene[1]	Propene[2] ("propylene")	2-Pentene (*cis* or *trans*)

Exercise 3.3 Write condensed structures for each of the following:
(a) 4-Methyl-2-pentene (b) 3-Propyl-1-heptene
(c) 3,3-Dimethyl-4-chloro-1-butene (d) 2,3-Dimethyl-2-butene

Exercise 3.4 Write IUPAC names and the common names, if they have already been mentioned, for each of the following. Ignore *cis* or *trans* designations if they are applicable.

```
(a) H3C   CH3                 (b)        CH3                  CH3
       \ /                               |                    |
        C                          CH3—CH—CH2—C—CH2—CH—CH3
        ||                                      ||
        CH2                                     C
                                               / \
                                            H3C   CH2—CH3

(c) CH3—CH=CH—Cl              (d) Br—CH2—CH=CH2

         CH2—CH3
         |
(e) CH3—CH—CH2—CH=CH2         (f) If [cyclohexene ring] is called cyclohexene
                                  what is its full structure? What
                                  is the structure of cyclopentene?
```

[1] This nonsystematic ("common") name is retained in the IUPAC system for the unsubstituted compound (1957 Rules, paragraph A-3.1). Most chemists, however, do not object to calling it ethene, and it must be so named in substituted ethenes.

[2] Whenever the double bond cannot be located differently to make an isomer, its position need not be designated by a number.

(g) If [3-methylcyclohexene ring structure, positions numbered 1–6, CH_3 at C-3] is 3-methylcyclohexene,[3] what is [cyclohexene ring with CH_3 groups on two adjacent carbons] ?

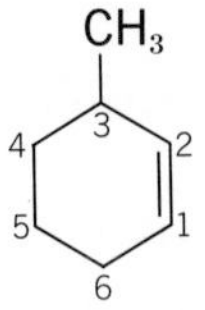

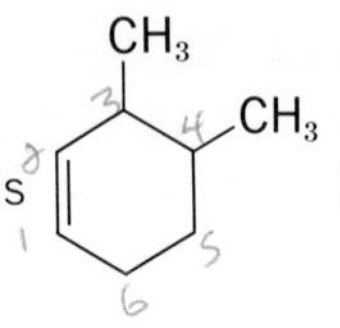

3.9 Chemical Properties of the Carbon-Carbon Double Bond

Addition reactions and oxidation reactions are two of the common types of chemical changes of a carbon-carbon double bond. An **addition reaction** is just that. Something adds to the double bond and, when finished, each of its carbons has another atom or group. Only a carbon-carbon single bond remains, and the two carbons of the original double bond have become saturated carbons. We shall study the addition of hydrogen, chlorine (and bromine), sulfuric acid, and water.

3.10 Hydrogenation of a Double Bond

Hydrogen, H_2, adds to a double bond only in the presence of a powdered metal catalyst (e.g., powdered nickel or platinum), and generally only under pressure. The alkene is put in a heavy-walled glass or steel bottle together with the catalyst, and the bottle is rocked to keep the catalyst suspended while hydrogen is let in under pressure. The reaction in general is called **hydrogenation;** the double bond is reduced.

In general:

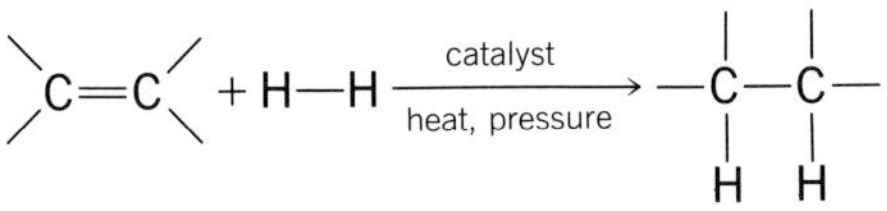

$$>C{=}C< + H{-}H \xrightarrow[\text{heat, pressure}]{\text{catalyst}} -\overset{|}{\underset{H}{C}}-\overset{|}{\underset{H}{C}}-$$

$$\underset{\text{Ethylene}}{CH_2{=}CH_2} + H_2 \xrightarrow[\text{heat, pressure}]{\text{Ni}} \underset{\text{Ethane}}{CH_3{-}CH_3}$$

$$\underset{\text{Isobutylene}}{(H_3C)_2C{=}CH_2} + H_2 \xrightarrow[\text{heat, pressure}]{\text{Ni}} \underset{\text{Isobutane}}{(H_3C)_2CH{-}CH_3}$$

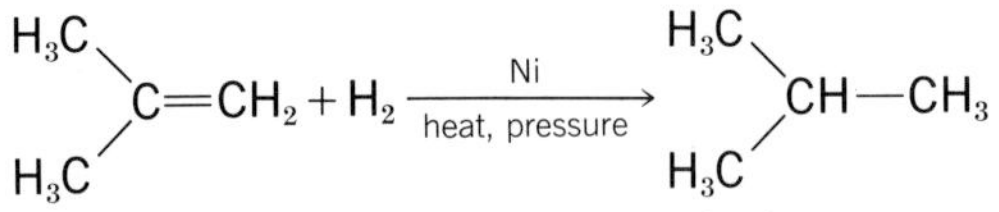

Specific examples:

$$\text{3-Methylcyclopentene} + H_2 \xrightarrow[\text{heat, pressure}]{\text{Ni}} \text{Methylcyclopentane}$$

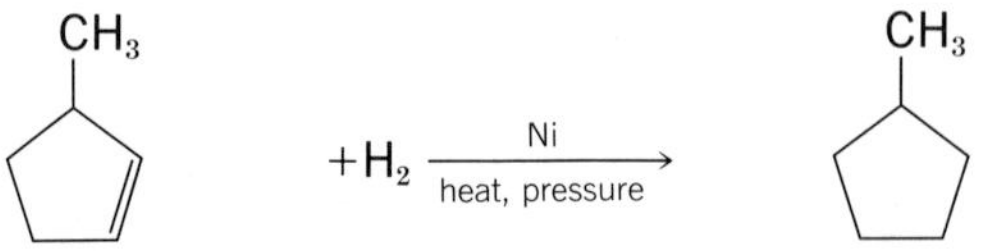

3-Methylcyclopentene — Methylcyclopentane

[3] The double bond in a cyclic system is always regarded as starting from position 1. Numbering then proceeds *through the double bond* around the ring in the direction that will give the lower set of numbers to the substituents.

3.11 Addition of Chlorine or Bromine to a Double Bond

Either of these chemicals will add to a double bond, exactly as does hydrogen, except that neither a catalyst nor pressure is needed. The reactions go smoothly in high yield at room temperature and below. Usually, the alkene is dissolved in some inert solvent such as carbon tetrachloride. Chlorine, a gas, is bubbled into the solution. Bromine, a corrosive liquid, is usually added in the form of a dilute solution in carbon tetrachloride. Because dibromoalkanes are essentially colorless compounds, when bromine is added a dramatic change in color is instantly observed. The dark brown of the bromine solution disappears as a nearly colorless solution forms. This change is the basis of an old and common test for telling an alkene from an alkane that you may perform in the laboratory.[4]

In general:

$$\gt C{=}C\lt + X{-}X \longrightarrow \gt \underset{X}{\underset{|}{C}}{-}\underset{X}{\underset{|}{C}}\lt \quad (\text{or } X{-}\overset{|}{\underset{|}{C}}{-}\overset{|}{\underset{|}{C}}{-}X)$$

(X = Cl or Br)

Specific examples:

$$CH_3CH{=}CHCH_2CH_3 + Br_2 \longrightarrow CH_3\underset{Br}{\underset{|}{C}}H{-}\underset{Br}{\underset{|}{C}}HCH_2CH_3$$

2-Pentene → 2,3-Dibromopentane (94%)

$$\text{Cyclohexene} + Br_2 \longrightarrow \text{1,2-Dibromocyclohexane}$$

Cyclohexene → 1,2-Dibromocyclohexane (*trans* isomer forms)

$$CH_3CH{=}CHCH_3 + Cl_2 \longrightarrow CH_3\underset{Cl}{\underset{|}{C}}H{-}\underset{Cl}{\underset{|}{C}}HCH_3$$

2-Butene → 2,3-Dichlorobutane (81%)[5]

The dichloro- and dibromo compounds are oily liquids. In this sense, therefore, ethylene was early regarded as an oil-forming gas or olefiant gas. From this came the nickname of all alkenes. Chemists commonly call them *olefins*.

[4] If the test is done in sunlight on an alkane, the color of brominc may disappear anyway because sunlight will catalyze the bromination of alkanes. In this substitution reaction, hydrogen bromide also evolves, however. It is easily observed because it fumes in moist air as it leaves the test tube.

[5] A percent in parentheses represents the mole percent yield. The percentages cited in nearly all examples are yields actually reported and then tabulated in *Synthetic Organic Chemistry* by R. B. Wagner and H. D. Zook, John Wiley & Sons, Inc., New York, 1953.

3.12 Addition of Hydrogen Bromide, Hydrogen Chloride, or Hydrogen Sulfate

Hydrogen bromide and hydrogen chloride are both gases; if either is bubbled into an alkene, which may or may not be dissolved in an inert solvent (e.g., carbon tetrachloride), it adds to the double bond. Unlike chlorine or bromine, H—Cl and H—Br are not symmetrical reagents. Therefore, different products may form depending on which of the two carbons of the double bond gets the hydrogen and which the halogen in HCl or HBr.

When ethylene reacts with hydrogen chloride, only one product is possible:

$$\underset{\text{Ethylene}}{CH_2{=}CH_2} + H{-}Cl \longrightarrow \underset{\text{Ethyl chloride}}{CH_3CH_2Cl}$$

When propylene reacts with hydrogen chloride, however, two products are possible, at least in principle: either

$$\underset{\text{Propylene}}{CH_3{-}CH{=}CH_2} + H{-}Cl \longrightarrow \underset{\substack{n\text{-Propyl chloride} \\ \text{(does not form)}}}{CH_3{-}\underset{H}{\underset{|}{CH}}{-}\underset{Cl}{\underset{|}{CH_2}}} \quad (\text{or } CH_3CH_2CH_2Cl)$$

or

$$CH_3{-}CH{=}CH_2 + H{-}Cl \longrightarrow \underset{\text{Isopropyl chloride}}{CH_3{-}\underset{Cl}{\underset{|}{CH}}{-}\underset{H}{\underset{|}{CH_2}}} \quad (\text{or } CH_3\underset{Cl}{\underset{|}{C}}HCH_3)$$

The actual product is largely isopropyl chloride. Very little *n*-propyl chloride forms. The reagent adds selectively. Propylene is an unsymmetrical alkene, one in which the carbons of the double bond hold different numbers of hydrogens. Ethylene, 2-butene, and 2-pentene are symmetrical alkenes; each carbon at their double bonds has as many hydrogens as the other carbon. As a general rule when an unsymmetrical reagent (H—Cl, H—Br, H_2SO_4, and H_2O) adds to an unsymmetrical double bond, the carbon of that bond with the greater number of hydrogens gets one more hydrogen. ("Them that has, gits.") This consistency was first noticed by a Russian chemist, Vladimer Markovnikov (1838–1904), and has since been called **Markovnikov's rule.**[6] The following examples are further illustrations:

$$\underset{\text{Isobutylene}}{CH_3\underset{CH_3}{\underset{|}{C}}{=}CH_2} + H{-}Cl \longrightarrow \underset{t\text{-Butyl chloride}}{CH_3{-}\overset{Cl}{\overset{|}{\underset{CH_3}{\underset{|}{C}}}}{-}CH_3} \quad \left(\text{not } CH_3{-}\underset{CH_3}{\underset{|}{CH}}{-}CH_2{-}Cl\right)$$

[6] When hydrogen bromide is used, it is important that no peroxides, compounds of the type R—O—O—H or R—O—O—R, be present. They catalyze a different mechanism, which we shall not discuss, with the result that in their presence hydrogen bromide (and only this hydrogen halide) adds itself to unsymmetrical alkenes against the Markovnikov rule. Traces of peroxides form when organic chemicals remain exposed to atmospheric oxygen for long periods.

$$\text{1-Methylcyclohexene} + H{-}Cl \longrightarrow \text{1-Chloro-1-methylcyclohexane} \quad (\text{not 2-chloro-1-methylcyclohexane})$$

1-Methylcyclohexene

1-Chloro-1-methylcyclohexane

Hydrogen sulfate is actually concentrated sulfuric acid, and if an alkene is shaken with this reagent, heat evolves and sulfuric acid adds to the double bond. A very polar compound forms and dissolves in the remaining sulfuric acid. Since an alkane does not behave this way but merely forms a separate layer, an alkene may be quickly distinguished from an alkane by shaking it with concentrated sulfuric acid. The alkene will dissolve exothermically; the alkane will not. (See also chapter-opening pictures.)

$$CH_3{-}CH{=}CH_2 + H{-}O{-}SO_2{-}O{-}H \xrightarrow[0\,^\circ C]{} CH_3{-}CH(CH_3){-}O{-}SO_2{-}O{-}H$$

Propylene — Sulfuric acid — Isopropyl hydrogen sulfate

If an alkene is symmetrical in the sense used in Markovnikov's rule, then the addition of an unsymmetrical reagent may occur in either of two directions. For example,

$$CH_3CH_2CH{=}CHCH_3 + H{-}Br \longrightarrow CH_3CH_2CHBrCH_2CH_3 + CH_3CH_2CH_2CHBrCH_3$$

2-Pentene — 3-Bromopentane — 2-Bromopentane

Both products form approximately equally.

3.13 Addition of Water

Ordinarily, an alkene can be shaken with water indefinitely and nothing will happen. If an acid is present and the temperature is adjusted, however, a water molecule can add to the double bond. The product is a member of the family of alcohols.

In general:

$$>C{=}C< + H{-}OH \xrightarrow{H^+} -\underset{H}{\overset{|}{C}}-\underset{OH}{\overset{|}{C}}-$$

Alkene — Alcohol

Specific examples:

$$CH_2{=}CH_2 + H{-}OH \xrightarrow[\substack{240\ ^\circ C \\ \text{(closed vessel)}}]{10\%\ H_2SO_4} CH_3{-}CH_2{-}OH$$

Ethylene → Ethyl alcohol

$$(H_3C)_2C{=}CH_2 + H{-}OH \xrightarrow[25\ ^\circ C]{10\%\ H_2SO_4} (H_3C)_2C(OH){-}CH_3 \quad \left(\text{not} \quad (H_3C)_2CH{-}CH_2{-}OH\right)$$

Isobutylene → *t*-Butyl alcohol

Markovnikov's rule applies to this reaction in which the H in H—OH goes to the carbon of the double bond that has the greater number of hydrogens; the —OH goes to the other carbon.

Exercise 3.5 Write the structures for the product(s) that would form under the conditions shown.

(a) $CH_2{=}CHCH_2CH_3 + HCl \longrightarrow$ [handwritten: $CH_3{-}CH(Cl)CH_2CH_3$]

(b) $CH_2{=}C(CH_3){-}CH_3 + HBr \longrightarrow$ [handwritten: $CH_3{-}C(Br)(CH_3){-}CH_3$]

(c) $CH_3{-}CH{=}C(CH_3){-}C_6H_{11} + H_2O \xrightarrow{H^+}$ (cyclohexyl ring)

(d) 3-methylcyclohexene (H_3C on ring) $+ H_2O \xrightarrow{H^+}$ (What mixture forms?)

Exercise 3.6 Write the structures and names of alkenes that are best used to prepare each of the following halides. Avoid selecting those yielding isomeric halides that are difficult to separate. If the halide cannot be prepared by the action of HCl or HBr on an alkene (given the limitations of Markovnikov's rule), state this fact.

(a) $CH_3CH_2CHCH_3$ with Cl on the third carbon ($CH_3CH_2CH(Cl)CH_3$)

(b) $CH_3CH_2CH_2Cl$

(c) Bromocyclopentane (cyclopentane ring with Br)

(d) $CH_3{-}C(CH_3)(Br){-}CH_2{-}CH_3$

(e) $Cl{-}CH_2{-}CH(CH_3){-}CH_2{-}CH_3$

3.14 How the Addition of Acids Occurs

Hydrogen chloride, hydrogen bromide, and sulfuric acid (by itself or as the catalyst in the addition of water) are all acids; all are able to donate protons. At a double bond, the pi electrons are less firmly held than sigma electrons, and because a proton is an electron-seeking reagent, alkenes behave as proton acceptors. In a confrontation between an alkene and a proton donor, we may represent the acid as G—H, where G = Cl, Br, OSO_3H, or even H_2O (in which case G—H is the hydronium ion, H_3O^+). The proton-transfer step, step 1, may be written as follows, using propylene as the alkene.

Step 1

$$:\ddot{G}—H + CH_2{=}CH—CH_3 \longrightarrow :\ddot{G}:^- + H—CH_2—\overset{+}{C}H—CH_3$$

Propylene — Isopropyl cation (a carbonium ion)

The organic product is a positively charged ion in which one carbon has around it only six electrons, not an octet. Called, in general, **carbonium ions** or **carbocations,** these ions are very unstable and have only the most fleeting existence. In step 2 they take anything in the mixture that might help them recover an outer octet for carbon. By taking the other product formed in step 1, $:\ddot{G}:^-$, the carbonium ion easily satisfies this need.

Step 2

$$:\ddot{G}:^- + CH_3—\overset{+}{C}H—CH_3 \longrightarrow CH_3—\underset{}{\overset{G}{\overset{|}{C}H}}—CH_3$$

For example,

$$CH_3—\overset{+}{C}H—CH_3 \xrightarrow{:\ddot{G}:^-}$$

if $:\ddot{G}: = :\ddot{Cl}:^-$ → $CH_3—\overset{Cl}{\overset{|}{C}H}—CH_3$ Isopropyl chloride

if $:\ddot{G}: = :\ddot{Br}:^-$ → $CH_3—\overset{Br}{\overset{|}{C}H}—CH_3$ Isopropyl bromide

if $:\ddot{G}: = {}^-O—SO_3H$ → $CH_3—\overset{O—SO_3H}{\overset{|}{C}H}—CH_3$ Isopropyl hydrogen sulfate

if $:\ddot{G}: = H—\ddot{O}—H$ → $CH_3—\overset{^+OH_2}{\overset{|}{C}H}—CH_3$ $\xrightarrow{-H^+}$ $CH_3—\overset{OH}{\overset{|}{C}H}—CH_3$ Isopropyl alcohol

These products illustrate Markovnikov's rule. Had the *n*-propyl cation formed in step 1, different products would have formed. This primary cation, the *n*-propyl cation, does not form because it is less stable than the isopropyl cation, a secondary cation. For carbonium ions in general, the order of their relative stability is

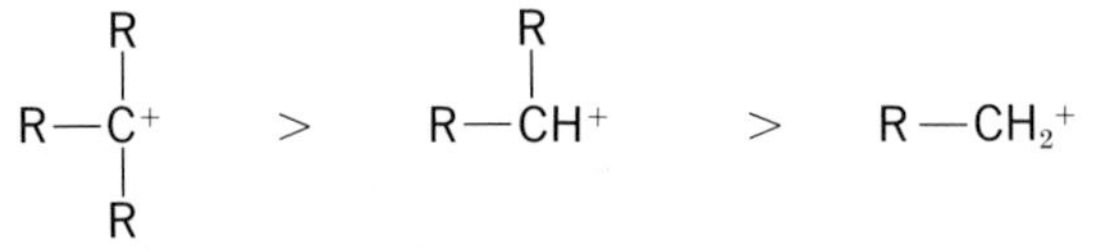

$$\underset{\substack{|\\ R}}{\overset{\substack{R\\ |}}{R{-}C^{+}}} \quad > \quad \overset{\substack{R\\ |}}{R{-}CH^{+}} \quad > \quad R{-}CH_2^{+}$$

Tertiary...(more stable than) ... secondary...(more stable than) ... primary

(To review the meaning of tertiary, secondary, and primary, see Section 2.7 under Rule 5.)

The alkyl groups joined to the carbon having the positive charge are responsible for this order of stability. With all the bonding electrons in them, alkyl groups furnish some electron density near the site of the positive charge, and this electron density helps to make the positively charged site more stable. The greater the number of alkyl groups at the site of positive charge, the more stable the system.

Exercise 3.7 Write the condensed structures for the two carbonium ions that can conceivably form if a proton becomes attached to each of the following alkenes. Circle the one that is preferred. Where both are reasonable, state that they are. Write the structures of the alkyl chlorides that will form by the addition of hydrogen chloride to each.

(a) $CH_3{-}CH_2{-}CH{=}CH_2$ (b) 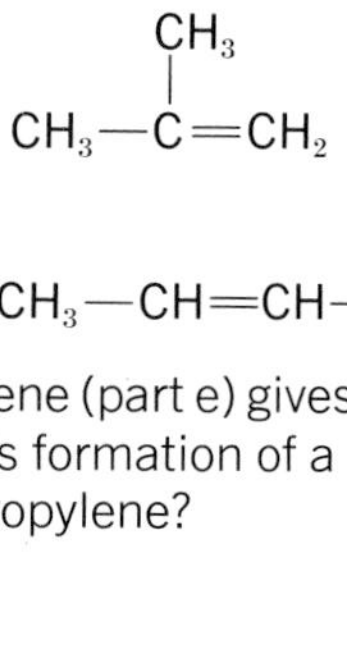 $CH_3{-}\overset{\substack{CH_3\\ |}}{C}{=}CH_2$ (c) 1-methylcyclohexene (six-membered ring with CH_3 on a double-bonded carbon)

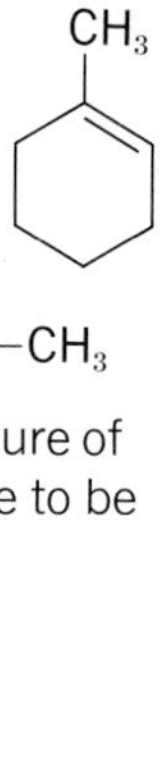

(d) $CH_3{-}CH{=}CH{-}CH_3$ (e) $CH_3{-}CH{=}CH{-}CH_2{-}CH_3$

(f) The addition of water to 2-pentene (part e) gives a mixture of alcohols. What are they? Why is formation of a mixture to be expected here but not from propylene?

3.15 Oxidation of Alkenes

Both ozone and hot solutions of potassium permanganate ($KMnO_4$) or potassium dichromate ($K_2Cr_2O_7$) vigorously oxidize carbon-carbon double bonds. The alkene is broken apart. When potassium permanganate is used, its deep red-purple color changes as it is consumed. The solution becomes less and less colored and a brown precipitate of manganese dioxide forms. The organic products depend on the particular alkene and may be ketones, carboxylic acids, carbon dioxide, or a mixture of these. We shall not study the details any further. Our interest is simply to learn that a carbon-carbon double bond makes a substance especially vulnerable to oxidation.

Mild oxidation of a double bond gives the net effect of the addition of hydrogen peroxide, HO—OH, and 1,2-dihydroxy compounds called glycols form. Either hydrogen peroxide (in an organic acid) or cold, aqueous potassium permanganate will work.

In general:

$$\text{>C=C<} \xrightarrow[\text{KMnO}_4,\ \text{H}_2\text{O}]{\text{H}_2\text{O}_2\text{ in formic acid, or}} \text{—}\underset{\text{OH}}{\underset{|}{\overset{|}{\text{C}}}}\text{—}\underset{\text{OH}}{\underset{|}{\overset{|}{\text{C}}}}\text{—}$$

A glycol

A specific example:

$$CH_2{=}CH(CH_2)_5CH_3 + H_2O_2 \xrightarrow{\text{Formic acid}} \underset{\text{OH}}{\underset{|}{CH_2}}\text{—}\underset{\text{OH}}{\underset{|}{CH}}(CH_2)_5CH_3$$

1-Octene — Hydrogen peroxide — 1,2-Octanediol (58%)

Alkynes

3.16 The Triple Bond

We understand the formation and structure of the carbon-carbon triple bond as we did the double bond. At the triple bond, carbon is bound to just two groups, and we invoke still another mode of hybridization: ***sp* hybridization.** Figure 3.5 illustrates how the hybrid orbitals are formed. Figure 3.6 shows how they participate in forming the bonds in acetylene. The triple bond forces the atoms immediately around it into a line and the bond angle is 180°. Therefore, there can be no geometric isomerism about this functional group.

Acetylene may be made by the action of water on calcium carbide:

$$CaC_2 + 2H_2O \longrightarrow H\text{—}C{\equiv}C\text{—}H + Ca(OH)_2$$

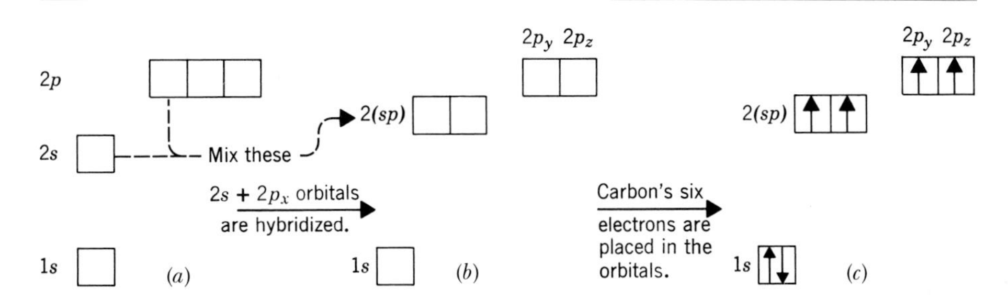

Figure 3.5 Forming *sp* hybrid orbitals. (*a*) The boxes represent atomic orbitals of carbon at levels 1 and 2. For the moment they are empty. (*b*) The orbitals when a carbon atom becomes *sp*-hybridized, when the 2*s* and one of the 2*p* orbitals "mix." (*c*) The valence state of *sp*-hybridized carbon with carbon's electrons distributed in available orbitals: $1s^2 2(sp)2(sp)2p_y 2p_z$.

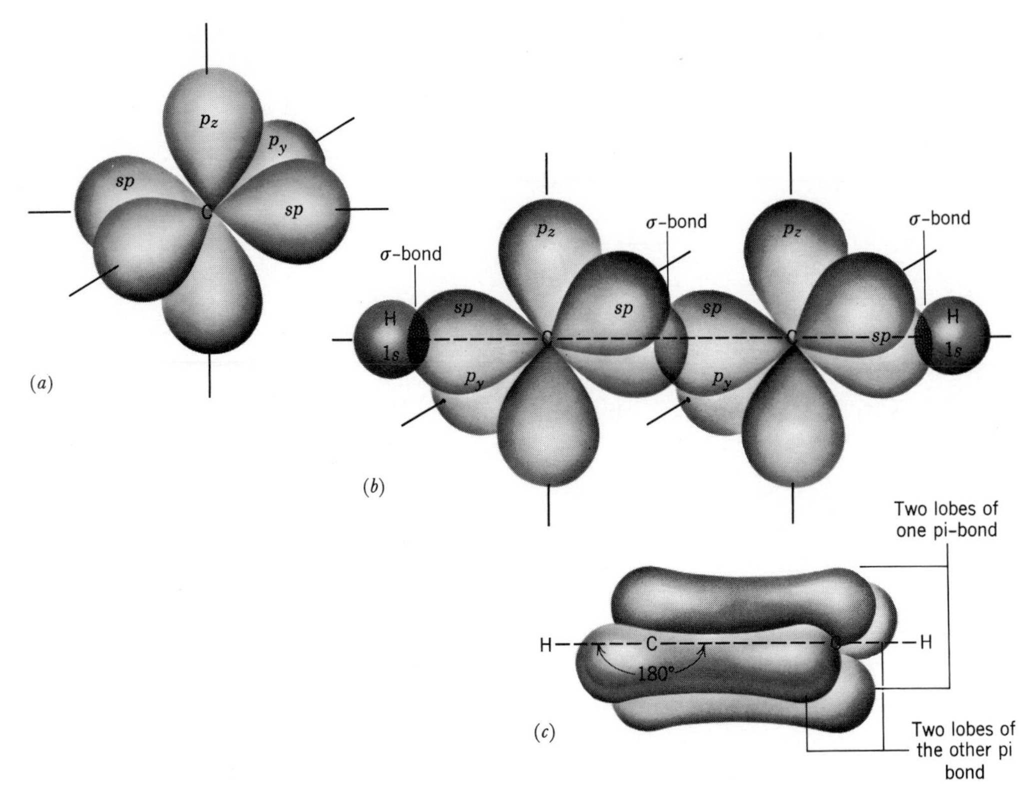

Figure 3.6 The triple bond in acetylene. (*a*) The orbitals of carbon in *sp* hybridization; two *sp* hybrid orbitals, one p_y and one p_z orbitals. (Only the large lobe of each *sp* orbital is shown.) (*b*) Sigma bonds of acetylene: *s-sp, sp-sp* and *sp-s,* left to right. (*c*) Pi bonds of acetylene: side to side overlap of p_y to p_y and p_z to p_z to give two pi bonds.

3.17 Nomenclature

The name "acetylene" has been adopted by the IUPAC system for the simplest alkyne. For higher members of the series we use essentially the same kinds of rules we used for alkenes, except that we use "-yne" for the word ending. These examples illustrate application of these rules.

$$CH_3—C{\equiv}C—H \qquad CH_3—\underset{}{\overset{CH_3}{\overset{|}{C}H}}—C{\equiv}C—H \qquad CH_3—C{\equiv}C—CH_3$$

Propyne 3-Methyl-1-butyne 2-Butyne

3.18 Chemical Properties of the Triple Bond

The triple bond is attacked by essentially the same reagents that attack a double bond: hydrogen, halogens (chlorine and bromine), hydrogen halides, and oxidizing agents. The following reactions illustrate some of these.

Hydrogenation

$$\underset{\text{Acetylene}}{H{-}C{\equiv}C{-}H} + H_2 \xrightarrow[\text{heat, pressure}]{\text{catalyst}} \underset{\text{Ethylene}}{H_2C{=}CH_2} \xrightarrow[\text{(same conditions)}]{H_2} \underset{\text{Ethane}}{CH_3CH_3}$$

Addition of Halogen (chlorine or bromine)

$$\underset{\text{Propyne}}{CH_3{-}C{\equiv}C{-}H} + Cl_2 \longrightarrow \underset{\text{1,2-Dichloropropene}}{CH_3{-}C(Cl){=}CH{-}Cl} \xrightarrow{Cl_2} \underset{\text{1,1,2,2-Tetrachloropropane}}{CH_3{-}CCl_2{-}CHCl_2}$$

Addition of Hydrogen Halide
Notice that Markovnikov's rule operates.

$$\underset{\text{Propyne}}{CH_3{-}C{\equiv}C{-}H} + H{-}Cl \longrightarrow \underset{\text{2-Chloropropene}}{CH_3{-}C(Cl){=}CH_2} \xrightarrow{H{-}Cl} \underset{\text{2,2-Dichloropropane}}{CH_3{-}CCl_2{-}CH_3}$$

Aromatic Hydrocarbons

3.19 The Benzene Ring

A large number of substances consist of molecules containing rings of six carbons but with only one substituent per carbon. These rings are called benzene rings after the simplest example, the hydrocarbon benzene, C_6H_6. Because the first compounds found to have this feature often had fragrant, aromatic odors, they were all called aromatic compounds. Today that name means any substance whose molecules contain the benzene ring or a similar structural feature.

Under extreme conditions, much more extreme than those needed to reduce ethylene, cyclohexene, or cyclohexadiene, benzene can be hydrogenated to give cyclohexane. From that and similar facts came the conclusion that benzene molecules have rings of six carbons. The six hydrogens in benzene are equivalent because only one product forms when benzene, C_6H_6, is changed to some monosubstituted compound, e.g., chlorobenzene, C_6H_5Cl. No isomers of chlorobenzene have ever been found. To put these facts together and write a structure for benzene is not, however, easy. The only way the hydrogens can be equivalent is to have one on each carbon, as in the following incomplete structure, where each carbon has only three bonds. The most nat-

Benzene, C_6H_6
(Structure is incomplete)

"Cyclohexatriene"
(unknown; is not benzene)

ural way to "give" each carbon four bonds would be to write alternating double bonds, that is, write a structure for 1,3,5-cyclohexatriene. The trouble with that solution is that benzene does not behave at all as a triene or any other unsaturated compound. All efforts to make cyclohexatriene from elimination reactions carried out on substituted cyclohexanes lead, instead, to benzene, not to a typical triene. Benzene simply does not behave chemically as an unsaturated compound except under the most powerful forcing conditions.

Hot potassium permanganate leaves benzene entirely alone; at room temperature so do concentrated sulfuric acid, chlorine, bromine, and the hydrogen halides. Hydrogenation is one of a very few addition reactions of benzene. Its normal reactions are **aromatic substitution reactions,** the substitution of a hydrogen on an aromatic ring by another group. The following are typical of benzene, and, as types of reactions, are typical of all aromatic compounds.

Nitration

$$C_6H_6 + HO{-}NO_2 \xrightarrow[50\text{–}55\ ^\circ C]{H_2SO_4(\text{concd})} C_6H_5{-}NO_2 + H_2O$$

Benzene (bp 80 °C) | Nitric acid | Nitrobenzene (85%) (bp 211 °C)

(Under these conditions alkenes undergo extensive oxidation and decomposition.)

Sulfonation

$$C_6H_6 + SO_3 \xrightarrow[\text{room temperature}]{H_2SO_4(\text{concd})} C_6H_5{-}S(\rightarrow O)_2{-}O{-}H$$

Benzenesulfonic acid (56%)
(mp 50–51 °C)

Sulfuric acid adds to double bonds in alkenes to form alkyl hydrogen sulfates.

Halogenation

$$C_6H_6 + Cl_2 \xrightarrow[FeCl_3]{\text{Fe or}} C_6H_5{-}Cl + H{-}Cl$$

Chlorobenzene (90%)
(bp 132 °C)

$$C_6H_6 + Br_2 \xrightarrow{Fe} \underset{\substack{\text{Bromobenzene (59\%)} \\ \text{(bp 155 °C)}}}{C_6H_5\text{—}Br} + H\text{—}Br$$

Chlorine and bromine add to an alkene with no catalyst required.

Oxidation

So stable is the benzene ring toward oxidation that if it holds an alkyl group, as in 1-phenylpropane, that group is oxidized and torn right down to the ring, and the ring is not attacked.

$$\underset{\substack{\text{1-Phenylpropane} \\ (n\text{-propylbenzene})}}{C_6H_5\text{—}CH_2CH_2CH_3} \xrightarrow{\text{hot } KMnO_4} C_6H_5\text{—}\overset{\overset{\displaystyle O}{\|}}{C}\text{—}OH \; (+CO_2 + H_2O + MnO_2)$$

What has been said here about the benzene ring largely applies to other substituted benzenes, those named, for example, in Section 3.23. Depending on the substituent, however, the ring may be much more reactive or much less reactive than the ring in benzene itself. Moreover, the ring is destroyed by oxidizing agents if it holds an amino group (—NH_2) or a hydroxyl group (—OH). We shall not need more details for our study, but too much of a simplification would be left by omitting the fact that aromatic compounds are not all alike. For most of them it is safe to say that in spite of the appearance of much unsaturation, aromatic compounds tend to give substitution reactions rather than addition reactions.

3.20 Kekulé Structure for Benzene

A German scientist, August Kekulé (1829–1896), was the first to propose the cyclohexatriene structure for benzene, **1**.

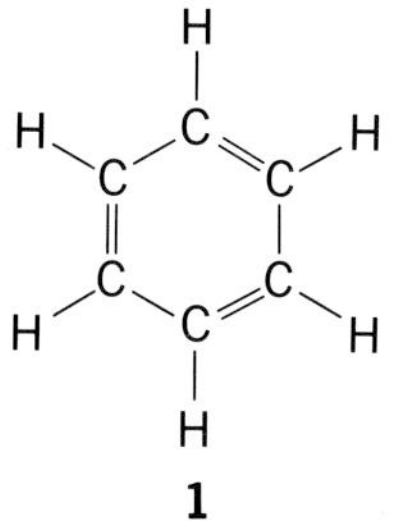

1 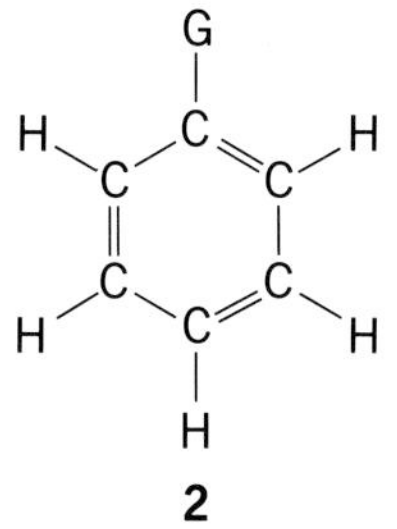 **2**

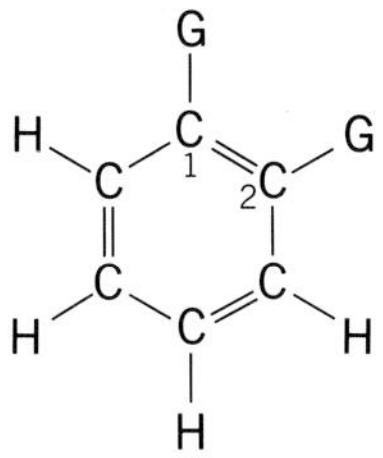

3 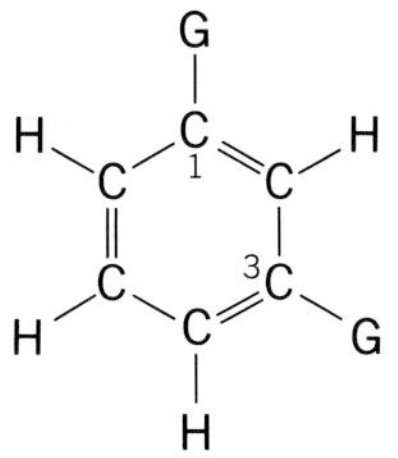 **4** 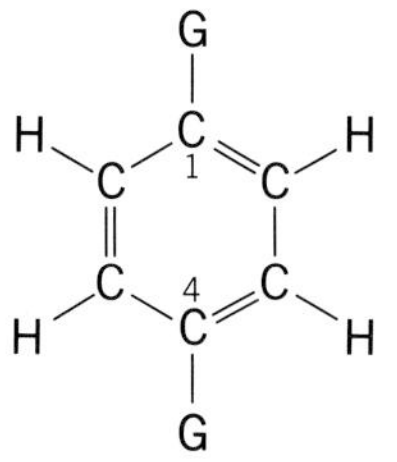 **5** 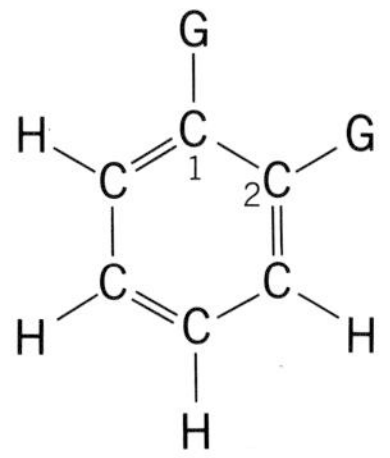 **6**

All six hydrogens in **1** are equivalent, and the structure of a monosubstituted benzene would be of type **2,** where G is some group or atom. Possible disubstituted products would be **3, 4,** and **5.** Kekulé's critics pointed out, however, that if his structure for benzene were correct, the 1,2-disubstituted product **3** should have an isomer **6** that differed from it only in the location of the double bonds. (Compare sites 1 to 2 in **3** and **6.**) No such isomers, **3** and **6,** have ever been found. Kekulé therefore modified his theory. He suggested that the double bonds so rapidly shift back and forth that **3** and **6,** while isomers, are in such rapid and mobile equilibrium that they can never be separated. His theory did not solve the whole problem, however. It failed completely to explain the remarkable fact that benzene is not an alkene, but nothing better could be used and the benzene "problem" persisted into the 1930s.

3.21 Molecular Orbital Model of Benzene

Kekulé came very close; the basic framework given by **1** is correct. Kekulé, who knew that benzene was not like an alkene, still included the alternating double bonds because he had to do something to preserve a valence of four for carbon. With molecular orbital theory, however, we can do something else as seen in Figure 3.7. The basic framework of sigma bonds in benzene is given in part (*a*). Each carbon is sp^2 hydridized as in ethylene. Each carbon has an unhybridized p_z orbital whose axis sticks in and out of the plane of the paper, seen in part (*b*). The side-to-side overlap of a p_z orbital is not limited to just one side. The six p_z orbitals interact with their neighbors all around the ring producing a circular, double-doughnut-shaped molecular orbital illustrated in part (*c*) of Figure 3.7. The six p_z electrons are distributed among three different energy levels enclosed by this general space. We need no details about these levels; for our purposes we shall simply consider the p_z electrons to be within the double-doughnut-shaped space. This space opens up room for electrons. The more electrons can spread out in space, the more stable is the system, since electrons naturally tend to repel each other. We say that the electrons in benzene are delocalized. In the ring-enclosing molecular orbital, the pi electrons of benzene spread out. This delocalization stabilizes

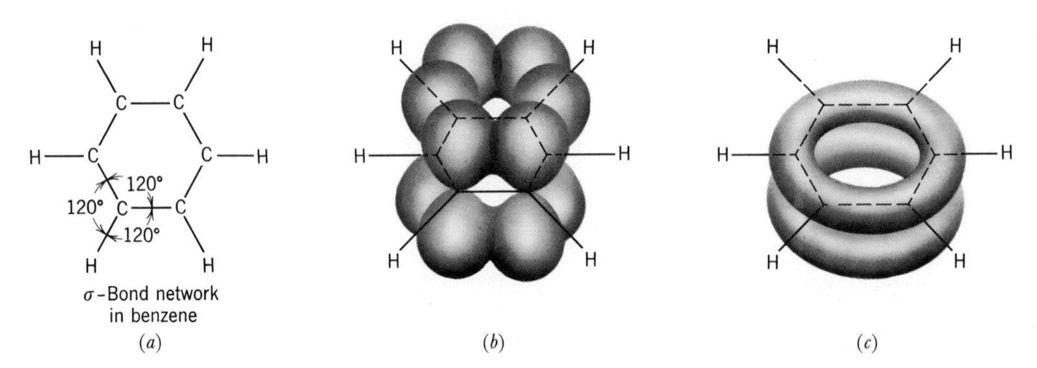

Figure 3.7 Molecular orbital model for benzene. (*a*) The sigma bond framework. All atoms lie in the same plane. (*b*) The p_z orbitals at the ring carbon atoms prior to side-to-side overlap. (*c*) The double-doughnut-shaped molecular orbital of benzene supplying a closed circuit for pi electrons.

benzene against any reaction, except under very unusual circumstances, that would permanently break up the closed-circuit. Benzene does not give addition reactions, because they would make one or more parts of the ring saturated. The circuit would no longer be closed, and the system would be less stable than before. The substitution reactions of benzene, however, do not permanently destroy the closed circuit.

3.22 Resonance in Benzene

Another term used for delocalization is **resonance.** We say that a condition of resonance exists in benzene and that benzene is stabilized by resonance. The theory of resonance for benzene emerged a few years prior to the molecular orbital treatment, and its language and symbols are still widely used. Pictures such as those in Figure 3.7 take time to draw and are not drawn equally well by all scientists. Resonance theory lets us use a simpler system of symbols. These symbols are Kekulé-type structures, and they let the mind of the reader mentally blend them into the view now given by the molecular orbital theory of benzene. In resonance theory benzene is represented by the pair of structures given by **7a** and **7b.** They differ only in the location of electrons. The representation given by **7c** is often substituted for the compos-

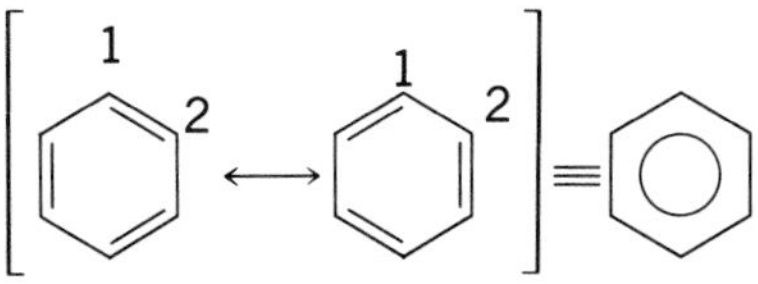

1 2 1 2

(a) (b) (c)

7

ite of **7a** and **7b,** but either of the latter will often be found in references as the structure of benzene. Once one is aware of the true nature of benzene, which structure is used, **7a** or **7b** or **7c,** is no longer an important issue.

The single arrow with two heads is the special symbol in resonance theory that informs us that none of the structures connected by that symbol is completely correct and that the mind must mentally blend them. Had two oppositely pointing arrows been used ($\rightleftarrows$), then a condition of equilibrium involving two real molecules would have been described. But benzene is not **7a** part of the time and then becomes **7b.** The bond from 1 to 2 in **7** is not a full double bond part of the time (as in **7a**) and a full single bond part of the time (as in **7b**). There is just one benzene molecule, but the mind is aided in picturing its structure by using **7a** and **7b,** structures with the virtue of obeying rules of valence, and saying that while they are not real they contribute to our picture of benzene. We call them **resonance contributors.** The mind looks at the bond from 1 to 2, or between any two adjacent corners, and sees that each such bond is a "partial double bond." While more than a simple single bond, it is less than a fully localized double bond. Instead it has characteristics of both types of bonds, and we call it a partial double bond.

The central postulate of the theory of resonance is that whenever, without violating rules of valence, two or more reasonable structures can be drawn that differ only in the location of electrons and not in any way in the location of atomic nuclei, the condition of resonance exists. In that circumstance, to continue the theory, the system's real structure is a blend, a resonance hybrid, of the two or more contributing structures. The substance will have more stability than we might expect of any one contributor. The most common examples of resonance stabilized substances are aromatic compounds where rings contain, in Kekulé-type structures, alternating double and single bonds.

Resonance exists in any planar six-membered ring, whether the ring members are all carbons or some members are other atoms (e.g., nitrogen or oxygen), where one can write alternating double and single bonds as in the Kekulé structure of benzene. Pyridine and quinoline are two of hundreds of examples.

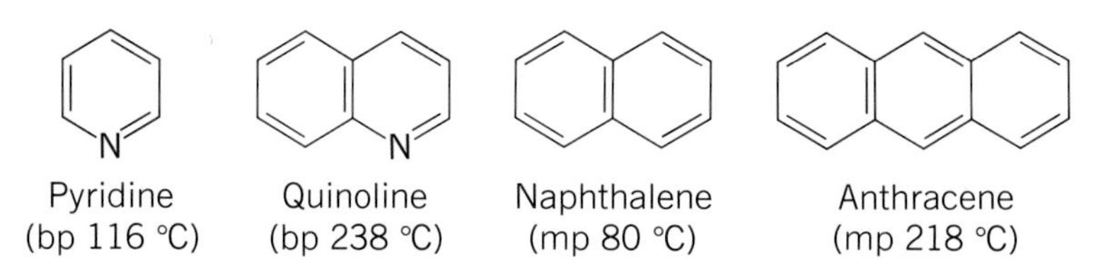

Pyridine (bp 116 °C) Quinoline (bp 238 °C) Naphthalene (mp 80 °C) Anthracene (mp 218 °C)

A number of fused ring systems such as the polynuclear aromatic hydrocarbons naphthalene and anthracene are also aromatic compounds. While seemingly very unsaturated, when viewed as Kekulé-type structures, their pi electrons are delocalized over large areas.

In many situations the benzene ring is represented as a hexagon with a circle within it as seen in several examples in the next section. While that symbol obscures adherance to rules of valence, it alerts us to the existence of an aromatic substance with **aromatic properties,** meaning substitution rather than addition reactions in spite of considerable unsaturation.

Naming Derivatives of Benzene

3.23 Monosubstituted Derivatives of Benzene

For a few compounds the nomenclature has some system. Prefixes naming the substituent are joined to the word *benzene*:

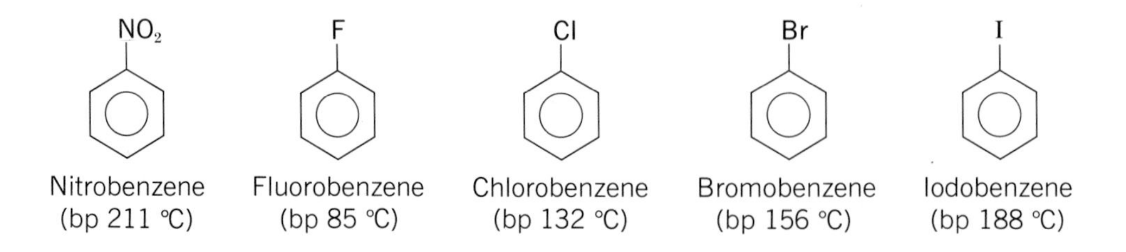

Nitrobenzene (bp 211 °C) Fluorobenzene (bp 85 °C) Chlorobenzene (bp 132 °C) Bromobenzene (bp 156 °C) Iodobenzene (bp 188 °C)

Other derivatives have common names that are always used, even though systematic names are possible:

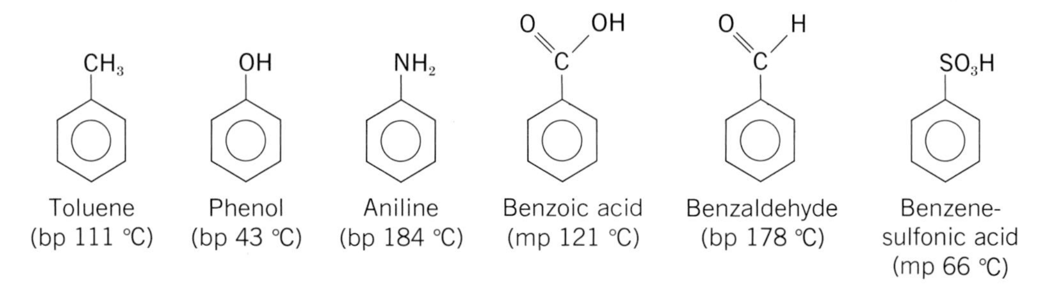

As a substituent on a hydrocarbon chain the benzene ring, C_6H_5—, is called the *phenyl* group; for example, C_6H_5—$CH_2CH_2CH_2CH_3$ is 1-phenylbutane.

3.24 Disubstituted Compounds

When two or more groups are attached to the same benzene ring, both their nature and their relative locations must be specified in the name. The prefixes "ortho," "meta," and "para" are used to distinguish, respectively, the 1,2-, the 1,3-, and the 1,4-relations. These prefixes are usually abbreviated *o*-, *m*-, and *p*-; in the following examples both substituent groups are the same.

Cl Cl Cl Cl Cl Cl

ortho
o-Dichlorobenzene
(bp 181 °C)

meta
m-Dichlorobenzene
(bp 173 °C)

para
p-Dichlorobenzene
(bp 174 °C)

If one of the two groups would give the compound a common name, were one group alone present, that name would be used, and the second group would be designated as a substituent. The following examples illustrate this point. These are named from left to right, respectively, as derivatives of toluene, aniline, benzoic acid, and phenol.

CH_3 / NO_2 — *p*-Nitrotoluene (mp 55 °C)

NH_2 / Br — *o*-Bromoaniline (bp 229 °C)

O OH C / Cl — *m*-Chlorobenzoic acid (mp 158 °C)

OH / NO_2 — o-Nitrophenol (mp 45 °C)

If neither group would be associated with a common name, were it alone on the benzene ring, both groups would be named and located in the name. For example,

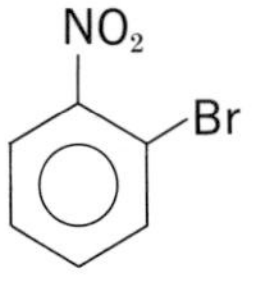

o-Bromonitrobenzene (mp 43 °C)

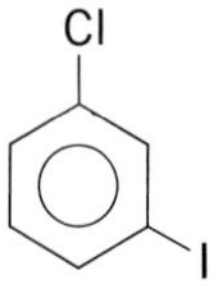

m-Chloroiodobenzene (bp 230 °C)

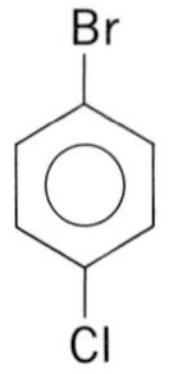

p-Chlorobromobenzene (mp 68 °C)

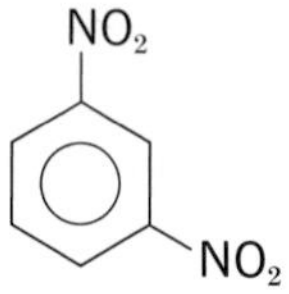

m-Dinitrobenzene (mp 90 °C)

3.25 Polysubstituted Benzenes

If three or more groups are on a benzene ring, the ring positions must be numbered. When one group is associated with a common name, its position is number 1. For example,

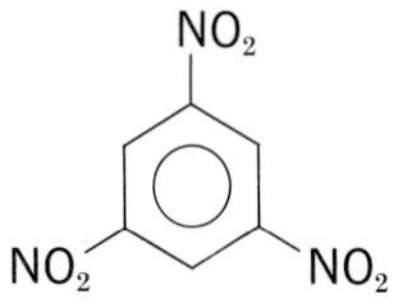

1,3,5-Trinitrobenzene (TNB) (mp 121 °C)

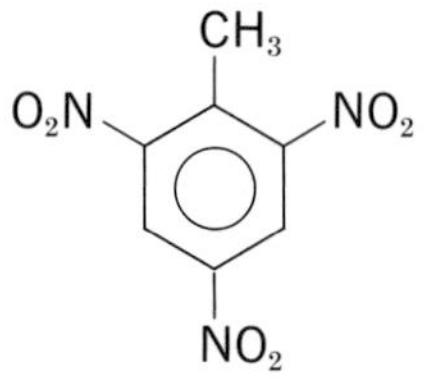

2,4,6-Trinitrotoluene (TNT) (mp 82 °C)

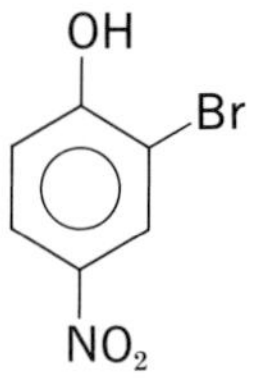

2-Bromo-4-nitrophenol (mp 114 °C)

Both TNB and TNT are explosives.

3.26 How Benzene Undergoes Substitution Reactions

The only kind of reagent with much chance of attacking the benzene ring is one that has a site that is positively charged, because the ring system is very electron-rich. The reagent must also be capable of forming a strong bond to carbon. An electrically neutral chlorine molecule, Cl—Cl, has a huge electron cloud, and cannot therefore penetrate the cloud on benzene and successfully attack a ring position. Without the iron catalyst, chlorine or bromine do not react with benzene under ordinary conditions. The iron catalyst, however, is first changed by some of the chlorine to iron(III) chloride, $FeCl_3$. This salt has binding ability for a chloride ion, and it reacts slightly with still unchanged chlorine to get that chloride ion as follows:

$$:\ddot{\underset{..}{Cl}}-\ddot{\underset{..}{Cl}}: + FeCl_3 \rightleftharpoons :\ddot{\underset{..}{Cl}}^+ + FeCl_4^-$$

The product includes a chlorine in an unusual state that lacks an outer octet and is positively charged—just the particle that might in the next step successfully attack an electron-rich benzene ring:

$$:\ddot{\underset{..}{Cl}}^{+} + C_6H_6 \longrightarrow \textbf{8}$$

We use a Kekulé strtucture for benzene to make it easier to write a structure for the intermediate carbonium ion, **8,** and we show one of the hydrogens in anticipation of the next step. Ion **8** is stabilized by some delocalization of the pi electrons among some of the ring atoms, but we shall not examine those details. In **8** the closed-circuit pi-electron network of benzene is broken but only temporarily. In the next step a proton drops off the ring to restore the ring's closed circuit, and the product, chlorobenzene (**9**),

$$\textbf{8} \longrightarrow \textbf{9} + H^+$$

forms. Overall, the system replaces H^+ by Cl^+, and it takes a special catalyst to generate the Cl^+. Bromination occurs by essentially an identical series of steps. We shall not examine the details of how nitration or sulfonation occur.

Brief Summary

Alkenes. The pair of carbons at a double bond are sp^2 hybridized and are joined by one sigma bond and one pi bond. One of the two is weaker than the other. The pi bond is strong enough to prevent free rotation at the double bond, and suitably substituted alkenes exhibit *cis-trans* isomerism. The pi bond undergoes addition reactions with hydrogen halides, sulfuric acid, chlorine, bromine, hydrogen (with a metal catalyst), water (with an acid catalyst), and oxidizing agents. See top of next page. Markovnikov's Rule usually applies.

Aromatic Hydrocarbons. When molecules are cyclic, planar, quite unsaturated and yet give substitution reactions rather than additions, the substance is an aromatic compound. Two typical hydrocarbons, benzene and toluene, are not affected at room temperature by reagents that typically react readily with alkenes—hydrogen, water, sulfuric acid, strong alkalies, hydrogen halides, chlorine or bromine (in the absence of sunlight, it must be said), or potassium permanganate. Hot permanganate, which is a typical, strong oxidizing agent, does react with alkyl groups on the ring producing benzoic acid, but the ring is left intact.

Resonance. Aromatic compounds are stabilized to avoid addition reactions by a

Alkenes ($>C{=}C<$)

- H_2, catalyst, heat, pressure → $-\overset{|}{\underset{H}{C}}-\overset{|}{\underset{H}{C}}-$ Alkanes
- X_2 → $-\overset{|}{\underset{X}{C}}-\overset{|}{\underset{X}{C}}-$ 1,2-Dihalo compounds
- H—X → $-\overset{|}{\underset{H}{C}}-\overset{|}{\underset{X}{C}}-$ Alkyl halides
- H_2O, H^+ → $-\overset{|}{\underset{H}{C}}-\overset{|}{\underset{OH}{C}}-$ Alcohols
- MnO_4^-, H_2O, cold → $-\overset{|}{\underset{OH}{C}}-\overset{|}{\underset{OH}{C}}- + MnO_2$ Glycols

Markovnikov's rule applies (alkyl halides, alcohols)

condition of resonance, which means that their molecules are hybrids or blends of two (or more) contributing structures. Each is a traditional, Kekulé-type system obeying rules of valence but differing in locations of electrons. Each does not actually exist, but the system is viewed as a hydrid of them all. An essentially identical picture emerges from the molecular orbital view of benzene in which the partial double bonds predicted by resonance theory are actually a ring-encircling molecular orbital holding the pi electrons in an unusually stable, closed circuit. Because this circuit breaks up only at great cost of energy, reactions of aromatic systems generally leave it alone and limit themselves to substitutions rather than additions.

Alkynes. The pair of carbons of a triple bond are in an *sp*-hybridized state and are joined by one sigma bond and two pi bonds. Alkynes give addition reactions essentially identical in type with those given by alkenes. A triple bond, however, can add two molecules of some reagent, instead of only one.

Questions and Exercises

1. Write the structure of each compound.
   (a) Propylene
   (b) Acetylene
   (c) *cis*-2-Hexene
   (d) 3-Bromo-1-butyne
   (e) *m*-Bromophenol
   (f) *p*-Chloroaniline
   (g) *o*-Nitrotoluene
   (h) 3,5-Dinitrobenzenesulfonic acid

2. Write condensed structures of all possible isomeric pentenes, C_5H_{10}. Include *cis* and *trans* isomers. Give IUPAC names. (Remember that the double bond is never "understood" in condensed structures.)

3. Write condensed structures of all isomeric pentynes, C_5H_8, and give IUPAC names.

4. Write IUPAC names for each of the following:

   (a) $CH_3(CH_2)_7CH{=}CH_2$

   (b) 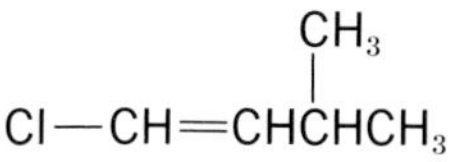 $Cl{-}CH{=}CHCH(CH_3)CH_3$ (CH₃ on the third carbon)

   (c) 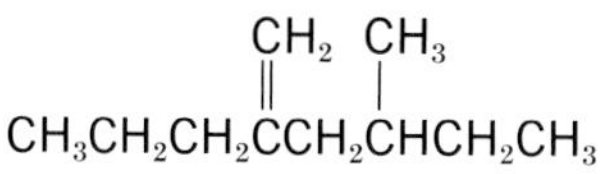 $CH_3CH_2CH_2C(={CH_2})CH_2CH(CH_3)CH_2CH_3$

   (d) 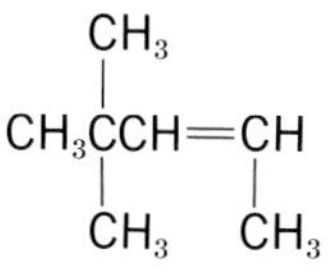 $CH_3C(CH_3)_2CH{=}CH{-}CH_3$

5. Predict which of the following would be able to exist as *cis-trans* isomers.

   (a) $Br{-}CH{=}CH{-}Br$

   (b) $Br_2C{=}CH_2$

   (c) 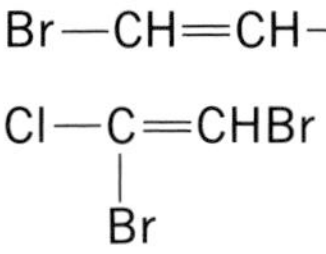 $Cl{-}C(Br){=}CHBr$

   (d) $CH_3CH{=}CHCH_2CH_3$

   (e) 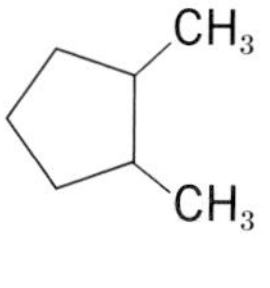 1,2-dimethylcyclopentane (cyclopentane ring with CH_3 on two adjacent carbons)

   (f) $(CH_3)_2C{=}CH_2$

6. Write equations for the reaction of isobutylene with each of the following reagents: (a) cold, concentrated H_2SO_4, (b) H_2(Ni, heat, pressure), (c) H_2O (H^+ catalysis), (d) H—Cl, and (e) H—Br.

7. Write equations for the reaction of 1-methylcyclopentene with each of the reagents of Exercise 6.

8. When 1-butene reacts with hydrogen chloride, the product is 2-chlorobutane. The isomer, 1-chlorobutane, does not form. Explain.

9. Ethane is insoluble in concentrated sulfuric acid. Ethylene dissolves readily. Write an equation to show how ethylene is converted into a substance polar enough to dissolve in the highly polar concentrated sulfuric acid.

10. Describe three simple chemical tests that could be used to distinguish between 2-butene and butane. Describe what you would do and see.

11. Write the structures for the products to be expected of the following situations. If no reaction is to be expected, so state. This exercise will help you to learn how to apply the chemical facts about functional groups you studied in this

chapter. To work this kind of exercise, you have to be able to do three things:

(a) Classify a specific organic compound into its proper family (or, in many cases in later chapters, several families).

(b) Recall the short list of chemical facts about the family to see if the reagent given in the question is on that list. If the reagent given is not on the list we assume that no reaction occurs.

(c) On the basis of the chemical fact recalled, apply the fact by writing the specific reaction.

Example

$$CH_3CH_2CH_2CH_3 + H_2SO_4 \longrightarrow ?$$

Step (a). The organic compound is an alkane.

Step (b). On the list of reactions or chemical facts about alkanes we have . . . nothing, except that they will burn and will react with chlorine (or bromine). We assume, therefore, that alkanes do not react specifically with sulfuric acid.

Step (c). We write "no reaction" as the answer.

Example

$$CH_3CH{=}CHCH_3 + H_2O \xrightarrow{H^+} ?$$

Step (a). The organic compound is an alkene.

Step (b). On the list of reactions or chemical facts about carbon-carbon double bonds are the following:

(1) They add water (in the presence of an acid) to form alcohols.

(2) They add hydrogen in the presence of a metal catalyst at high pressure and temperature to change to a saturated system (an alkane).

(3) They add X_2, HX (X = Cl, Br), and H_2SO_4.

Step (c). Since water is the given reagent, together with an acid catalyst, we know we have to write an alcohol for our answer. The skeleton of the alcohol will be identical with the skeleton of the alkene (at least in all our examples), but the double bond will disappear and at one of its carbons will be an —OH group and at the other carbon of the former double bond, another —H.

Answer: $CH_3CH_2\underset{|}{\overset{}{C}}HCH_3$ with OH on the third carbon

$$CH_3CH_2\underset{\displaystyle OH}{\underset{|}{C}}HCH_3$$

(a) $CH_3CH_2CH{=}CH_2 + H_2 \xrightarrow[\text{pressure}]{\text{Ni, heat}}$

(b) $CH_2{=}CHCH_3 + H_2O \xrightarrow[\text{heat}]{H^+}$

(c) $+ Br_2 \xrightarrow{\text{Fe}}$

(d) $CH_3CH_2CH_3 + \text{concd } H_2SO_4 \longrightarrow$

(e) $+ H_2O \xrightarrow{H^+}$

(f) $CH_3CH{=}CHCH_2CH_3 + H_2 \xrightarrow[\text{pressure}]{\text{Ni, heat}}$

(g) $+ \text{concd } H_2SO_4 \longrightarrow$

(h) $-CH{=}CH- \quad + H_2O \xrightarrow{H^+}$

(i) $+ H_2 \xrightarrow[\text{pressure}]{\text{Ni, heat}}$

(j) $CH_3CH_2CH_2CH_3 + O_2 \xrightarrow[\text{combustion}]{\text{complete}}$
(Balance the equation)

(k) $+ H_2O \xrightarrow{H^+}$

12. Write equations for the steps in the bromination of benzene, catalyzed by iron(III) bromide.

chapter 4

Alcohols, Ethers, and Phenols

The demonstration that compelled belief in ether as a successful anesthetic in surgery occurred on October 16, 1846, at Massachusetts General Hospital. This painting recreates the dramatic moment when surgeon John C. Warren put the scalpel to patient Gilbert Abbott anesthetized with ether administered by a dentist, William T. G. Morton. (Courtesy Massachusetts General Hospital, Boston.)

4.1 Occurrence

The **alcohol group** is one of nature's most widely occurring functional groups. It occurs in all sugars and starches, cellulose (cotton), proteins, many natural and synthetic drugs, hormones, flavoring agents, and some vitamins.

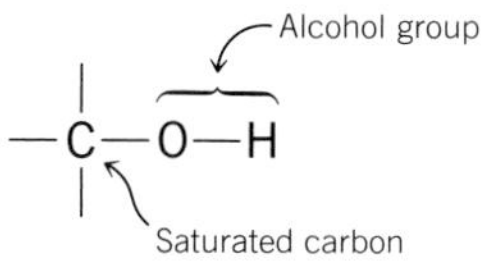

Alcohol group

Saturated carbon

Essential features of all alcohols

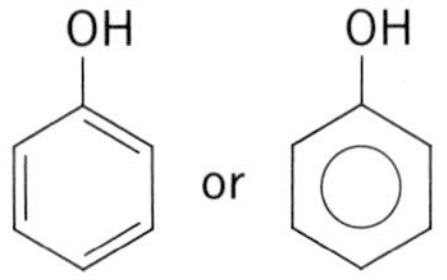

Phenol (two accepted symbols)

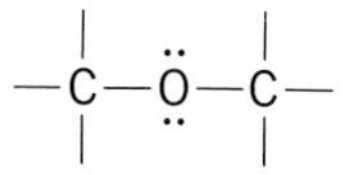

Ether group (Neither carbon may have a double bond to another oxygen.)

Phenols are also widespread both in nature and in commerce. One of the amino acid "building blocks" of proteins—tyrosine—contains a phenolic system. Commercial uses of phenol include the synthesis of aspirin, bakelite plastics and several bactericides, fungicides, herbicides, detergents, and softening agents (called plasticizers) for certain plastics.[1]

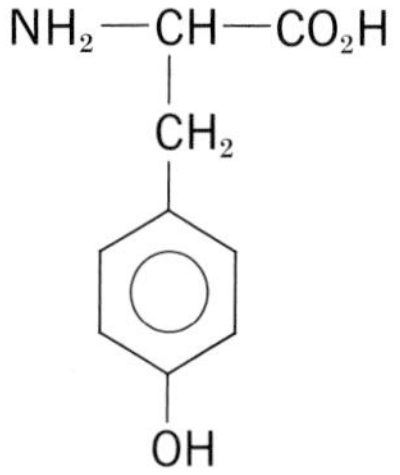

Tyrosine (an amino acid)

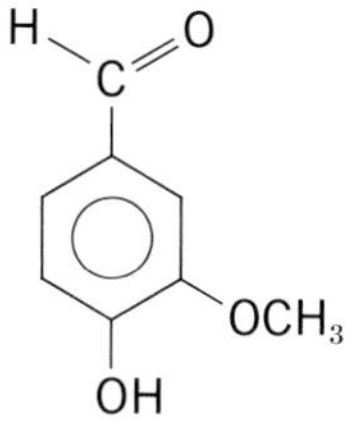

Vanillin (a flavoring agent)

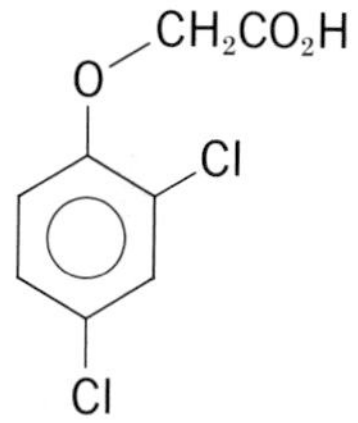

2,4-D (a herbicide)

Structural Features and Types

4.2 Subclasses of Alcohols

We classify an alcohol as primary (1°), secondary (2°), or tertiary (3°) according to the condition of the alcohol carbon, the carbon holding the —OH group. If that carbon has just one other carbon directly attached to it, the alcohol is a **primary alcohol.** If there are two carbons joined directly to the carbon bearing the —OH group, the alcohol is **secondary;** if three, **tertiary.** Any alcohol with only one —OH group per molecule is called a **monohydric alcohol.**

[1] Bactericides are chemicals that kill bacteria. Fungicides are chemicals used to control fungi, especially on seed grains. Herbicides are chemicals used to control weeds and brush.

$$\begin{array}{ccc} \text{H} & \text{R}' & \text{R}' \\ | & | & | \\ \text{R—C—OH} & \text{R—C—OH} & \text{R—C—OH} \\ | & | & | \\ \text{H} & \text{H} & \text{R}'' \end{array}$$

Primary alcohol (1°) | Secondary alcohol (2°) | Tertiary alcohol (3°)

As the primes of the R-groups indicate, these groups need not be identical. Several alcohols are listed in Table 4.1.

4.3 Polyhydric Alcohols

Compounds having two or more hydroxyl groups per molecule are quite common. To be stable, the —OH groups must be on separate carbons. With few exceptions, 1,1-dihydroxy compounds (diols) are unstable.

Dihydric alcohols, those whose molecules have two —OH groups, are commonly called **glycols.** Glycerol is a common and important trihydric alcohol. Carbohydrates are polyhydric compounds.

TABLE **4.1**
Some Common Alcohols

Common Name	Structure	Subclass	Boiling Point, in °C
Methyl alcohol	CH_3OH	—	65
Ethyl alcohol	CH_3CH_2OH	1°	78.5
n-Propyl alcohol	$CH_3CH_2CH_2OH$	1°	97
Isopropyl alcohol	$CH_3CH(OH)CH_3$	2°	82
n-Butyl alcohol	$CH_3CH_2CH_2CH_2OH$	1°	117
Secondary butyl alcohol (*sec*-butyl alcohol)	$CH_3CH_2CH(OH)CH_3$	2°	100
Isobutyl alcohol	$(H_3C)_2CHCH_2OH$	1°	108
Tertiary butyl alcohol (*t*-butyl alcohol)	$(CH_3)_3C{-}OH$	3°	83

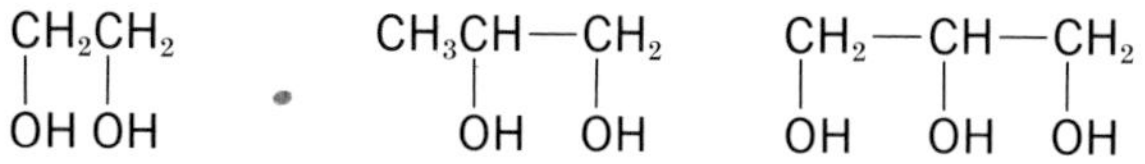

CH_2CH_2 (OH OH)	$CH_3CH{-}CH_2$ (OH OH)	$CH_2{-}CH{-}CH_2$ (OH OH OH)
Ethylene glycol (bp 197 °C)	Propylene glycol (bp 189 °C)	Glycerol (bp 290 °C)

4.4 Phenols

Phenols are compounds whose molecules contain a hydroxyl group attached directly to a carbon in an aromatic ring. In Table 4.2 some common phenols are listed.

4.5 Ethers

The **ethers** are substances whose molecules are of any of the generic formulas:

TABLE 4.2
Some Phenols

Name	Structure	Some Uses
Phenol (mp, 41 °C)	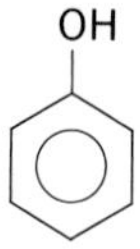 benzene ring with OH	Raw material for plastics, drugs, dyes, and other commodities
o-Cresol (mp, 30 °C)	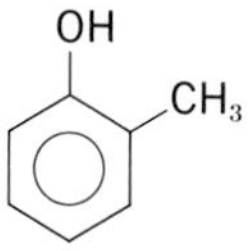 benzene ring with OH and ortho CH_3	
m-Cresol (mp, 11 °C)	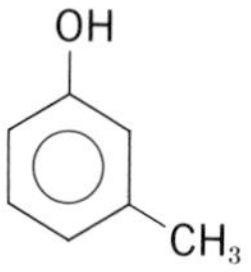 benzene ring with OH and meta CH_3	The cresols are raw materials for other chemicals and, as a mixture, used in disinfectants
p-Cresol (mp 36 °C)	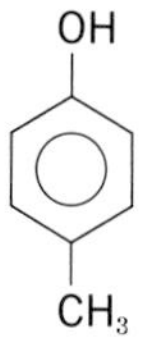 benzene ring with OH and para CH_3	
Hydroquinone (mp, 170 °C)	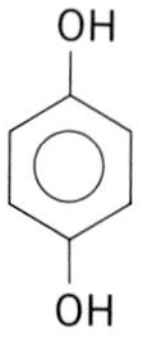 benzene ring with OH and para OH	Photographic developer; chemical intermediate
Pyrogallol (mp, 133 °C)	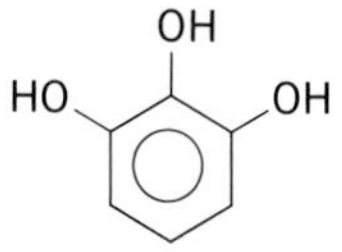 benzene ring with HO, OH, OH (adjacent)	Photographic developer; agent for combining with and removing oxygen from gas mixtures

R—O—R′ R—O—Ar Ar—O—Ar

where R is an aliphatic hydrocarbon group and Ar is an aromatic ring. A few ethers are listed in Table 4.3.

The ether group occurs widely in nature and in commercial products. (Both vanillin and 2,4-D, above, have ether linkages.) One of the very simple ethers, diethyl ether, is the widely used anesthetic commonly called simply "ether."

Exercise 4.1 Classify each as an ether, a phenol, or a monohydric or dihydric alcohol. If the structure is of a monohydric alcohol, classify it further as a 1°, 2°, or 3° alcohol. If none of these, state so.

(a) $CH_3—O—CH_2CH_3$ (b) cyclohexene ring—OH

(c) $CH_3—C(=O)—OH$ (d) cyclohexane ring bearing CH_3 and OH on the same carbon

(e) cyclohexane ring with OH on two adjacent carbons (f) $CH_3—C(=O)—O—CH_3$

(g) CH_3—benzene ring—OH (h) benzene ring—$CH_2—OH$

TABLE 4.3
Some Ethers

Name	Structure	Mp (°C)	Bp (°C)	Solubility in Water
Dimethyl ether	CH_3OCH_3	−138	−23	37 volumes gas dissolved in 1 volume water at 18 °C
Methyl ethyl ether	$CH_3OCH_2CH_3$	−116	11	
Diethyl ether	$CH_3CH_2OCH_2CH_3$	−116	34.5	8 g/100 ml (16 °C)
Di-*n*-propyl ether	$CH_3CH_2CH_2OCH_2CH_2CH_3$	−122	91	Slightly soluble
Methyl phenyl ether	$CH_3OC_6H_5$	−38	155	Insoluble
Diphenyl ether	$C_6H_5OC_6H_5$	28	259	Insoluble
Divinyl ether	$CH_2═CHOCH═CH_2$		29	

Nomenclature of Alcohols, Phenols, and Ethers

4.6 Common Names for Alcohols

Common names are used almost exclusively for simple alcohols. If we have a name for the alkyl group joined to the —OH in an alcohol, we form the name of the alcohol by simply writing the word "alcohol" after the name of the alkyl group, as seen in the examples of Table 4.1 (and which should be learned).

4.7 IUPAC Names for Alcohols

The rules for naming alcohols are similar to those for alkanes. In both selecting the main chain and numbering it, the alcohol group dominates.

1. For the parent structure, select the longest continuous chain of carbons that includes the —OH group. Name the alkane that corresponds to that chain. Then drop the final "-e" from that name and replace it by "-ol." For example,

CH_3OH	CH_3CH_2OH	$CH_3CH(CH_3)CH_2CH_2OH$
Methanol	Ethanol	Butanol (incomplete; see under rule 3)

2. When the —OH group can be variously located on the chain, specify its location by numbering the carbons of the parent chain from whichever end will give the location of the —OH group the lower number. For example,

$CH_3CH_2CH_2OH$	$CH_3CH(CH_3)OH$	$CH_3CH_2CH(CH_3)OH$	$CH_3CH(CH_3)CH_2CH_2OH$
1-Propanol	2-Propanol	2-Butanol	1-Butanol (incomplete; see under rule 3)

3. Note the names and locations of any groups on the parent chain. Assemble these names and numbers before the parent name developed thus far. These examples show how it is done. Special attention should be paid to the use of commas and hyphens to make the final name one word.

$CH_3CH(CH_3)CH_2CH_2OH$	$CH_3CH_2CH(CH_3)CH_2OH$	$CH_3CH_2C(CH_3)_2CH_2OH$
3-Methyl-1-butanol (complete)	2-Methyl-1-butanol	2,2-Dimethyl-1-butanol

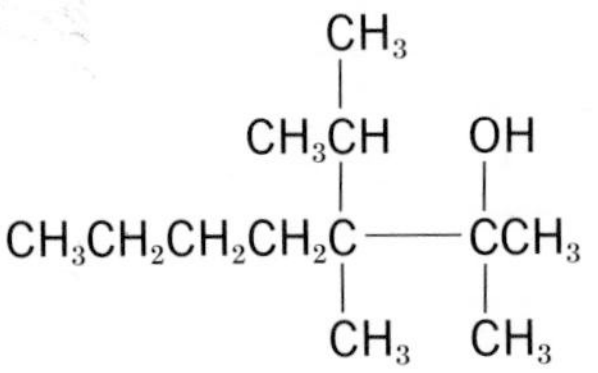

```
                 CH3
                 |
               CH3CH    OH
                 |      |
CH3CH2CH2CH2C ------- CCH3
                 |      |
                CH3    CH3
```

3-Isopropyl-2,3-dimethyl-2-heptanol

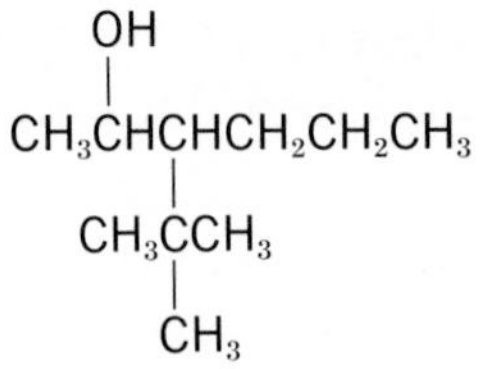

```
      OH
      |
CH3CHCHCH2CH2CH3
        |
      CH3CCH3
        |
       CH3
```

3-*t*-Butyl-2-hexanol

4. If other atoms or groups are attached to or incorporated into the parent chain, first apply rules 1-3 and then work the other groups into the name. For example,

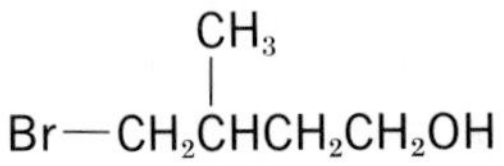

```
           CH3
           |
Br—CH2CHCH2CH2OH
```

4-Bromo-3-methyl-1-butanol
(not 1-Bromo-2-methyl-4-butanol)

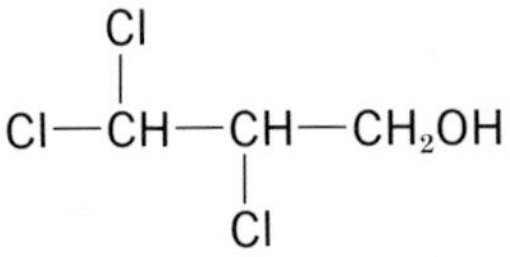

```
    Cl
    |
Cl—CH—CH—CH2OH
        |
        Cl
```

2,3,3-Trichloro-1-propanol

5. In a polyfunctional compound having an —OH group, where it may be awkward to include the ending "-ol", the —OH group may be named as a substituent, "hydroxy." For example, *p*-hydroxybenzoic acid.

Exercise 4.2 Name each by the IUPAC system.

```
        CH3
        |
(a) CH3—CH—CH2—CH2—CH2—OH
```

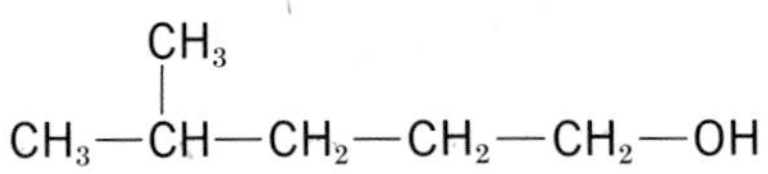

```
           CH3
           |
(b) CH3—C—OH
           |
           CH3
```

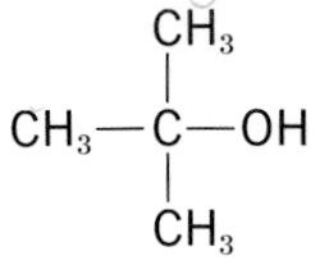

```
               CH3
               |
(c) CH3—CH2—C—CH2—OH
               |
               CH2—CH2—CH3
```

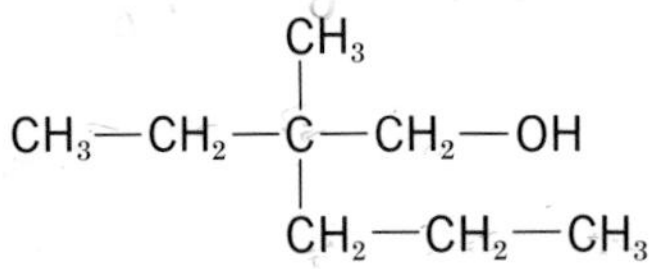

(d) Br—CH_2—CH_2—CH_2—OH

Exercise 4.3 If the IUPAC name of $HOCH_2CH_2OH$ is 1,2-ethanediol, what is the IUPAC name for glycerol: $HOCH_2CHCH_2OH$?

```
              |
              OH
```

Exercise 4.4 If the IUPAC name for **X** is 1,3-cyclohexanediol, what is the IUPAC name for **Y**?

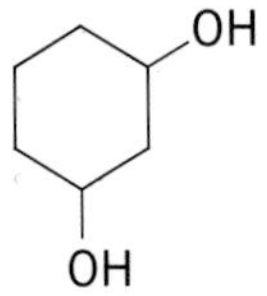

OH
OH
X

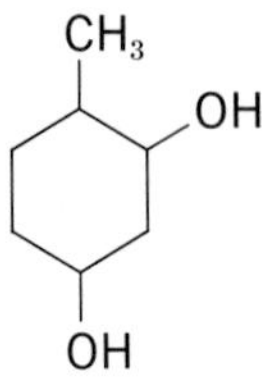

CH3
OH
OH
Y

4.8 Nomenclature of Phenols

"Phenol" is a parent name; it is usually but not always used in naming substituted phenols. Halogen or nitro derivatives of phenol are named with this as the basis as seen in the examples.

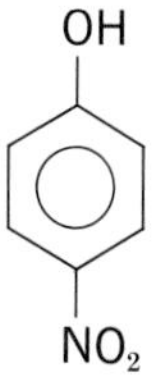

OH benzene ring NO_2	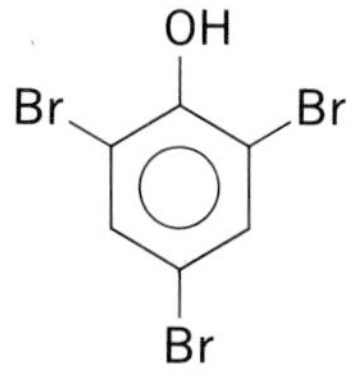 OH, Br, Br benzene ring Br	OH, NH_2 benzene ring	O=C–OH benzene ring OH
p-Nitrophenol	2,4,6-Tribromophenol	*o*-Aminophenol	*p*-Hydroxybenzoic acid

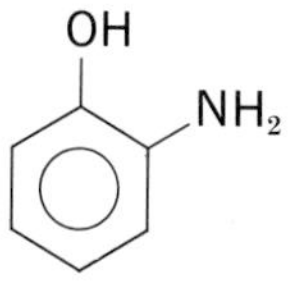

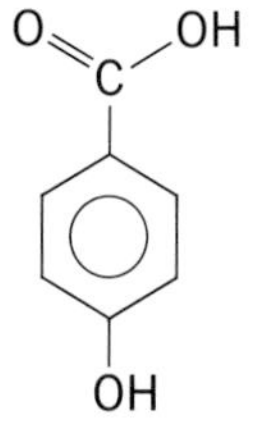

Many phenols, however, have common names that are used almost exclusively. The methylphenols, for example, are called cresols. See Table 4.2.

4.9 Ethers. Common Names

The two groups joined to oxygen are named, and the word *ether* is placed last. If the two groups are identical, the prefix "di-" is used (but it is often omitted by chemists).

$CH_3—O—CH_3$	$CH_3CH_2—O—CH_2CH_3$	$CH_3OCH_2CH_2CH_2CH_3$
Dimethyl ether	Diethyl ether "Ether"	Methyl *n*-butyl ether (Note: The parts of the name are not run together.)

IUPAC rules exist for naming ethers but we shall have no need for them.

Exercise 4.5 Give a common name to each compound.
(a) $CH_3—CH_2—O—CH_3$ (b) $CH_3—CH_2—CH_2—O—CH_2—CH_2—CH_3$

Physical Properties

4.10 Hydrogen Bonds

Ethane and methyl alcohol have similar formula weights (30 vs. 32), yet ethane boils 152° below methyl alcohol (−88.6° versus 64.5°). Far more energy is evidently needed to separate molecules of the alcohol from each other than to separate those of the alkane. Alcohol molecules cling to each other by **hydrogen bonds** just as do water molecules (Figure 4.1). This electrical force of attraction between partial charges, from a $\delta+$ on a hydrogen held covalently at one highly electronegative atom (F, O, or N) to the $\delta-$ on another such atom, affects boiling points, solubilities, and other properties of a number of organic families. The —OH group can donate a hydrogen bond because it has the hydrogen covalently attached to an electronegative atom. This group in

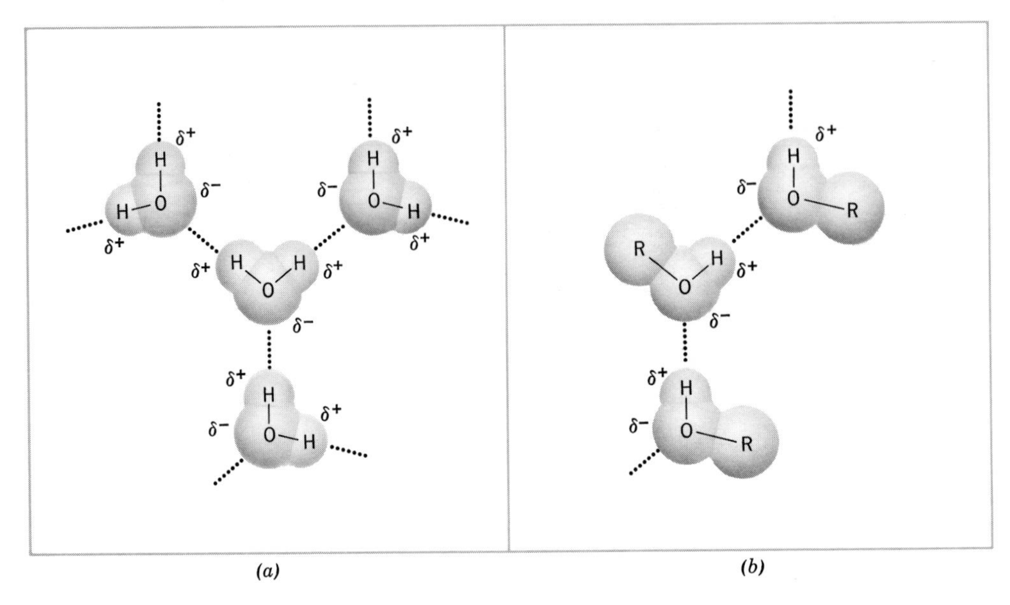

Figure 4.1 Hydrogen bonding in water (*a*) and alcohols (*b*). Hydrogen bonding in water may occur from three sides of the molecule (*a*). In liquid alcohols hydrogen bonding is limited to two sides of the molecule. Low-formula-weight alcohols, therefore, have lower boiling points than water. (Although the drawings suggest considerable order and rigidity in these liquids, their molecules move about constantly as hydrogen bonds break and reform.)

another molecule or elsewhere on the same molecule can accept hydrogen bonds, too, because it has an oxygen function.

Exercise 4.6 Propylene glycol and *n*-butyl alcohol have similar formula weights. Hydrogen bonding between molecules in one of these is more extensive and stronger than in the other. Which? What data given in this chapter support the answer?

4.11 Solubility

Substances in which molecules are attracted to each other by hydrogen bonds are said to be associated. They often tend to dissolve in each other. The mutual solubility of water in methyl alcohol is an example.

Hydrocarbons are nonassociated substances. In order for an alkane or any hydrocarbon to dissolve in water, hydrogen bonds between water molecules would have to be broken and reduced in number. The formation of new hydrogen bonds to replace the old ones cannot occur because hydrocarbons do not have their own —OH groups. In contrast, methyl alcohol molecules can slip into the network of hydrogen bonds (or vice versa) in a water solution, Figure 4.2, and low-formula-weight alcohols and most di- or polyhydric alcohols are soluble in water in all proportions.

Straight-chain alcohols having five or more carbons are too hydrocarbonlike, and they are virtually insoluble in water.

Since all alcohols are partly hydrocarbonlike, it is possible for most of them

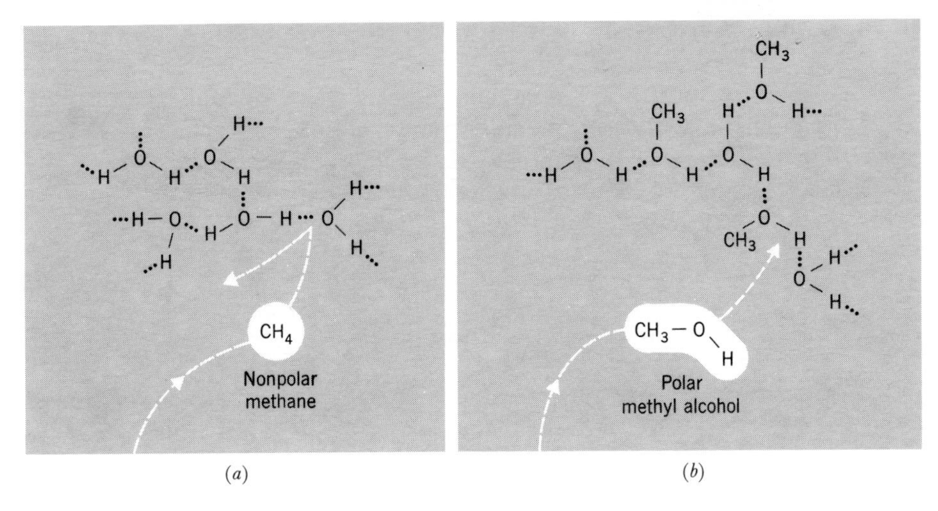

Figure 4.2 The solubility of short-chain alcohols in water. Because an alkane or any hydrocarbon cannot replace hydrogen bonds that must be broken to dissolve something in water (*a*), it cannot dissolve. An alcohol can replace some of these hydrogen bonds (*b*) and can therefore dissolve in water.

to dissolve in typical "hydrocarbon solvents": benzene, ether, carbon tetrachloride, chloroform, gasoline, petroleum ether, and methylene chloride.

To recapitulate, water tends to dissolve waterlike molecules; nonpolar solvents tend to dissolve hydrocarbonlike substances. Polar solvents dissolve polar solutes; nonpolar solvents dissolve nonpolar (or moderately polar) solutes; "likes dissolve likes."

While the ether group cannot donate a hydrogen bond, it can accept one, as seen in Figure 4.3. Therefore, ethers are more soluble in water than comparable alkanes while having comparable boiling points. Both *n*-butyl alcohol and diethyl ether have about the same solubility in water, about 8 g in 100 ml of water, whereas the butanes and pentanes are insoluble. Diethyl ether, however, boils at 35 °C, *n*-pentane at 36 °C, and *n*-butyl alcohol at 117 °C. Ether molecules are not polar enough to attract each other any more than those of *n*-pentane.

Chemical Properties of Alcohols, Phenols, and Ethers

4.12 Relative Acidity of Alcohols and Phenols

Alcohols are neither acids nor bases in the Arrhenius sense in water. They cannot donate protons and neutralize sodium or potassium hydroxide or oth-

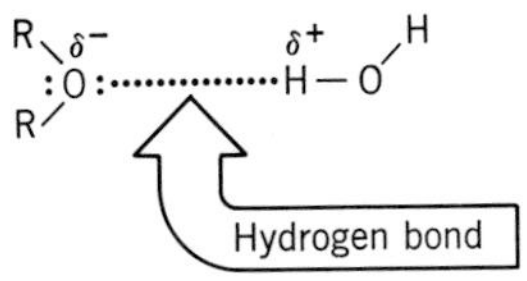

$R_2\overset{\delta-}{O}$:··············$\overset{\delta+}{H}$—O—H

Hydrogen bond

Figure 4.3 An ether molecule accepts a hydrogen bond. (The δ^- end of a hydrogen bond is the "acceptor" end.)

erwise react with alkali at room temperature. Alcohols cannot release hydroxide ions in water, either.

One reason for creating separate classes for phenols and alcohols is the much higher acidity of phenols. Phenol, although a very weak acid compared to hydrochloric or sulfuric acids, is a million times stronger an acid than ethyl alcohol.[2] Phenol is not strong enough an acid to neutralize the weak base, sodium bicarbonate, but it reacts quantitatively with sodium hydroxide, a strong base:

$$C_6H_5\text{—}O\text{—}H + NaOH \longrightarrow C_6H_5\text{—}O^-Na^+ + H_2O$$

Phenol $\qquad\qquad$ Salt of phenol

Exercise 4.7 Write equations for the reactions, if any, between (a) ethyl alcohol and sodium hydroxide, and (b) *p*-cresol and sodium hydroxide.

4.13 Dehydration of Alcohols

Alcohols can be changed to alkenes or to ethers by the catalytic action of strong acids and heat. The catalyst is essential. A molecule of water also forms, and these reactions are therefore called **dehydrations.** Let us study the formation of alkenes first.

The pieces of a water molecule, H— and —OH, can be pulled from an alcohol molecule to give an alkene if the hydrogen is on the carbon next to the alcohol carbon. The pieces of a water molecule, in other words, must be on adjacent carbons. In general:

$$\text{—}\underset{\text{H}}{\overset{|}{\underset{|}{C}}}\text{—}\underset{\text{OH}}{\overset{|}{\underset{|}{C}}}\text{—} \xrightarrow[\text{heat}]{H^+} \;\rangle C{=}C\langle\; + H\text{—}OH$$

Alcohol $\qquad\qquad$ Alkene

Specific examples:

$$\underset{\text{Ethyl alcohol}}{CH_3CH_2\text{—}OH} \xrightarrow[170\text{–}180\ ^\circ C]{\text{concd } H_2SO_4} \underset{\text{Ethylene}}{CH_2{=}CH_2} + H_2O$$

$$\underset{sec\text{-Butyl alcohol}}{CH_3CH_2\underset{\text{OH}}{\underset{|}{C}}HCH_3} \xrightarrow[100\ ^\circ C]{60\%\ H_2SO_4} \underset{\substack{\text{2-Butene}\\(\text{principal product})}}{CH_3CH{=}CHCH_3} + \underset{\text{1-Butene}}{CH_3CH_2CH{=}CH_2} + H_2O$$

[2] Respective K_a-values at or near room temperature are: for most alcohols, 10^{-16} to 10^{-18}; and for phenol, 10^{-10}; where $K_a = [H^+][A^-]/[HA]$, HA being the weak acid and A^- its anion.

$$CH_3-\underset{CH_3}{\overset{CH_3}{\underset{|}{\overset{|}{C}}}}-OH \xrightarrow[80\text{–}90\ ^\circ C]{20\%\ H_2SO_4} CH_2{=}C\genfrac{}{}{0pt}{}{\diagup CH_3}{\diagdown CH_3} + H_2O$$

t-Butyl alcohol — Isobutylene

4.14 Ethers from Alcohols

Besides changing to alkenes, some alcohols may change to ethers when they are heated in the presence of a mineral acid. The elements of water may form from two different molecules, instead of from within one molecule. In general:

$$R-O-H + H-O-R \xrightarrow[\Delta]{\text{concd } H_2SO_4} R-O-R + H-OH$$

Specific examples:

$$CH_3CH_2-O-H + H-O-CH_2CH_3 \xrightarrow[140^\circ]{\text{concd } H_2SO_4} CH_3CH_2-O-CH_2CH_3 + H_2O$$

Ethyl alcohol — Diethyl ether

$$CH_3-O-H + H-O-CH_3 \xrightarrow[\Delta]{\text{concd } H_2SO_4} CH_3-O-CH_3 + H_2O$$

Methyl alcohol — Dimethyl ether

At a temperature of 140 °C in the presence of concentrated sulfuric acid, ethyl alcohol may be converted to diethyl ether. At a higher temperature, 170 to 180 °C, ethyl alcohol is dehydrated internally by acid, and ethylene is the product. When one system of reagents can lead to two or more products, the experimental conditions determine which forms in the major amount.

Exercise 4.8 Write equations for the acid-catalyzed dehydration of these compounds.

(a) cyclohexane ring—OH (b) benzene ring—$\underset{OH}{\underset{|}{CH}}-CH_3$

Exercise 4.9 What alcohol would be needed to make each ether? (Write its structure.)

(a) $CH_3CH_2CH_2OCH_2CH_2CH_3$ (b) benzene ring—CH_2-O-CH_2—benzene ring

4.15 Oxidation

A number of oxidizing agents will convert 1° alcohols to **aldehydes** or **carboxylic acids;** 2° alcohols are oxidized to **ketones.**

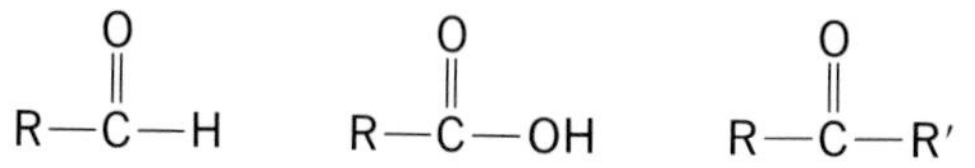

$$\underset{\text{Aldehyde}}{\mathrm{R{-}\overset{\overset{\displaystyle O}{\|}}{C}{-}H}} \qquad \underset{\text{Carboxylic acid}}{\mathrm{R{-}\overset{\overset{\displaystyle O}{\|}}{C}{-}OH}} \qquad \underset{\text{Ketone}}{\mathrm{R{-}\overset{\overset{\displaystyle O}{\|}}{C}{-}R'}}$$

As a reminder, an **oxidation reaction** is any reaction that results in either a gain in the number of oxygens or a loss in the number of hydrogens in a molecule. A **reduction reaction** is one in which a molecule suffers either a gain in hydrogens (e.g., hydrogenation of alkenes) or a loss in oxygens.

Controlled oxidation of alcohols (i.e., not combustion) removes the pieces of the element hydrogen, H—H, one —H from the —OH group and the other —H from the alcohol carbon.[3]

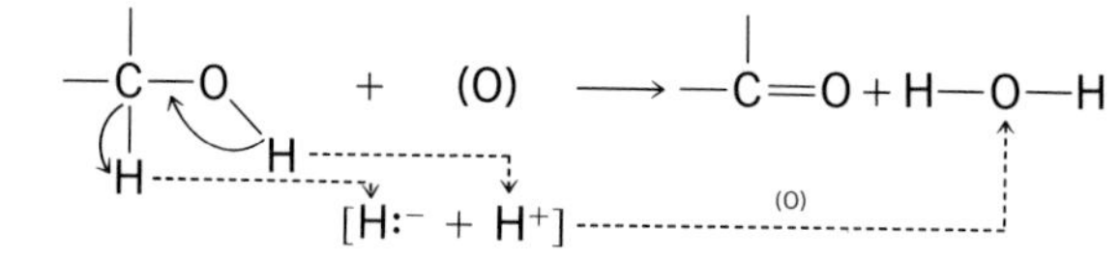

$$\underset{\text{(1° or 2° alcohol)}}{-\overset{|}{\underset{\underset{\displaystyle H}{|}}{C}}-O-H} + (O) \longrightarrow -\overset{|}{C}{=}O + H{-}O{-}H$$

$$[H:^- + H^+] \xrightarrow{(O)} H{-}O{-}H$$

By "pieces of the element hydrogen" we mean enough particles to make up one molecule of hydrogen, for example: $H:^-$ and H^+. Potassium dichromate, $K_2Cr_2O_7$, and potassium permanganate, $KMnO_4$, are two inorganic oxidizing agents that can oxidize alcohols. They remove the pieces of elemental hydrogen and convert them into a molecule of water. In living systems special acceptors of hydrogen are used. Some of them are molecules of vitamins or their near relatives, and it is to that system we are leading.

Primary alcohols are oxidized to aldehydes. If an inorganic oxidizing agent is used, the aldehyde has to be removed almost as rapidly as it forms. Otherwise, the aldehyde will be oxidized by unchanged oxidizing agent to the carboxylic acid because aldehydes are even more readily oxidized than alcohols. Since aldehydes have lower boiling points than their parent alcohols, we can boil them off and thereby remove them as they form. (In the examples that follow, the equations are not balanced.) In general, for 1° alcohols:

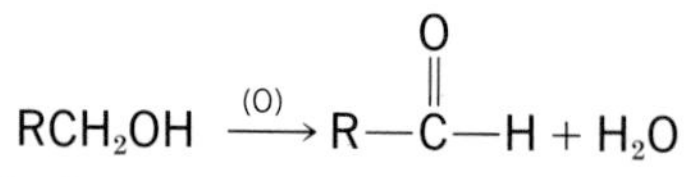

$$\underset{\text{1° Alcohol}}{RCH_2OH} \xrightarrow{(O)} \underset{\text{Aldehyde}}{\mathrm{R{-}\overset{\overset{\displaystyle O}{\|}}{C}{-}H}} + H_2O$$

[3] The symbol (O) represents any chemical oxidizing agent capable of effecting the reaction. The "equations" shown should be spoken of as "reaction sequences." Specific equations are difficult to balance without devoting more space than is intended.

$$\underset{}{RCH_2OH} + \underset{\substack{\text{Potassium}\\\text{permanganate}\\\text{(deep purple)}}}{KMnO_4} \longrightarrow \underset{\text{Aldehyde}}{R-\overset{\overset{\displaystyle O}{\|}}{C}-H} \xrightarrow{KMnO_4} \underset{\substack{\text{Salt of carboxylic}\\\text{acid}}}{R-\overset{\overset{\displaystyle O}{\|}}{C}-O^-K^+} + \underset{\substack{\text{Manganese}\\\text{dioxide}\\\text{(brown sludge)}}}{MnO_2} + KOH$$

$$\xrightarrow{H_3O^+} \underset{\text{Carboxylic acid}}{R-\overset{\overset{\displaystyle O}{\|}}{C}-O-H} + H_2O + K^+$$

$$RCH_2OH + \underset{\substack{\text{Sodium}\\\text{dichromate}\\\text{(bright orange)}}}{Na_2Cr_2O_7} \xrightarrow{H^+} R-\overset{\overset{\displaystyle O}{\|}}{C}-H \xrightarrow{Cr_2O_7^{2-}} R-\overset{\overset{\displaystyle O}{\|}}{C}-OH + \underset{\substack{\text{Chromium}\\\text{ion}\\\text{(bright green)}}}{Cr^{3+}}$$

Specific examples:

$$\underset{\substack{n\text{-Propyl alcohol}\\\text{(bp 97 °C)}}}{CH_3CH_2CH_2OH} \xrightarrow[\text{(49\%)}]{Cr_2O_7^{2-},H^+} \underset{\substack{\text{Propionaldehyde}\\\text{(bp 55 °C)}}}{CH_3CH_2\overset{\overset{\displaystyle O}{\|}}{C}H}$$

$$\underset{\substack{n\text{-Butyl alcohol}\\\text{(bp 118 °C)}}}{CH_3CH_2CH_2CH_2OH} \xrightarrow[\text{(72\%)}]{Cr_2O_7^{2-},H^+} \underset{\substack{n\text{-Butyraldehyde}\\\text{(bp 82 °C)}}}{CH_3CH_2CH_2\overset{\overset{\displaystyle O}{\|}}{C}H}$$

Carboxylic acid from 1° alcohol:

$$\underset{n\text{-Propyl alcohol}}{CH_3CH_2CH_2OH} \xrightarrow[\text{(65\%)}]{Cr_2O_7^{2-},H^+} \underset{\text{Propionic acid}}{CH_3CH_2\overset{\overset{\displaystyle O}{\|}}{C}OH}$$

Secondary alcohols are oxidized to ketones. In sharp contrast to aldehydes, ketones are quite resistant to oxidation. They need not be removed as they form, and yields are usually higher than in the syntheses of aldehydes. In general, for 2° alcohols:

$$\underset{2°\text{ Alcohol}}{R-\overset{\overset{\displaystyle OH}{|}}{C}H-R} + (O) \longrightarrow \underset{\text{Ketone}}{R-\overset{\overset{\displaystyle O}{\|}}{C}-R} + H_2O$$

Specific examples:

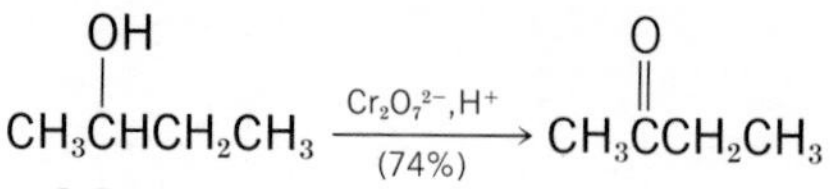

$$CH_3CH(OH)CH_2CH_3 \xrightarrow[(74\%)]{Cr_2O_7^{2-}, H^+} CH_3C(=O)CH_2CH_3$$

2-Butanol → 2-Butanone

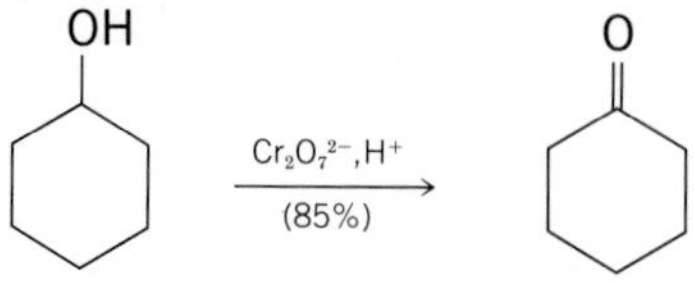

$$\text{Cyclohexanol} \xrightarrow[(85\%)]{Cr_2O_7^{2-}, H^+} \text{Cyclohexanone}$$

Cyclohexanol → Cyclohexanone

Tertiary alcohols such as *t*-butyl alcohol cannot be oxidized in these ways because their molecules lack the hydrogen on the alcohol carbon.

Ethers cannot be oxidized in these ways, either, because they lack the hydrogen on an —OH group.

Exercise 4.10 Write the structure of the product of oxidation in each case. (Take 1° alcohols to the aldehyde stage only.) If oxidation cannot occur (as studied in this section), state so.

(a) CH_3OH (b) CH_3CH_2OH

(c) 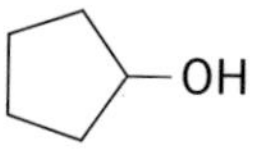 Cyclopentanol (cyclopentane ring with —OH) (d) 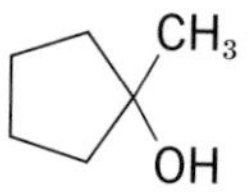 1-Methylcyclopentanol (cyclopentane ring bearing CH_3 and OH on the same carbon)

4.16 Reactions of Phenols

In sharp contrast to the rings of benzene, alkylbenzenes (e.g., toluene), and their halogenated or nitrated derivatives, the rings in phenols are very easily attacked by oxidizing agents. The reaction is seldom useful for making other compounds, however. Instead, deeply colored dyes and tars form. Even exposure to air is enough to oxidize crystals of phenol and give them a reddish-purplish coating. Pigments of the skin are formed by the oxidation of a phenolic side chain contributed by the amino acid tyrosine. If the enzyme for catalyzing and controlling this oxidation is absent or defective, the individual is an albino.

One phenolic system important in living systems that can be oxidized in a controlled way is the phenolic system in hydroquinone.

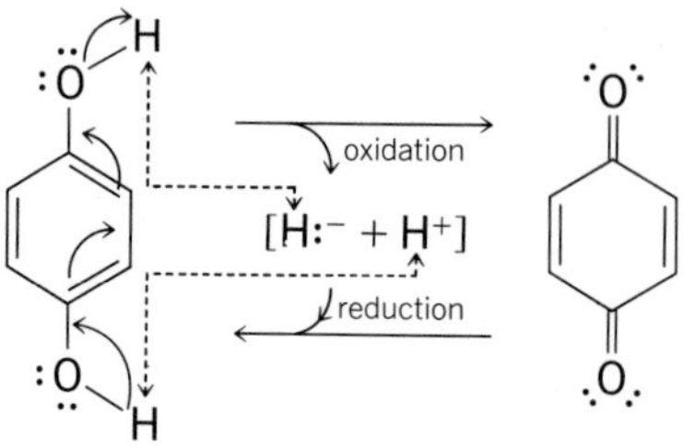

$$\text{Hydroquinone} \underset{\text{reduction}}{\overset{\text{oxidation}}{\rightleftharpoons}} p\text{-Quinone} \quad [H:^- + H^+]$$

Hydroquinone   *p*-Quinone

As the curved arrows indicate, an easy path is open for the loss of the pieces of hydrogen; H:⁻ (a hydride ion) comes from one oxygen and H^+ from the other. Although the change to quinone appears to break up the pi-electron network of the ring in hydroquinone, a new network of almost the same degree of electron delocalization forms. Quinone is easily reduced by mild reducing agents back to hydroquinone. The quinone-hydroquinone system, therefore, can shuttle a pair of electrons (the pair on the hydride ion, $H:^-$) back and forth depending on the chemical environment. The quinone-hydroquinone electron shuttle is an integral part of a major biological network, the respiratory chain, for oxidizing molecules of food to obtain energy for living. The chain or series of enzymes includes quinonelike compounds called the ubiquinones in one enzyme. Vitamin K_2, phylloquinone, is also a quinone.

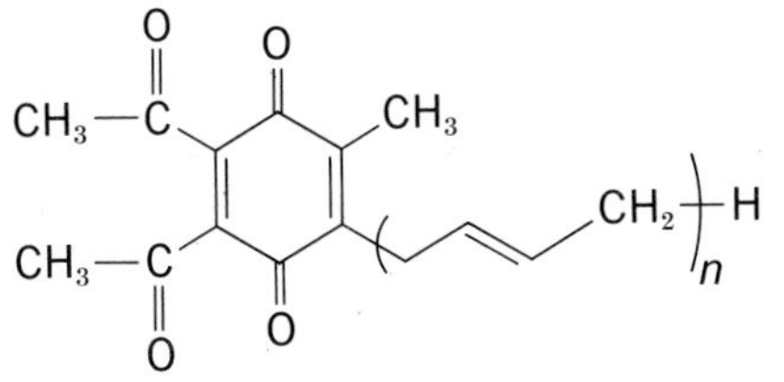

CH_3—C(=O) CH_3—C(=O) CH_3 CH_2—H n

Ubiquinones (n = 1,2,...10)
(coenzyme Q)

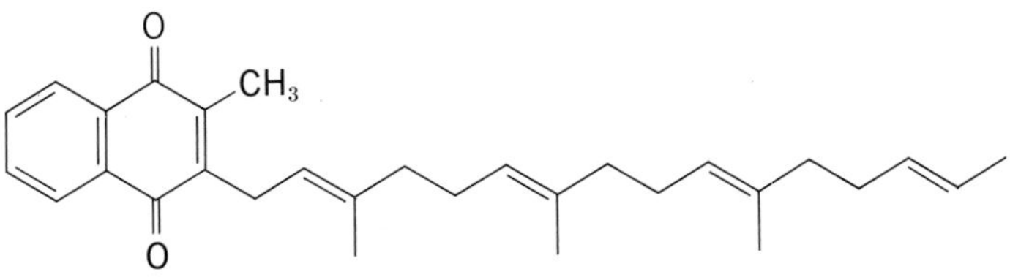

CH_3

Vitamin K_2
(phylloquinone)

The phenol ring is easily attacked by other kinds of electron-poor reagents besides oxidizing agents. Where benzene requires an iron or iron(III) salt as a catalyst for chlorination or bromination, phenol needs no catalyst and the reaction cannot be controlled until three halogens have entered the system. The —OH group on a benzene ring is thus a powerful **activating group.** We cannot examine a detailed explanation for that behavior, but we may note that the —OH group has pairs of electrons adjacent to the pi-electron network of

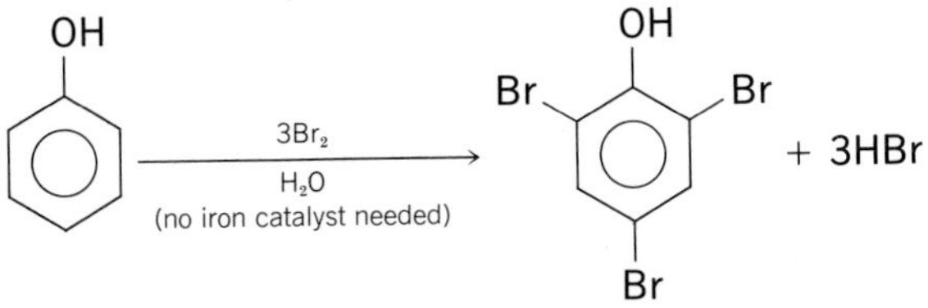

OH —[$3Br_2$, H_2O (no iron catalyst needed)]→ OH, Br, Br, Br + 3HBr

the ring. As some attacking particle, for example, Cl^+ or Br^+, removes pi-electron density from the ring breaking up the closed circuit, temporarily, electron density from oxygen can flow in to replace lost electron density temporarily. (Conversely, of course, if a substituent is a strong electron-withdrawing group it ought to act as a **deactivating group,** i.e., make the ring less reactive than that of benzene if our explanation for the —OH group is correct. The nitro group is an electron-withdrawing, deactivating group; the ring in nitrobenzene is unusually resistant to further substitution and to oxidation, tending to confirm the theory.)

4.17 Reactions of Ethers

At room temperature, the ether function is stable to acids, bases, oxidizing agents, reactive metals (e.g., sodium), and reactive nonmetals (e.g., bromine). Ethers are highly flammable. Diethyl ether is particularly dangerous, since its vapors are denser than air and tend to accumulate along the floor, where chance sparks (e.g., from shoe cleats) can detonate the ether-air mixture.

What few reactions ethers do undergo are of small consequence in the chemistry of health. Since the ether function does occur in some types of molecules in the body, it is important to know that this group is chemically very stable.

Important Individual Compounds

4.18 Alcohols

Methyl Alcohol.
("Wood Alcohol," Methanol)

Most methyl alcohol is made by the reaction of carbon monoxide with hydrogen under extremely high pressure and temperature.

$$2H_2 + CO \xrightarrow[\substack{\text{temp.} = 350\text{–}400\ ^\circ\text{C} \\ \text{catalyst} = ZnO\text{—}Cr_2O_3}]{3000\ \text{lb/in.}^2} CH_3\text{—}OH$$

It acquired its nickname, wood alcohol, from the fact that it is obtainable when wood is heated in the absence of air during the manufacture of charcoal.

Taken internally in sufficient quantity, methyl alcohol produces either blindness or death. It is used primarily as the raw material for the industrial synthesis of formaldehyde, as a solvent, and as a denaturant (poison) for ethyl alcohol.

Ethyl Alcohol.
("Grain Alcohol," Ethanol)

Some ethyl alcohol is obtained by the fermentation of sugars, but most is synthesized by the hydration of ethylene in the presence of a catalyst. A 70% (vol/vol) solution of ethyl alcohol is used as a disinfectant. Industrially, it is used as a solvent and in the compounding of pharmaceuticals, perfumes, lotions, tonics, and rubbing compounds. For these purposes, it is adulterated (denatured) by poisons that are very difficult to remove. This is done to prevent its use as a beverage; for nearly all governments derive considerable revenue by taxing potable alcohol.

The drinkability of (dilute) ethyl alcohol is unique among the alcohols. The body possesses enzymes that act to destroy it rapidly by oxidation. In excess it causes permanent damage to the liver. The illusion that alcohol is a stimulant derives from the fact that its first effect is to depress activity in the uppermost level of the brain, the center of judgment, inhibition, and restraint. Alcohol abuse and alcoholism constitute the major drug problem in the United States and many other countries.

Isopropyl Alcohol

Isopropyl alcohol is a common substitute for ethyl alcohol as a rubbing compound. It is twice as toxic as ethyl alcohol. Isopropyl alcohol is also used as a disinfectant in concentrations from 50 to 99% (vol/vol).

Ethylene Glycol. Propylene Glycol

These two glycols serve as the base for all permanent-type antifreezes. Their great solubility in water and their very high boiling points make them ideal for this purpose.

Glycerol (Glycerin)

Glycerol, a colorless, syrupy liquid with a sweet taste, is freely soluble in water and insoluble in nonpolar solvents. It is a product of the digestion of simple fats and oils.

Sugars

All carbohydrates consist of polyhydroxy compounds. They will be studied in a later chapter.

4.19 Phenols

Phenol ("Carbolic Acid")

Phenol is a general protoplasmic poison and was the first antiseptic to be employed by Joseph Lister (1827–1912), an English surgeon. Because it is dangerous to healthy tissue, other antiseptic agents have since been developed, but phenol and the cresols (the phenolic toluenes) are still used in aqueous solutions as disinfectants for floors, drains, walls, cesspools, and toilets. Phenol should not be allowed to come in contact with bare hands. Industrially, phenol is a chemical of major importance being used to make drugs (e.g., aspirin), dyes, and plastics.

***n*-Hexylresorcinol**

This phenol is used in dilute solution as an antiseptic. In concentrated form it is an anthelmintic, an agent for destroying intestinal worms.

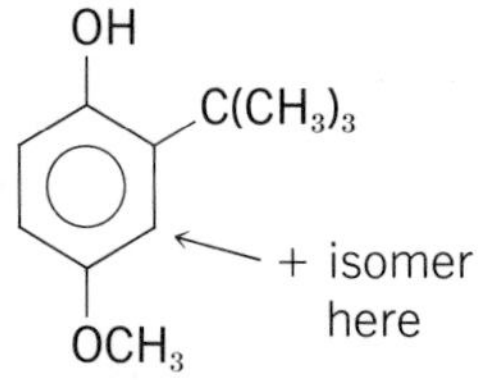

OH, $C(CH_3)_3$, + isomer here, OCH_3
BHA (Butylated hydroxy anisole)

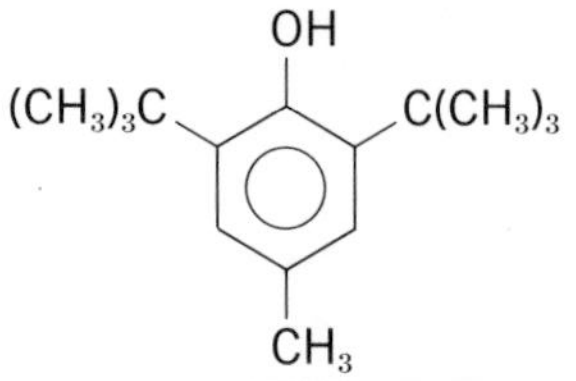

OH, $(CH_3)_3C$, $C(CH_3)_3$, CH_3
BHT (Butylated hydroxy toluene)

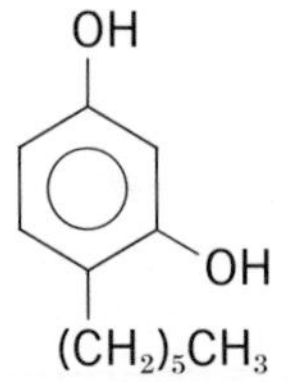

OH, OH, $(CH_2)_5CH_3$
n-Hexylresorcinol

BHA and BHT

These are widely used as antioxidants in gasoline, lubricating oils, rubber, edible fats and oils, and materials used for packaging foods that might turn rancid. Being phenols and easily oxidized, they act by interfering with oxidizing reactions of the materials they are designed to protect. In food or container uses they are limited to concentrations of 200 ppm (0.02%) based on the fat or oil content of the food.

4.20 Ethers

Diethyl Ether ("Ether")

Diethyl ether is a colorless, volatile liquid with a pungent, somewhat irritating odor, widely used as an anesthetic. Although it was first used for this purpose in 1842 by Crawford Long in Georgia, word of the results was not published, and the medical profession did not profit from his discovery. Four years later, in 1846, a Boston dentist, William T. G. Morton, rediscovered the anesthetic properties of ether and successfully demonstrated them in an actual operation.

Diethyl ether is a depressant for the central nervous system and is, at the same time, somewhat of a stimulant for the sympathetic system. It exerts an effect on nearly all tissues of the body.

Divinyl Ether ("Vinethene")

This ether is another anesthetic, being more rapid in its action than ethyl ether. It forms an explosive mixture with air.

Brief Summary

Structural Features. Alcohols may be represented as R—OH where the carbon holding the —OH group must be saturated. The alcohol carbon may be 1°, 2°, or 3°. Alcohols may have two or more —OH groups. Phenols may be symbolized as

Ar—OH, where Ar— is an aromatic ring directly holding the —OH group. Ethers may be represented as R—O—R where the R— groups may be alike or different, aliphatic or aromatic.

Names. Common names of alcohols are made by writing the name of the R— group followed by "alcohol" as a separate word. Common names of ethers are devised by naming each group joined to the oxygen followed by "ether" as a separate word. Common names of phenols simply have to be learned. The IUPAC names of alcohols are made in the same way we make names of alkanes except that the —OH group dominates both the choice of the chain and its numbering. The main chain must include the alcohol carbon, and the chain is numbered to give the location of the —OH group the lower number. (We did not study IUPAC ways to name ethers or phenols.)

Hydrogen Bonds. Opportunities for hydrogen bonding greatly affect the relative boiling points and water solubilities of substances; the more hydrogen bonding possible, the higher the boiling point and water solubility.

Chemical Properties of Alcohols. Alcohols do not ionize as either acids or bases. Water, dilute acids, dilute bases, or reducing agents do not react with alcohols at room temperature. Hot acids dehydrate alcohols producing either alkenes or ethers, depending on the conditions. (Usually, some of both form, and conditions are adjusted to maximize the formation of the desired product.) Hot oxidizing agents change 1° alcohols to aldehydes, which may be further oxidized to carboxylic acids. Secondary alcohols are oxidized to ketones, which resist further oxidation. (Reactions of alcohols with organic compounds will be studied in succeeding chapters.)

Chemical Properties of Phenols. Phenols are much stronger acids than alcohols, strong enough to neutralize sodium hydroxide but are much less strong than carboxylic acids or mineral acids. Even though they are very weak acids, they are damaging to tissue and sometimes are used to sterilize equipment. The —OH group on a benzene ring powerfully activates the ring to further substitution, and the ring is much more susceptible to oxidation (to colored products). The special phenolic system in hydroquinone is widely found in living systems as part of a vitamin and an enzyme.

Chemical Properties of Ethers. Ethers give no reaction with water, dilute acids or bases, and dilute oxidizing or reducing agents, at least at room temperature. (However, they do burn well.)

Syntheses. Alkenes and ethers can be made from alcohols. Aldehydes and carboxylic acids can be made from 1° alcohols, ketones from 2° alcohols.

Questions and Exercises

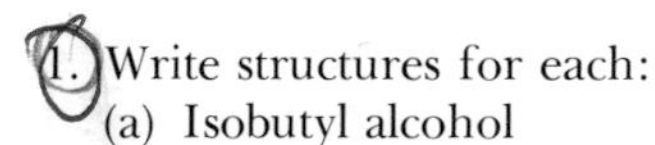

1. Write structures for each:
   (a) Isobutyl alcohol
   (b) Glycerol
   (c) Isopropyl alcohol
   (d) Propylene glycol
   (e) *t*-Butyl alcohol
   (f) *n*-Heptyl alcohol
   (g) *o*-Nitrophenol
   (h) *m*-Cresol
   (i) Diisopropyl ether
   (j) Ethyl isobutyl ether
2. Write common names for each.

(a) $(CH_3)_2CHOH$

(b)
```
CH3CHCH2CH3
    |
    OH
```

(c)
```
CH3—O—CHCH2CH3
      |
      CH3
```

(d) CH_3—C₆H₄—OH (para-substituted benzene ring)

(e) C₆H₄(OH)(NO₂) (benzene ring with —OH and meta —NO_2)

(f) $(CH_3)_2CHCH_2OH$

3. Write the IUPAC name for the following compound.

```
        CH3 OH    CH2CH2CH3
         |   |     |
CH3CH2CCH2CHCH2CHCHCH3
         |          |
         CH         CH3
        /  \
     CH3    CH3
```

4. Write an equation (not necessarily balanced) for the reaction of *n*-propyl alcohol with each of the following reagents.
(a) Potassium permanganate (excess)
(b) Concentrated sulfuric acid, heat. (Show two kinds of reactions.)

5. You are given an unknown liquid and told that it is either *t*-butyl alcohol or *n*-butyl alcohol. Moreover, you are told that when a few drops of the unknown are shaken with dilute, potassium permanganate, the purple permanganate color disappears and a brown precipitate forms. Which alcohol is it? Explain in detail.

6. Write the structure of the principal organic product that, on the basis of what has been studied, would be expected to form in each of the following situations. If no reaction would occur, state "no reaction." (*Note:* To differentiate between alkene formation and ether formation, a coefficient of 2 will be placed before the structure of an alcohol if it is intended that you write the structure of the ether that could form. Otherwise, it is assumed that you would write reactants and products before you try to balance equations.) To review how to handle problems such as these see Exercise 11, Chapter 3. Some equations will be reviews of reactions of previous chapters.

(a) $(CH_3)_2CHOH \xrightarrow[H^+]{Cr_2O_7^{2-}}$

(b) $CH_3CH_2CH_2CH_3 + H_2SO_4 \longrightarrow$ no because no OH

(c) $CH_3CH_2CH_2C(CH_3)_2 \xrightarrow{MnO_4^-}$ (with OH on the C bearing the two CH_3 groups)

(d) $CH_3CH_2CH_2OH + NaOH_{aq} \longrightarrow$ no, alcohols dont react with NaOH

(e) $2CH_3OH \xrightarrow[\Delta]{H_2SO_4}$ CH_3-O-CH_3

(f) $CH_3CH_2CHCH_2CH_3 + (O) \longrightarrow$ (with OH on C-3) CH_3-CH_2-C(=O)-CH_2-CH_3

(g) $CH_3—O—CH_3 + (O) \longrightarrow$ ethers dont accept Os

(h) $CH_3CH{=}CHCH_3 + H_2O \xrightarrow{H^+}$ CH_3-CH(OH)-CH(H)-CH_3

(i) $CH_3CH_2CH_2OH \xrightarrow[\Delta]{H_2SO_4}$

(j) $CH_3CHCH_2CH_3 + (O) \longrightarrow$ (with OH on C-2)

(k) $CH_3CH{=}CH_2 + H_2 \xrightarrow[\text{pressure}]{\text{Ni, heat}}$

(l) $CH_3{-}\overset{\overset{\displaystyle OH}{|}}{\underset{\underset{\displaystyle CH_3}{|}}{C}}CH_2CH_3 + (O) \longrightarrow$

(m) 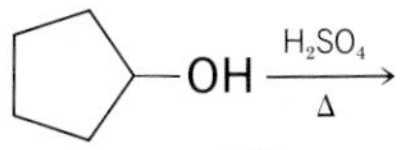 cyclopentyl$-OH \xrightarrow[\Delta]{H_2SO_4}$

(n) $CH_3CH_2OCH_3 + HCl_{aq} \longrightarrow$

(o) 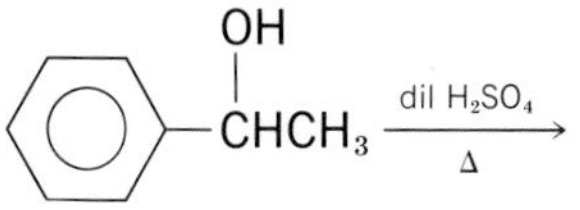 $C_6H_5{-}\overset{\overset{\displaystyle OH}{|}}{C}HCH_3 \xrightarrow[\Delta]{\text{dil } H_2SO_4}$

(p) 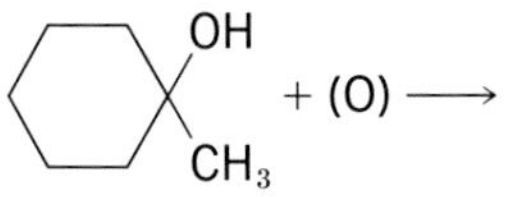 1-methylcyclohexanol (cyclohexane ring bearing OH and CH_3 on the same carbon) $+ (O) \longrightarrow$

(q) $CH_3{-}O{-}CH_2CH{=}CH_2 + H_2 \xrightarrow[\text{pressure}]{\text{Ni, heat}}$

(r) 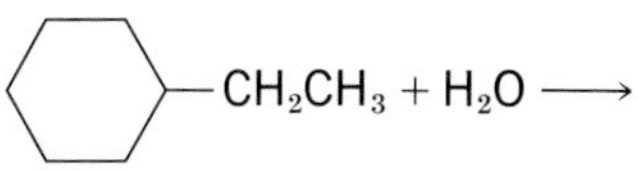 cyclohexyl$-CH_2CH_3 + H_2O \longrightarrow$

(s) $CH_3\overset{\overset{\displaystyle CH_3}{|}}{\underset{\underset{\displaystyle OH}{|}}{C}}CH_3 \xrightarrow[\Delta]{H_2SO_4}$

(t) $CH_3CH_2CH_2OH + (O) \longrightarrow$

(u) $CH_3{-}O{-}CH_2\overset{\overset{\displaystyle OH}{|}}{C}HCH_3 + (O) \longrightarrow$

7. What alcohols would be needed to make the following compounds by oxidation? Write their structures.

(a) $CH_3\overset{\overset{\displaystyle O}{\|}}{C}CH_2CH_3$

(b) 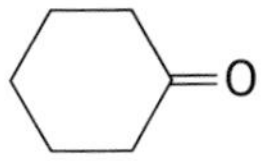 cyclohexanone (cyclohexane ring $=O$)

(c) $CH_3CH_2\overset{\overset{\displaystyle O}{\|}}{C}OH$

(d) $H{-}\overset{\overset{\displaystyle O}{\|}}{C}{-}CH_2CH(CH_3)_2$

(e) $C_6H_5\overset{\overset{\displaystyle O}{\|}}{C}{-}H$

8. Account for the great differences in boiling points as you go from *n*-butane (bp 0 °C, formula weight 60), to *n*-propyl alcohol (bp 97 °C, formula weight 60), to ethylene glycol (bp 197 °C, formula weight 62).

9. *n*-Propyl alcohol is very soluble in all nonpolar solvents, whereas ethylene glycol and glycerol are insoluble. Explain.

chapter 5

Halides, Mercaptans, and Amines

Serenity sometimes comes in the possession of a safe, relatively harmless and altogether effective defense. The skunk's reservoir of mercaptans, studied briefly in this chapter, may not give it serenity but it surely provides security. (Photo by Ed Cesar/The National Audubon Society Collection-Photo Researchers.)

Organohalogen Compounds

5.1 Types and Occurrence

The halogens—fluorine, chlorine, bromine, and iodine—occur only very rarely within organic compounds of living cells. One hormone, thyroxin, contains iodine, and without iodide ion in the diet people have a tendency to develop a goiter. A number of drugs contain one or another of the halogens, and many industrial and agricultural compounds are **organohalogens.**

With few exceptions compounds can exist in which a hydrogen attached to almost any kind of carbon has been replaced by any of the halogens, whether that carbon is saturated or unsaturated, aliphatic or aromatic. The two most common types of organohalogen compounds are the alkyl halides and aryl halides. Examples are given in Table 5.1. Direct halogenation of alkanes or aromatic hydrocarbons can be used to make many organochlorine and organobromine compounds, and alkenes add hydrogen chloride, bromide, or iodide. Organofluorine compounds usually require different methods of syntheses. We shall not deal further with the fluorine compounds except for a brief mention of Freons. The nomenclature of halides has been given in earlier chapters and should be reviewed. Some important individual organochlorines are listed in Table 5.2.

TABLE 5.1
Some Organohalogen Compounds

	Boiling Points (in °C) X =			Densities g/ml (at 20 °C) X =		
Type	Cl	Br	I	Cl	Br	I
Alkyl halides						
R—*X*						
Methyl	−24	5	43	(gas)	(gas)	2.28
Ethyl	13	38	72	(gas)	1.44	1.93
Propyl	47	71	102	0.89	1.34	1.75
Butyl	79	102	130	0.88	1.22	1.52
Cyclohexyl	143	165	180 (dec)	1.00	1.32 (15°)	1.62 (15°)
Aryl halides						
C_6H_5—X	132	156	189	—	—	—
CH_3—C_6H_4—X	162	184	211	—	—	—
Cl—C_6H_4—X	175	196	226	—	—	—

TABLE 5.2
Some Important Organohalogen Compounds

Name	Structure	Boiling Point, in °C	Uses
Ethyl chloride	$CH_3—CH_2—Cl$	12.5	External local anesthetic ("freeze" technique); total anesthetic if breathed (may cause death)
Chloroform	$CHCl_3$	61	Powerful, rapid-acting anesthetic; nonflammable
Carbon tetrachloride	CCl_4	77	Nonflammable dry-cleaning solvent (ample ventilation is essential)
Iodoform	CHI_3	(mp 119)	Antiseptic wound dressing (releases iodine slowly)

5.2 Physical Properties

Table 5.1 gives a few halides and some of their physical properties. The data illustrate the general rule that, all other factors being equal, a rise in formula weight means a rise in boiling point and a rise in density. Compare, for example, any horizontal row of Table 5.1: alkyl chloride, bromide, or iodide. We note, too, that compounds with either bromine or iodine are all more dense than water.

Unable to donate or accept hydrogen bonds, all organohalogen compounds are insoluble in water and soluble in the nonpolar solvents. Many are absorbed directly through the skin, and some enter the body in trace amounts in food and air. They migrate to fatty tissue where they have greater solubility, which means an extra load of alien chemicals for the liver to handle. One of the many functions of this vital organ is the processing of poisons to get them into a form that can be excreted, generally via the urine. Unless a halogen can be replaced by a hydroxyl group, it is hard for a living system to get the chemical moving in aqueous media on a course that leads to its elimination. Some organochlorine compounds simply cannot be changed chemically, at least at rates fast enough. They are not biodegradable either in our bodies or by organisms or chemicals in the environment. They tend to build up in fatty tissue and liver as exposure continues. Such compounds include carbon tetrachloride (CCl_4), some organochlorines once widely used as pesticides (e.g., aldrin, dieldrin, DDT), and members of a family called the polychlorinated biphenyls or PCBs.

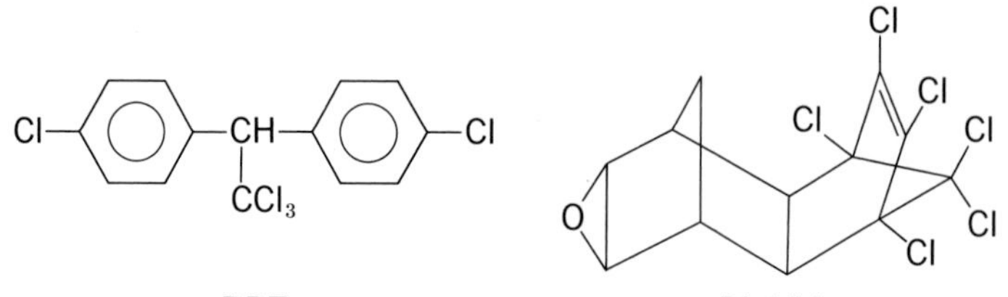

DDT

Dieldrin

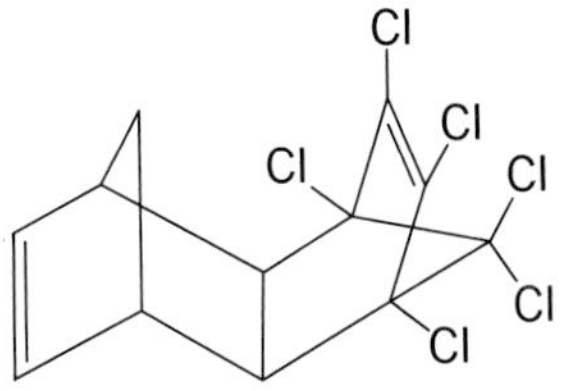

Aldrin

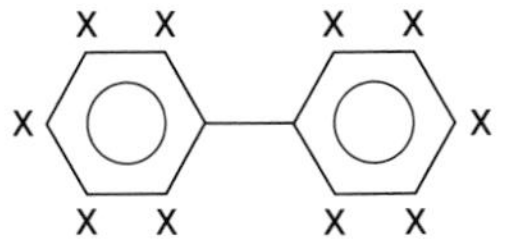

Biphenyl

Polychlorinated biphenyls; sites marked (x) may hold —Cl, and 210 possible compounds are possible. Commercial products are mixtures.

PCBs

While in the liver organochlorine compounds may activate processes that cause harm, or their mere physical presence may upset delicate balances in other chemical systems. Thus the simple physical property of being relatively nonpolar means that organohalogen compounds can be taken in selectively by fatty tissue where, being foreign molecules, they may lead to damage.

5.3 Some Chemical Properties

To understand how some organohalogens are not biodegradable, we have to see how others can react with various chemicals. Because the halogens are more electronegative than carbon, the carbon-halogen bond is somewhat polar. The carbon bears a small, partial positive charge; the halogen a small, partial negative charge. The carbon, therefore, can be attacked by electron-rich particles such as molecules of ammonia or ions such as the hydrogen sulfide or the hydroxide ion. The reactions of methyl bromide with these reagents illustrate what can happen if special conditions (not to be described) are satisfied. These reactions are fairly general methods to make alcohols, amines, and mercaptans.

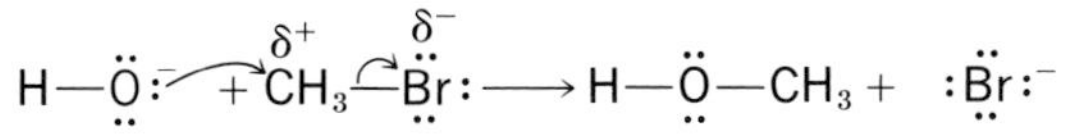

$$H{-}\ddot{O}{:}^{-} + \overset{\delta+}{C}H_3{-}\overset{\delta-}{Br}{:} \longrightarrow H{-}\ddot{O}{-}CH_3 + {:}\ddot{Br}{:}^{-}$$

Hydroxide ion | Methyl bromide | Methyl alcohol | Bromide ion

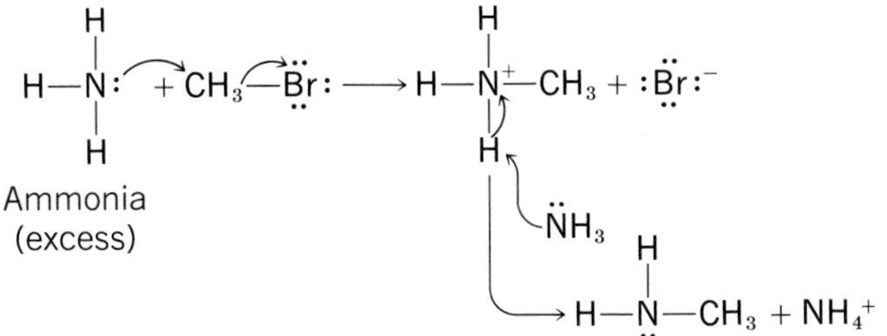

$$H_3N{:} + CH_3{-}Br{:} \longrightarrow H_3\overset{+}{N}{-}CH_3 + {:}\ddot{Br}{:}^{-}$$

Ammonia (excess)

$$H_3\overset{+}{N}{-}CH_3 + {:}NH_3 \longrightarrow H_2\ddot{N}{-}CH_3 + NH_4^{+}$$

Methylamine

$$\text{H—}\ddot{\text{S}}\text{:}^- + \text{CH}_3\text{—}\ddot{\text{Br}}\text{:} \longrightarrow \text{H—}\ddot{\text{S}}\text{—CH}_3 + \text{:}\ddot{\text{Br}}\text{:}^-$$

Hydrogen sulfide ion — Methyl mercaptan

These reactions are examples of **substitution reactions.** In compounds with molecules of longer chains, only the carbon holding the halogen can be successfully attacked because success depends on ejection of the halide ion from one side as the new bond forms on the other. The stabilities of the halogens as halide ions, which also form, contribute to success, too.

Another factor in successful substitutions of these types is simply the access to the carbon holding the halogen. If the approach to it is blocked in any way and the attacking particle cannot get to it, the reaction either cannot occur at all or it takes place in an entirely different way (which we shall not pursue). For example, aryl halides are generally inert toward water, ammonia, the hydroxide ion, or the hydrogen sulfide ion. The electron cloud on the benzene ring repels incoming molecules of the reactant, also electron-rich. That is why the polychlorinated biphenyls or PCBs are not biodegradable. The organochlorine pesticides such as aldrin and dieldrin have cagelike structures that make it difficult for some attacking group to reach a carbon holding a chlorine.

The carbon atom in the carbon tetrachloride molecule is engulfed in the electron clouds of the chlorine substituents, and the chlorines cannot get replaced by —OH groups.

Freon 12, CCl_2F_2 (bp −30 °C), is one of several substances consisting of carbon, fluorine, and chlorine used as propellants in aerosol cans and refrigerants in air conditioners, freezers and refrigerators. Freons provide propellant force without changing the odor, taste, or color of the other contents of aerosols. Almost totally untouched by any chemical in the environment, the Freons have accumulated over the years and gradually have migrated into the stratosphere where they are interfering with the stratospheric ozone shield.

Exercise 5.1 Write the structures of the principal organic products that would form in a substitution reaction between ethyl bromide and each reagent:
(a) NaOH (b) NaSH (c) NH_3 (Give the structure of the product before and after the proton leaves.)

Another reaction of alkyl halides is **elimination.** Action of base on an alkyl halide, under suitable conditions of solvent and temperature, causes the alkyl halide to lose the halogen together with a proton from an adjacent carbon. Isopropyl bromide, for example, can react as follows:

$$\text{H}\ddot{\text{O}}\text{:}^- + \text{H—CH}_2\text{—}\underset{\text{Br}}{\underset{|}{\text{CH}}}\text{—CH}_3 \xrightarrow[\text{alcohol}]{} \text{HO—H} + \text{CH}_2\text{=CH—CH}_3 + \text{:}\ddot{\text{Br}}\text{:}^-$$

Isopropyl bromide — Propylene

Some isopropyl alcohol also forms because some attacks by the hydroxide ion occur at carbon-2 holding the bromine atom.

Exercise 5.2 In the environment DDT is partly degraded to DDE by the elimination of HCl. Using the structure of DDT given earlier in this chapter, what is the structure of DDE (*d*ichlorodiphenyl*d*ichloro*e*thylene)? (Unhappily, DDE is physiologically active like DDT in the environment.)

Mercaptans and Disulfides

5.4 Structural Features

Sulfur analogs of alcohols are important in the chemistry of proteins. Generally, they are called thio alcohols or **mercaptans,** and they have the general structural formula: R—S—H. The general structure for **disulfide** is R—S—S—R.

5.5 Naming the Mercaptans

The common names and structures of a few simple mercaptans are listed in Table 5.3. The word mercaptan specifies the —SH group; preceding this word is the name of the alkyl group to which it is attached. Another name for the —SH group itself is **sulfhydryl group.**

5.6 Properties

It is rather astonishing that the simple, formal replacement of oxygen by sulfur in an alcohol should produce such a dramatic change in odor. Mercaptans possess the most disagreeable odors of all organic compounds. The fluid ejected by a skunk when defending itself contains mercaptans.

The one important reaction of the —SH group is its behavior toward mild oxidizing agents, a reaction that is important in protein chemistry. These oxi-

TABLE 5.3
Common Mercaptans

Common Name	Structure	Boiling Point, in °C
Methyl mercaptan	CH_3—SH	6
Ethyl mercaptan	CH_3CH_2—SH	36
n-Propyl mercaptan	$CH_3CH_2CH_2$—SH	68
n-Butyl mercaptan	$CH_3CH_2CH_2CH_2$—SH	98
Cysteine (an amino acid, a building block of proteins)	HS—$CH_2CH(NH_2)C(=O)OH$	Solid

dizing agents remove the elements of hydrogen from two —S—H groups:

$$CH_3—S—H + H—S—CH_3 + (O) \longrightarrow CH_3—S—S—CH_3 + H_2O$$

The product is a disulfide. Mild reducing agents easily convert the disulfide back into separate —S—H groups. We may let the symbol (H) represent any mild reducing agent.

$$CH_3—S—S—CH_3 \xrightarrow[\text{(H)}]{\text{reducing agents}} 2CH_3—S—H$$

Exercise 5.3 Write the structure of the disulfide that would form by oxidation of each mercaptan.
(a) Ethyl mercaptan (b) Cyclohexyl mercaptan

Exercise 5.4 Write the structure of the product(s) of the reduction of each.
(a) $CH_3CH_2CH_2—S—SCH_2CH_2CH_3$ (b) $CH_3—S—SCH_2CH_3$

Amines

5.7 Structural Features

The amino group, $—NH_2$, is one of the characteristic groups in molecules of amino acids, the building blocks of proteins. Simple **amines** are alkyl or aryl derivatives of ammonia. One, two, or all three of the hydrogens on a molecule of ammonia may be replaced by such groups. For example,

CH_3NH_2	$(CH_3)_2NH$	$(CH_3)_3N$	$CH_3NHCH_2CH_3$
Methylamine	Dimethylamine	Trimethylamine	Methylethylamine

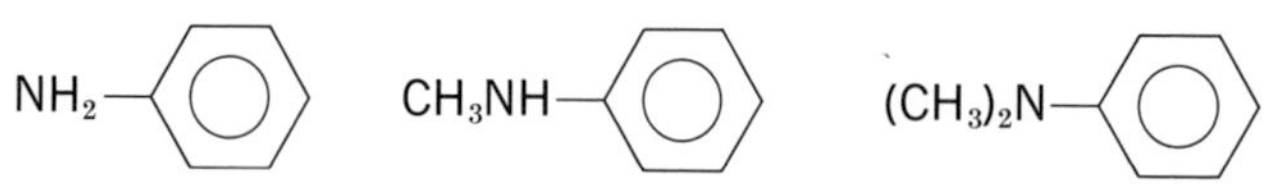

$NH_2—C_6H_5$	$CH_3NH—C_6H_5$	$(CH_3)_2N—C_6H_5$
Aniline	*N*-Methylaniline	*N,N*-Dimethylaniline

Aliphatic amines are those in which the carbon(s) attached directly to the nitrogen has (or have) only single bonds to other groups. Thus **1** is not an amine, but **2** has an amine function (and a keto group). If even one of the groups attached directly to the nitrogen is aromatic, the amine is classified as an aromatic amine. Several amines are listed in Table 5.4.

$$\underset{\mathbf{1}}{R—\overset{O}{\overset{\|}{C}}—NH_2} \qquad \underset{\mathbf{2}}{R—\overset{O}{\overset{\|}{C}}—CH_2—NH_2}$$

TABLE 5.4
Some Common Amines

Common Name	Structure	Boiling Point, in °C	Solubility in Water	K_b (at 25 °C)
Methylamine	CH_3-NH_2	−8	Very soluble	4.4×10^{-4}
Dimethylamine	$CH_3-\underset{H}{\underset{\vert}{N}}-CH_3$	8	Very soluble	5.2×10^{-4}
Trimethylamine	$CH_3-\underset{CH_3}{\underset{\vert}{N}}-CH_3$	3	Very soluble	0.5×10^{-4}
Ethylamine	$CH_3CH_2-NH_2$	17	Very soluble	5.6×10^{-4}
Diethylamine	$CH_3CH_2-\underset{H}{\underset{\vert}{N}}-CH_2CH_3$	55	Very soluble	9.6×10^{-4}
Triethylamine	$CH_3CH_2-\underset{CH_2CH_3}{\underset{\vert}{N}}-CH_2CH_3$	89	14 g/100 g H_2O	5.7×10^{-4}
n-Propylamine	$CH_3CH_2CH_2-NH_2$	49	Very soluble	4.7×10^{-4}
Aniline	[benzene ring]$-NH_2$	184	4 g/100 g H_2O	3.8×10^{-10}
Glycine	$NH_2-CH_2\overset{O}{\overset{\Vert}{C}}-OH$	mp 233	Very soluble	—

5.8 Some Important Heterocyclic Amines

Cyclic compounds (for instance, benzene and cyclopropane), in which all the members of the ring are alike (e.g., all are carbons) are **homocyclic compounds.** Ring compounds in which all the ring members are not alike are **heterocyclic compounds.**

Among the most important heterocyclic compounds are those whose molecules contain one or more nitrogens in a ring. For example,

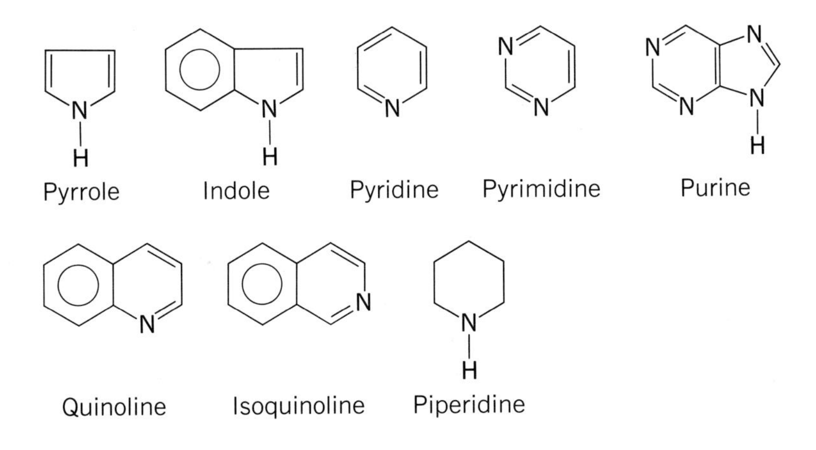

The pyrimidine and purine systems are important in the molecules of genes. Piperidine behaves as an ordinary aliphatic amine, but the others are aromatic compounds.

5.9 Naming the Amines

In the common system of nomenclature, aliphatic amines are named by designating the alkyl groups attached to the nitrogen and following this series by the word-part "-amine."

$(CH_3)_2CHCH_2—NH_2$
Isobutylamine

$(CH_3)_2CH—NH—CH_2CH_3$
Ethylisopropylamine

$CH_3CH_2CH_2—N(CH_3)—CH_2CH_2CH_2CH_3$
Methyl-*n*-propyl-*n*-butylamine

Where this system breaks down, the IUPAC system can be employed. In this system the $—NH_2$ group is named the amino group, or *N*-alkylamino or *N,N*-dialkylamino, and the specific groups are, of course, named.

$CH_3—CH(NH_2)—CH_2—OH$
2-Amino-1-propanol

$CH_3CH_2NH—CH_2CH_2CH_2CH_2CH_2CH_2CH_2CH_3$
1-(*N*-Ethylamino)octane (That is, at the 1 position in octane an *N*-ethylamino group is attached.)

The capital *N* designation means that the group immediately following it is attached to nitrogen. (See also the earlier examples of *N*-methylaniline and *N,N*-dimethylaniline.)

The simplest aromatic amine is always called aniline.

Exercise 5.5 Write structures for the following compounds.

(a) Triethylamine
(b) Dimethylethylamine
(c) *t*-Butyl-*sec*-butyl-*n*-butylamine
(d) *p*-Nitroaniline
(e) *p*-Aminophenol
(f) Cyclohexylamine

Exercise 5.6 Write the names for each of the following according to the common system whenever possible, otherwise according to the IUPAC system.

(a) $CH_3CH_2N(CH_2CH_2CH_2CH_3)_2$
(b) $NH_2CH_2CH_2NH_2$
(c) $C_6H_5—N(CH_2CH_3)_2$
(d) cyclopentyl—NH_2

5.10 Physical Properties

An amino group is ammonialike, that is, moderately polar and capable of giving and receiving hydrogen bonds. As a result, when compounds of similar formula weights are compared, amines have higher boiling points than alkanes but lower boiling points than alcohols.

	Formula Wt.	Bp (°C)
CH_3CH_3	30	−89
CH_3NH_2	31	−6
CH_3OH	32	65

Like the —OH group in alcohols, the $—NH_2$ group in amines helps make them soluble in water (Figure 5.1). Amines are also soluble in less polar solvents.

The lower-formula-weight amines (e.g., methylamine and ethylamine) have odors very much like that of ammonia. At higher formula weights the odors become "fishy." Aromatic amines have moderately pleasant, pungent odors, but they are at the same time very toxic. They are absorbed directly through the skin with sometimes fatal results.

5.11 Basicity of Amines

Much of the chemistry of amines can be understood by comparing them to ammonia. Amines are simply derivatives of ammonia. Reactions with water and with dilute mineral acids illustrate the similarities.

Equilibrium between ammonia and water:

$$H_3N: + H—\ddot{O}—H \rightleftharpoons NH_4^+ + :\ddot{O}H^-$$

Equilibrium between an amine and water:

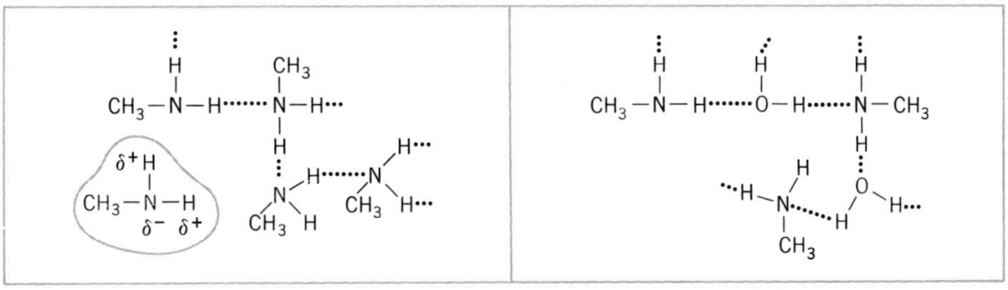

Figure 5.1 Hydrogen bonding in amines. (*a*) Between molecules of an amine. (*b*) Between molecules of an amine and molecules of water in an aqueous solution. Low-formula-weight amines are very soluble in water because they can slip into the hydrogen-bonding network in water.

$$R—\ddot{N}H_2 + H—\ddot{O}—H \rightleftharpoons R—NH_3^+ + :\ddot{O}H^-$$

The relative ability of ammonia to take a proton from a water molecule is given by the value of its base dissociation constant, $K_b = 1.8 \times 10^{-5}$, where $K_b = \dfrac{[NH_4^+][OH^-]}{[NH_3]}$. The greater the value of K_b, the stronger the base. As K_b values of Table 5.4 show, aliphatic amines are usually slightly stronger bases than ammonia and like ammonia react quantitatively with mineral acids, reagents that supply the hydronium ion.

Reaction of ammonia with hydronium ion:

$$H_3N: + H—\overset{+}{O}(H)_2: \longrightarrow NH_4^+ + :O(H)_2:$$

Reaction between an amine and hydronium ion:

$$R—\ddot{N}H_2 + H—\overset{+}{O}(H)_2: \longrightarrow R—NH_3^+ + :O(H)_2:$$

Aromatic amines such as aniline (Table 5.4) are about 10 thousand times weaker bases than aliphatic amines. For aniline, $K_b = 3.8 \times 10^{-10}$. When the amino group is joined directly to a benzene ring, the pair of electrons on nitrogen is actually partially delocalized into the pi-electron network of the ring that lets some of nitrogen's electron density into the ring. Hence nitrogen's electron pair is not actually as available to form a strong bond to a proton as the conventional structure indicates. Since some delocalization of the amino group's electron pair into the ring occurs, the ring in aniline is much more reactive than benzene toward electron-seeking reagents. Aniline, like phenol, is unusually reactive in chlorination and bromination, for example. Like phenol, aniline picks up three bromine substituents per ring before the reaction stops:

$$C_6H_5NH_2 \xrightarrow[\text{(no catalyst)}]{3Br_2 \text{ in } H_2O} C_6H_2Br_3NH_2 \quad (+\ 3HBr)$$

2,4,6-Tribromoaniline

Unlike the bromination of benzene, no iron or iron salt catalyst is needed. Moreover, like phenol, the ring in aniline is torn apart into tars and colored products by oxidizing agents. Normally a colorless liquid when pure, aniline gradually turns black when stored exposed to air, because even air slowly oxidizes this compound.

5.12 Properties of Amine Salts

Table 5.5 contains a list of a few representative salts of amines. These are salts in the true sense; that is, they are typical ionic compounds. All are solids and all have relatively high melting points, especially compared with the melting points of their corresponding amines. Like ammonium salts, amine salts are soluble in water but are not soluble in nonpolar solvents. The amino group is one of the great **"solubility switches"** of organic and biochemical molecules. When it accepts a proton and becomes positively charged, it switches on the water solubility of molecules with even long hydrocarbon groups. For example,

$$CH_3(CH_2)_8CH_2NH_2 + HCl_{aq} \longrightarrow CH_3(CH_2)_8CH_2\overset{+}{N}H_3Cl^-$$

Unprotonated amine, Insoluble in water — Protonated amine, Soluble in water

The water solubility of a **protonated amine**—that is, a substituted ammonium ion—can be switched off just as easily as it is turned on, simply by changing the pH to more basic values. The ammonium ion itself easily and quantitatively gives up a proton to the hydroxide ion:

$$H_3N^{+}\text{—}H + {:}\ddot{O}\text{—}H^- \longrightarrow H_3N{:} + H\text{—}\ddot{O}\text{—}H$$

Also, protonated amines neutralize hydroxide ion:

$$CH_3(CH_2)_6CH_2\overset{+}{N}H_2\text{—}H + {:}\ddot{O}\text{—}H^- \longrightarrow CH_3(CH_2)_6CH_2NH_2{:} + H\text{—}\ddot{O}\text{—}H$$

Soluble in water — Insoluble in water

TABLE 5.5

Amine Salts

Name	Structure	Mp (°C)
Methylammonium chloride	$CH_3NH_3^+Cl^-$	232
Dimethylammonium chloride	$(CH_3)_2NH_2^+Cl^-$	171
Dimethylammonium bromide	$(CH_3)_2NH_2^+Br^-$	134
Dimethylammonium iodide	$(CH_3)_2NH_2^+I^-$	155
Tetramethylammonium hydroxide (base as strong as KOH)	$(CH_3)_4N^+OH^-$	130–135 (decomposes)

A number of amines are beneficial drugs, e.g., quinine, codeine, and morphine. To make them more soluble than they are as free amines in the aqueous medium used to administer them, they are converted to protonated amines by reaction with sulfuric, phosphoric, or hydrochloric acid. The examples given are sold as morphine sulfate, quinine sulfate, and codeine phosphate.

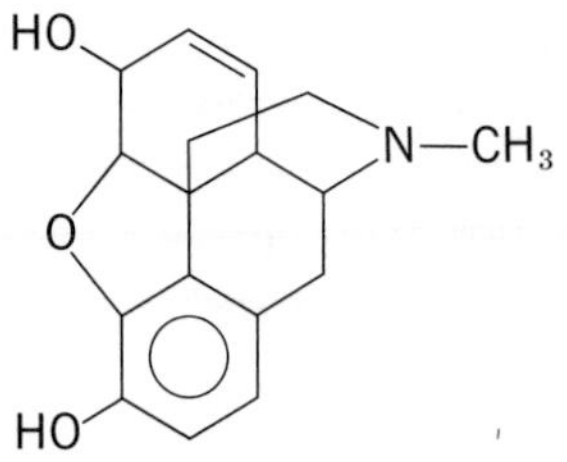

Morphine 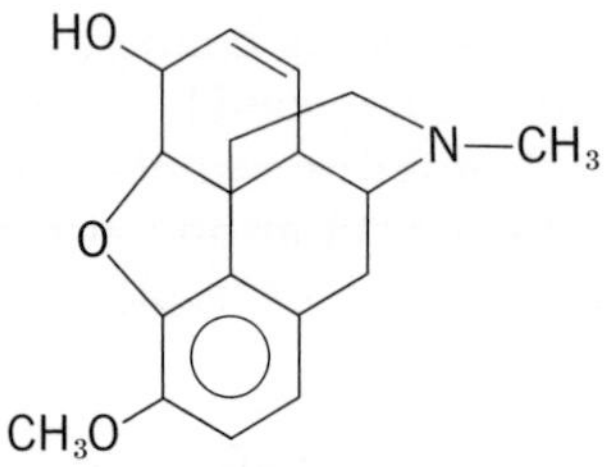 Codeine 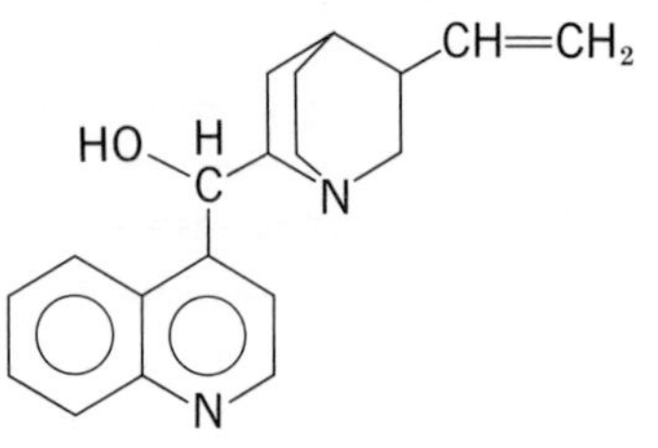 Quinine

Exercise 5.7 Write the equation for the reaction of each compound with HCl_{aq}.
(a) Methylamine (b) Methylethylamine (c) Trimethylamine

Exercise 5.8 Write the equation for the reaction of each compound with $NaOH_{aq}$.
(a) Ethylammonium chloride (b) Dimethylammonium bromide

5.13 Important Individual Amines

Dimethylamine $(CH_3)_2NH$

One of the constituents that gives herring brine its distinctive odor, dimethylamine is synthesized and used industrially in the manufacture of fungicides and accelerators for the vulcanization of rubber.

Several aliphatic amines (e.g., ethyl-, isobutyl-, and β-phenylethylamine) are synthesized in animal cells. They stimulate the central nervous system.

Invert Soaps. $CH_3(CH_2)_n\overset{+}{N}(CH_3)_3Cl^-$

Quaternary ammonium salts in which one of the alkyl groups is long (e.g., $n = 15$) have detergent and germicidal properties. In ordinary soaps the detergent action is associated with a negatively charged organic ion. In an invert soap, a positive ion has the detergent property. (Detergent action is discussed in Section 11.19.)

Aniline. $C_6H_5NH_2$

The annual production of aniline in the United States is well over 100,000 tons. Virtually all of it is used in the manufacture of aniline dyes, pharmaceuticals, and chemicals for the plastics industry.

Brief Summary

Organohalogen Compounds. The organohalogens illustrate the relations between formula weight and boiling point and between polarity and solubility. They show how the chemistry of a carbon chain is controlled in part by the polarity of a substituent and by the geometric accessibility of the carbon holding the substituent. Alkyl halides can undergo both substitution and elimination reactions. By varying the reagents and conditions, we can change alkyl halides to alcohols, amines, mercaptans, or alkenes. Aryl halides generally do not give these substitution or elimination reactions.

Mercaptans. The presence of an —SH group makes the substance a reducing agent. Proteins generally include this group or its oxidized form, the disulfide group. Oxidizing agents change mercaptans to disulfides, and reducing agents reverse that change.

Amines. The amino group, like ammonia, is a proton-accepting or basic group. Low-formula-weight amines are soluble in water (via hydrogen bonding to water), and high-formula-weight amines are made soluble when they accept protons. Protonated amines will neutralize hydroxide ions and change back to free amines. The amino group is one of nature's important solubility switches and exhibits this behavior at the molecular level of life among proteins.

Reagents. Water has no effect on mercaptans or disulfides. Amines undergo slight ionization in water giving trace concentrations of hydroxide ions and protonated amines. Aqueous alkali is neutralized by protonated amine. Aqueous acids have no special effect on mercaptans or disulfides, but are promptly neutralized by amines. Mild reducing agents affect only disulfides, of all the groups studied in this chapter, and change disulfides to mercaptans. Mild oxidizing agents affect mercaptans, changing them to disulfides, and react with aromatic amines to give colored, sometimes tarlike products.

Questions and Exercises

1. Write common names for each compound.

(a) $CH_3CHCH_2CH_3$ with Br on C-2 (i.e., $CH_3CH(Br)CH_2CH_3$)

(b) Br—C_6H_4—NH_2 (para)

(c) cyclopentyl—SH

(d) $CH_3NCH_2CH_3$ with CH_3CHCH_3 on N (i.e., $CH_3N(CH(CH_3)_2)CH_2CH_3$)

(e) $(CH_3)_3CNH_2$

(f) Cl—C_6H_4—OH (para)

2. Arrange these compounds in the order of increasing boiling points.

CH_3CH_2OH	$CH_3CH_2CH_3$	$CH_3CH_2NH_2$
(a)	(b)	(c)

(You should not have to consult tables to answer this question.)

3. A solution of dimethylamine in water tests basic to litmus. Account for the obvious excess of hydroxide ions over hydrogen ions in this solution. (Write an equation showing how dimethylamine would interact with water.)

4. If you add a few drops of isobutylamine to a small volume of water, the solution will smell like overripe fish. If you now add a bit more than a molar equivalent of hydrochloric acid and mix the contents thoroughly, the odor disappears. Explain the disappearance of the odor, writing an equation as part of your answer. (Remember, salts in general are not volatile.)

5. Arrange the following compounds into what would be the most reasonable order of their increasing solubility in water. Give the number 1 to the letter of the least soluble and number 4 to the most soluble.

$CH_3CH_2\overset{+}{N}H_3Cl^-$	$CH_3(CH_2)_6CH_2OH$
(a)	(b)
$CH_3CH_2CH_2CH_3$	$CH_3CH_2CH_2—O—CH_2CH_2CH_3$
(c)	(d)

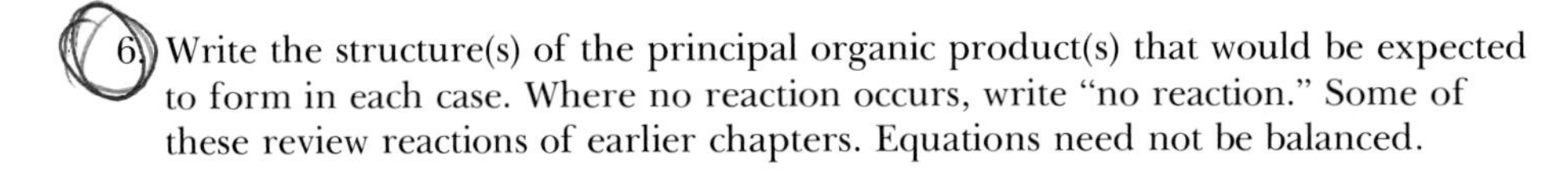

6. Write the structure(s) of the principal organic product(s) that would be expected to form in each case. Where no reaction occurs, write "no reaction." Some of these review reactions of earlier chapters. Equations need not be balanced.

(a) $CH_3CH_2CH_3 + H_2O \xrightarrow{\text{heat}}$

(b) $CH_3CH_2CH_2Br + H_2O \xrightarrow{\text{heat}}$

(c) $CH_3CH_2CH_2SH + H_2O \xrightarrow{\text{heat}}$

(d) $CH_3CH_2NH_2 + HCl_{aq} \longrightarrow$

(e) $CH_3\underset{\displaystyle NH_3^+Cl^-}{\underset{|}{C}}HCH_3 + NaOH_{aq} \longrightarrow$

(f) (cyclopentyl)—$SH + (O) \longrightarrow$

(g) $CH_3CH_2\underset{\displaystyle Br}{\underset{|}{C}}HCH_2CH_3 + NaOH \xrightarrow[\text{tion)}]{\text{(elimina-}}$

(h) $CH_3CH_2\overset{+}{N}H(CH_3)_2 + NaOH_{aq} \longrightarrow$

(i) $CH_2{=}CHCH_2NH_2 + H_2 \xrightarrow[\text{heat, pressure}]{\text{metal}}$

(j) 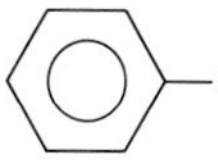 —Cl + $NaOH_{aq}$ $\longrightarrow$

(k) 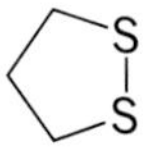 S—S + (H) $\longrightarrow$

(l) 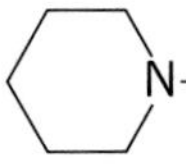 N—H + HCl_{aq} $\longrightarrow$

(m) $CH_3CH_2O—CH_2CH_2Cl + NH_3 \xrightarrow[\text{by } NaOH_{aq})]{}$ (followed by $NaOH_{aq}$)

(n) $Cl^{-}\overset{+}{N}H_3CH_2CH_2NH_2 + NaOH_{aq} \longrightarrow$

(o) $CH_3—S—S—CH_2CH_2NH_2 + HCl_{aq} \longrightarrow$

chapter 6

Aldehydes and Ketones

Both the ring stand and the meter stick are reflected by this test tube because its inner surface has been silvered by the Tollens' reaction, which we study in this chapter. (Photo by K. T. Bendo.)

6.1 The Carbonyl Group

The carbon-oxygen double bond or **carbonyl group,** C=O, occurs in a wide variety of nature's most important substances—in all foods, all enzymes, most

δ− O ⟵ Carbonyl oxygen

δ+ C ⟵ Carbonyl carbon

120°

Carbonyl group

The two carbon-oxygen bonds in the carbonyl group consist of one σ-bond and one π-bond.

TABLE 6.1
The Carbonyl Group in Families of Organic Compounds

Family Name	Generic Family Structure[a]
Aldehydes	(H)R—C(=O)—H
Ketones	R—C(=O)—R′
Carboxylic acids	(H)R—C(=O)—OH
Derivatives of carboxylic acids	
Acid chlorides	R—C(=O)—Cl
Anhydrides	R—C(=O)—O—C(=O)—R′
Esters	(H)R—C(=O)—O—R′
Amides	(H)R—C(=O)—NH_2 (simple amides)
	(H)R—C(=O)—NH—R′ (*N*-alkyl amides)
	(H)R—C(=O)—N(R″)—R′ (*N,N*-dialkyl amides)

[a] When the symbol for hydrogen is placed in parentheses before (or after) R [e.g., (H)R— or —R(H)], the specific R— group may be either an alkyl group or hydrogen. In these compounds, R— may also be aryl. When R— is primed (e.g., R′— or R″—), the two or more R— groups in a given symbol may be different.

hormones, all genes, and many vitamins. A number of families of compounds have this group, each defined according to the other groups present on the carbon. See Table 6.1. We shall study two of them, aldehydes and ketones, in this chapter.

6.2 Structural Features of Aldehydes and Ketones

$$\textbf{Aldehydes:}\quad (H)R-\overset{\overset{\large O}{\|}}{C}-H \qquad \textbf{Aldehyde group:}\quad -\overset{\overset{\large O}{\|}}{C}-H$$

To be classified as an **aldehyde,** the molecules of the substance must have a hydrogen attached to the carbonyl carbon. The other group must be joined by a bond from one of its carbons to the carbonyl carbon. (This other group may also be hydrogen as in formaldehyde.) The aldehyde group is often written as —CHO, and the double bond to oxygen is "understood." Several examples are given in Table 6.2.

$$\textbf{Ketones:}\quad R-\overset{\overset{\large O}{\|}}{C}-R' \qquad \textbf{Keto group:}\quad -\underset{|}{\overset{|}{C}}-\overset{\overset{\large O}{\|}}{C}-\underset{|}{\overset{|}{C}}-$$

Molecules in **ketones** must have a carbonyl joined on both sides by bonds to other carbons. Specific examples are given in Table 6.3. In condensed structures the keto group may be written as RCOR′ and the carbon-oxygen double bond is "understood."

6.3 Naming Aldehydes and Ketones

The common names of aldehydes are made from two word parts, "aldehyde" as a suffix and another as a prefix. The prefixes for the aldehydes having one to four carbons are as follows, and each is associated with the entire molecule except the aldehyde's hydrogen:

$$\underset{\text{Form-}}{H-\overset{\overset{\large O}{\|}}{C}-} \quad \underset{\text{Acet-}}{CH_3-\overset{\overset{\large O}{\|}}{C}-} \quad \underset{\text{Propion-}}{CH_3-CH_2-\overset{\overset{\large O}{\|}}{C}-} \quad \underset{\text{Butyr-}}{CH_3-CH_2-CH_2-\overset{\overset{\large O}{\|}}{C}-} \quad \underset{\text{Isobutyr-}}{CH_3-\overset{\overset{\large CH_3}{|}}{CH}-\overset{\overset{\large O}{\|}}{C}-}$$

These prefixes are also used in giving common names to organic acids. Since we shall encounter some simple acids as we study aldehydes and ketones, we shall look at their names now. The following are the common names of the al-

TABLE 6.2
Aldehydes

Common Name	Structure	Mp (°C)	Bp (°C)	Solubility in water
Formaldehyde	$CH_2{=}O$	−92	−21	Very soluble
Acetaldehyde	$CH_3CH{=}O$	−125	21	Very soluble
Propionaldehyde	$CH_3CH_2CH{=}O$	−81	49	16 g/100 ml (25°)
n-Butyraldehyde	$CH_3CH_2CH_2CH{=}O$	−99	76	4 g/100 ml
Isobutyraldehyde	$(CH_3)_2CHCH{=}O$	−66	65	9 g/100 ml
Valeraldehyde	$CH_3CH_2CH_2CH_2CH{=}O$	−92	102	Slightly soluble
Caproaldehyde	$CH_3CH_2CH_2CH_2CH_2CH{=}O$	−56	128	Very slightly soluble
Acrolein	$CH_2{=}CHCH{=}O$	−87	53	40 g/100 ml
Crotonaldehyde	$CH_3CH{=}CHCH{=}O$	−77	104	Moderately soluble
Benzaldehyde	$C_6H_5{-}CH{=}O$	−56	178	0.3 g/100 ml
Salicylaldehyde	$C_6H_4(OH){-}CH{=}O$ (OH ortho)	−10	197	Slightly soluble
Vanillin	$HO{-}C_6H_3(OCH_3){-}CH{=}O$ (CH₃O meta to CH=O)	81	285	1 g/100 ml
Cinnamaldehyde (*trans*)	$C_6H_5{-}CH{=}CHCH{=}O$	−8	253	Insoluble
Furfural	furan ring (O) $-CH{=}O$	−31	162	9 g/100 ml
Glucose[a] (dextrose)	$CH_2(OH){-}CH(OH){-}CH(OH){-}CH(OH){-}CH(OH){-}CH{=}O$	146 (decomposes)	—	Very soluble

[a] Only the general structural features of the open-chain form of glucose are shown. Additional details are discussed in a later chapter.

dehydes and carbonxylic acids having one to four carbons. (The formal or IUPAC names are in parentheses, but they are seldom used.)

In the IUPAC system aldehydes are named by changing the final "-e" in the name of the corresponding alkane with the longest chain that includes the carbonyl group to "-al." The corresponding acids have IUPAC names made by changing the final "-e" in the name of the alkane corresponding to the

$H{-}\overset{\overset{\large O}{\|}}{C}{-}H$	$H{-}\overset{\overset{\large O}{\|}}{C}{-}OH$
Formaldehyde (Methanal)	Formic acid (Methanoic acid)
$CH_3{-}\overset{\overset{\large O}{\|}}{C}{-}H$	$CH_3{-}\overset{\overset{\large O}{\|}}{C}{-}OH$
Acetaldehyde (Ethanal)	Acetic acid (Ethanoic acid)
$CH_3{-}CH_2{-}\overset{\overset{\large O}{\|}}{C}{-}H$	$CH_3{-}CH_2{-}\overset{\overset{\large O}{\|}}{C}{-}OH$
Propionaldehyde (Propanal)	Propionic acid (Propanoic acid)
$CH_3{-}CH_2{-}CH_2{-}\overset{\overset{\large O}{\|}}{C}{-}H$	$CH_3{-}CH_2{-}CH_2{-}\overset{\overset{\large O}{\|}}{C}{-}OH$
Butyraldehyde (Butanal)	Butyric acid (Butanoic acid)
$CH_3{-}\overset{\overset{\large CH_3}{\|}}{CH}{-}\overset{\overset{\large O}{\|}}{C}{-}H$	$CH_3{-}\overset{\overset{\large CH_3}{\|}}{CH}{-}\overset{\overset{\large O}{\|}}{C}{-}OH$
Isobutyraldehyde (2-Methylpropanal)	Isobutyric acid (2-Methylpropanoic acid)

longest chain that includes the carbonyl group to "oic acid." In IUPAC names the carbon chain is always numbered to give the carbonyl carbon number 1. (Compare the IUPAC names for isobutyraldehyde and isobutyric acid.) The following names illustrate further applications of this rule:

$$\underset{4}{CH_3}{-}\underset{3}{\overset{\overset{\large CH_3}{\|}}{CH}}{-}\underset{2}{CH_2}{-}\underset{1}{\overset{\overset{\large O}{\|}}{C}}{-}H$$

3-Methylbutanal

(Not: 3-methylbutyraldehyde. Do not mix the systems.)

$$\overset{6}{CH_3}{-}\overset{5}{\underset{\underset{\large Br}{|}}{CH}}{-}\overset{4}{\underset{\underset{\large Br}{|}}{CH}}{-}\overset{3}{CH_2}{-}\overset{2}{CH_2}{-}\underset{1}{\overset{\overset{\large O}{\|}}{C}}{-}OH$$

4,5-Dibromohexanoic acid

In the common systems the names for slightly more complicated compounds are made by designating positions on the main chain by Greek letters. The carbonyl carbon is *not* lettered. Other carbons are lettered going away from the carbonyl carbon:

$$\underset{\delta}{C}{-}\underset{\lambda}{C}{-}\underset{\beta}{C}{-}\underset{\alpha}{C}{-}\overset{\overset{\large O}{\|}}{C}{-}$$

TABLE 6.3
Ketones

Name	Structure	Mp (°C)	Bp (°C)	Solubility in Water
Acetone	$CH_3C(=O)CH_3$	−95	56	Very soluble
Methyl ethyl ketone	$CH_3C(=O)CH_2CH_3$	−87	80	33 g/100 ml (25°)
2-Pentanone	$CH_3C(=O)CH_2CH_2CH_3$	−84	102	6 g/100 ml
3-Pentanone	$CH_3CH_2C(=O)CH_2CH_3$	−40	102	5 g/100 ml
2-Hexanone	$CH_3C(=O)CH_2CH_2CH_2CH_3$	−57	128	1.6 g/100 ml
3-Hexanone	$CH_3CH_2C(=O)CH_2CH_2CH_3$	—	124	1.5 g/100 ml
Cyclopentanone	(cyclopentane ring)=O	−53	129	Slightly soluble
Cyclohexanone	(cyclohexane ring)=O	—	156	Slightly soluble
Camphor	bicyclic structure with H_3C, CH_3 on bridging C, and =O	176	209	Insoluble

The alpha position (α) is always the carbon attached directly to the carbonyl group. The following examples illustrate how the common system of nomenclature may be extended by the use of Greek letters. (The IUPAC names are in parentheses.)

$$CH_3-\underset{\substack{|\\ Br}}{CH}-\overset{\substack{O\\ \|}}{C}-H$$

α-Bromopropionaldehyde
(2-Bromopropanal)

$$CH_3-\underset{\substack{|\\ Br}}{CH}-\underset{\substack{|\\ Br}}{CH}-\overset{\substack{O\\ \|}}{C}-OH$$

α,β-Dibromobutyric acid
(2,3-Dibromobutanoic acid)

To give ketones common names, use the word "ketone" for the carbonyl group and name the groups attached to it. These examples illustrate this procedure.

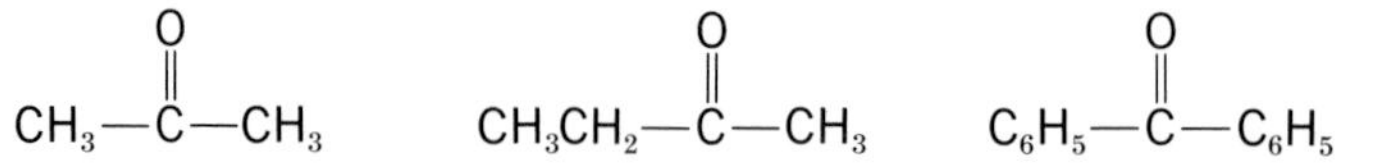

$$CH_3-\overset{\overset{\displaystyle O}{\|}}{C}-CH_3 \qquad CH_3CH_2-\overset{\overset{\displaystyle O}{\|}}{C}-CH_3 \qquad C_6H_5-\overset{\overset{\displaystyle O}{\|}}{C}-C_6H_5$$

(Dimethyl ketone) Acetone — Ethyl methyl ketone — (Diphenyl ketone) Benzophenone

Although names in parentheses in these examples are acceptable, the common names, those beneath them, are even more "common"; they are the only names used for their respective compounds by chemists. As a minimum you should learn the structure and name for acetone.

The IUPAC names for ketones are made by selecting the longest carbon chain that includes the carbonyl group, numbering it so as to give the carbonyl carbon the lower of two possible numbers, and then changing the final "-e" in the corresponding alkane to "-one" (pronounced to rhyme with "bone"). For example,

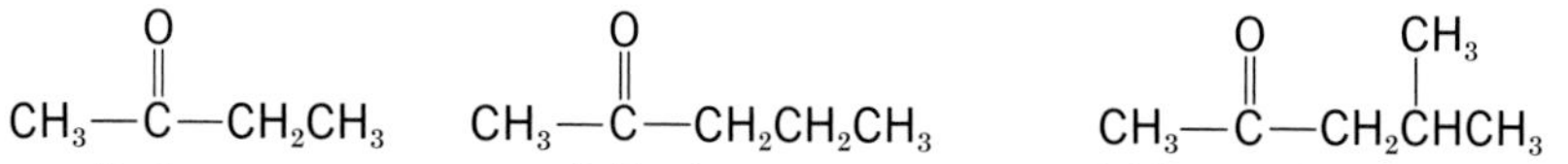

$$CH_3-\overset{\overset{\displaystyle O}{\|}}{C}-CH_2CH_3 \qquad CH_3-\overset{\overset{\displaystyle O}{\|}}{C}-CH_2CH_2CH_3 \qquad CH_3-\overset{\overset{\displaystyle O}{\|}}{C}-CH_2\overset{\overset{\displaystyle CH_3}{|}}{C}HCH_3$$

Butanone (no number needed) — 2-Pentanone — 4-Methyl-2-pentanone (Not: 2-Methyl-4-pentanone)

Exercise 6.1 Write condensed structures for each compound.
(a) α-Chloropropionaldehyde (b) Diisopropyl ketone
(c) β-Hydroxyisobutyraldehyde (d) Phenyl ethyl ketone

Exercise 6.2 Write common names for each compound.

(a) $H-\overset{\overset{\displaystyle O}{\|}}{C}-\overset{\overset{\displaystyle CH_3}{|}}{C}HCH_3$ (b) $CH_3-\overset{\overset{\displaystyle O}{\|}}{C}-\overset{\overset{\displaystyle CH_3}{|}}{C}HCH_3$

(c) $Cl-CH_2CH_2CH_2\overset{\overset{\displaystyle O}{\|}}{C}-H$ (d) $HO-\overset{\overset{\displaystyle O}{\|}}{C}-\underset{\underset{\displaystyle OH}{|}}{C}HCH_3$

(e) $CH_3\overset{\overset{\displaystyle Br}{|}}{C}H\underset{\underset{\displaystyle CH_3}{|}}{C}H\overset{\overset{\displaystyle O}{\|}}{C}-H$ (f) $(CH_3)_3C-\overset{\overset{\displaystyle O}{\|}}{C}-C(CH_3)_3$

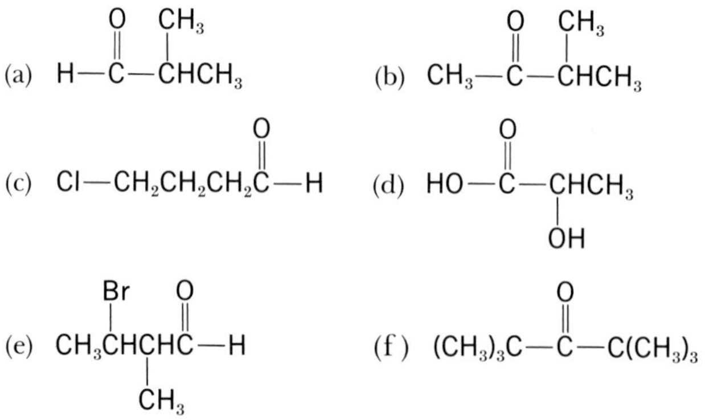

Exercise 6.3 Give the IUPAC names for each compound of Exercise 6.2.

Exercise 6.4 Each of these names is wrong in some way. Can you describe the error?

(a) 4-Methyl-3-pentanone (b) 1-Methylethanal
(c) α-Bromobutanal (d) 2-Chloroacetaldehyde

6.4 Physical Properties of Aldehydes and Ketones

Aldehydes and ketones are moderately polar compounds. We see how their polarities relate to those of other families by the boiling point data for substances of closely similar formula weights in Table 6.4.

Low-formula-weight aldehydes and ketones are soluble in water, but by the time the molecule has five carbons, this solubility has become quite low. The carbonyl group cannot act as a hydrogen bond donor. It has no —H on oxygen (or nitrogen). However, the carbonyl oxygen is, of course, a hydrogen bond acceptor. Trends in typical properties influenced by polarity and formula weight are seen in the data of Tables 6.2 and 6.3.

6.5 Preparation of Aldehydes and Ketones

In Chapter 4 we saw that careful oxidation (i.e., dehydrogenation) of a 1° alcohol yields an aldehyde and the oxidation of a 2° alcohol gives a ketone:

$$\underset{\text{1° Alcohol}}{R{-}CH_2{-}OH} + \underset{\text{Oxidizing agent}}{(O)} \longrightarrow \underset{\text{Aldehyde}}{R{-}\overset{\overset{\displaystyle O}{\|}}{C}{-}H} + H_2O$$

$$\underset{\text{2° Alcohol}}{R{-}\overset{\overset{\displaystyle OH}{|}}{CH}{-}R'} + (O) \longrightarrow \underset{\text{Ketone}}{R{-}\overset{\overset{\displaystyle O}{\|}}{C}{-}R'} + H_2O$$

TABLE 6.4
Structure and Boiling Point

Compound	Structure	Formula Weight	Bp
Propane	$CH_3CH_2CH_3$	44	−45 °C
Dimethyl ether	CH_3OCH_3	46	−25
Methyl chloride	CH_3Cl	50	−24
Ethylamine	$CH_3CH_2NH_2$	45	17
Acetaldehyde	$CH_3CH{=}O$	44	21
Ethyl alcohol	CH_3CH_2OH	46	78.5

Chemical Properties of Aldehydes and Ketones

6.6 Oxidation

The difference in their ease of oxidation is the principal reason why aldehydes and ketones are in separate classes. Aldehydes are extremely easy to oxidize to carboxylic acids without losing any carbons from the chain. In fact, aldehydes are not easy to store for any length of time even in stoppered bottles. Air slowly oxidizes them. Ketones, in sharp contrast, are hard to oxidize.

6.7 Tollens' Test

Silvering Mirrors

The ease with which the aldehyde group is oxidized by even mild reagents makes it possible to detect it with reagents that leave other groups alone. Tollens' reagent is one example. It consists of a solution of the diammonia complex of the silver ion, $Ag(NH_3)_2^+$, dissolved together with the nitrate ion in dilute ammonia. Dilute ammonia contains some hydroxide ion, which would form a precipitate of silver oxide, Ag_2O, except for the fact that ammonia molecules form this complex with silver ion. Hydroxide ions cannot get at silver ions to form the precipitate when the silver ions are complexed with ammonia. The general equation for the action of Tollens' reagent on an aldehyde is as follows; the salt of the carboxylic acid, RCO_2^-, forms from the aldehyde and silver metal appears:

$$\underset{\text{Aldehyde}}{RCH{=}O} + 2Ag(NH_3)_2^+ + 3OH^- \longrightarrow \underset{\text{Carboxylate ion}}{RCO_2^-} + \underset{\text{Silver}}{2Ag\downarrow} + 2H_2O + 4NH_3$$

A specific example is

$$\underset{\text{Acetaldehyde}}{CH_3CH{=}O} + 2Ag(NH_3)_2^+ + 3OH^- \longrightarrow \underset{\text{Acetate ion}}{CH_3CO_2^-} + \underset{\text{Silver}}{2Ag\downarrow} + 2H_2O + 4NH_3$$

In a positive **Tollens' test** metallic silver appears in a previously clear, colorless solution. If the inner wall of the test vessel is clean and grease-free, the silver comes out as a beautiful mirror, a reaction that serves as the basis for silvering mirrors. If the glass surface is not clean, the silver separates as a gray, finely divided, powdery precipitate.

6.8 Benedict's Test and Fehling's Test

Benedict's solution and Fehling's solution are also mild oxidizing agents. Both are alkaline and contain the copper(II) ion, Cu^{2+}, stabilized by a complexing agent. This agent is the citrate ion in Benedict's solution and the tartrate ion in Fehling's solution. Without these complexing agents, copper(II) ions would precipitate as copper(II) hydroxide in the presence of the alkali.

Benedict's solution is more frequently used because it stores well. Fehling's solution must be prepared just before use.

With either solution the Cu^{2+} ion is the oxidizing agent. In the presence of certain easily oxidized groups it is reduced to the copper(I) state, but copper(I) ions cannot be kept in solution by either citrate or tartrate ions. A precipitate forms consisting of copper(I) oxide, Cu_2O. [Copper(I) hydroxide, CuOH, is unstable and changes to the oxide.] What we see in a positive test, then, is a change from the brilliant blue color of the test solution to the bright orange-red color of precipitated copper(I) oxide. The principal systems that give this result are α-hydroxy aldehydes, α-keto aldehydes, and α-hydroxy ketones. The first two, α-hydroxy aldehydes and α-hydroxy ketones, are

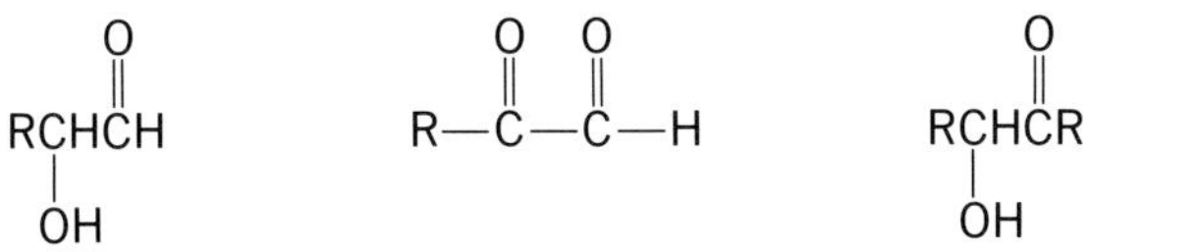

$$\underset{\text{OH}}{\text{RCH}}\text{—}\overset{\text{O}}{\overset{\|}{\text{C}}}\text{H} \qquad \text{R—}\overset{\text{O}}{\overset{\|}{\text{C}}}\text{—}\overset{\text{O}}{\overset{\|}{\text{C}}}\text{—H} \qquad \underset{\text{OH}}{\text{RCH}}\text{—}\overset{\text{O}}{\overset{\|}{\text{C}}}\text{R}$$

α-Hydroxy aldehyde   α-Keto aldehyde   α-Hydroxy ketone

common to the sugar family. (Glucose is an α-hydroxy aldehyde; fructose is an α-hydroxy ketone.) **Benedict's test** is a common method for detecting the presence of glucose in urine. Normally, urine does not contain glucose, but in certain conditions, such as diabetes, it does. A positive test for glucose varies from a bright green color (0.25% glucose), to yellow-orange (1% glucose), to brick red (over 2% glucose).

6.9 Strong Oxidizing Agents

Aldehydes are, of course, readily oxidized to carboxylic acids by permanganate ion, dichromate ion, and other oxidizing agents. Ketones also eventually yield to such reagents under more drastic conditions but except in isolated instances the reaction is not particularly useful. Ketones tend to fragment somewhat randomly on both sides of the carbonyl carbon:

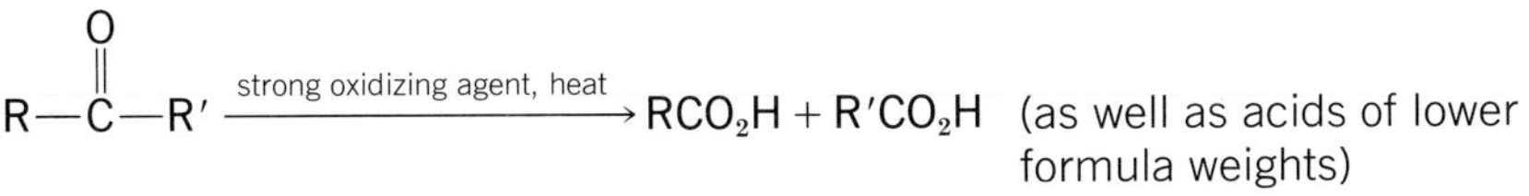

$$\text{R—}\overset{\text{O}}{\overset{\|}{\text{C}}}\text{—R}' \xrightarrow{\text{strong oxidizing agent, heat}} \text{RCO}_2\text{H} + \text{R}'\text{CO}_2\text{H} \quad \text{(as well as acids of lower formula weights)}$$

6.10 Reduction

Several methods are available for reducing aldehydes and ketones to their corresponding primary and secondary alcohols. We shall study two: **direct hydrogenation** and **hydride-ion transfer.**

6.11 Catalytic Hydrogenation

Hydrogen adds to the carbonyl group as it does to an alkene.

In general:

$$\underset{\text{Aldehyde}}{R{-}\overset{\overset{\displaystyle O}{\|}}{C}{-}H} + H_2 \xrightarrow[\text{pressure, heat}]{\text{Ni}} \underset{1^\circ\ \text{Alcohol}}{R{-}CH_2{-}OH}$$

$$\underset{\text{Ketone}}{R{-}\overset{\overset{\displaystyle O}{\|}}{C}{-}R'} + H_2 \xrightarrow[\text{pressure, heat}]{\text{Ni}} \underset{2^\circ\ \text{Alcohol}}{R{-}\overset{\overset{\displaystyle OH}{|}}{C}H{-}R'}$$

Specific examples:

$$\underset{n\text{-Butyraldehyde}}{CH_3CH_2CH_2CH{=}O} + H_2 \xrightarrow[\text{pressure, heat}]{\text{Ni}} \underset{n\text{-Butyl alcohol (85\%)}}{CH_3CH_2CH_2CH_2OH}$$

$$\underset{\text{Acetone}}{CH_3\overset{\overset{\displaystyle O}{\|}}{C}CH_3} + H_2 \xrightarrow[\text{pressure, heat}]{\text{Ni}} \underset{\text{Isopropyl alcohol (100\%)}}{CH_3\overset{\overset{\displaystyle OH}{|}}{C}HCH_3}$$

6.12 Reduction by Hydride-Ion Transfer

The hydride ion, H:⁻, is available from certain inorganic compounds, the metal hydrides (e.g., NaH, BH_3, AlH_3, or $LiAlH_4$). It reacts avidly with electron-poor sites, as illustrated by its behavior toward water:

$$\underset{\text{Lithium hydride}}{Li^+H:^-} + \underset{\text{Water}}{\overset{\delta+}{H}{-}\overset{\delta-}{\ddot{O}}{-}H} \longrightarrow \underset{\text{Hydrogen}}{H{-}H} + \underset{\text{Lithium hydroxide}}{Li^+ + :\ddot{O}H^-}$$

Nearly all metal hydrides give this type of reaction, and therefore metal hydrides cannot exist in the aqueous medium of the body. However, body cells contain a number of substances whose molecules can donate the hydride ion provided the right catalyst, an enzyme, is present. The aldehyde and ketone groups are both good acceptors of the hydride ion because of the partial positive charge on the carbonyl carbon. Letting M:H be a molecule of hydride ion donor, or a metal hydride, we may represent the first step in **hydride-ion transfer** as follows:

$$\underset{\text{Hydride donor}}{M{:}H} + \underset{\text{Aldehyde or ketone}}{>C{=}\ddot{O}:} \longrightarrow \underset{\text{Salt of an alcohol}}{H{-}\overset{|}{\underset{|}{C}}{-}\ddot{O}:^-\ M^+}$$

Since the negative ion of an alcohol is a powerful proton-acceptor (a powerful base), it takes a proton from water (or any other acid) as follows:

$$H-\overset{|}{\underset{|}{C}}-\ddot{\underset{..}{O}}:^- + H-\ddot{\underset{..}{O}}-H \longrightarrow H-\overset{|}{\underset{|}{C}}-\ddot{\underset{..}{O}}-H + {}^-:\ddot{\underset{..}{O}}-H$$

Anion of alcohol — Water — Alcohol — Hydroxide ion

Addition of Water and Alcohols

6.13 Hydrates

Most aldehydes react in aqueous solutions with water to establish an equilibrium mixture with the hydrated form:

$$R-\overset{O}{\overset{\|}{C}}-H + H-OH \overset{H^+}{\rightleftharpoons} R-CH(OH)_2$$

Aldehyde — Aldehyde hydrate

Formaldehyde is especially prone to form its hydrate. Water probably adds as follows:

$$H-\overset{:O:}{\overset{\|}{C}}-H + \underset{\text{Catalyst}}{H^+} \rightleftharpoons H-\overset{H-\ddot{O}:}{\overset{|}{\underset{+}{C}}}-H$$

$$H-\overset{H-\ddot{O}:}{\overset{|}{\underset{+}{C}H}} + H-\ddot{\underset{..}{O}}-H \rightleftharpoons H-CH(\ddot{O}-H)(^+\ddot{O}H_2) \xrightleftharpoons[\text{proton loss}]{} H-CH(\ddot{O}H)_2 + \underset{\text{Recovery of catalyst}}{H^+}$$

Aldehyde hydrates are actually 1,1-diols, and these compounds are generally too unstable to isolate and purify. They easily break back down to the aldehyde. However, the oxidation of an aldehyde to a carboxylic acid in an aqueous medium very likely proceeds by way of the aldehyde hydrate. In this "alcohol" form, the oxidation occurs.

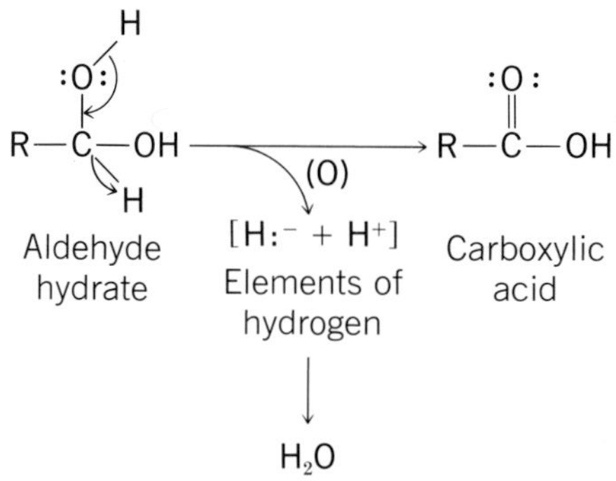

$$R-\underset{H}{\overset{OH}{C}}-OH \xrightarrow{(O)} R-\overset{O}{\overset{\|}{C}}-OH$$

Aldehyde hydrate

$[H:^- + H^+]$ Elements of hydrogen → H_2O

Carboxylic acid

6.14 Hemiacetals and Acetals Hemiketals and Ketals

Most aldehydes dissolved in alcohols consist of equilibrium mixtures:

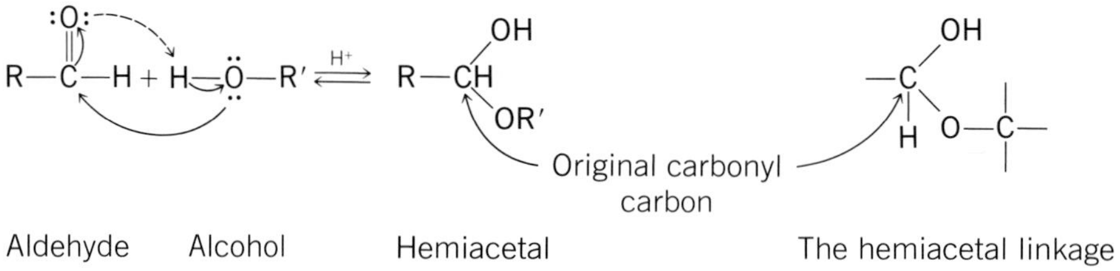

$$R-\overset{O}{\overset{\|}{C}}-H + H-\ddot{O}-R' \overset{H^+}{\rightleftharpoons} R-\underset{OR'}{\overset{OH}{CH}}$$

Aldehyde — Alcohol — Hemiacetal — The hemiacetal linkage

Original carbonyl carbon

From the alcohol the hydrogen goes to the carbonyl oxygen; the alcohol oxygen and its alkyl group go to the carbonyl carbon. The product with an —OH group and an —OR group joined to the same carbon is a **hemiacetal,** and most cannot be isolated. Efforts to do so convert them back to the original aldehyde and alcohol. In fact, the hemiacetal group is often called a **potential aldehyde group.** The hemiacetals that can be isolated are, however, very important, for they occur among carbohydrates.

What we have said about aldehydes applies to some extent to ketones. The main difference is that ketones are less prone to add water or alcohols. In ketones there is one more bulky group at the carbonyl carbon, a group that discourages the addition of water or alcohols. It gets in the way. We do not even consider hydrates of ketones. However, we may speak of **hemiketals,** which are just like hemiacetals except that they come from ketones. We shall encounter one that is stable when we study the structure of fructose, one of the important sugars.

Ordinary alcohols interact and change to ordinary ethers only under rather extreme conditions, as we learned in Chapter 4. As we shall see next, however, the —OH group in a hemiacetal or hemiketal is easily converted into a second ether linkage.

The "hemi-" part of hemiacetal suggests that we are halfway to something, and we are. Hemiacetals, in the presence of an acid catalyst, react with alcohols and form **acetals.** Ketones can be changed to acetal-like compounds called **ketals** via hemiketals.

In general:

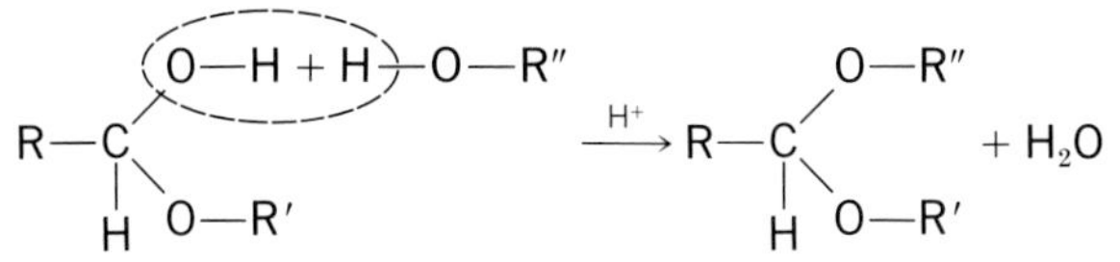

$$R\text{—}\underset{\substack{|\\H}}{C}(O\text{—}H)(O\text{—}R') + H\text{—}O\text{—}R'' \xrightarrow{H^+} R\text{—}\underset{\substack{|\\H}}{C}(O\text{—}R'')(O\text{—}R') + H_2O$$

Hemiacetal Alcohol Acetal

Specific example:

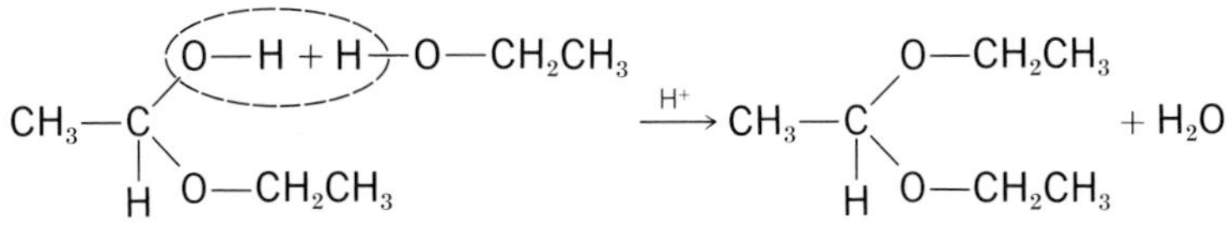

$$CH_3\text{—}\underset{\substack{|\\H}}{C}(O\text{—}H)(O\text{—}CH_2CH_3) + H\text{—}O\text{—}CH_2CH_3 \xrightarrow{H^+} CH_3\text{—}\underset{\substack{|\\H}}{C}(O\text{—}CH_2CH_3)(O\text{—}CH_2CH_3) + H_2O$$

Acetal or ketal formation is very similar to ether formation. However, the latter requires high temperatures; the former occurs at room temperature because the —OH in a hemiacetal is activated by the presence of the ether linkage attached to the same site. It is generally true among organic compounds that if different functional groups are situated close to each other on the same molecule, one modifies the properties of the other. This generalization also applies to acetals and ketals. The characteristic feature of an acetal or ketal is the presence of two ether linkages joining to the same carbon (bold-faced), the carbon that was the carbonyl carbon of the original aldehyde or ketone.

$$\text{—}\underset{\substack{|\\H}}{\mathbf{C}}(O\text{—}\overset{|}{\underset{|}{C}}\text{—})(O\text{—}\overset{|}{\underset{|}{C}}\text{—}) \qquad \text{—}\underset{|}{\mathbf{C}}(O\text{—}\overset{|}{\underset{|}{C}}\text{—})(O\text{—}\overset{|}{\underset{|}{C}}\text{—})$$

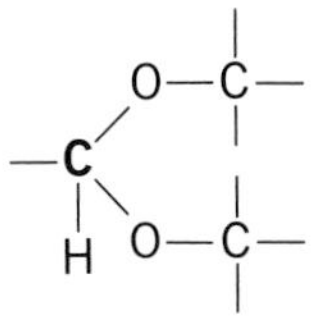

Acetal 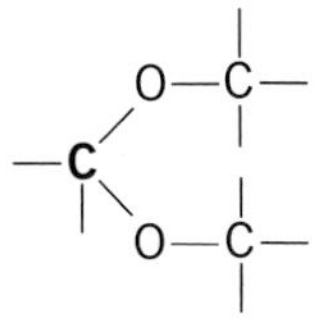 Ketal

As a result, the two adjacent ether linkages do not behave as simple, ordinary ethers (characterized by great resistance to chemical attack). In the presence of a trace of acid, acetals and ketals are broken up by water. They undergo easy hydrolysis (*hydro-* water; *lysis*, breaking) all the way back to the original aldehyde (or ketone) and alcohol. However, acetals and ketals are stable in base. These two facts about acetals and ketals are important in sugar chemistry. Both the hemiacetal and the acetal groups are features in the structures of carbohydrates. Table sugar, sucrose, has a ketal group. The enzyme-catalyzed hydrolysis of acetal and ketal linkages is the basic chemistry of the digestion of carbohydrates.

In general:

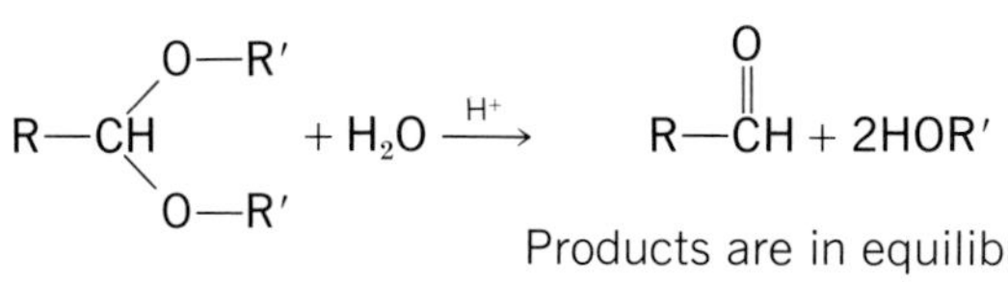

$$R\text{—}CH(O\text{—}R')(O\text{—}R') + H_2O \xrightarrow{H^+} R\text{—}\overset{\overset{O}{\|}}{C}H + 2HOR'$$

Products are in equilibrium with the hemiacetal.

Specific example:

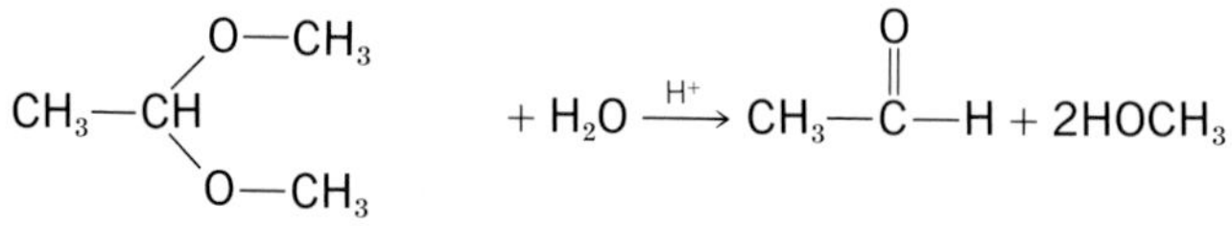

$$CH_3{-}CH(O{-}CH_3)_2 + H_2O \xrightarrow{H^+} CH_3{-}\overset{O}{\overset{\|}{C}}{-}H + 2HOCH_3$$

Acetaldehyde dimethylacetal (bp 65 °C) — Acetaldehyde — Methyl alcohol

Exercise 6.5 Write the structure of the hemiacetal that will exist in equilibrium with each of the following pairs of compounds.

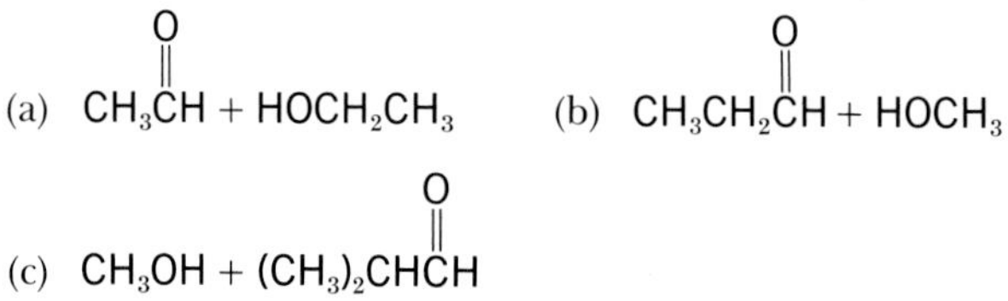

(a) $CH_3\overset{O}{\overset{\|}{C}}H + HOCH_2CH_3$ (b) $CH_3CH_2\overset{O}{\overset{\|}{C}}H + HOCH_3$

(c) $CH_3OH + (CH_3)_2CH\overset{O}{\overset{\|}{C}}H$

Exercise 6.6 Write the structure of the acetal that will form if the aldehyde in each of the parts of Exercise 6.5 combines with two molecules of the alcohol that is shown with it.

Exercise 6.7 Write the structure of the original aldehyde and the original alcohol that will form if the following acetals are hydrolyzed. One example is not an acetal. Which one is it?

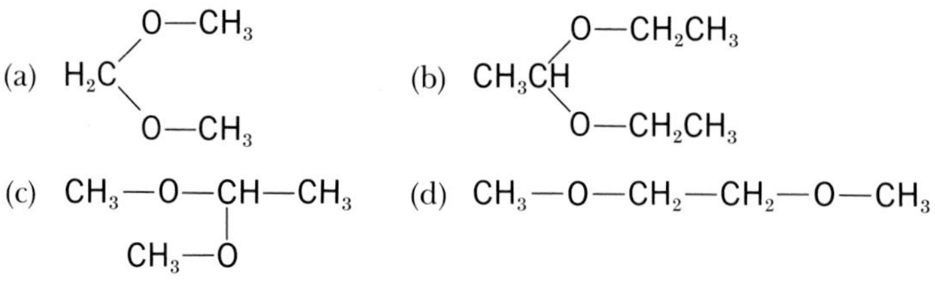

(a) $H_2C(O{-}CH_3)_2$ (b) $CH_3CH(O{-}CH_2CH_3)_2$

(c) $CH_3{-}O{-}CH(O{-}CH_3){-}CH_3$ (d) $CH_3{-}O{-}CH_2{-}CH_2{-}O{-}CH_3$

6.15 Aldol Condensation

Not only will one functional group alter properties of another group very near it (as with the acetals and ketals), it may also change the nearby hydrogens. Hydrogens on the carbon atom next to a carbonyl group, in the alpha position, are more acidic than hydrogens in a normal alkanelike environment. Not acidic enough to be classed as acids, aldehydes and ketones with alpha hydrogens nonetheless can interact with strong base (or the equivalent in an enzyme). A reaction known as the **aldol condensation** ("aldol," because the product is an aldehyde-alcohol or ketone-alcohol and "condensation" because two molecules condense into one) can occur in alkali (usually 10% NaOH) or in living cells where the enzyme exists. The aldol condensation is one of the great molecule-building reactions of living things. We shall encounter it, as well as the reverse reaction, in the biochemistry of carbohydrates. We introduce it here with the simplest example possible, the aldol condensation of acetaldehyde.

When heated in the presence of 10% sodium hydroxide, acetaldehyde reacts to form β-hydroxybutyraldehyde.

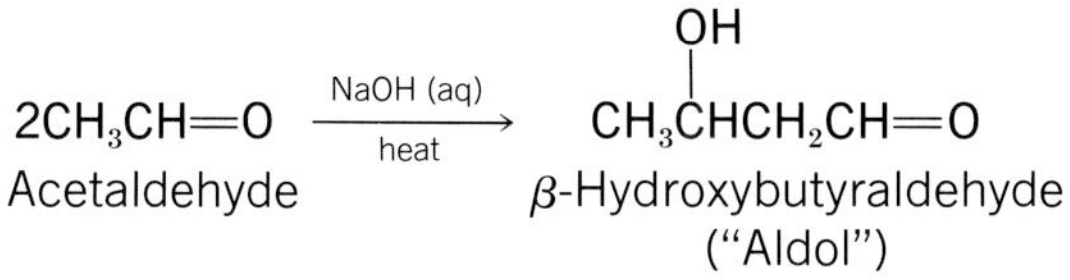

$$2CH_3CH{=}O \xrightarrow[\text{heat}]{\text{NaOH (aq)}} CH_3CH(OH)CH_2CH{=}O$$

Acetaldehyde → β-Hydroxybutyraldehyde ("Aldol")

In general:

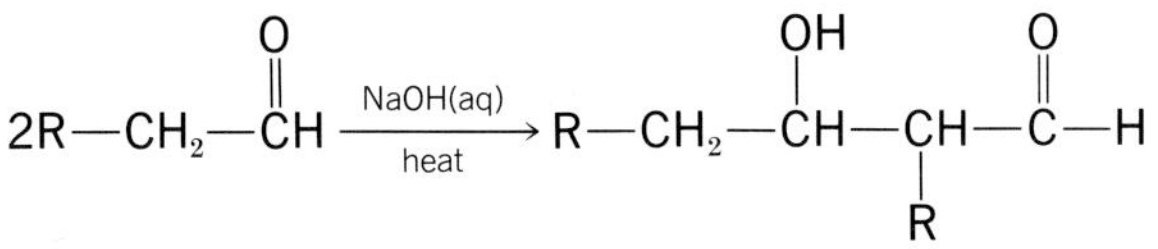

$$2R{-}CH_2{-}\overset{O}{\overset{\|}{C}}H \xrightarrow[\text{heat}]{\text{NaOH(aq)}} R{-}CH_2{-}\overset{OH}{\overset{|}{C}}H{-}\underset{R}{\underset{|}{C}}H{-}\overset{O}{\overset{\|}{C}}{-}H$$

This reaction is illustrative of a large number that have the following feature in common:

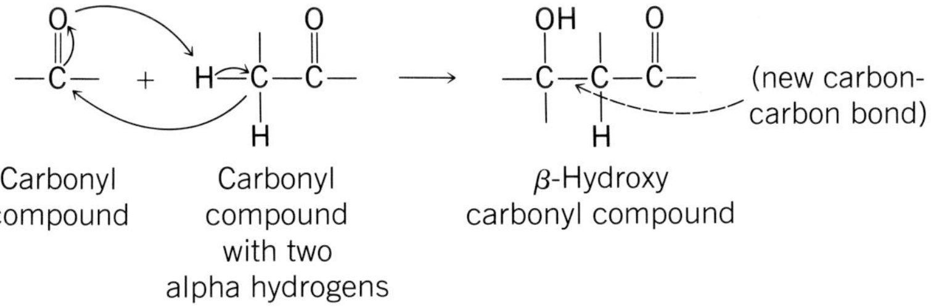

$$-\overset{O}{\overset{\|}{C}}- \;+\; H{-}\underset{H}{\underset{|}{\overset{|}{C}}}{-}\overset{O}{\overset{\|}{C}}- \longrightarrow -\underset{|}{\overset{OH}{\overset{|}{C}}}{-}\underset{H}{\underset{|}{\overset{|}{C}}}{-}\overset{O}{\overset{\|}{C}}-$$

(new carbon-carbon bond)

Carbonyl compound + Carbonyl compound with two alpha hydrogens → β-Hydroxy carbonyl compound

In words, one aldehyde adds across the carbonyl group of another.

We may illustrate how this reaction happens in alkali by the example of acetaldehyde.

Step 1. The aldehyde gives an alpha hydrogen to the base:

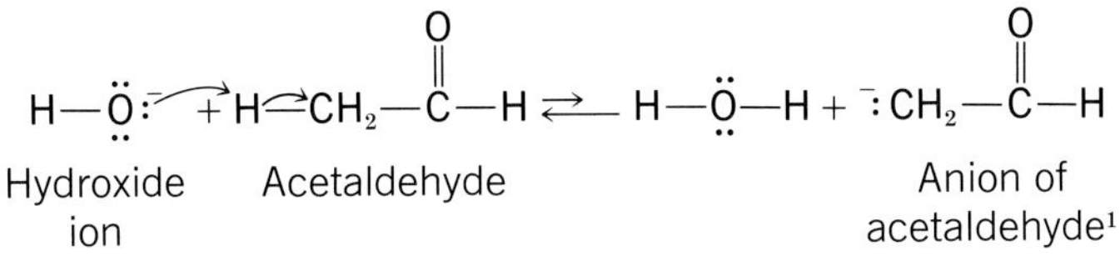

$$H{-}\ddot{\underset{..}{O}}{:}^- + H{-}CH_2{-}\overset{O}{\overset{\|}{C}}{-}H \rightleftharpoons H{-}\ddot{\underset{..}{O}}{-}H + {}^-{:}CH_2{-}\overset{O}{\overset{\|}{C}}{-}H$$

Hydroxide ion + Acetaldehyde ⇌ Anion of acetaldehyde[1]

Step 2. The anion of acetaldehyde is attracted to the carbonyl carbon of another molecule of acetaldehyde:

[1] This anion is stabilized by the delocalization of the negative charge via resonance:

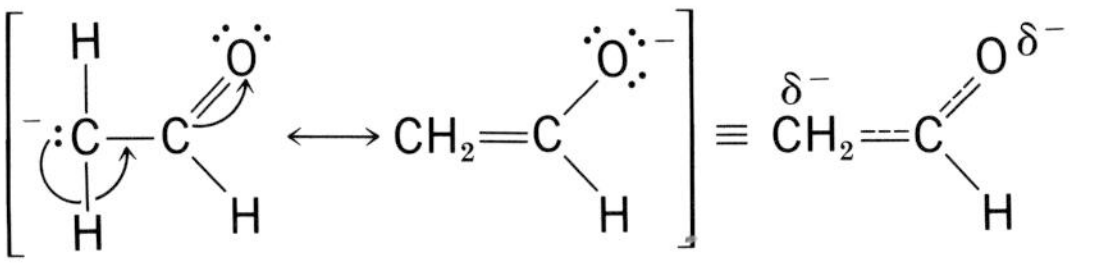

$$\left[{}^-{:}CH_2{-}\underset{H}{C}{=}\ddot{O}{:} \longleftrightarrow CH_2{=}\underset{H}{C}{-}\ddot{O}{:}^- \right] \equiv \overset{\delta^-}{CH_2}{\cdots}\underset{H}{C}{\cdots}O^{\delta^-}$$

Delocalization via resonance means that the negative charge can spread out, and that makes the system more stable.

$$CH_3-\overset{\overset{:O:}{\|}}{C}-H + {}^{-}:CH_2-\overset{\overset{O}{\|}}{C}-H \rightleftharpoons CH_3-\overset{\overset{:\ddot{O}:^{-}}{|}}{CH}-CH_2-\overset{\overset{O}{\|}}{C}-H$$

Acetaldehyde — Anion of acetaldehyde — Anion of aldol

Step 3. The anion of the aldol, a strong base like the hydroxide ion (except much stronger), takes a proton from a water molecule:

$$H-\ddot{O}-H + CH_3-\overset{\overset{:\ddot{O}:^{-}}{|}}{CH}-CH_2-\overset{\overset{O}{\|}}{C}-H \rightleftharpoons H\ddot{O}:^{-} + CH_3-\overset{\overset{H-\ddot{O}:}{|}}{CH}-CH_2-\overset{\overset{O}{\|}}{C}-H$$

Recovered base catalyst — Aldol

As the double arrows indicate, each step can be reversed. Aldols will break apart if heated with dilute base or in the presence of appropriate enzymes; that is, they undergo a reverse aldol condensation.

Exercise 6.8 Write the structure of the product of the aldol condensation involving each of the following aldehydes:
(a) Propionaldehyde (b) Butyraldehyde
(c) Phenylacetaldehyde ($C_6H_5CH_2CH{=}O$)

6.16 Important Individual Aldehydes and Ketones

Formaldehyde

At room temperature formaldehyde is a gas with a very irritating and distinctive odor. It is commonly marketed as Formalin, an approximately 37% solution of formaldehyde in water and methyl alcohol. Formaldehyde, along with its Formalin solution, is used as a disinfectant and as a preservative for biological samples and specimens. The largest amounts of Formalin, however, are consumed in the manufacture of various resins and plastics (e.g., Bakelite).

Acetone

One of the most important organic solvents, acetone not only dissolves a wide variety of organic substances but is also miscible with water in all proportions. Industrially, acetone is used as a solvent. It is made in many ways, including the air oxidation of isopropyl alcohol at elevated temperatures.

Brief Summary

Carbonyl Compounds. The carbon-oxygen double bond can itself be attacked, and it makes a hydrogen on its alpha position more acidic than is normal for a C—H system. What reactions the carbonyl group has are dependent altogether on the other groups at the carbonyl carbon.

Aldehydes. Aldehydes, like mercaptans, are easily oxidized. Made by the oxidation of 1° alcohols, aldehydes can also be reduced back to 1° alcohols. In water, aldehydes form equilibrium mixtures with their hydrated forms, 1,1-diols, and in alcohols, aldehydes form equilibrium mixtures with corresponding hemiacetals. The latter, if a trace of acid and more alcohol are present, can be changed to 1,1-diethers or acetals. In strong alkali, aldehydes (and to some extent, ketones) double in size as aldols form. The aldol condensation is reversible under suitable conditions.

Ketones. Unlike aldehydes, ketones are quite stable to oxidizing agents. Made by oxidation of 2° alcohols, ketones can be reduced back to 2° alcohols. Ketones, like aldehydes, can be changed to hemiketals and ketals.

Hemiacetals and Hemiketals. Formed in an equilibrium between an alcohol and either an aldehyde or a ketone, these compounds represent the simple addition of the alcohol to the carbonyl group. Normally not isolable because they too easily revert back to original alcohol and aldehyde or ketone, some stable examples occur among the carbohydrates.

Acetals and Ketals. These are formed via hemiacetals or hemiketals when an aldehyde or ketone is heated with an alcohol in the presence of a trace of acid catalyst. They can be isolated and purified, but if warmed with aqueous acid, they hydrolyze back to the original alcohol and aldehyde or ketone. Acetals and ketals occur among carbohydrates, and carbohydrate digestion is simply hydrolysis of these groups.

Reagents. Chemical properties of aldehydes and ketones may be summarized according to their responses to reagent systems as follows:

	Chemical Behavior	
Reagent System	Aldehydes	Ketones
Oxidation	Easily oxidized to carboxylic acids (compare also Tollens', Benedict's, and Fehling's tests)	Much more stable than aldehydes; when oxidized, undergo fragmentation
Reduction	Change to 1° alcohols	Change to 2° alcohols
Water	No major reaction; unstable hydrates appear	No major reaction; essentially no hydrates form
Alcohols	Unstable hemiacetals, then (if H^+ is present) acetals	Unstable hemiketals then (if H^+ is present) ketals
Aqueous alkali	Aldol condensations	Aldol condensations

Questions and Exercises

1. Write the structures for each of the following.
   (a) 2,3-Dimethylhexanal (b) Chloroacetic acid
   (c) di-*n*-Butyl ketone (d) Diisopropyl ketone
2. If the common name for $CH_3CH_2CH_2CH_2CO_2H$ is valeric acid, write the common and IUPAC names for $CH_3CH_2CH_2CH_2CH{=}O$.
3. If the structure of benzoic acid is $C_6H_5{-}\overset{\overset{\displaystyle O}{\|}}{C}{-}OH$, write the structure of benzaldehyde.
4. Write common names for each of the following compounds.

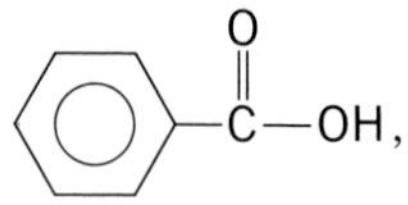

   (a) $CH_3CH_2\overset{\overset{\displaystyle O}{\|}}{C}H$ (b) $(CH_3)_2CH\overset{\overset{\displaystyle O}{\|}}{C}H$

   (c) $BrCH_2CH_2CH_2\overset{\overset{\displaystyle O}{\|}}{C}H$ (d) $CH_3\overset{\overset{\displaystyle CH_3}{|}}{C}HCH_2\overset{\overset{\displaystyle O}{\|}}{C}\underset{\underset{\displaystyle CH_3}{|}}{C}HCH_3$

   (e) $CH_3\overset{\overset{\displaystyle OH}{|}}{C}HCH_2\overset{\overset{\displaystyle O}{\|}}{C}OH$ (f) 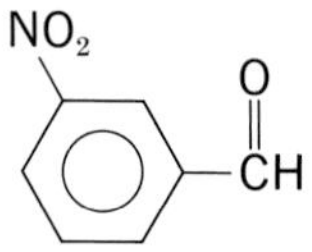 3-NO_2-$C_6H_4{-}\overset{\overset{\displaystyle O}{\|}}{C}H$ (benzene ring with NO_2 substituent meta to the CHO group)

5. Write the structure(s) of the principal organic product(s) that might reasonably be expected to form if acetaldehyde were subjected to each of the following reagents and conditions.
   (a) H_2, Ni, pressure, heat (b) Excess CH_3OH, dry HCl
   (c) CH_3OH (as a solvent) (d) 10% NaOH, heat
   (e) Tollens' reagent (f) $Cr_2O_7^{2-}$, H^+
6. Repeat Exercise 5 using acetone instead of acetaldehyde.
7. Write the structures of the principal organic products that would form in each of the following reactions.

   (a) $(CH_3)_2CHOH \xrightarrow{Cr_2O_7^{2-},H^+}$

   (b) $CH_3\overset{\overset{\displaystyle O}{\|}}{C}H + H_2 \xrightarrow[\text{heat, pressure}]{\text{catalyst}}$

   (c) $CH_3CH_2CH_2CH_2OH \xrightarrow[\text{(1st stage only)}]{Cr_2O_7^{2-},H^+}$

   (d) $C_6H_5{-}\underset{\underset{\displaystyle OH}{|}}{C}H{-}C_6H_5 \xrightarrow{Cr_2O_7^{2-},H^+}$

8. Which of the following would be expected to give a positive Benedict or Fehling test?

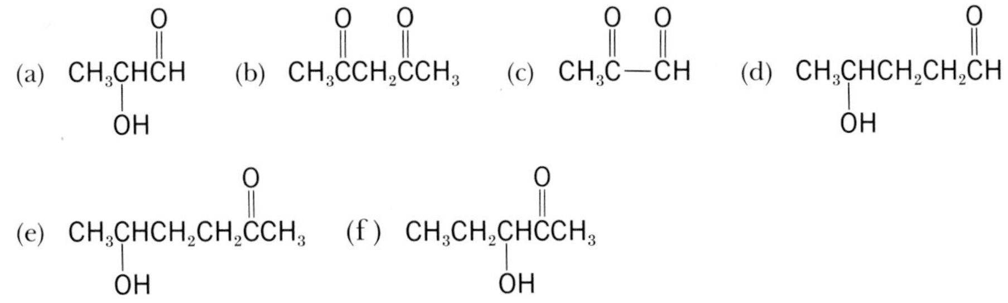

(a) $CH_3CH(OH)CHO$ (drawn as $CH_3\underset{OH}{CH}\overset{O}{\overset{\|}{C}}H$) (b) $CH_3\overset{O}{\overset{\|}{C}}CH_2\overset{O}{\overset{\|}{C}}CH_3$ (c) $CH_3\overset{O}{\overset{\|}{C}}-\overset{O}{\overset{\|}{C}}H$ (d) $CH_3\underset{OH}{CH}CH_2CH_2\overset{O}{\overset{\|}{C}}H$

(e) $CH_3\underset{OH}{CH}CH_2CH_2\overset{O}{\overset{\|}{C}}CH_3$ (f) $CH_3CH_2\underset{OH}{CH}\overset{O}{\overset{\|}{C}}CH_3$

9. The structure of a hypothetical molecule follows:

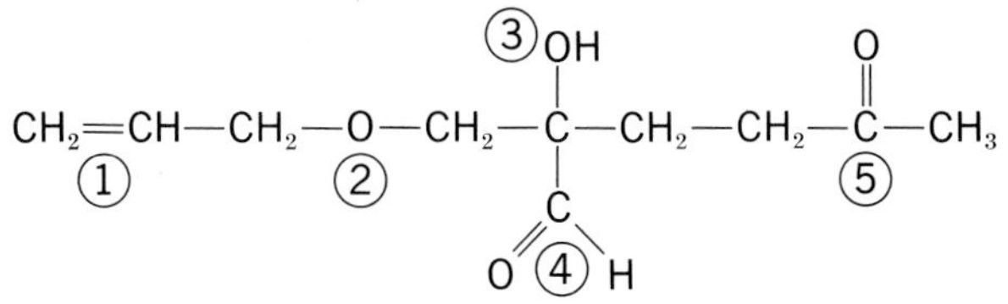

$$\underset{①}{CH_2{=}CH}-CH_2-\underset{②}{O}-CH_2-\overset{③OH}{\underset{\overset{|}{C(=O)H\ ④}}{C}}-CH_2-CH_2-\underset{⑤}{\overset{O}{\overset{\|}{C}}}-CH_3$$

Predict the chemical reactions it would probably undergo if it were subject to the reagents and conditions listed. To simplify the writing of the equations illustrating these reactions, isolate the portion of the molecule that would be involved, replace the other noninvolved portions by symbols as G, G′, G″, and so forth, and then write the equation (see example). You are expected to make reasonable predictions based only on the reactions we have studied thus far. If no reaction is to be predicted, write "None."

Example Reagent: excess H_2, Ni, heat, and pressure. We approach this problem in the following steps.

(a) First identify by name the functional groups in the molecule. (By doing this by name, you take advantage of the way you store information about the reactions of functional groups. Thus the name for —CHO is aldehyde or aldehyde group. In your mental "file" on this group should be the statement "aldehydes can be reduced to 1° alcohols" or "aldehydes can be easily oxidized to acids," and so forth.

In the structure given, you should recognize ① alkene, ② ether, ③ 3° alcohol, ④ aldehyde, ③ and ④ together the α-hydroxyaldehyde system and ⑤ ketone.

(b) To continue the example, ask yourself how each group would respond to the reagent given, the hydrogenation done catalytically.
   (1) Alkenes will add hydrogen.
   (2) Ethers do not react with hydrogen.
   (3) 3° Alcohols do not react with hydrogen.
   (4) Aldehydes can be reduced to 1° alcohols.
   (5) Ketones can be reduced to 2° alcohols.

(c) Third, write the equations. For example,

$$CH_2{=}CH{-}G + H_2 \xrightarrow[\text{heat, pressure}]{\text{Ni}} CH_3{-}CH_2{-}G$$

where G in this reaction is:

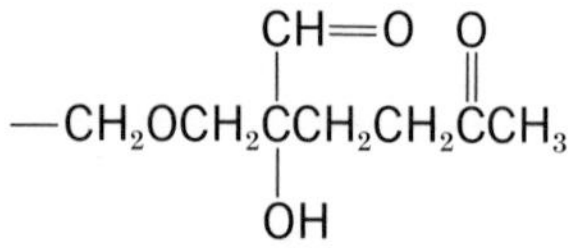

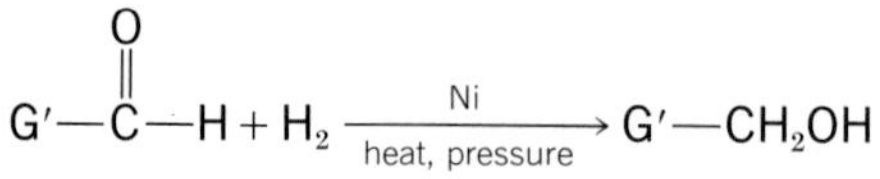

$$-CH_2OCH_2\overset{\overset{\displaystyle CH=O}{|}}{\underset{\underset{\displaystyle OH}{|}}{C}}CH_2CH_2\overset{\overset{\displaystyle O}{\|}}{C}CH_3$$

$$G'{-}\overset{\overset{\displaystyle O}{\|}}{C}{-}H + H_2 \xrightarrow[\text{heat, pressure}]{\text{Ni}} G'{-}CH_2OH \qquad \text{(What is } G'\text{?)}$$

$$G''{-}\overset{\overset{\displaystyle O}{\|}}{C}{-}CH_3 + H_2 \xrightarrow[\text{heat, pressure}]{\text{Ni}} G''{-}\overset{\overset{\displaystyle OH}{|}}{C}H{-}CH_3$$

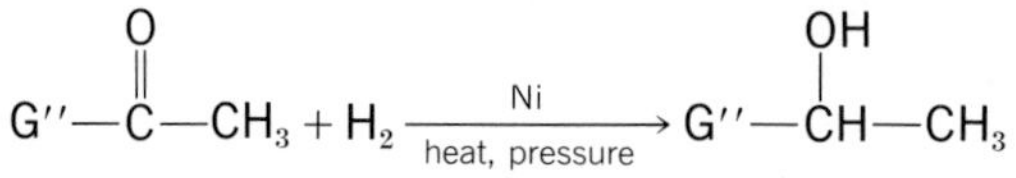

The reagents and conditions for this exercise are
(1) Br_2 in CCl_4 solution, room temperature.
(2) Excess CH_3OH in the presence of dry HCl.
(3) Tollens' reagent.
(4) Benedict's reagent (do not attempt to write an equation).
(5) Dilute aqueous sodium hydroxide at *room* temperature.

10. Repeat the directions given in Exercise 9 for the following hypothetical structure and the reagents given.

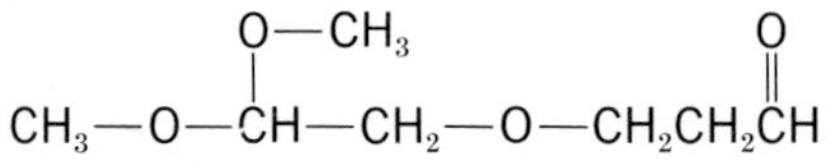

$$CH_3{-}O{-}\overset{\overset{\displaystyle O{-}CH_3}{|}}{C}H{-}CH_2{-}O{-}CH_2CH_2\overset{\overset{\displaystyle O}{\|}}{C}H$$

(a) Tollens' reagent (which is a basic solution).
(b) Excess water, H^+ catalyst.
(c) MnO_4^-, OH^-.

chapter 7

Carboxylic Acids and Their Derivatives

This cascade of aspirin tablets is a mere trickle when compared with the billions that are manufactured each year in the United States. Molecules of aspirin have two functional groups that we study in this chapter, the carboxylic acid and the ester groups. (Photo by Erich Hartmann/Magnum.)

7.1 Occurrence of Acids

The two principal types of organic acids are the **carboxylic acids** and the sulfonic acids. The sulfonic acids are much less common than the carboxylic acids, and we shall not continue their study.

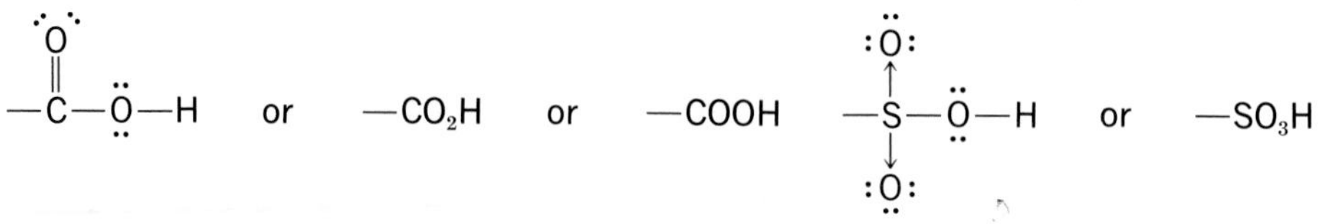

$$-\overset{\overset{\displaystyle \ddot{O}}{\|}}{C}-\ddot{\underset{\cdot\cdot}{O}}-H \quad \text{or} \quad -CO_2H \quad \text{or} \quad -COOH \qquad -\overset{\overset{\displaystyle :\ddot{O}:}{\uparrow}}{\underset{\underset{\displaystyle :\underset{\cdot\cdot}{O}:}{\downarrow}}{S}}-\ddot{\underset{\cdot\cdot}{O}}-H \quad \text{or} \quad -SO_3H$$

Carboxyl group (carbonyl + hydroxyl)

Sulfonic acid group

The molecules in substances classified as carboxylic acids must have a hydroxyl group attached to the carbonyl carbon. The other bond from the carbonyl carbon must be attached to another carbon (or to hydrogen in formic acid). The name of the group is either the carboxylic acid group or carboxyl group, and it is often condensed to —COOH or to —CO_2H. When you see it either way, remember the presence of the carbon-oxygen double bond and the hydroxyl group. Although their molecules possess hydroxyl groups, carboxylic acids definitely are not alcohols. The carbonyl group changes the properties of the —OH attached to it (and vice versa) too much. Specific examples of carboxylic acids are listed in Table 7.1. The straight-chain acids are commonly called fatty acids because many are obtained by the hydrolysis of animal fats (or vegetable oils).

Formic acid is a sharp-smelling, irritating liquid responsible for the sting of certain ants and the nettle.

Acetic acid gives tartness to vinegar, where its concentration is 4 to 5%. Because pure acetic acid (mp 16.6 °C or 63 °F) congeals to an icelike solid in a cool room, it is often called glacial acetic acid.

Butyric acid causes the odor of rancid butter. What is "strong" about valeric acid (Latin *valerum,* to be strong) is its odor. Caproic, caprylic, and capric acids (Latin, *caper,* goat) also have disagreeable odors.

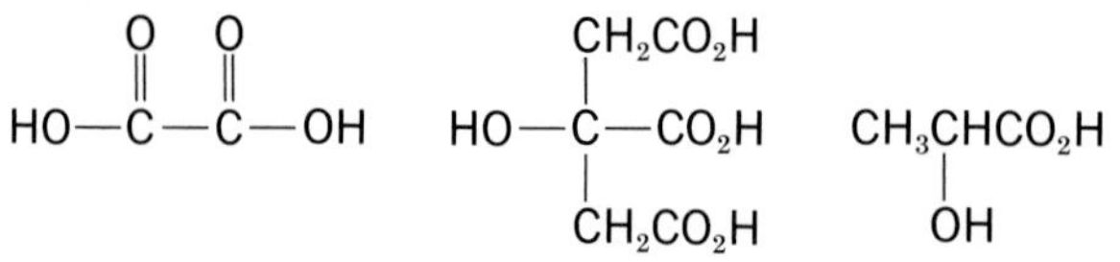

$$HO-\overset{\overset{\displaystyle O}{\|}}{C}-\overset{\overset{\displaystyle O}{\|}}{C}-OH \qquad HO-\overset{\overset{\displaystyle CH_2CO_2H}{|}}{\underset{\underset{\displaystyle CH_2CO_2H}{|}}{C}}-CO_2H \qquad CH_3\underset{\underset{\displaystyle OH}{|}}{C}HCO_2H$$

Oxalic acid — Citric acid — Lactic acid

Oxalic acid gives the tart flavor to rhubarb. Because it is both an acid and a reducing agent, aqueous solutions of oxalic acid will remove the stains caused by inks based on iron compounds.

The tart taste of citrus fruits is caused by citric acid. Its salt, sodium citrate, is sometimes added to drawn, whole blood to keep it from clotting. So important is citric acid in metabolism that one major biochemical sequence is called the "citric acid cycle."

TABLE 7.1
Carboxylic Acids

	n	Structure	Name	Origin of Name	Mp (°C)	Bp (°C)	Solubility (in g/100 g water a 20 °C)	K_a (at 25 °C)
Straight-chain saturated acids $C_nH_{2n}O_2$	1	HCO_2H	Formic acid (methanoic)	L.*formica*, ant	8	101	∞	1.77×10^{-4} (20 °C)
	2	CH_3CO_2H	Acetic acid (ethanoic)	L.*acetum*, vinegar	17	118	∞	1.76×10^{-5}
	3	$CH_3CH_2CO_2H$	Propionic acid (propanoic)	Gr.*proto*, first *pion*, fat	−21	141	∞	1.34×10^{-5}
	4	$CH_3CH_2CH_2CO_2H$	Butyric acid (butanoic)	L.*butyrum*, butter	−6	164	∞	1.54×10^{-5}
	5	$CH_3(CH_2)_3CO_2H$	Valeric acid (pentanoic)	L.*valere*, to be strong (valerian root)	−35	186	4.97	1.52×10^{-5}
	6	$CH_3(CH_2)_4CO_2H$	Caproic acid (hexanoic)	L.*caper*, goat	−3	205	1.08	1.31×10^{-5}
	7	$CH_3(CH_2)_5CO_2H$	Enanthic acid (heptanoic)	Gr.*oenanthe*, vine blossom	−9	223	0.26	1.28×10^{-5}
	8	$CH_3(CH_2)_6CO_2H$	Caprylic acid (octanoic)	L.*caper*, goat	16	238	0.07	1.28×10^{-5}
	9	$CH_3(CH_2)_7CO_2H$	Pelargonic acid (nonanoic)	pelargonium (geranium family)	15	254	0.03	1.09×10^{-5}
	10	$CH_3(CH_2)_8CO_2H$	Capric acid (decanoic)	L.*caper*, goat	32	270	0.015	1.43×10^{-5}
	12	$CH_3(CH_2)_{10}CO_2H$	Lauric acid (dodecanoic)	Laurel	44	–	0.006	–
	14	$CH_3(CH_2)_{12}CO_2H$	Myristic acid (tetradecanoic)	Myristica (nutmeg)	54	–	0.002	–
	16	$CH_3(CH_2)_{14}CO_2H$	Palmitic acid (hexadecanoic)	Palm oil	63	–	0.0007	–
	18	$CH_3(CH_2)_{16}CO_2H$	Stearic acid (octadecanoic)	Gr.*stear*, tallow	70	–	0.0003	–
Miscellaneous carboxylic acids		$C_6H_5CO_2H$	Benzoic acid	Gum benzoin	122	249	0.34 (25 °C)	6.46×10^{-5}
		$C_6H_5CH{=}CHCO_2H$	Cinnamic acid (*trans*)	Cinnamon	42	–	0.04	3.65×10^{-5}
		$CH_2{=}CHCO_2H$	Acrylic acid	L.*acer*, sharp	13	141	soluble	5.6×10^{-5}
		o-$HOC_6H_4CO_2H$ (benzene ring with CO_2H and OH)	Salicylic acid	L.*salix*, willow	159	211	0.22 (25 °C)	1.1×10^{-3} (19 °C)

Lactic acid gives the tart taste to sour milk. During periods of strenuous exercise, glucose units in the body are changed to lactic acid, and this acid is believed to be responsible for the soreness of muscles.

In all these examples of naturally occurring acids the common names were given. Several additional examples are given in Table 7.1, where the IUPAC names for most of the acids are shown in parentheses. Because the common names of biologically important acids are either shorter than IUPAC names or are more solidly entrenched in the scientific literature (or both), we emphasize them here. Be sure you know the common names for the acids from one through four carbons in length.

7.2 Synthesis of Carboxylic Acids

In earlier chapters we studied two ways to introduce the carboxyl group, by the oxidation of 1° alcohols directly to acids or by the oxidation of aldehydes.

$$\underset{\text{1° Alcohol}}{RCH_2OH} \xrightarrow{(O)} RCO_2H \qquad (O) = \text{Any strong oxidizing agent (e.g., } CrO_3,\ MnO_4^-\text{)}$$

$$\underset{\text{Aldehyde}}{RCHO} \xrightarrow{(O)} RCO_2H \qquad (O) = \text{Any moderate or strong oxidizing agent}$$

7.3 Physical Properties of Carboxylic Acids

The carboxyl group confers considerable polarity to molecules. A carboxylic acid has a higher boiling point than an alcohol of the same formula weight, as these data show:

CH_3CH_2OH	(form. wt. 46), bp 78 °C	$CH_3CH_2CH_2OH$	(form. wt. 60), bp 97 °C
HCO_2H	(form. wt. 46), bp 101 °C	CH_3CO_2H	(form. wt. 60), bp 118 °C

The reason is that pairs of carboxylic acid molecules can be held together by two hydrogen bonds:

$$R-C\begin{matrix}\overset{\delta-}{O}\cdot\cdot\overset{\delta+}{H}-O \\ \\ O-\underset{\delta+}{H}\cdot\cdot\underset{\delta-}{O}\end{matrix}C-R$$

Therefore, a carboxylic acid has an effective formula weight higher than the structural formula implies. Its boiling point is thus higher than we would otherwise expect.

The first four members of the homologous series of carboxylic acids are soluble in water. The carboxyl group can both donate and accept hydrogen bonds to and from water molecules. As the hydrocarbon chain in the acid lengthens, the solubility of the acid in water falls off sharply. The long-chain acids (C_{12}—C_{18}) are normal products of the digestion of fats and oils. The body has a mechanism for dissolving fatty acids in its aqueous fluids (e.g., blood and lymph), but if this mechanism breaks down, fats in the diet present serious problems, especially when their fatty acids tend to clog the circulatory system.

The low-formula-weight carboxylic acids with three to eight carbons, like the corresponding aldehydes, have extremely nauseous odors.

7.4 Acidity of Carboxylic Acids

In an aqueous medium, molecules of carboxylic acids interact with water molecules and a slight ionization occurs that produces hydronium ions and **carboxylate ions,** RCO_2^-.
In general:

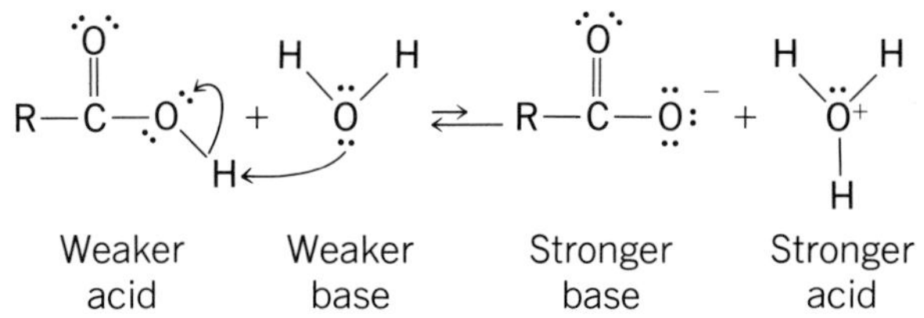

$$R{-}C({=}O){-}O{-}H + H_2O \rightleftarrows R{-}C({=}O){-}O^- + H_3O^+$$

Weaker acid — Weaker base — Stronger base — Stronger acid

A 1-*M* solution of acetic acid is ionized only to about 0.5% at room temperature. Carboxylic acids are relatively weak proton donors; they are weak acids. Compared with alcohols, however, which also have the hydroxyl group, carboxylic acids are stronger acids by factors of several billion.

The K_a values for several acids are given in Table 7.1, where $K_a = \frac{[RCO_2^-][H_3O^+]}{[RCO_2H]}$. The quantities in brackets are the concentrations in moles per liter of the ions or molecules present in the equilibrium given above. The higher the K_a value, the stronger the acid. All K_a values of carboxylic acids are on the order of 10^{-5}, compared to alcohols of 10^{-16} and phenols of about 10^{-10}. The relatively higher acidity of the carboxylic acids comes partly from the second oxygen, the carbonyl oxygen. Near the hydrogen that can ionize, it provides a second electron-withdrawing group that helps to pull the electrons of the O—H bond away from the hydrogen. Hence the hydrogen is taken off from a molecule of a carboxylic acid by the colliding water molecule more readily than from an alcohol or a phenol.

Another factor making carboxylic acids stronger acids than alcohols or phenols is that the negative charge in the carboxylate ion delocalizes into the pi bond of the carbonyl group. This delocalization of charge stabilizes the ion.

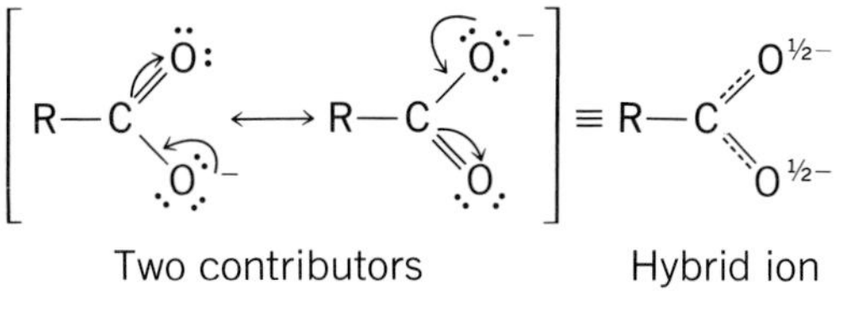

$$\left[R{-}C({=}O){-}O^- \longleftrightarrow R{-}C(O^-){=}O\right] \equiv R{-}C(O^{1/2-})(O^{1/2-})$$

Two contributors — Hybrid ion

7.5 Neutralization of Carboxylic Acids

Water is too weak a proton-acceptor—too weak a base—to convert acid molecules quantitatively to carboxylate ions. If hydroxide ions are present, however, the carboxlyic acid is quantitatively neutralized, because the hydroxide ion is a strong base, to give a salt and water.
In general,

$$\underset{\text{Stronger acid}}{R{-}\overset{\overset{\displaystyle O}{\|}}{C}{-}O{-}H} + \underset{\text{Stronger base}}{Na^+OH^-} \rightleftharpoons \underset{\text{Weaker base}}{R{-}\overset{\overset{\displaystyle O}{\|}}{C}{-}O^-Na^+} + \underset{\text{Weaker acid}}{H{-}OH}$$

Specific examples are

$$\underset{\text{Acetic acid}}{CH_3{-}\overset{\overset{\displaystyle O}{\|}}{C}{-}O{-}H} + \underset{\text{Sodium hydroxide}}{Na^+OH^-} \rightleftharpoons \underset{\text{Sodium acetate (a salt)}}{CH_3{-}\overset{\overset{\displaystyle O}{\|}}{C}{-}O^-Na^+} + \underset{\text{Water}}{H{-}OH}$$

$$\underset{\text{Stearic acid (insoluble in water)}}{CH_3(CH_2)_{16}\overset{\overset{\displaystyle O}{\|}}{C}{-}O{-}H} + Na^+OH^- \rightleftharpoons \underset{\text{Sodium stearate (soluble in water, a soap)}}{CH_3(CH_2)_{16}\overset{\overset{\displaystyle O}{\|}}{C}{-}O^-Na^+} + H{-}OH$$

$$\underset{\text{Benzoic acid (insoluble in water)}}{C_6H_5\overset{\overset{\displaystyle O}{\|}}{C}{-}O{-}H} + Na^+OH^- \rightleftharpoons \underset{\text{Sodium benzoate (soluble in water)}}{C_6H_5\overset{\overset{\displaystyle O}{\|}}{C}{-}O^-Na^+} + H{-}OH$$

Exercise 7.1 Write structures of the organic products of the reaction of each compound with aqueous sodium hydroxide at room temperature.

(a) $CH_3CH_2CO_2H$ (b) $CH_3{-}O{-}C_6H_4{-}CO_2H$ (para-substituted benzene ring) (c) $CH_3CH{=}CHCO_2H$

The purified salts of carboxylic acids are genuine salts—assemblies of oppositely charged ions—and therefore are all solids at room temperature. Table 7.2 gives a few examples of sodium salts. Like all sodium salts, these are soluble in water but completely insoluble in nonpolar solvents such as ether.

The salts of acids are but one of several types of derivatives of carboxylic acids. Let us turn next to a survey of the structural features and names of all these derivatives.

TABLE 7.2
Salts of Carboxylic Acids

Name	Structure	Mp (°C)	Solubility: Water	Solubility: Ether
Sodium formate	$HCO_2^-Na^+$	253	Soluble	Insoluble
Sodium acetate	$CH_3CO_2^-Na^+$	323	Soluble	Insoluble
Sodium propionate	$CH_3CH_2CO_2^-Na^+$	—	Soluble	Insoluble
Sodium benzoate	$C_6H_5CO_2^-Na^+$	—	66 g/100 ml	Insoluble
Sodium salicylate	$C_6H_4(OH)CO_2^-Na^+$ (benzene ring with $CO_2^-Na^+$ and ortho OH)	—	111 g/100 ml	Insoluble

7.6 Derivatives of Carboxylic Acids

The formation of carboxylate ions by the neutralization of organic acids introduces us to the first of several acid derivatives, compounds that may be both made from acids and also converted back to acids relatively easily. These derivatives are acid salts, esters, amides, acid chlorides, and acid anhydrides.

Carboxylic Acid Salts

$$R-\overset{\displaystyle O}{\overset{\|}{C}}-O^-M^+ \qquad \text{("}M^+\text{" is some metal ion.)}$$

The metal ion may be from any metal, but the sodium salts are the most common commercially.

Carboxylic Acid Esters (Esters)

$$(H)R-\overset{\displaystyle O}{\overset{\|}{C}}-O-R' \qquad \text{or} \qquad (H)RCO_2R'$$

The ester group has the features shown by **1.** The bond drawn with a heavier line is called the ester linkage. To be an ester, the molecule must have a carbonyl-oxygen-carbon network. Ester groups are sometimes written as RCO_2R'.

$$-\overset{\displaystyle O}{\overset{\|}{C}}-O-\overset{|}{\underset{|}{C}}-$$

Ester linkage

Ester group

1

Carboxylic Acid Amides ("Amides")

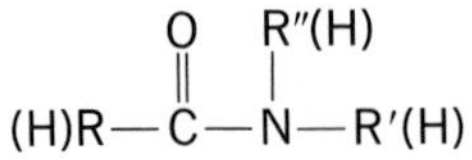

$$\text{(H)R—}\overset{\overset{\large O}{\|}}{C}\text{—}\overset{\overset{R''(H)}{|}}{N}\text{—R'(H)}$$

So-called "simple amides" are ammonia derivatives of carboxylic acids. Although not easily made this way, amides may be regarded as having been formed by the splitting out of water between an acid and ammonia:

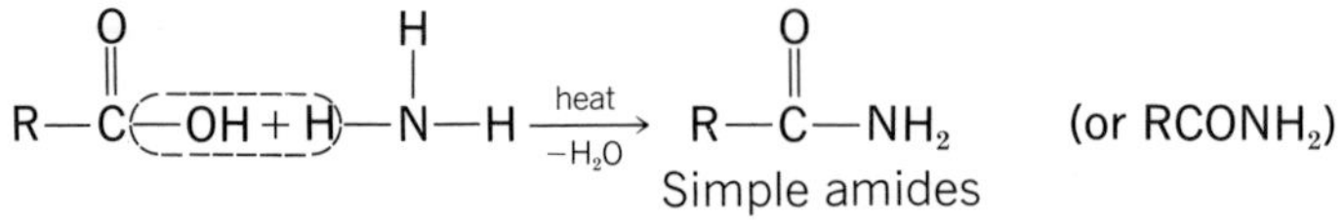

$$\text{R—}\overset{\overset{O}{\|}}{C}\text{—OH} + \text{H—}\overset{\overset{H}{|}}{N}\text{—H} \xrightarrow[-H_2O]{\text{heat}} \text{R—}\overset{\overset{O}{\|}}{C}\text{—NH}_2 \qquad \text{(or RCONH}_2\text{)}$$

Simple amides

Other amides, the so-called *N*-alkyl or *N,N*-dialkyl amides, where *N* denotes attachment to the nitrogen of a simple amide, are also well-known types of compounds:

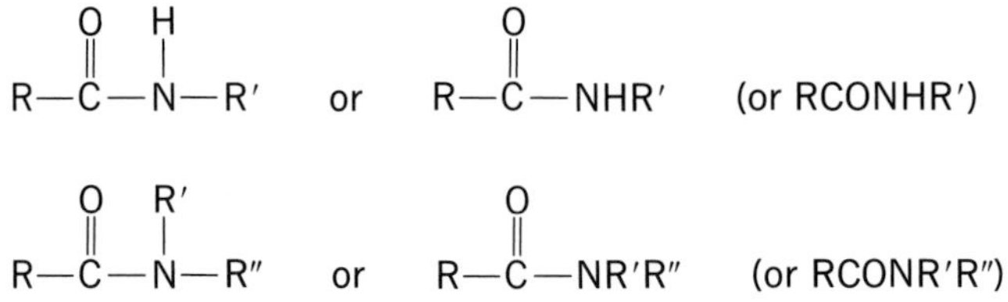

$$\text{R—}\overset{\overset{O}{\|}}{C}\text{—}\overset{\overset{H}{|}}{N}\text{—R'} \quad \text{or} \quad \text{R—}\overset{\overset{O}{\|}}{C}\text{—NHR'} \quad \text{(or RCONHR')}$$

$$\text{R—}\overset{\overset{O}{\|}}{C}\text{—}\overset{\overset{R'}{|}}{N}\text{—R''} \quad \text{or} \quad \text{R—}\overset{\overset{O}{\|}}{C}\text{—NR'R''} \quad \text{(or RCONR'R'')}$$

Any or all of the R-groups may be aromatic, too.

To be classified as an amide, the molecule must have a carbonyl-to-nitrogen bond, called the amide linkage. See **2.** In proteins the name peptide bond is synonymous with amide linkage.

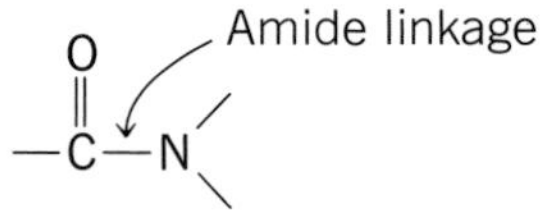

$$\text{—}\overset{\overset{O}{\|}}{C}\text{—N}\langle$$

Amide linkage

Amide group

2

Carboxylic Acid Chlorides (Acid Chlorides)

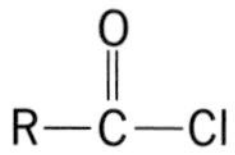

$$\text{R—}\overset{\overset{O}{\|}}{C}\text{—Cl}$$

The molecules of these compounds have carbonyl-to-chlorine bonds. Some of the corresponding acid fluorides, bromides, and iodides are known, but they are seldom used. (We shall ignore them.)

Carboxylic Acid Anhydrides (Anhydrides)

$$R{-}\overset{\displaystyle O}{\overset{\|}{C}}{-}O{-}\overset{\displaystyle O}{\overset{\|}{C}}{-}R$$

In the molecules of these compounds an oxygen atom is flanked on both sides by carbonyl groups. They are called anhydrides (meaning "not hydrated") because the dehydration or loss of water between two molecules of a carboxylic acid would give an anhydride:

$$R{-}\overset{\displaystyle O}{\overset{\|}{C}}{-}O{-}H + H{-}O{-}\overset{\displaystyle O}{\overset{\|}{C}}{-}R \longrightarrow R{-}\overset{\displaystyle O}{\overset{\|}{C}}{-}O{-}\overset{\displaystyle O}{\overset{\|}{C}}{-}R + H{-}OH$$

The R-groups may be aromatic, but they may not be replaced by hydrogens in either acid chlorides or anhydrides because such substances are too unstable to exist.

7.7 Naming Derivatives of Carboxylic Acids

We base both the common and the IUPAC names for acid derivatives on the name of the acid itself. See Table 7.3. To name any of these derivatives, first write the name of the parent acid in the system chosen, common or IUPAC. Then drop the "-ic acid" suffix portion and replace it by the suffix specified for the derivative in the second column of Table 7.3. These suffixes are the same in both the common and the IUPAC systems. For example,

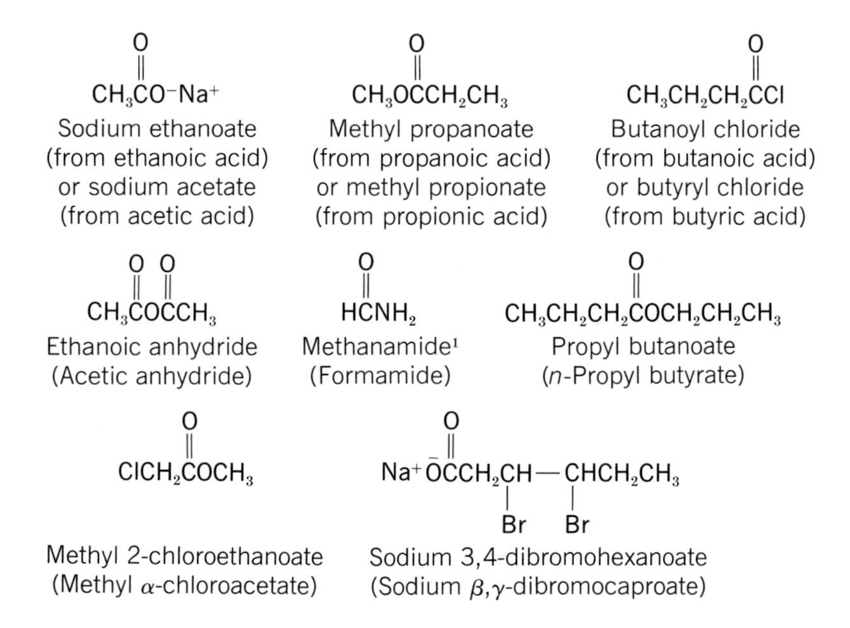

[1] Note that the "o" is also dropped for phonetic reasons. The name is not methanoamide.

TABLE 7.3
Common Names for Acids and Acid Derivatives

Class	Characteristic Suffix for the Name	Characteristic Prefix in the Carbonyl Portion of the Name			
		C_1 form-	C_2 acet-	C_3 propion-	C_4 butyr-[a]
Carboxylic acids	-ic acid	HCO_2H Formic acid	CH_3CO_2H Acetic acid	$CH_3CH_2CO_2H$ Propionic acid	$CH_3CH_2CH_2CO_2H$ Butyric acid
Carboxylic acid salts	-ate (with name of positive ion written first)	$HCO_2^-Na^+$ Sodium formate	$CH_3CO_2^-Na^+$ Sodium Acetate	$CH_3CH_2CO_2^-K^+$ Potassium propionate	$CH_3CH_2CH_2CO_2^-NH_4^+$ Ammonium butyrate
Acid chlorides	-yl chloride	(unstable)	$CH_3\overset{\overset{O}{\|}}{C}Cl$ Acetyl chloride	$CH_3CH_2\overset{\overset{O}{\|}}{C}Cl$ Propionyl chloride	$CH_3CH_2CH_2\overset{\overset{O}{\|}}{C}Cl$ Butyryl chloride
Anhydrides	-ic anhydride	(unstable)	$CH_3\overset{\overset{O}{\|}}{C}O\overset{\overset{O}{\|}}{C}CH_3$ Acetic anhydride	$CH_3CH_2\overset{\overset{O}{\|}}{C}O\overset{\overset{O}{\|}}{C}CH_2CH_3$ Propionic anhydride	$(CH_3CH_2CH_2\overset{\overset{O}{\|}}{C})_2O$ Butyric anhydride
Esters	-ate (with name of alkyl group on oxygen written first)	$H\overset{\overset{O}{\|}}{C}OCH_2CH_3$ Ethyl formate	$CH_3\overset{\overset{O}{\|}}{C}OCH(CH_3)_2$ Isopropyl acetate	$CH_3CH_2\overset{\overset{O}{\|}}{C}OCH_2CH_3$ Ethyl propionate	$CH_3CH_2CH_2\overset{\overset{O}{\|}}{C}OCH_3$ Methyl butyrate
Simple amides	-amide	$H\overset{\overset{O}{\|}}{C}NH_2$ Formamide	$CH_3\overset{\overset{O}{\|}}{C}NH_2$ Acetamide	$CH_3CH_2\overset{\overset{O}{\|}}{C}NH_2$ Propionamide	$CH_3CH_2CH_2\overset{\overset{O}{\|}}{C}NH_2$ Butyramide

[a] It is also common practice to write "*n*-" before "butyr-" in all these names. Thus "*n*-butyric acid," meaning the *normal* isomer. This practice, however, is not necessary.

Always figure out the name of the acid first and then operate on that to write a name for the acid derivative. To identify the parent acid mentally, draw a line around the "acid portion" of the structure: that part of the *carbon* skeleton with the carbonyl group but excluding anything such as oxygen, nitrogen, or chlorine also bound to the carbonyl carbon and any alkyl groups bound to oxygen or nitrogen.

For example,

$CH_3-\overset{\overset{O}{\|}}{C}-O-CH_3$

Carbonyl portion
(Parent acid is acetic acid or ethanoic acid.)

$CH_3-NH-\overset{\overset{O}{\|}}{C}-CH_2CH_3$

Carbonyl portion
(Parent acid is propionic acid or propanoic acid.)

$H-\overset{\overset{O}{\|}}{C}-O-CH_3$

Carbonyl portion
(Parent acid is formic acid or methanoic acid.)

$CH_3-\overset{\overset{O}{\|}}{C}-Cl$

Carbonyl portion
(Parent acid is acetic acid or ethanoic acid.)

Exercise 7.2 Write the IUPAC names for all the compounds in Table 7.3.

7.8 Carboxylate Ions

Since they form from weak acids, carboxylate ions must be relatively strong bases. In other words, they are quite good proton acceptors, especially toward a good proton donor such as the hydronium ion.
In general:

$$R{-}\overset{\overset{\displaystyle O}{\|}}{C}{-}O^- + H{-}\overset{+}{O}H_2 \longrightarrow R{-}\overset{\overset{\displaystyle O}{\|}}{C}{-}O{-}H + H{-}OH$$

Carboxylate ion (stronger base)	Hydronium ion (stronger acid)	Carboxylic acid (weaker acid)	Water (weaker base)

Specific examples:

$$C_6H_5CO_2^-Na^+ + HCl \xrightarrow[\text{water}]{} C_6H_5CO_2H + Na^+Cl^-$$

Sodium benzoate (soluble in water)	Hydrochloric acid	Benzoic acid (insoluble in water)	Sodium chloride

$$CH_3(CH_2)_{16}\overset{\overset{\displaystyle O}{\|}}{C}{-}O^-Na^+ + HCl \longrightarrow CH_3(CH_2)_{16}\overset{\overset{\displaystyle O}{\|}}{C}{-}O{-}H + NaCl$$

Sodium stearate	Stearic acid (insoluble in water)

We learn from these reactions that water-insoluble carboxylic acids can be made soluble by converting them into their sodium or potassium salts. The free acids may be recovered from their salts by the action of any strong acid. The carboxyl group, therefore, is another of nature's important solubility "switches." Through it we switch on or off the solubility of a long-chain acid merely by adjusting the pH of the medium.

Exercise 7.3 Write structures of the organic products of the reaction of each compound with dilute hydrochloric acid at room temperature.

(a) $CH_3{-}O{-}C_6H_4{-}CO_2^-K^+$ (para-substituted benzene ring) (b) $CH_3CH_2CO_2^-Li^+$ (c) $(CH_3CH{=}CHCO_2^-)_2Ca^{2+}$

7.9 Reactions of Carboxylic Acids

Besides forming salts, the carboxylic acids may be changed into other acid derivatives, esters, amides, acid chlorides, and anhydrides. Acid chlorides and acid anhydrides are not directly important at the molecular level of life because they cannot last long in an aqueous environment. We shall not study how to make them, but we shall still learn a little of their chemistry because living systems contain substances similar to them, mixed anhydrides with phosphoric acid. A number of simple esters are listed in Table 7.4, and several amides in Table 7.5.

7.10 Synthesis of Esters. Esterification

When a carboxylic acid and an alcohol are boiled together with a trace of a mineral acid catalyst, they react to form an ester and water. The reaction is

TABLE 7.4
Esters of Carboxylic Acids

	Name	Structure	Mp (°C)	Bp (°C)	Solubility (in g/100 g water at 20 °C)
Ethyl esters of straight-chain carboxylic acids, $RCO_2C_2H_5$	Ethyl formate	$HCO_2C_2H_5$	−79	54	Soluble
	Ethyl acetate	$CH_3CO_2C_2H_5$	−82	77	7.39 (25 °C)
	Ethyl propionate	$CH_3CH_2CO_2C_2H_5$	−73	99	1.75
	Ethyl butyrate	$CH_3(CH_2)_2CO_2C_2H_5$	−93	120	0.51
	Ethyl valerate	$CH_3(CH_2)_3CO_2C_2H_5$	−91	145	0.22
	Ethyl caproate	$CH_3(CH_2)_4CO_2C_2H_5$	−68	168	0.063
	Ethyl enanthate	$CH_3(CH_2)_5CO_2C_2H_5$	−66	189	0.030
	Ethyl caprylate	$CH_3(CH_2)_6CO_2C_2H_5$	−43	208	0.007
	Ethyl pelargonate	$CH_3(CH_2)_7CO_2C_2H_5$	−45	222	0.003
	Ethyl caproate	$CH_3(CH_2)_8CO_2C_2H_5$	−20	245	0.0015
Esters of acetic acid, CH_3CO_2R	Methyl acetate	$CH_3CO_2CH_3$	−99	57	24.4
	Ethyl acetate	$CH_3CO_2CH_2CH_3$	−82	77	7.39 (25 °C)
	n-Propyl acetate	$CH_3CO_2CH_2CH_2CH_3$	−93	102	1.89
	n-Butyl acetate	$CH_3CO_2CH_2CH_2CH_2CH_3$	−78	125	1.0 (22 °C)
Miscellaneous esters	Methyl acrylate	$CH_2{=}CHCO_2CH_3$		80	5.2
	Methyl benzoate	$C_6H_5CO_2CH_3$	−12	199	Insoluble
	Methyl salicylate	$C_6H_4(OH)$—CO_2CH_3 (benzene ring with CO_2CH_3 and ortho OH)	−9	223	Insoluble
	Acetylsalicylic acid	benzene ring with CO_2H and ortho $OCCH_3$ (C=O)	135		
	"Waxes"	$CH_3(CH_2)_nCO_2(CH_2)_nCH_3$	n = 23–33: carnauba wax; = 25–27: beeswax; = 14–15: spermaceti		

TABLE 7.5
Amides of Carboxylic Acids

Name	Structure	Mp (°C)	Bp (°C)	Solubility in Water
Formamide	$HCONH_2$	2.5	210 (decomposes)	∞
N-Methylformamide	$HCONHCH_3$	−5	131 (at 90 mm pressure)	Very soluble
N,N-Dimethylformamide	$HCON(CH_3)_2$	−61	153	∞
Acetamide	CH_3CONH_2	82	222	Very soluble
N-Methylacetamide	$CH_3CONHCH_3$	28	206	Very soluble
N,N-Dimethylacetamide	$CH_3CON(CH_3)_2$	−20	166	∞
Propionamide	$CH_3CH_2CONH_2$	79	213	Soluble
Butyramide	$CH_3(CH_2)_2CONH_2$	115	216	Soluble
Valeramide	$CH_3(CH_2)_3CONH_2$	106	Sublimes	Soluble
Caproamide	$CH_3(CH_2)_4CONH_2$	100	Sublimes	Slightly soluble
Benzamide	$C_6H_5-CONH_2$	133	290	1.5 g/100 g H_2O

called **esterification,** and we say that the acid (or the alcohol) has been esterified.

In general:

$$R-\overset{O}{\overset{\|}{C}}-OH + H-O-R' \underset{\Delta}{\overset{H^+}{\rightleftharpoons}} R-\overset{O}{\overset{\|}{C}}-O-R' + H-OH$$

Carboxylic acid — Alcohol — Ester

Specific examples:

$$CH_3-\overset{O}{\overset{\|}{C}}-O-H + H-O-CH_2CH_3 \underset{\Delta}{\overset{H^+}{\rightleftharpoons}} CH_3-\overset{O}{\overset{\|}{C}}-O-CH_2CH_3 + H_2O$$

Acetic acid — Ethyl alcohol — Ethyl acetate

$$o\text{-}HOC_6H_4-C(=O)-O-H + H-O-CH_3 \underset{\Delta}{\overset{H^+}{\rightleftharpoons}} o\text{-}HOC_6H_4-C(=O)-O-CH_3 + H_2O$$

Salicylic acid — Methyl alcohol — Methyl salicylate (oil of wintergreen)

$$\underset{\text{Benzoic acid}}{C_6H_5\overset{\overset{\displaystyle O}{\|}}{C}OH} + \underset{\text{Methyl alcohol}}{CH_3OH} \underset{\text{heat}}{\overset{H^+}{\rightleftharpoons}} \underset{\text{Methyl benzoate}}{C_6H_5\overset{\overset{\displaystyle O}{\|}}{C}OCH_3} + H_2O$$

In esterification the pieces of a water molecule are split out from the acid and alcohol, with the —OH group coming from the acid and the —H from the alcohol. (As time permits, your instructor may show the details of how the reaction happens.) Steps that may be used to write the formation of an ester between an acid and any alcohol are as follows, illustrated for the reaction of propionic acid and methyl alcohol.

Step 1. Set down the correct structures of the reactants first (you cannot do anything else until this is done correctly) and blacken or erase the *hydrogen* atom on the —OH function of the alcohol. Do the same to the *hydroxyl* group on the acid:

$$\underset{\text{Propionic acid}}{CH_3CH_2\overset{\overset{\displaystyle O}{\|}}{C}—OH} + \underset{\text{Methyl alcohol}}{CH_3—O—H}$$

Step 2. Link the remaining fragments—the oxygen of the alcohol to the carbonyl carbon of the acid:

$$CH_3CH_2\overset{\overset{\displaystyle O}{\|}}{C} + CH_3—O— \longrightarrow CH_3CH_2\overset{\overset{\displaystyle O}{\|}}{C}—O—CH_3$$

Step 3. Having determined the correct structure of the ester, write a neat and orderly equation for its formation.

Exercise 7.4 Write the structure and the common name of the ester that could be prepared from each of the following pairs of compounds.
(a) Acetic acid and methyl alcohol
(b) Benzoic acid and isopropyl alcohol
(c) Isobutyric acid and propyl alcohol
(d) Propionic acid and *n*-butyl alcohol

7.11 Synthesis of Amides

Ammonia, and monoalkyl and dialkyl derivatives of ammonia react with carboxylic acids to form amides if the reactants are heated strongly together. (At room temperature ammonium salts form.)

In general:

$$\underset{\text{Carboxylic acid}}{R—\overset{\overset{\displaystyle O}{\|}}{C}—OH} + \underset{NH_3,\ NH_2R',\ \text{or } NHR'_2}{H—N\langle} \underset{\Delta}{\rightleftharpoons} \underset{\text{Amide}}{R—\overset{\overset{\displaystyle O}{\|}}{C}—N\langle} + H—OH$$

Specific examples:

$$\underset{\text{Acetic acid}}{CH_3CO_2H} + \underset{\text{Ammonia}}{NH_3} \rightleftharpoons \underset{\text{Ammonium acetate}}{CH_3CO_2^-NH_4^+} \xrightarrow[\text{acetic acid, 110 °C}]{\text{Excess}} \underset{\text{Acetamide (90\%)}}{CH_3\overset{O}{\overset{\|}{C}}NH_2} + H_2O$$

$$H_2N\text{—}H + \underset{\text{Benzoic acid}}{H\text{—}O\text{—}\overset{O}{\overset{\|}{C}}\text{—}C_6H_5} \xrightarrow{\Delta} \underset{\text{Benzamide}}{H\text{—}\overset{H}{\overset{|}{N}}\text{—}\overset{O}{\overset{\|}{C}}\text{—}C_6H_5} + H_2O$$

$$\underset{\text{Acetic acid}}{CH_3\text{—}\overset{O}{\overset{\|}{C}}\text{—}O\text{—}H} + \underset{\text{Methylamine}}{H\text{—}\overset{H}{\overset{|}{N}}\text{—}CH_3} \xrightarrow{\Delta} \underset{N\text{-Methylacetamide}}{CH_3\text{—}\overset{O}{\overset{\|}{C}}\text{—}\overset{H}{\overset{|}{N}}\text{—}CH_3} + H_2O$$

$$\underset{\text{Dimethylamine}}{CH_3\text{—}\overset{CH_3}{\overset{|}{N}}\text{—}H} + \underset{\text{Benzoic acid}}{H\text{—}O\text{—}\overset{O}{\overset{\|}{C}}\text{—}C_6H_5} \xrightarrow{\Delta} \underset{N,N\text{-Dimethylbenzamide}}{CH_3\text{—}\overset{H_3C}{\overset{|}{N}}\text{—}\overset{O}{\overset{\|}{C}}\text{—}C_6H_5} + H_2O$$

Trisubstituted amines (e.g., trimethylamine) obviously cannot enter into amide formation. They possess no N—H bond.

The interaction of an amine with a carboxylic acid is the essential feature of protein synthesis. Proteins are huge molecules whose "backbones" involve amide linkages.

Exercise 7.5 Write the structure and common name of the amide that would be expected to form (if any) from the following pairs of compounds.
(a) Acetic acid and dimethylamine
(b) Butyric acid and diethylamine
(c) Benzoic acid and aniline
(d) Propionic acid and trimethylamine

7.12 Esters and Amides from Acid Chlorides or Anhydrides

Esters and amides are made more rapidly and in generally higher yields if, instead of using the free acid, the acid chloride or anhydride is used.[2]
In general, for esterification:

[2] While general methods for making anhydrides exist, a chemist usually is interested in only one or two, e.g., acetic anhydride and benzoic anhydride, and these are readily purchased at a chemical supply house. Acid chlorides may also be purchased or they may be made by a reaction between the carboxylic acid and one of three inorganic reagents: thionyl chloride ($SOCl_2$), phosphorus pentachloride (PCl_5), and phosphorus trichloride (PCl_3). We omit further details, but some mention of the source of anhydrides and acid chlorides was considered appropriate. Our interest is eventually in specific anhydrides of biochemical importance.

$$\underset{\text{Acid chloride}}{R{-}\overset{O}{\overset{\|}{C}}{-}Cl} + \underset{\text{Alcohol}}{H{-}O{-}R'} \longrightarrow \underset{\text{Ester}}{R{-}\overset{O}{\overset{\|}{C}}{-}O{-}R'} + \underset{\text{Hydrogen chloride}}{H{-}Cl}$$

$$\underset{\text{Acid anhydride}}{R{-}\overset{O}{\overset{\|}{C}}{-}O{-}\overset{O}{\overset{\|}{C}}{-}R} + \underset{\text{Alcohol}}{H{-}O{-}R'} \longrightarrow \underset{\text{Ester}}{R{-}\overset{O}{\overset{\|}{C}}{-}O{-}R'} + \underset{\text{Carboxylic acid}}{H{-}O{-}\overset{O}{\overset{\|}{C}}{-}R}$$

An example:

$$\underset{\text{Salicylic acid}}{o\text{-}HO_2C{-}C_6H_4{-}O{-}H} + \underset{\text{Acetic anhydride}}{CH_3\overset{O}{\overset{\|}{C}}{-}O{-}\overset{O}{\overset{\|}{C}}CH_3} \xrightarrow[\text{heat}]{H^+} \underset{\text{Aspirin (Acetylsalicylic acid)}}{o\text{-}HO_2C{-}C_6H_4{-}O{-}\overset{O}{\overset{\|}{C}}CH_3} + \underset{\text{Acetic acid}}{HO\overset{O}{\overset{\|}{C}}CH_3}$$

In general, for making amides:

$$\underset{\text{Acid chloride}}{R{-}\overset{O}{\overset{\|}{C}}{-}Cl} + \underset{\text{Ammonia (or amine) (excess)}}{H{-}N(H(R))(H(R'))} \longrightarrow \underset{\text{Amide}}{R{-}\overset{O}{\overset{\|}{C}}{-}N(H(R))(H(R'))} + HCl \text{ (as its salt with the excess ammonia or amine)}$$

$$\underset{\text{Acid anhydride}}{R{-}\overset{O}{\overset{\|}{C}}{-}O{-}\overset{O}{\overset{\|}{C}}{-}R} + \underset{\text{Ammonia (or amine) (excess)}}{H{-}N(H(R))(H(R'))} \longrightarrow \underset{\text{Amide}}{R{-}\overset{O}{\overset{\|}{C}}{-}N(H(R))(H(R'))} + HO\overset{O}{\overset{\|}{C}}R \text{ (as a salt)}$$

All these reactions are quite exothermic. The acid chloride of acetic acid, acetyl chloride, reacts violently with an alcohol or ammonia. The reason is partly because the expelled group, the chloride ion, is stable and poised to leave and partly because the carbonyl carbon is wide open to attack, as we shall soon study.

7.13 Reactions of Acid Derivatives

All of the acid derivatives can be made to react with water, with alcohols, or with ammonia (or amines). The site of attack is the electron-poor carbonyl carbon; what attacks is the electron-rich oxygen in a molecule of water or alcohol or the electron-rich nitrogen in a molecule of ammonia (or an amine). All the reactions are characterized by the following general change, in which we shall let :Z represent the electron-rich attacking particle:

$$R-\overset{\delta+}{C}(=\overset{\delta-}{O})-G + :Z \longrightarrow R-C(-O^{-})(-Z)-G \longrightarrow R-C(=O)-Z + :G$$

All other factors being equal, the larger the partial positive charge on the carbonyl carbon the more attractive it is to an electron-rich particle (:Z) and the faster the reaction. Acid chlorides and anhydrides have carbonyl carbons with relatively high, partial positive charges, because they have good electron-withdrawing groups (—Cl or —OCOR); therefore, these families are very reactive toward water. When G = —O—R′ in an ester, it is also somewhat electron-withdrawing at the carbonyl carbon—more so than the nitrogen in the amide. (Why?) Therefore, amides have small partial positive charges on their carbonyl carbons, and amides are the least reactive of the acid derivatives toward water. Thus proteins, whose molecules have many amide groups, can endure in the aqueous fluids of living cells.

7.14 Hydrolysis of Esters

The **hydrolysis** of an ester is simply the reverse of its formation—the reverse of esterification.

In general:

$$R-\overset{O}{\overset{\|}{C}}-O-R' + H-OH \xrightarrow{H^+} R-\overset{O}{\overset{\|}{C}}-O-H + H-O-R'$$

Specific examples:

$$CH_3-\overset{O}{\overset{\|}{C}}-O-CH_2CH_3 + H_2O \xrightarrow{H^+} CH_3-\overset{O}{\overset{\|}{C}}-OH + HO-CH_2CH_3$$

Ethyl acetate → Acetic acid + Ethyl Alcohol

$$CH_3-O-\overset{O}{\overset{\|}{C}}-C_6H_5 + H_2O \xrightarrow{H^+} CH_3-O-H + H-O-\overset{O}{\overset{\|}{C}}-C_6H_5$$

Methyl benzoate → Methyl alcohol + Benzoic acid

Note carefully, in the hydrolysis of an ester, that the carbonyl-to-oxygen bond is the only bond that breaks in the ester molecule. To be able to write the structures of the acid and the alcohol formed by the hydrolysis of an ester, use the following steps (illustrated for *n*-propyl acetate).

Step 1. Set down the structure of the ester and break it at the carbonyl-to-oxygen bond:

$$CH_3CH_2CH_2{-}O{\wr}\overset{\overset{\large O}{\|}}{C}CH_3 \longrightarrow CH_3CH_2CH_2{-}O{-} + \overset{\overset{\large O}{\|}}{C}CH_3$$

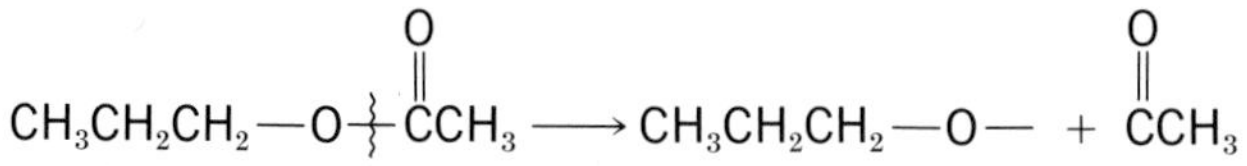

Propyl acetate

Step 2. Attach the elements of water to these fragments: H— goes to the oxygen of the developing alcohol; —OH goes to the carbonyl carbon:

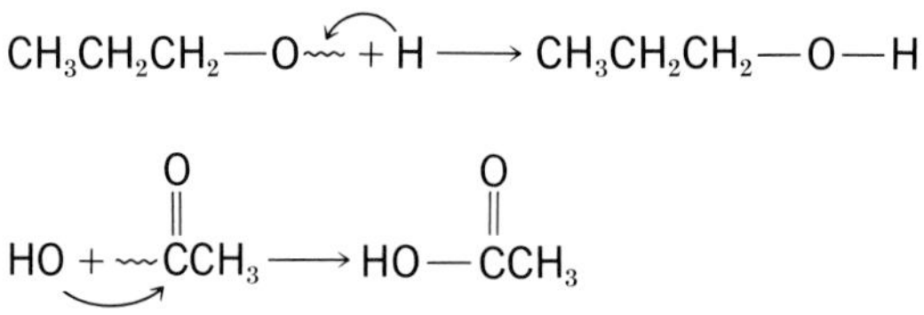

$$CH_3CH_2CH_2{-}O{\sim} + H \longrightarrow CH_3CH_2CH_2{-}O{-}H$$

$$HO + {\sim}\overset{\overset{\large O}{\|}}{C}CH_3 \longrightarrow HO{-}\overset{\overset{\large O}{\|}}{C}CH_3$$

These operations enable you to write the correct structures for the products and the equation for the reaction.

Animal fats and vegetable oils are special kinds of esters, and the hydrolysis of their ester linkages occurs during their digestion.

Exercise 7.6 Write the structures of the organic products that would be expected to occur in each of the following situations. If no reaction is to be expected, write "No reaction."

(a) 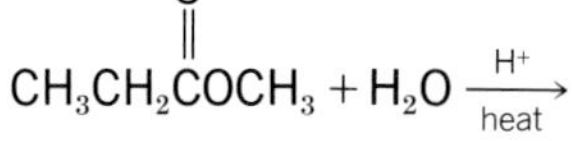 $CH_3CH_2\overset{\overset{\large O}{\|}}{C}OCH_3 + H_2O \xrightarrow[\text{heat}]{H^+}$

(b) $CH_3CH_2O\overset{\overset{\large O}{\|}}{C}CH_3 + H_2O \xrightarrow[\text{heat}]{H^+}$

(c) $C_6H_5\overset{\overset{\large O}{\|}}{C}OCH(CH_3)_2 + H_2O \xrightarrow[\text{heat}]{H^+}$

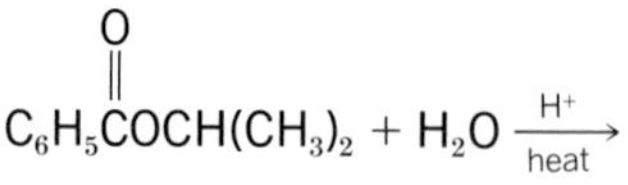

(d) 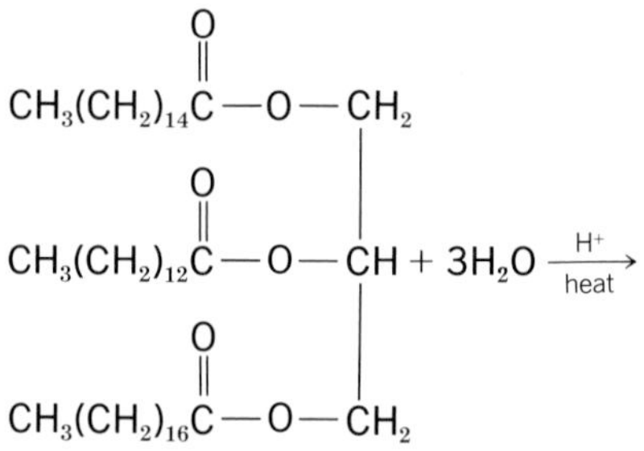
$$\begin{array}{l} CH_3(CH_2)_{14}\overset{\overset{\large O}{\|}}{C}{-}O{-}CH_2 \\ \qquad\qquad\qquad\qquad\quad | \\ CH_3(CH_2)_{12}\overset{\overset{\large O}{\|}}{C}{-}O{-}CH + 3H_2O \xrightarrow[\text{heat}]{H^+} \\ \qquad\qquad\qquad\qquad\quad | \\ CH_3(CH_2)_{16}\overset{\overset{\large O}{\|}}{C}{-}O{-}CH_2 \end{array}$$

(e) 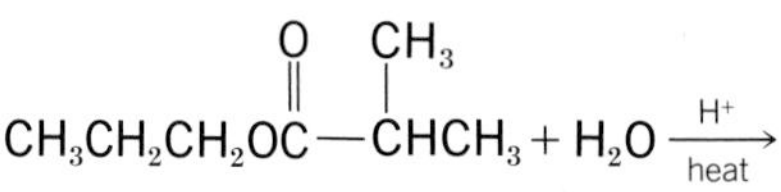 $CH_3CH_2CH_2O\overset{\overset{\large O}{\|}}{C}{-}\overset{\overset{\large CH_3}{|}}{C}HCH_3 + H_2O \xrightarrow[\text{heat}]{H^+}$

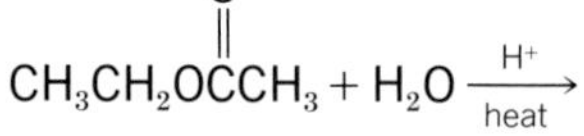

7.15 Saponification of Esters

This reaction of esters is merely a slight variation of ester hydrolysis. Ester hydrolysis occurs in the presence of an acid catalyst or an enzyme. **Saponification** occurs in aqueous sodium or potassium hydroxide. Essentially the same products form; the difference is that in saponification the carboxylic acid appears not as a free acid but as its sodium or potassium salt. The alcohol, however, emerges as the free alcohol in both hydrolysis and saponification.
In general:

$$R—\overset{O}{\overset{||}{C}}—O—R' + NaOH\ (aq) \xrightarrow[\Delta]{} R—\overset{O}{\overset{||}{C}}—O^-Na^+ + H—O—R'$$

Specific examples:

$$\underset{\text{Ethyl acetate}}{CH_3—\overset{O}{\overset{||}{C}}—O—CH_2CH_3} + NaOH \xrightarrow[\Delta]{H_2O} \underset{\text{Sodium acetate}}{CH_3—\overset{O}{\overset{||}{C}}—O^-Na^+} + \underset{\text{Ethyl alcohol}}{H—O—CH_2CH_3}$$

$$\underset{\text{Methyl benzoate}}{CH_3—O—\overset{O}{\overset{||}{C}}—C_6H_5} + NaOH \xrightarrow[\Delta]{H_2O} \underset{\text{Methyl alcohol}}{CH_3—O—H} + \underset{\text{Sodium benzoate}}{Na^+O^-—\overset{O}{\overset{||}{C}}—C_6H_5}$$

Saponification begins with a collision between a hydroxide ion, OH^-, and the carbonyl carbon with its partial positive charge.

$$R—\overset{:O:}{\overset{||}{C}}{}^{\delta+}—\ddot{O}R' + {}^-:\ddot{O}H \longrightarrow R—\overset{:\ddot{O}:^-}{\overset{|}{C}}(—\ddot{O}—R')(—\ddot{O}—H) \longrightarrow R—\overset{:O:}{\overset{||}{C}}—\ddot{O}:^- + H—\ddot{O}—R'$$

"Saponification" comes from the Latin *sapo,* soap, and it means to make soap. Ordinary soap is a mixture of sodium salts of long-chain carboxylic acids.

Exercise 7.7 Write the structures of the products of the saponification of each ester given in Exercise 7.6. Assume the presence of enough aqueous sodium hydroxide to react with all ester links.

7.16 Hydrolysis of Amides

Just as ester hydrolysis is actually the reverse of esterification, so amide hydrolysis is the exact reverse of amide formation.
In general:

$$R{-}\overset{\displaystyle O}{\overset{\|}{C}}{-}\overset{\displaystyle H}{\overset{|}{N}}{-}H + H{-}OH \xrightarrow[\Delta]{} R{-}\overset{\displaystyle O}{\overset{\|}{C}}{-}O{-}H + H{-}\overset{\displaystyle H}{\overset{|}{N}}{-}H$$

Amide — Water — Original carboxylic acid — Original amine (ammonia)

Note carefully what has happened structurally. First, the carbonyl-to-nitrogen bond (and only this bond) has ruptured:

$$R{-}\overset{\displaystyle O}{\overset{\|}{C}}{\nmid}\overset{\displaystyle H}{\overset{|}{N}}{-}H \dashrightarrow R{-}\overset{\displaystyle O}{\overset{\|}{C}}{-} + {-}\overset{\displaystyle H}{\overset{|}{N}}{-}H$$

Second, a hydroxyl group from water has become attached to the carbonyl carbon. The hydrogen left over from water has become attached to nitrogen:

$$R{-}\overset{\displaystyle O}{\overset{\|}{C}}{-} \quad H{-}O{\nmid}H \quad \overset{\displaystyle H}{\overset{|}{N}}{-}H \longrightarrow R{-}\overset{\displaystyle O}{\overset{\|}{C}}{-}\mathit{OH} + \mathit{H}{-}\overset{\displaystyle H}{\overset{|}{N}}{-}H$$

Specific examples:

$$CH_3{-}\overset{\displaystyle O}{\overset{\|}{C}}{-}\overset{\displaystyle H}{\overset{|}{N}}{-}H + H{-}OH \xrightarrow[\Delta]{} CH_3{-}\overset{\displaystyle O}{\overset{\|}{C}}{-}O{-}H + H{-}\overset{\displaystyle H}{\overset{|}{N}}{-}H$$

Acetamide — Acetic acid — Ammonia

$$C_6H_5{-}\overset{\displaystyle O}{\overset{\|}{C}}{-}\overset{\displaystyle H}{\overset{|}{N}}{-}H + H{-}OH \xrightarrow[\Delta]{} C_6H_5{-}\overset{\displaystyle O}{\overset{\|}{C}}{-}O{-}H + H{-}\overset{\displaystyle H}{\overset{|}{N}}{-}H$$

Benzamide — Benzoic acid

For all practical purposes, the only significant reaction that occurs to proteins when they are digested is just this one—amide hydrolysis. That is why mastery of this reaction now will make your later study easier.

Acids catalyze amide hydrolysis. Alkalies also can be used as catalysts. (Depending on whether you use acid or alkali, you get different initial products. However, the solution is normally made neutral before the products are isolated, and it is these isolated products that have been given.)

Exercise 7.8 Write the structures of the products of the hydrolysis of each of the following.

(a) $CH_3\overset{\displaystyle O}{\overset{\|}{C}}NH_2$ (b) $CH_3NH\overset{\displaystyle O}{\overset{\|}{C}}CH_2CH_3$ (c) $C_6H_5NH\overset{\displaystyle O}{\overset{\|}{C}}CH_3$

(d) 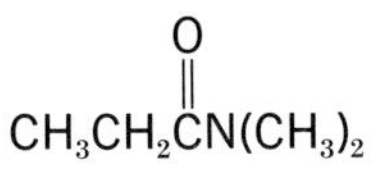 $CH_3CH_2\overset{O}{\overset{\|}{C}}N(CH_3)_2$ (e) 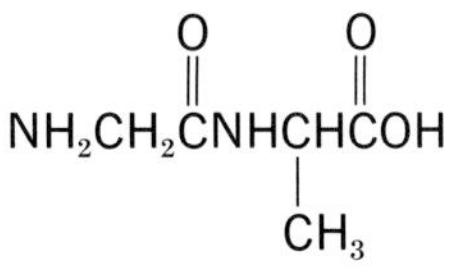 $NH_2CH_2\overset{O}{\overset{\|}{C}}NH\underset{CH_3}{\underset{|}{C}}H\overset{O}{\overset{\|}{C}}OH$

7.17 Amides as Neutral Compounds

In marked contrast to amines, amides are not proton-acceptors in the usual sense of being bases in water. The carbonyl group in the amide drains too much electron density away from the nitrogen; not enough is left at nitrogen to form a good bond to a proton. Amines are basic; amides are neutral.

Another way of understanding how amides are neutral compared to amines is by resonance. The unshared pair of electrons on nitrogen in the amide is delocalized as shown by these resonance contributors:

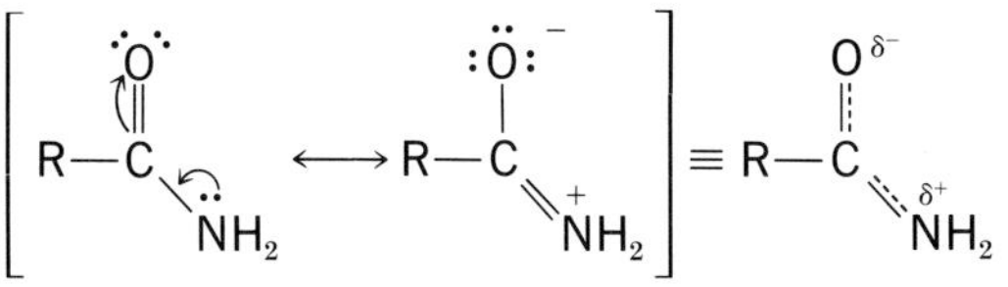

$$\left[R-\overset{\ddot{O}:}{\overset{\|}{C}}-\ddot{N}H_2 \longleftrightarrow R-\overset{:\ddot{O}:^-}{\overset{|}{C}}=\overset{+}{N}H_2 \right] \equiv R-\overset{O^{\delta-}}{\overset{\|}{C}}\cdots\overset{\delta+}{N}H_2$$

Resonance contributors for an amide — Hybrid

The same pair on the nitrogen of an amine cannot be delocalized; therefore, it is more available to form a bond to some hydrogen ion and change to a substituted ammonium ion.

7.18 Physical Properties of Acid Derivatives

A number of physical properties were assembled in Tables 7.4 and 7.5 for esters and amides. The usual trends in boiling points and solubilities with increasing molecular size may be noted there. Hydrogen bonding between amide molecules, illustrated in Figure 7.1, evidently is very strong, because all amides with at least one hydrogen on nitrogen, except the simplest amide (formamide), are solids at room temperature. The trend of a rapid drop in the melting point among the substituted formamides or acetamides as the hydrogens on nitrogen are replaced by alkyl groups (Table 7.5) is a striking indication that the hydrogen bond makes a considerable difference in physical properties. We may see this difference also in comparing propionic acid with methyl acetate, two isomers. Propionic acid, which has an —OH group, boils at 141 °C; methyl acetate, which has no —OH group, boils at 57 °C. Esters are only moderately polar.

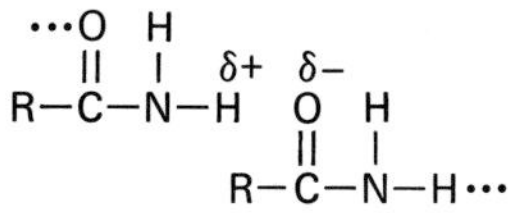

$$\cdots O\;\;H$$
$$R-\overset{\|}{C}-\overset{|}{N}-\overset{\delta+}{H}\cdots\overset{\delta-}{O}\;\;H$$
$$R-\overset{\|}{C}-\overset{|}{N}-H\cdots$$

Figure 7.1 Hydrogen bonding between molecules of amides.

The low-formula-weight esters have unusually pleasant odors and flavors, in huge contrast with their parent acids (or alcohols), as seen in the information of Table 7.6. Amides generally have no noteworthy odors.

Organic Derivatives of Inorganic Oxyacids

7.19 Anhydrides and Esters of Phosphoric Acid

Whereas simple anhydrides of carboxylic acids react too readily with water to be part of living systems, mixed derivatives between phosphoric acid (or its close relatives) and carboxylic acids or alcohols are involved in a number of metabolic pathways. Let us first see what these systems are and then look at some biologically important examples.

If a molecule of water splits out between two molecules of phosphoric acid, H_3PO_4, diphosphoric acid, $H_4P_2O_7$, results.

$$\underset{\text{Phosphoric acid}}{HO{-}\overset{\overset{O}{\|}}{\underset{\underset{OH}{|}}{P}}{-}O{-}H} + H{-}O{-}\overset{\overset{O}{\|}}{\underset{\underset{OH}{|}}{P}}{-}OH \xrightarrow[215\ ^\circ C]{} \underset{\text{Diphosphoric acid (mp 61 °C)}}{HO{-}\overset{\overset{O}{\|}}{\underset{\underset{OH}{|}}{P}}{-}O{-}\overset{\overset{O}{\|}}{\underset{\underset{OH}{|}}{P}}{-}OH} + H_2O$$

All four hydrogens in $H_4P_2O_7$ are replaceable. In addition, diphosphoric acid is also very much like an anhydride.

TABLE 7.6
Fragrances or Flavors of Some Esters

Name	Structure	Source or Flavor
Ethyl formate	$HCO_2CH_2CH_5$	Rum
Isobutyl formate	$HCO_2CH_2CH(CH_3)_2$	Raspberries
n-Pentyl acetate (*n*-amyl acetate)	$CH_3CO_2CH_2CH_2CH_2CH_2CH_3$	Bananas
Isopentyl acetate (isoamyl acetate)	$CH_3CO_2CH_2CH_2CH(CH_3)_2$	Pears
n-Octyl acetate	$CH_3CO_2(CH_2)_7CH_3$	Oranges
Ethyl butyrate	$CH_3CH_2CH_2CO_2CH_2CH_5$	Pineapples
n-Pentyl butyrate	$CH_3CH_2CH_2CO_2(CH_2)_4CH_3$	Apricots
Methyl salicylate	$C_6H_4(OH){-}CO_2CH_3$ (benzene ring with $-CO_2CH_3$ and ortho OH)	Oil of wintergreen

$$-\overset{\large O}{\overset{\|}{C}}-O-\overset{\large O}{\overset{\|}{C}}- \qquad -\overset{\large O}{\overset{\|}{\underset{|}{P}}}-O-\overset{\large O}{\overset{\|}{\underset{|}{P}}}-$$

Skeleton of a carboxylic acid anhydride

Skeleton of a phosphoric acid anhydride

The system occurs widely in living cells in the form of compounds of the following general type. They are shown in the acid forms, but usually they are present as singly or doubly ionized particles.

$$RO-P(=O)(OH)-O-P(=O)(OH)-OH$$

Monoester of diphosphoric acid (a diphosphate)[3]

Example: $(CH_3)_2C{=}CHCH_2O-P(=O)(OH)-O-P(=O)(OH)-OH$

Isopentenyl diphosphate (an intermediate in the biosynthesis of cholesterol)

Diphosphate esters are simultaneously esters, acids, and anhydrides. As anhydrides they participate in typical reactions of these compounds, for example, hydrolysis:

$$RO-P(=O)(OH)-O-P(=O)(OH)-OH + H_2O \longrightarrow RO-P(=O)(OH)-OH + HO-P(=O)(OH)-OH$$

Diphosphate ester — Monophosphate ester — Phosphoric acid

Triphosphate derivatives are also known and are essentially double anhydrides, plus being esters as well as acids. Adenosine triphosphate (ATP) is the

$$RO-P(=O)(OH)-O-P(=O)(OH)-O-P(=O)(OH)-OH$$

Triphosphate ester

$$HO-P(=O)(OH)-O-P(=O)(OH)-O-P(=O)(OH)-O-CH_2-\text{(ribose, 2',3'-OH; adenine with }NH_2\text{)}$$

Adenosine triphosphate, ATP

[3] Note carefully that in this and similar contexts the term *diphosphate* does not mean two separate PO_4^{3-} groups.

principal storehouse of the chemical energy that is most quickly available to living systems. The whole "purpose" of the breakdown of sugars or fats in the diet is to remake ATP when it has been used up as a source of energy.

7.20 Esters of Nitric and Nitrous Acids

Just as phosphoric acid looks like a phosphorus analog of a carboxylic acid, both nitric and nitrous acid seem to be nitrogen analogs. Many of their esters are known. Nitroglycerin (glyceryl trinitrate) is a powerful explosive. Isoamyl nitrite and nitroglycerin both have been used in controlling the severe pains of attacks of angina pectoris. These esters are covalent substances; they have no saltlike properties.

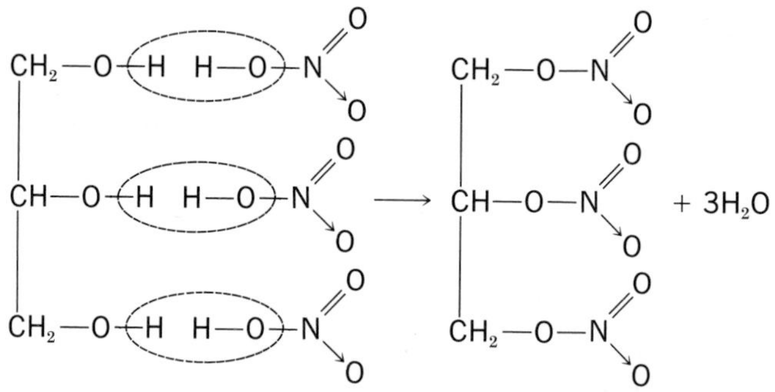

$$\begin{array}{l} CH_2-O-H \quad H-O-NO_2 \\ CH-O-H \quad H-O-NO_2 \\ CH_2-O-H \quad H-O-NO_2 \end{array} \longrightarrow \begin{array}{l} CH_2-O-NO_2 \\ CH-O-NO_2 \\ CH_2-O-NO_2 \end{array} + 3H_2O$$

Glycerol — Nitric acid (three molecules) — Glyceryl trinitrate ("Nitroglycerin")

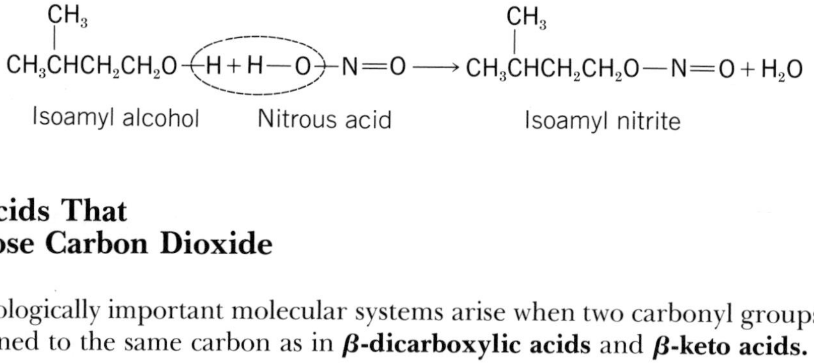

$$CH_3CH(CH_3)CH_2CH_2O-H + H-O-N=O \longrightarrow CH_3CH(CH_3)CH_2CH_2O-N=O + H_2O$$

Isoamyl alcohol — Nitrous acid — Isoamyl nitrite

7.21 Acids That Lose Carbon Dioxide

Biologically important molecular systems arise when two carbonyl groups are joined to the same carbon as in **β-dicarboxylic acids** and **β-keto acids.**

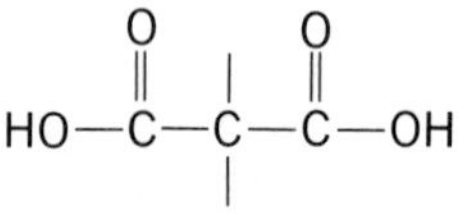

$$HO-\overset{O}{\overset{\|}{C}}-\underset{|}{\overset{|}{C}}-\overset{O}{\overset{\|}{C}}-OH \qquad \text{Example: } HO-\overset{O}{\overset{\|}{C}}-CH_2-\overset{O}{\overset{\|}{C}}-OH$$

β-Dicarboxylic acid — Malonic acid

$$R-\overset{O}{\overset{\|}{C}}-\underset{|}{\overset{|}{C}}-\overset{O}{\overset{\|}{C}}-OH \qquad \text{Example: } CH_3-\overset{O}{\overset{\|}{C}}-CH_2-\overset{O}{\overset{\|}{C}}-OH$$

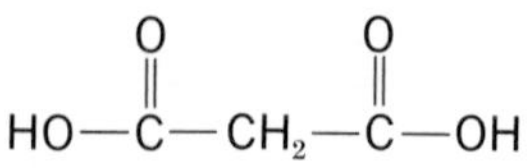

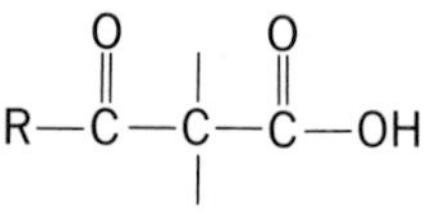

β-Keto acid — Acetoacetic acid

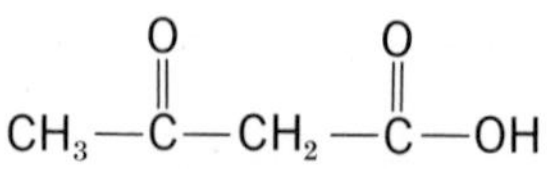

These compounds lose carbon dioxide very easily if heated; we say that they decarboxylate. The reason for this ease of **decarboxylation** is the second carbonyl group, which not only draws electrons to itself but provides a temporary home for an electron pair until a proton can get around to taking that pair for a new carbon-hydrogen bond. For example, the decarboxylation of a β-keto acid proceeds roughly as follows:

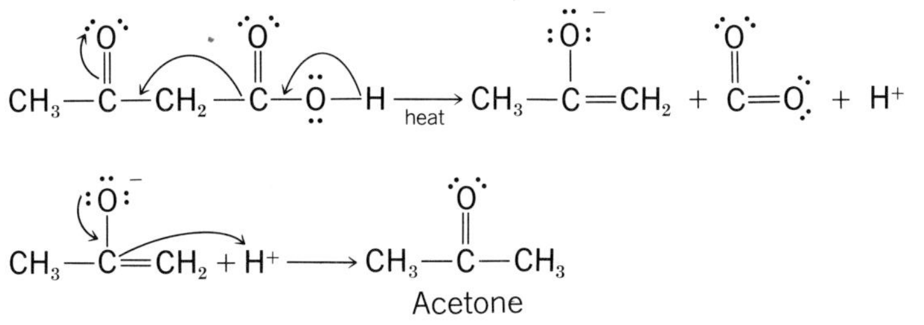

$$CH_3—\overset{O}{\overset{\|}{C}}—CH_2—\overset{O}{\overset{\|}{C}}—O—H \xrightarrow{heat} CH_3—\overset{O^-}{\overset{|}{C}}=CH_2 + \overset{O}{\overset{\|}{C}}=O + H^+$$

$$CH_3—\overset{O^-}{\overset{|}{C}}=CH_2 + H^+ \longrightarrow \underset{\text{Acetone}}{CH_3—\overset{O}{\overset{\|}{C}}—CH_3}$$

The decarboxylation of malonic acid or any β-dicarboxylic acid, occurs in a similar fashion producing acetic acid or a substituted acetic acid.

$$\underset{\text{A substituted malonic acid}}{R_2C(COOH)_2} \xrightarrow{heat} \underset{\text{A substituted acetic acid}}{R_2CH—\overset{O}{\overset{\|}{C}}—O—H} + CO_2$$

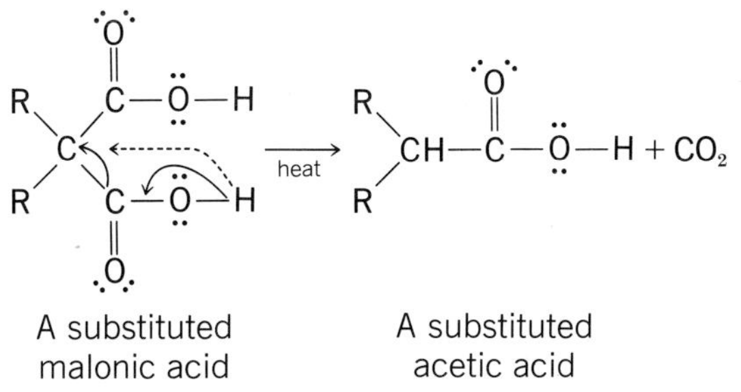

Both of these types of decarboxylations occur among metabolic pathways of the body. They are the principal ways whereby the carbon dioxide that we exhale forms.

Exercise 7.9 Which of these could decarboxylate and what product(s) (besides carbon dioxide) would form? (Write their structures.)

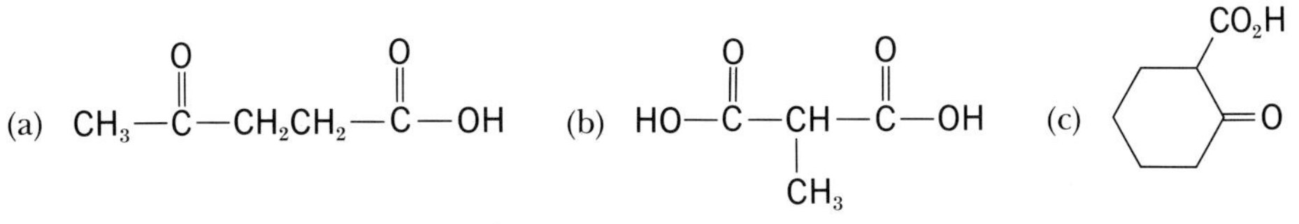

(a) $CH_3—\overset{O}{\overset{\|}{C}}—CH_2CH_2—\overset{O}{\overset{\|}{C}}—OH$ (b) $HO—\overset{O}{\overset{\|}{C}}—\underset{CH_3}{\underset{|}{CH}}—\overset{O}{\overset{\|}{C}}—OH$ (c) 2-oxocyclohexanecarboxylic acid (ring with CO_2H and $=O$ on adjacent carbons)

7.22 Synthesis of β-Keto Esters

We learned in the last chapter that one aldehyde molecule can add to another in an aldol condensation and that this reaction was an important type of molecule-building event in living systems. The reaction occurred partly because the carbonyl group activates its alpha hydrogens.

Simple esters also contain activated hydrogens on their alpha carbons. Something like an aldol condensation can also occur between two ester molecules (Step 1), but immediately after (Step 2), the initial product breaks down to a β-keto ester. The overall reaction is called the **Claisen condensation** (after Ludwig Claisen, a German chemist).

Step 1 (illustrated with ethyl propionate):

$$CH_3CH_2C(=O)OC_2H_5 + HCH(CH_3)C(=O)OC_2H_5 \rightleftharpoons \left[CH_3CH_2C(OH)(OC_2H_5){-}CH(CH_3)C(=O)OC_2H_5 \right]$$

Ethyl propionate — Ethyl propionate — (unstable)

Step 2

$$\left[CH_3CH_2C(OH)(OC_2H_5){-}CH(CH_3)C(=O)OC_2H_5 \right] \rightleftharpoons CH_3CH_2C(=O)CH(CH_3)C(=O)OC_2H_5 + C_2H_5OH$$

Ethyl α-methyl-β-ketovalerate — Ethyl alcohol

The steps are reversible. Therefore, this reaction under one set of conditions is an important molecule-building event, and under another set, a molecule breaking event. To make the Claisen condensation occur in the laboratory, an especially powerful base is needed, usually the sodium salt of ethyl alcohol ($Na^+ \ ^-OCH_2CH_3$). In the cell, special enzymes help the Claisen condensation happen. The reverse Claisen occurs in the breakdown of long-chain fatty acids during the metabolism of fats and oils.

Exercise 7.10 Write the structure of the β-keto ester that could be made from each by a Claisen condensation.
(a) $CH_3CO_2C_2H_5$ (b) $CH_3CH_2CH_2CO_2C_2H_5$

Important Individual Compounds

Acids and Salts

7.23 Acetic Acid

Over two billion pounds of this chemical are manufactured each year in the United States. Acetic acid is used principally in the production of a variety of acetate fibers and plastics—cellulose acetate, acetate rayon and others—and as a solvent. The acetate ion is an important intermediate in several metabolic pathways of the body.

7.24 2,4-Dichlorophenoxyacetic acid. 2,4-D

Cl (2-position), Cl (4-position) on benzene ring —O—CH_2CO_2H

This compound and several of its esters and salts are plant growth hormones that serve as herbicides (weed killers). When a plant takes any of them into its system, its growing patterns are so upset that it dies.

7.25 Sodium Benzoate

The sodium salt of benzoic acid is used as a food additive in foods and beverages that naturally have pH values in the range below 4.5 or 4.0. Acting to prevent the growth of yeasts and harmful bacteria—that is, serving as an antimicrobial agent—sodium benzoate is used in carbonated and still beverages, syrups, jams and jellies, pickles, salted margarine, syrups, fruit salads, icings and pie fillings. Concentrations are limited to 0.05 to 0.10%. Neither the benzoate ion nor benzoic acid accumulates in the body, and sodium benzoate has long been recognized as safe when used according to the limits given.

7.26 Propionates

The sodium and calcium salts of propionic acid, sodium propionate and calcium propionate, are widely used in baked goods and processed cheese to prevent the formation of molds and to inhibit the bacterium responsible for "rope" in bread. Propionic acid occurs naturally in Swiss cheese where its concentration may be as high as 1%. As food additives, the sodium and calcium salts are limited to 0.3 to 0.4%.

7.27 Sorbic Acid and the Sorbates

Sorbic acid is a white solid with two alkene double bonds per molecule. Both sorbic acid and its potassium and sodium salts are added to a variety of foods to inhibit the growth of molds and yeasts. These additives are effective up to pH values of 6.5, higher than the effective upper limit for benzoates and propionates. The products are used in fruit juices, fresh fruits, wines, soft drinks, sauerkraut and other pickled products, and some meat and fish products. With some foods, such as cheese, smoked fish, and dried fruit, solutions of sorbic acid salts are sprayed onto the surfaces or the wrappers.

$CH_3CH{=}CHCH{=}CHCO_2H$

Sorbic acid

HO—C_6H_4—CO_2R

Esters of *p*-hydroxybenzoic acid
(Parabens)

Esters

7.28 Parabens. Esters of *p*-Hydroxybenzoic Acid

The alkyl esters, particularly the methyl, ethyl, propyl, and butyl esters, are used as antimicrobial agents in cosmetics and pharmaceuticals as well as foods. They are active against molds and yeasts but less active against bacteria. Parabens may be used in baked goods, beer, cheese, fruit products, olives and pickles, and syrups.

7.29 Salicylates

Salicylic acid can function either as an acid or as a phenol, since it possesses both functional groups.

Salicylic acid, its esters, and its salts, taken internally, have both an analgesic effect (depressing sensitivity to pain) and an antipyretic action (reducing fever). As analgesics, they act to raise the threshold of pain by depressing pain centers in the thalamus region of the brain. Salicylic acid, itself, is too irritating to be used internally, but sodium salicylate and especially acetylsalicylic acid (aspirin) are widely used.

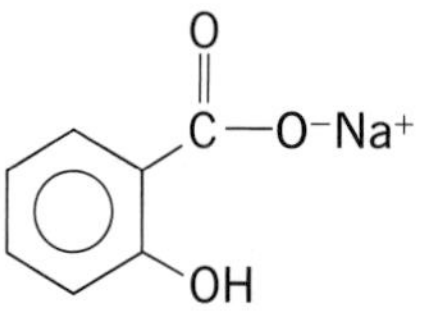

O ‖ C—O⁻Na⁺, OH

Sodium salicylate

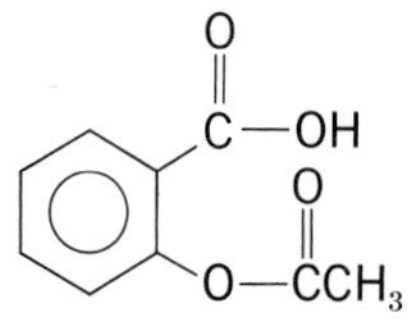

O ‖ C—OH, O—C(=O)CH$_3$

Acetylsalicylic acid
(Aspirin)

O ‖ C—OCH$_3$, OH

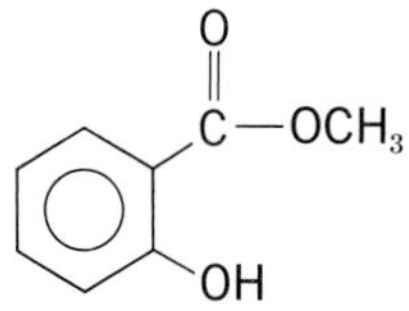

Methyl salicylate
(Oil of wintergreen)

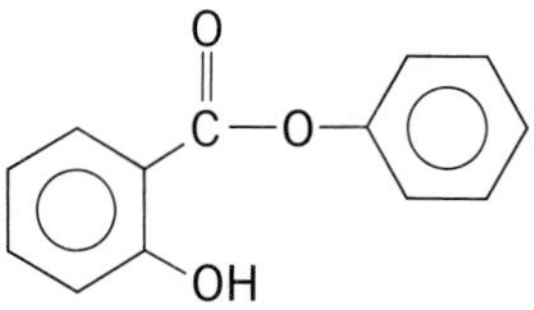

O ‖ C—O—C$_6$H$_5$, OH

Phenyl salicylate
(Salol)

Methyl salicylate is used in liniments. Phenyl salicylate has been used in ointments that protect the skin against ultraviolet rays.

Amides

7.30 Nicotinamide (Niacinamide)

Nicotinamide, the amide of nicotinic acid (niacin), has the physiological properties of niacin, a B-vitamin, and is the usual form in which niacin is administered. It bears some structural resemblance to the deadly poison, nicotine. Nicotine, in sufficient dosage, is lethal; niacin is essential to every cell in the human body.

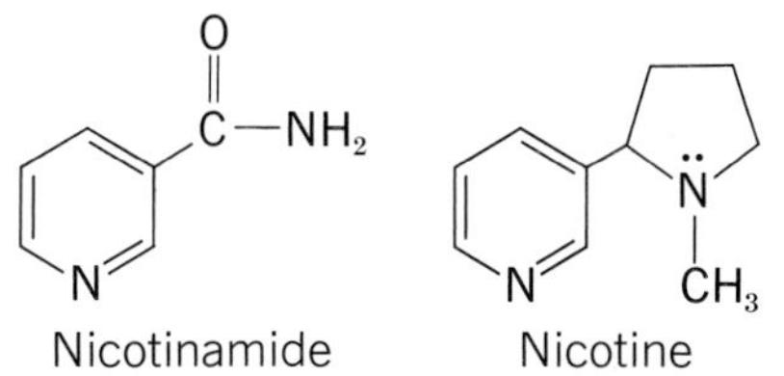

Nicotinamide Nicotine

Brief Summary

Acids and Their Salts. The carboxyl group, $—CO_2H$, is a polar group that confers moderate water solubility to a molecule while not preventing solubility in nonpolar solvents. The group is very resistant to oxidation and reduction. Carboxylic acids are proton donors toward hydroxide ions whereas alcohols are not. Toward water, carboxylic acids are weak acids; therefore, their carboxylate ions are proton acceptors (bases) toward mineral acids and their hydronium ions.

Salts of carboxylic acids are ionic compounds, and the potassium or sodium salts are very soluble in water. Hence the carboxyl group is another of nature's important "solubility switches."

The derivatives of acids—chlorides, anhydrides, esters, and amides—may all be made from the acids and converted back to the acids by hydrolysis.

Esters. Esters form by an acid-catalyzed reaction between an alcohol and a carboxylic acid. (Acid chlorides and anhydrides also react with alcohols to give esters, but these methods do not occur in living systems.) The action of water in the presence of an acid catalyst or an enzyme hydrolyzes esters back to the alcohol and the carboxylic acid. Esters are saponified by the action of hot alkali, which produces the alcohol and the salt of the acid. A near doubling of the ester molecule occurs in the Claisen condensation. The ester group is a key feature of molecules of animal fats and vegetable oils.

Amides. Amides form by strongly heating an acid together with ammonia or a suitable amine. They are hydrolyzed by the action of hot water and either an alkali or an acid. The amide bond is the principal bond that joins the individual units (amino acids) together in proteins. The amide group is neutral toward aqueous acids and bases.

Esters of Inorganic Acids. Esters of phosphoric, diphosphoric, and triphosphoric acid occur in living systems. The diphosphates and triphosphates are simultaneously esters, anhydrides, and acids; the triphosphates, in particular, are important energy-storage systems in cells. Esters of nitric and nitrous acids are also known and some serve as drugs.

Reactivities of Acid Derivatives. Toward water, with or without an acid or basic catalyst, the acid derivatives display the following order of reactivity:

$$\text{acid chlorides} > \text{acid anhydrides} > \text{esters} > \text{amides}$$

The carbonyl group in the acid chlorides is most cpen to attack by an electron-rich reagent (e.g., water or hydroxide ion); it has the largest partial positive charge of all the derivatives; and the group ejected (Cl^-) is the most stable that can be released from these derivatives.

Physical Properties. Where hydrogen bonding can be maximized, substances have the highest boiling points and, up to a point, the highest solubility in water. Hence all other factors being equal (especially the formula weights), amides have higher boiling points than carboxylic acids, and esters have relatively low boiling points.

Reagents. *Aqueous acids* attack and hydrolyze esters and amides, and change carboxylate ions to free carboxylic acids. *Aqueous alkali* saponifies esters and amides, but the acids that form emerge as their salts. Aqueous base also changes acids to their salts. *Oxidizing agents* generally do not attack acids or their derivatives. Mild *reducing agents* will not attack the carbonyl groups of acids or any derivatives. (The *Study Guide* has an extensive review of organic reactions.)

Acids That Decarboxylate. β-Keto acids and β-dicarboxylic acids are two types of acids that readily lose carbon dioxide.

Questions and Exercises

1. Write the structure of each of the following compounds.
(a) Acetamide (b) Methyl propionate
(c) Acetyl chloride (d) Acetic anhydride
(e) *N*-Methylformamide (f) Isopropyl *n*-butyrate
(g) Glycerol triacetate

2. Write the common names for each of the following compounds. (Some are for review.)

(a) $CH_3CH_2CH_2CH_2O\overset{O}{\overset{\|}{C}}CH_2CH_3$ (b) $CH_3CH_2CH_2\overset{O}{\overset{\|}{C}}NH_2$

(c) $CH_3\overset{O}{\overset{\|}{C}}CH_2\overset{O}{\overset{\|}{C}}OH$ (d) $CH_3CH_2\overset{O}{\overset{\|}{C}}-Cl$

(e) $C_6H_5\overset{O}{\overset{\|}{C}}OCH_2CH_3$ (f) $CH_3CH_2OCH_2CH_3$

(g) $CH_3CH_2CH_2CH_2NH_2$ (h) $CH_3CH_2CH_2CH_2CH_2\overset{O}{\overset{\|}{C}}OH$

(i) $(CH_3)_2CHO\overset{O}{\overset{\|}{C}}H$

3. Complete the following equations by writing the structure(s) of the principal

organic product(s) that are expected to form. If no reaction is expected, write "No reaction." (Many are for review.)

(a) $CH_3CHO \xrightarrow{K_2Cr_2O_7}$

(b) $CH_3OH + CH_3COOH \xrightarrow[heat]{H^+}$

(c) $CH_3CH_2CH_3 + H_2SO_4 \longrightarrow$

(d) $CH_3CH_2CH(OH)CH_3 \xrightarrow[heat]{KMnO_4}$

(e) $CH_3CH_2CONH_2 + H_2O \xrightarrow[heat]{H^+}$

(f) $CH_3CH(OH)CH_3 \xrightarrow[heat]{H_2SO_4}$

(g) $CH_3OCOCH(CH_3)_2 + H_2O \xrightarrow[heat]{H^+}$

(h) $CH_3CH_2CH{=}CH_2 + HCl \longrightarrow$

(i) $CH_3CH(OCH_3)_2 + H_2O \xrightarrow[heat]{H^+}$

(j) $CH_3CO_2^-Na^+ + HBr_{aq} \longrightarrow$

(k) $CH_3CH_2OCH_2CH_3 + H_2O \xrightarrow{^-OH}$

(l) $CH_3CH_2CHO + 2CH_3OH \xrightarrow{HCl}$

(m) $CH_3CH{=}CHCH_3 + H_2 \xrightarrow[heat,\ pressure]{Ni}$

(n) $CH_3C(OH)(CH_3)CH_2CH_3 \xrightarrow{K_2Cr_2O_7}$

(o) $(CH_3)_2CHCOOH + CH_3OH \xrightarrow{H^+}$

(p) $CH_3NHCOCH_3 + H_2O \xrightarrow[heat]{}$

(q) $CH_3CH_2OH \xrightarrow{KMnO_4}$

(r) $CH_3CH_2COCH_2COOH \xrightarrow[heat]{}$

(s) $C_6H_6 + Br_2 \xrightarrow{Fe}$

(t) $CH_3CH_2NHCOCH_3 + H_2O \xrightarrow[heat]{}$

(u) $CH_3COCH_3 \xrightarrow{KMnO_4}$

(v) $NH_2CH_2CH_2CH_3 + HCl_{aq} \longrightarrow$

(w) $CH_3CH_2CH_2CH_3 + H_2O \xrightarrow{H^+}$

(x) $CH_3OCH_2\overset{O}{\overset{\|}{C}}OCH_3 + H_2O \xrightarrow[heat]{H^+}$

(y) $CH_3OCH_2\overset{O}{\overset{\|}{C}}CH_3 + H_2O \xrightarrow[heat]{H^+}$

(z) $NH_2CH_2\overset{O}{\overset{\|}{C}}NHCH_2\overset{O}{\overset{\|}{C}}OH + H_2O \xrightarrow[heat]{}$

(aa) $CH_3\overset{O}{\overset{\|}{C}}OH + NaOH_{aq} \longrightarrow$

(bb) $CH_3CH_2\overset{CH_3}{\overset{|}{\underset{H}{\underset{|}{N^+}}}}{-}H\ Cl^- + NaOH_{aq} \longrightarrow$

(cc) $CH_3OCH_2\underset{CH_3O}{\underset{|}{C}}HCH_3 + H_2O \xrightarrow[heat]{}$

(dd) $CH_3NH_2 + CH_3\overset{O}{\overset{\|}{C}}OH \xrightarrow[heat]{}$

(ee) $CH_3C{\equiv}CCH_3 + \underset{(excess)}{H_2} \xrightarrow{Ni}$

(ff) $C_6H_5O\overset{O}{\overset{\|}{C}}CH_3 + H_2O \xrightarrow{H^+}$

(gg) $CH_3\overset{O}{\overset{\|}{C}}OCH_2CH_2O\overset{O}{\overset{\|}{C}}CH_3 + H_2O \xrightarrow[heat]{H^+}$

(hh) $CH_3O\overset{O}{\overset{\|}{C}}CH_2CH_2\overset{O}{\overset{\|}{C}}OCH_3 + H_2O \xrightarrow[heat]{H^+}$

4. Arrange the following compounds in the order of increasing boiling point. (You should be able to do this without consulting tables.) Explain your answer.

$CH_3CH_2CH_2CH_2NH_2$ (a) $CH_3CH_2CH_2CH_2OH$ (b) $CH_3CH_2CH_2CH_2CH_3$ (c) $CH_3CH_2\overset{O}{\overset{\|}{C}}OH$ (d)

5. Arrange the following compounds in the order of increasing solubility in water and outline the reasons for the order you select.

$CH_3CH_2CH_2CH_2CH_2\overset{O}{\overset{\|}{C}}O^-Na^+$ (a) $CH_3CH_2CH_2CH_2CH_2\overset{O}{\overset{\|}{C}}OH$ (b) $CH_3CH_2CH_2CH_2\overset{O}{\overset{\|}{C}}OCH_3$ (c)

$CH_3CH_2CH_2CH_2CH_2OCH_3$ (d) $CH_3CH_2CH_2CH_2CH_2CH_3$ (e)

6. Explain how *N,N*-dimethylacetamide melts at a temperature so much lower than that of acetamide. (See Table 7.5.)
7. Explain how the carbonyl group makes acetic acid a stronger acid than ethyl alcohol.
8. When aspirin is hydrolyzed, what forms? (Give structures.)

chapter 8

Synthetic Polymers

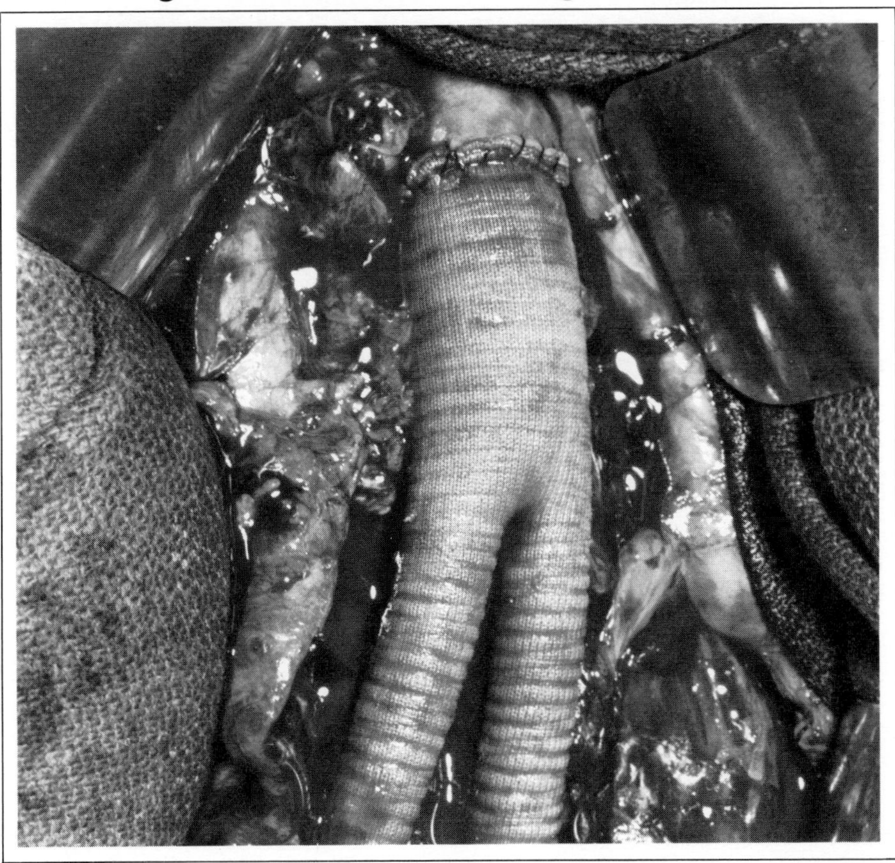

One of the very versatile synthetics is Dacron, seen here as a graft of knitted tubing in an operative site. Synthetics have many applications in medicine, and we shall learn about several in this chapter. (Courtesy the Department of Surgery, Baylor University College of Medicine.)

8.1 The World of Synthetics

Petroleum, coal, and natural gas have long been familiar to us as fuels to heat buildings and drive machinery. Much less understood by most people is the enormous importance of these natural substances as chemical raw materials for the world of synthetics.

About 80% of the gasoline used in cars and trucks is synthetic, being made from other parts of petroleum that do not work in gasoline engines.

Much of the rubber used to make tires and innertubes is synthetic, made from petroleum.

Drip-dry, wash and wear clothing is made wholly or mostly from synthetics, fabrics made from the chemicals wrung from coal and petroleum and natural gas and changed to nylon, Dacron, Orlon, and the acrylates.

The entire world of plastics—polyolefins, polyamides, polyesters, acrylates and silicones—emerged from the ingenuity of scientists and engineers; the daring of administrators, managers, and investors; and the availability of raw materials.

High-yield agriculture, without which there is not the slightest hope of feeding the world population, depends heavily, for good or ill, on considerable use of synthetic weed killers, bug killers, and fungi preventers.

High-quality medicine depends increasingly on synthetic drugs; disposable hospital items; and plastic fabric, rod, tubing, sheet, knitted mesh, and molding compounds for the surgical strengthening and repairing of damaged tissue.

Many hobbies and recreational activities depend on synthetics: strong, ultralight fabrics for tents and backpacks; sails; lines and lures for fishing; epoxy resins for fiber glass plastic boat hulls and the bodies of some racing vehicles, to say nothing of surfboards and skis and flame resistant backing for photographic film.

We could not hope in one chapter to survey all of the materials in the world of synthetics. We shall limit our study largely to polymeric substances and to those most useful in the general health field, but a number of common and popular uses will be described.

8.2 Some Terms in Polymer Science

A **polymer** is a very high-formula-weight substance consisting of molecules having in common a repeating structural unit, the word coming from the Greek *poly-*, "many," and *meros,* "parts." A polymer is one kind of **macromolecule** ("macro-," huge), one with repeating units. The starting material for making a polymer is called a **monomer.** If only one monomer is used, the polymer is sometimes called a homopolymer—all repeating units are identical and all come from one monomer. The reaction producing a polymer is called **polymerization.** Polyethylene is a homopolymer and ethylene is its monomer:

$$nCH_2{=}CH_2 \longrightarrow (CH_2CH_2)_n$$

Ethylene → Polyethylene

n may be thousands (Because n is large, its exact value is unimportant.)

In general, if we let "A" be the symbol for the monomer molecule, then its polymerization may be represented as, where n = several thousand:

$$nA \longrightarrow A—A—A—A—A—A—A—A—A—A—A—A—A—etc.$$

Sometimes polymerization generates long molecules with various branches. For example,

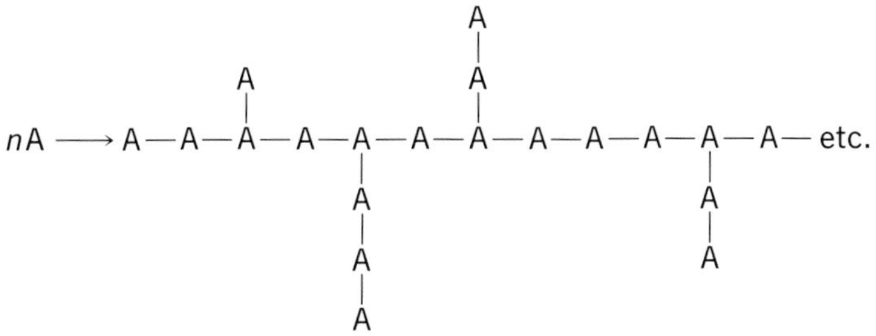

```
                                A
                                |
                  A             A
                  |             |
nA ──→ A—A—A—A—A—A—A—A—A—A—A—A— etc.
                      |               |
                      A               A
                      |               |
                      A               A
                      |
                      A
```

Often two or more monomers are mixed together and copolymerized. The product is called a **copolymer.** Symbolizing the second monomer by "B," then the possibilities are that A and B may alternate with each other; they may be grouped as blocks; they may go together in a wholly random way; or strands of B may be grafted onto strands of A. These possibilities are illustrated in Figure 8.1. All but the graft copolymer is a **linear polymer;** one unit follows another and the structure can be printed on one line. In the main chain the graft polymer is also linear. A number of important polymers are **cross-linked.** A second monomer reacts with parts of two linear molecules to make a molecular bridge; when cross-links occur in a number of places, the substance is an intricate, interlacing network (Figure 8.1). Polymer scientists and engineers can control copolymerization to give products of desired properties by selecting monomers and by changing conditions of temperature, pressure, catalyst, time, and order of mixing.

Not all polymers are plastics because the large family of polymers includes many natural substances fitting the definition: proteins, nucleic acids, polysaccharides (e.g., starch, cellulose), gums, and natural rubber.

A **plastic** is a finished or semifinished article or substance made from a resin by molding, casting, extruding, drawing, or laminating during or after polymerization of the raw materials.

A synthetic **resin** is the unfabricated polymeric material used to make all or most of a plastic article or to coat some surface. The finished plastic usually contains a number of other materials.

An inert **filler** may be added during polymerization to reduce the amount of shrinkage that occurs as the monomer mixture "sets." When thousands of molecules of monomer change to a relatively few molecules of polymer, the larger spaces used in nonbonding situations become the much smaller spaces determined by bonding distances. Silica, a form of hydrated silicon dioxide, is a filler in some dental restorative plastics, for example, where shrinkage must be completely overcome.

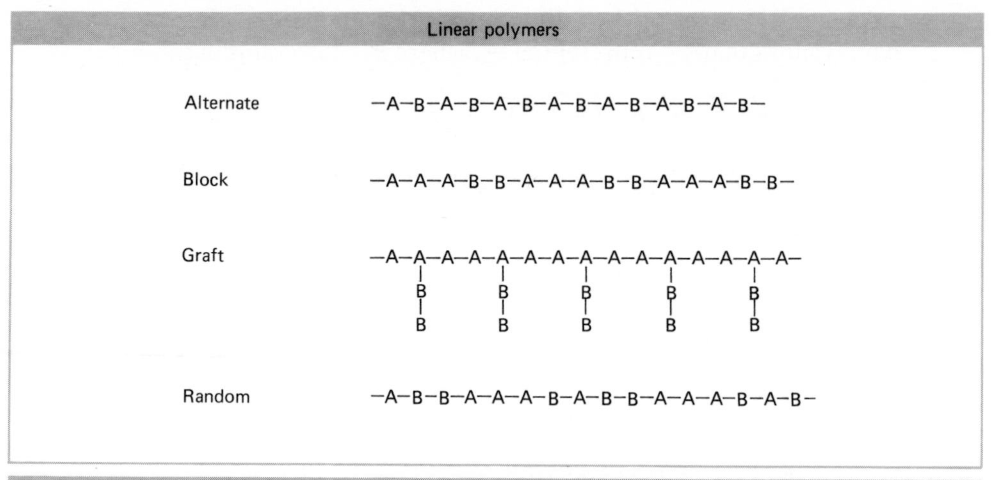

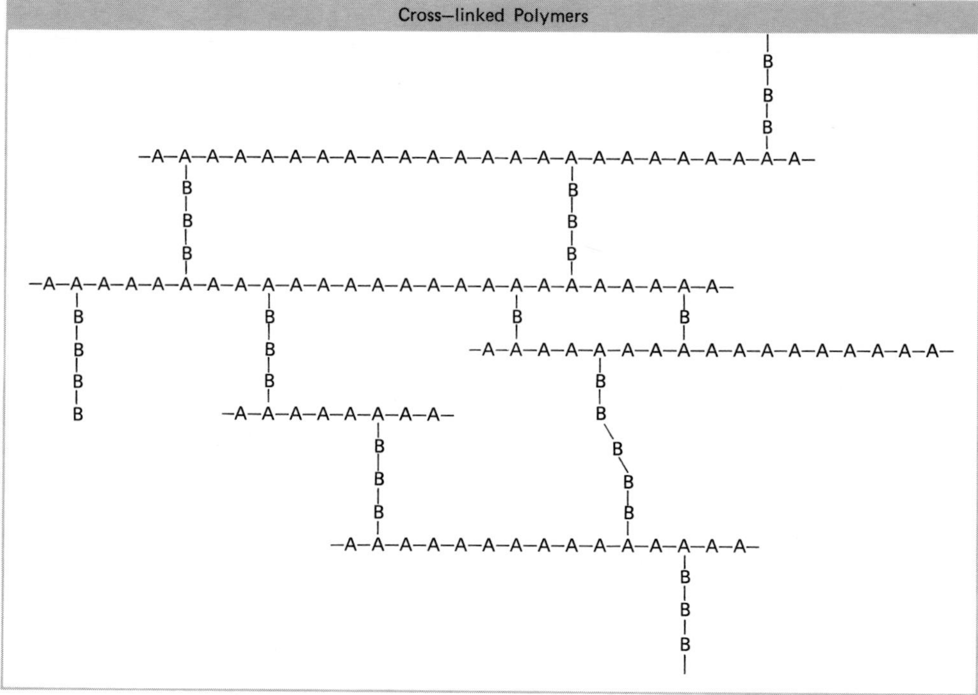

Figure 8.1 Some types of polymers. The cross-linked type shown here has cross-links of random length and location. In some cross-linked polymers much greater regularity is observed. Moreover, cross-linking occurs in all directions and the structural array is three-dimensional.

Traces of the catalyst used to induce polymerization will be present in the resin, some bound covalently at the ends of the polymer chains, some simply intermixed. The catalyst is more properly called the **initiator.** Unlike a true catalyst, an initiator of polymerization is not recovered unchanged. The words, however, are generally used interchangeably, as we shall do.

To keep the monomer from prematurely polymerizing, it may have been treated with a **stabilizer.** Hydroquinone is sometimes used because it scavenges stray oxidants that might initiate polymerization.

To reduce hardness and brittleness in the final product, a molecular lubricant called a **plasticizer** may be added. Some plasticizers are oily esters. Their molecules ride between neighboring polymer strands keeping the whole system more supple and flexible. Extra ingredients must be carefully controlled when medical applications are considered. Some plasticizers, for example, are harmful when released internally.

Some fillers or additives other than monomers can act to cross-link polymer molecules. Carbon black serves this function in synthetic rubber, which may contain as much as 20 to 30% carbon black.

One large group of polymers is **thermoplastic.** They soften when heated and can be remelted or redissolved and then recast or reformed into different plastic articles. Thermoplastic polymers generally have few or no cross-links. Heavily cross-linked polymers are **thermosetting.** Once formed and cast under heat, they set. The cross-linked molecular framework prevents remelting and reworking.

In terms of the broad chemical changes that produce polymers, the two chief types are **addition polymers** and **condensation polymers.** Addition polymers form as in the example of polyethylene discussed earlier. Monomer molecules simply join or add together. Condensation polymers form by the splitting out of small molecules between the monomers. In making a polyamide such as nylon or a polyester such as Dacron, molecules of water split out as new amide bonds or new ester bonds form.

Addition Polymers

8.3 Polyolefins

"Olefin" is the nickname of any alkene and **polyethylene** is the simplest **polyolefin.** When ethylene is polymerized under high pressure and high temperature, a low-density form of polyethylene is produced that softens at about 110 °C. We may use its formation to illustrate one of the ways that polymerization can occur—by a free-radical chain reaction.

The initiator for low-density polyethylene induces one-electron shifts (where we represent the initiator by IN·, the second of the two bonds in a double bond by two dots, and one-electron shifts by arrows with only one barb):

$$\mathrm{IN\cdot} \quad + CH_2 \overset{..}{-} CH_2 + CH_2 \overset{..}{-} CH_2 + CH_2 \overset{..}{-} CH_2 + CH_2 \overset{..}{-} CH_2 + \text{etc.}$$

$$\downarrow$$

$$\mathrm{IN{-}CH_2{-}CH_2{-}CH_2{-}CH_2{-}CH_2{-}CH_2{-}CH_2{-}CH_2{-}}\text{etc.} \quad (\text{or: } \mathrm{IN{+}CH_2{-}CH_2{+}_4}\text{etc.}$$

Polyethylene

As the chain develops, its growing end occasionally turns back on itself, throws out a hydrogen atom, and opens a site for the growth of a branch. Low-density polyethylene is considerably branched (see Section 8.2). The branches prevent the molecules from packing together tightly and developing a higher-density, mechanically strong product.

Special initiators such as the Ziegler-Natta[1] catalyst, a combination of titanium trichloride–aluminum triethyl, direct the polymerization of ethylene by a different route, one involving two-electron shifts. The product consists of molecules that are much longer and virtually unbranched. It has a much higher density and higher strength. Letting "IN" again be the symbol for the initiator, this time with an orbital that can accept a pair of electrons (and letting arrows with two barbs indicate two-electron shifts):

$$IN + CH_2{=}CH_2 \longrightarrow IN^{-}{-}CH_2{-}CH_2^{+} \xrightarrow{CH_2{=}CH_2} IN^{-}{-}CH_2{-}CH_2{-}CH_2{-}CH_2^{+}$$

$$\xrightarrow{CH_2{=}CH_2 \text{ etc. many times}} IN^{-}{+}CH_2{-}CH_2{+}_n^{+}$$

Eventually, the initiator is released by other reactions not shown and the polymer molecule, one without side branches, is free. This ultra-high-formula-weight, high-density polyethylene is presently the polymer of choice for replacement of the cup-shaped socket (the acetabulum) in a knee joint repair or other ball and socket joints. Polyethylene has a high resistance to chemical change; it resists penetration by fatty material but is quite permeable to both oxygen and carbon dioxide. Although many of its applications have been replaced by polypropylene, a close relative, polyethylene has been used for drains (as in middle ear vents, paranasal sinuses, and glaucoma), catheters, tubing, sutures, and for the wrapping of aneurysms.[2] Familiar household and laboratory articles made of polyethylene include pails and pans, funnels, icebox dishes, drinking glasses, syringe barrels and plungers, and other molded items (Figure 8.2).

Polypropylene is another useful polyolefin. Made by polymerizing propylene, this polymer is used to manufacture most of the items that are also made of polyethylene.

$$n\,CH_2{=}\underset{\displaystyle CH_3}{\underset{|}{CH}} \longrightarrow {+}CH_2{-}\underset{\displaystyle CH_3}{\underset{|}{CH}}{+}_n$$

Propylene → Polypropylene

One form of polypropylene has become one of the major synthetic fibers. Indoor-outdoor carpeting is made of it. If a polymer is used successfully in making fibers, its molecules must not only be long; they must also have uniformity and symmetry and yet have some features that make neighboring

[1] Karl Ziegler of Germany and Guilio Natta of Italy shared the 1963 Nobel Prize in Chemistry for their discoveries of catalysts that guide polymerizations along unique courses.

[2] It is assumed that the reader has access to other books, including a dictionary, in which medical and anatomical terms are defined and illustrated.

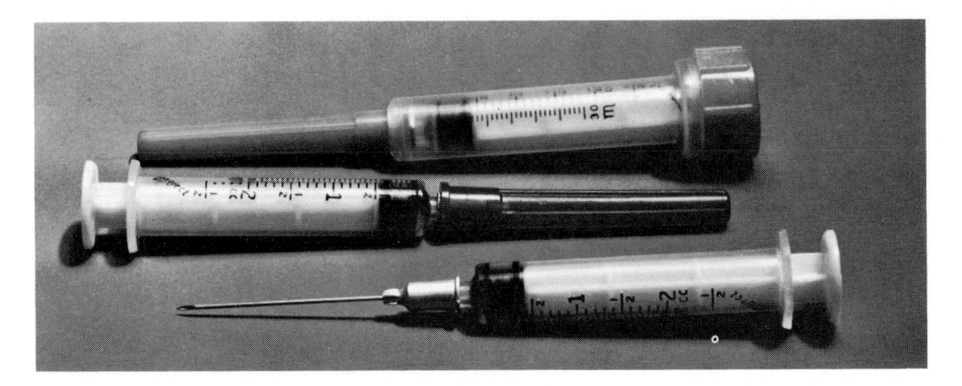

Figure 8.2 The costs of syringes used in clinics and hospitals are reduced by using presterilized, disposable plastic barrels and plungers made of polyethylene. (Courtesy Exxon Chemical Co.)

molecules line up side by side and stick to each other. Polyethylene molecules tangle well together but do not line up well side by side and stick. Polypropylene molecules have methyl groups on every other carbon. These help neighboring molecules catch on each other but they do not line up well side by side unless all the methyl groups are on the same side of the chain, as illustrated in Figure 8.3. If they alternate, first one on one side, then the next on the other side, and so forth, or if they are randomly oriented, good fibers cannot be made. Each of these three types of polypropylene—methyl groups all on the same side, or alternate, or random—can be made. The correct catalysts and conditions of temperature and pressure are known for each.

Polypropylene has outstanding flex life (Figure 8.4), making it useful surgically in integral hinge applications and heart valve leaflets. It has also been used in knitted surgical mesh, blood filters, and in hip joint repair. Most earlier medical uses of polyethylene are now better served by polypropylene.

8.4 Diene Polymers

Natural rubber, the coagulated milk juice (latex) of many tropical trees, is a polymer of isoprene. The isoprene units per molecule in natural rubber vary from 1500 to 15 000 for polymer formula weights of 100 000 to 1 million. Synthetic rubber with all the properties of natural rubber can be made from isoprene using a special catalyst. At each double bond in natural rubber the —CH_2— units forming part of the main chain are always *cis* to each other.

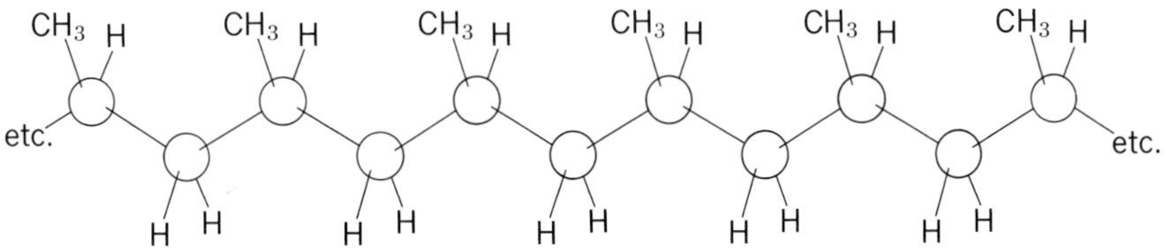

CH_3 H CH_3 H CH_3 H CH_3 H CH_3 H CH_3 H

etc. etc.

H H H H H H H H H H

Figure 8.3 Polypropylene in which all methyl side chains are on the same side of the polymer chain.

$$CH_2{=}\overset{CH_3}{\overset{|}{C}}{-}CH{=}CH_2 + CH_2{=}\overset{CH_3}{\overset{|}{C}}{-}CH{=}CH_2 + CH_2{=}\overset{CH_3}{\overset{|}{C}}{-}CH{=}CH_2 + CH_2{=}\overset{CH_3}{\overset{|}{C}}{-}CH{=}CH_2 + \text{etc.}$$

Isoprene

$$\downarrow \text{catalyst}$$

$$\cdots{-}CH_2{-}\underset{CH_3}{\underset{|}{C}}{=}CH{-}CH_2{-}CH_2{-}\overset{CH_3}{\overset{|}{C}}{=}CH{-}CH_2{-}CH_2{-}\underset{CH_3}{\underset{|}{C}}{=}CH{-}CH_2{-}CH_2{-}\overset{CH_3}{\overset{|}{C}}{=}CH{-}CH_2{-}\cdots$$

Natural Rubber
(Note the isoprene units
between the dashed lines.)

The isomer with these units all *trans* to each other is gutta percha. Gutta percha, less elastic than rubber, cannot substitute for rubber but it has had a few medical applications, for example, as a dental cement and as material for making fracture splints. Gutta percha is also obtained from various tropical trees.

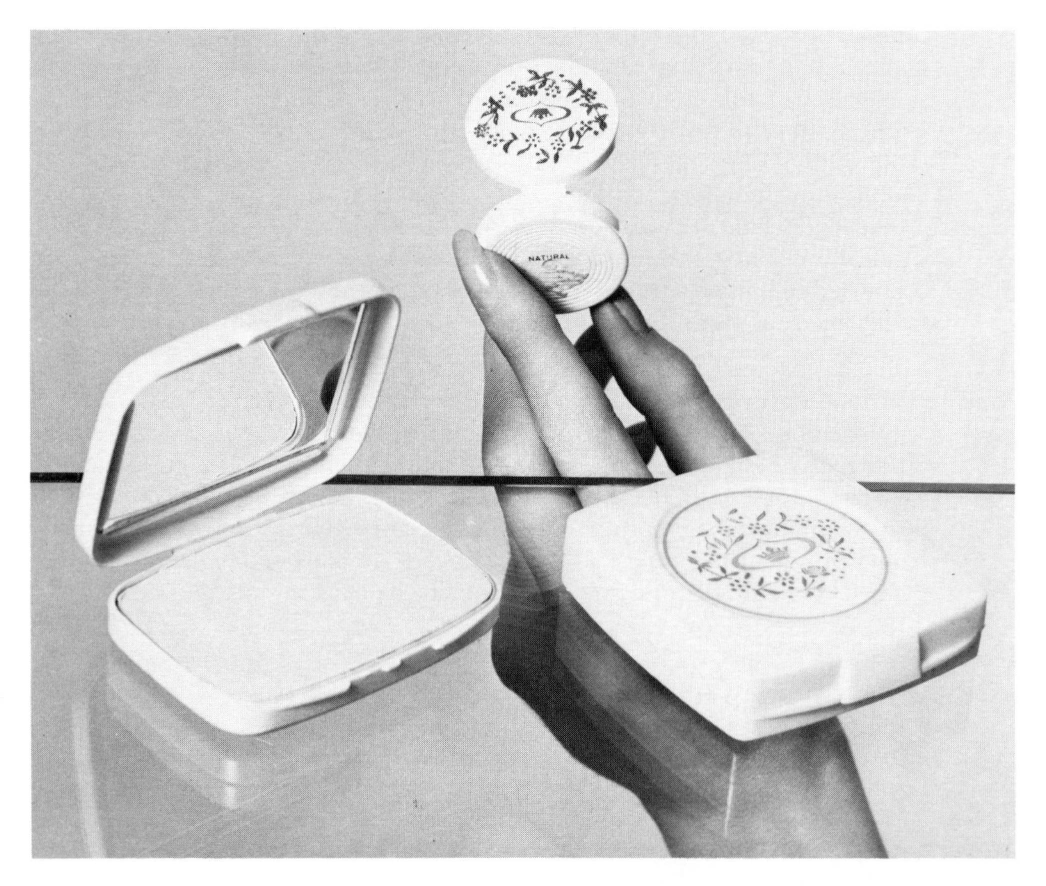

Figure 8.4 The exceptional flex strength of polypropylene contributes to its popularity as a resin for fabricating hinged articles. (Courtesy of ENJAY Chemical Company).

8.5 Elastomers

Polymers that are elastic are called **elastomers** (*elastic* poly*mers*). A number of olefin copolymers are elastomers, for example, the copolymer between isoprene and isobutylene.

Elastomers such as rubber may be made tougher and more resistant to abrasion by **vulcanization.** Rubber is mixed with a vulcanizing agent, such as sulfur and a catalyst, heated (hence "vulcan," god of fire), and numerous cross-links form.

8.6 Other Polyolefins and Halogenated Derivatives

Polystyrene, made from styrene, has phenyl groups on alternate carbons of the main chain. If an inert gas is blown into the polymerizing mixture, familiar polystyrene foam articles can be made.

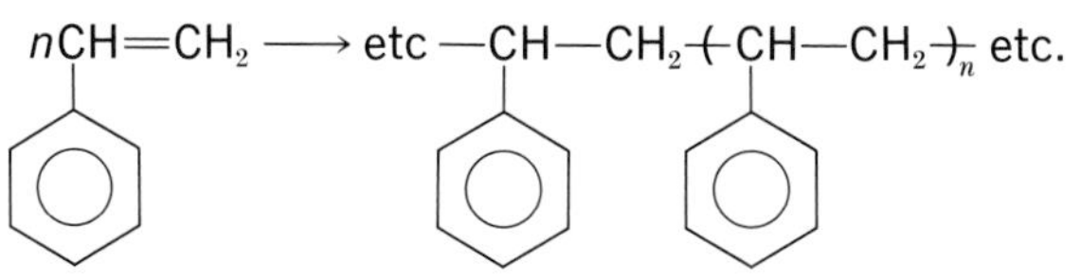

$$n\,CH(C_6H_5){=}CH_2 \longrightarrow \text{etc}\ {-}CH(C_6H_5){-}CH_2{+}CH(C_6H_5){-}CH_2{+}_n\ \text{etc.}$$

Styrene　　　Polystyrene

$CH_2{=}CHCl$	$CH_2{=}CCl_2$	$F_2C{=}CF_2$	$CF_3{-}CF{=}CF_2$
Vinyl chloride	Vinylidene chloride	Tetrafluoro-ethylene	Perfluoro-propylene

Polyvinyl chloride (PVC) is the principal resin used to make clear, hard plastic bottles. Its monomer, vinyl chloride, has repeatedly been implicated as a cause of a rare form of liver cancer, and the government has moved to specify rigid standards of air quality for the environment of workers using vinyl chloride. One common tubing, Tygon, is made of polyvinyl chloride. The polymer of vinylidene chloride is called Saran, and is used in making packaging film. The polymer of tetrafluoroethylene is **Teflon,** one of the most unusual of all the synthetic polymers. Only molten sodium or potassium will attack it chemically. Teflon is nonsticking and self-lubricating, and has been used medically as tips for hemodialysis shunts, coating of Dacron sutures, artery replacement segments, and some surgical drains and tubes. When copolymerized with some perfluoropropylene, tetrafluoroethylene forms a Teflonlike copolymer that is easier to work and make into articles.

8.7 Acrylates

A number of important polymers are made from acrylic acid or close relatives. **Polymethylmethacrylate,** the polymer of methyl methacrylate, when cast into sheets and molded items is marketed as Plexiglas or Lucite. The polymer has been used in making contact eye lenses, dentures, and in plastic surgery where a hard material is needed, as in correcting bony defects in the skull and in hip joint protheses.

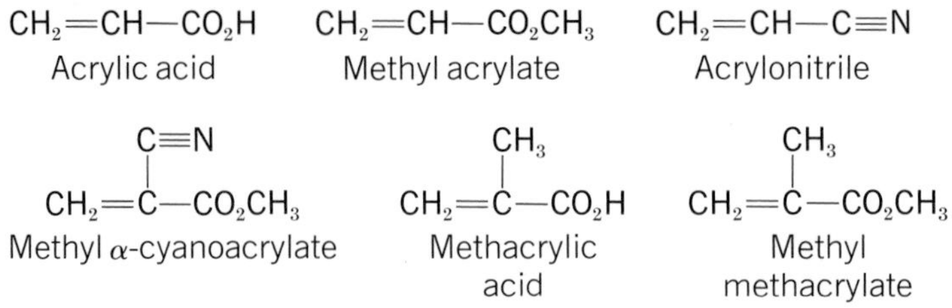

$CH_2{=}CH{-}CO_2H$ Acrylic acid

$CH_2{=}CH{-}CO_2CH_3$ Methyl acrylate

$CH_2{=}CH{-}C{\equiv}N$ Acrylonitrile

$CH_2{=}C(C{\equiv}N){-}CO_2CH_3$ Methyl α-cyanoacrylate

$CH_2{=}C(CH_3){-}CO_2H$ Methacrylic acid

$CH_2{=}C(CH_3){-}CO_2CH_3$ Methyl methacrylate

Exercise 8.1 Write enough of the structure of polyacrylic acid to show all the essential features.

Methyl α-cyanoacrylate is very unusual in that its polymerization is catalyzed by water. When spread on a surface in air of normal humidity, the monomer quickly sets to an astonishingly tough glue, "super glue." If you are not careful in using super glue, you can glue your thumb to your forefinger. (A small supply of acetone—fingernail polish remover—should be kept on hand to loosen nearly dried glue as quickly as possible from surfaces where it is unwanted.) Cyanoacrylates with higher alkyl groups on the ester link have been used as soft tissue adhesives.

8.8 Epoxies

Epoxies form a broad family of thermosetting polymers. Some formulations give us familiar epoxy glues, those that come in separate tubes or must be mixed just before use. Other epoxies are used to make dentures or in the direct filling of teeth. Some epoxies impregnated during setting with powdered aluminum oxide, have been used to make condyles in hip joint replacements.

The chemistry of epoxies begins with the **epoxide** functional group, a three-membered heterocyclic ether. We learned earlier that cyclopropane is unlike most other cyclic compounds; its ring is under strain and opens rather easily in the presence of bromine or hydrogen (under pressure and with a catalyst). The three-membered epoxide ring also opens easily in either acidic or basic media. For example, ethylene oxide, made from ethylene and the simplest of all epoxides, can open either to ethylene glycol or to a polymer, depending on the concentration of water.

In acid and an excess of water,

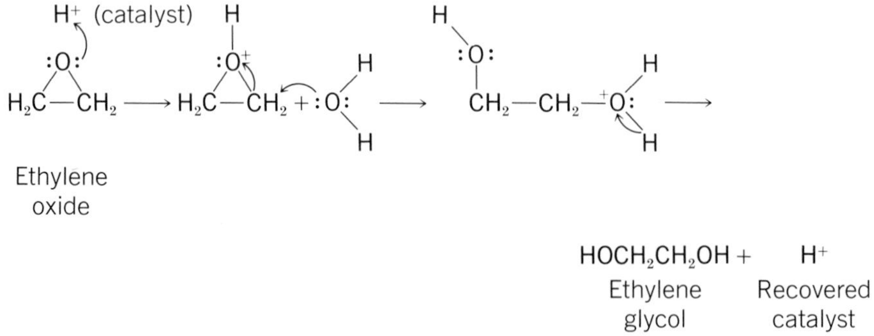

H^+ (catalyst)

$$H_2C{-}CH_2\ (\text{ring with } {:}O{:}) \longrightarrow H_2C{-}CH_2\ (\text{ring with } {:}O^+{-}H) + {:}OH_2 \longrightarrow HO{-}CH_2{-}CH_2{-}O^+H_2 \longrightarrow$$

Ethylene oxide

$$HOCH_2CH_2OH + H^+$$

Ethylene glycol   Recovered catalyst

In alkali and an excess of water,

$$H_2C\overset{O}{\frown}CH_2 + {}^-\!:\!\ddot{O}\!-\!H \longrightarrow H_2C(\!:\!\ddot{O}\!:^-)\!-\!CH_2\!-\!\ddot{O}\!-\!H \xrightarrow{H-\ddot{O}-H} HOCH_2CH_2OH + {}^-\!:\!\ddot{O}\!-\!H$$

Ethylene oxide; Hydroxide ion; Ethylene glycol; Recovered catalyst

If an excess of water is avoided, and an excess of ethylene oxide (by implication) is present, then polymerization may occur. By the careful control of all conditions, polymerization may be stopped at intermediate stages, at the dimer, trimer, or higher forms. Without attempting to describe all of the details, polymerization occurs roughly as follows:

$$HOCH_2CH_2OH + H_2C\overset{O}{\frown}CH_2 + H_2C\overset{O}{\frown}CH_2 + H_2C\overset{O}{\frown}CH_2 + \text{etc.} \longrightarrow$$

Ethylene glycol stage

$$HOCH_2CH_2O\!-\!CH_2CH_2O\!-\!CH_2CH_2O\!-\!CH_2CH_2O\!-\text{etc.}$$

Polyether

The popular elastic fibers, Spandex and Lycra, incorporate polyethers from ethylene oxide, as we shall see when we survey the polyurethanes.

One particular epoxide, epichlorohydrin, combined with a diol or a diphenol, such as bisphenol-A, forms the liquid "cement" for epoxy glue and some medically important epoxy resins. A large variety of other diphenols and diols may be used instead of bisphenol-A to make the liquid "cement."

$$H_2C\overset{O}{\frown}CH\!-\!CH_2Cl + {}^-\!:\!\ddot{O}\!-\!C_6H_4\!-\!C(CH_3)_2\!-\!C_6H_4\!-\!\ddot{O}\!:^- + ClCH_2\!-\!CH\overset{O}{\frown}CH_2 \xrightarrow{2Cl^-}$$

Epichlorohydrin; Bisphenol-A (di-sodium salt); Epichlorohydrin

$$H_2C\overset{O}{\frown}CH\!-\!CH_2\!-\!O\!-\!C_6H_4\!-\!C(CH_3)_2\!-\!C_6H_4\!-\!O\!-\!CH_2\!-\!CH\overset{O}{\frown}CH_2$$

An epoxy liquid "cementing" agent

The "curing agent" or "hardener" may be a triamino compound—one hardener used in epoxy glues—or the curing agent may itself be a low-formula-weight polymer such as polymethacrylic acid. The polymer structures are very complicated (see Figure 8.5), but they illustrate how cross-

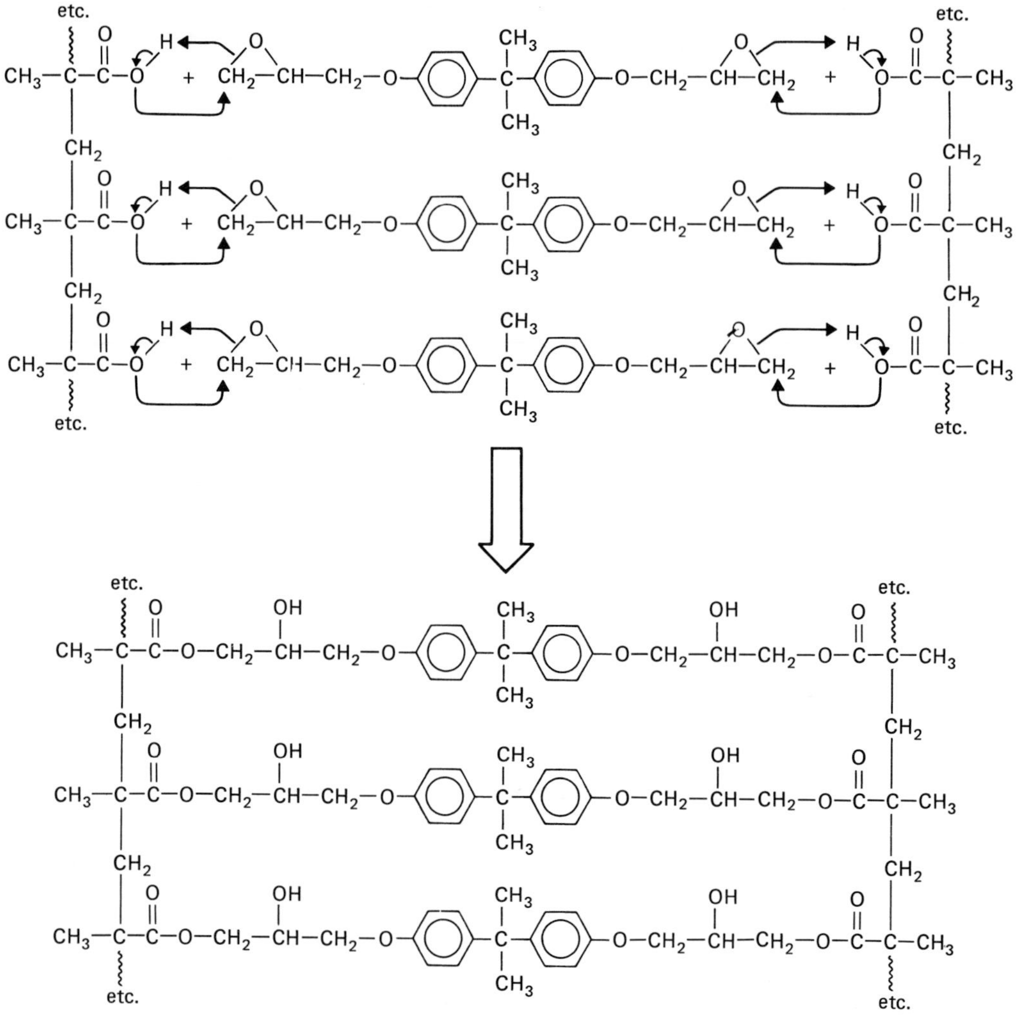

Figure 8.5 Idealized structure of an epoxy resin. Curing occurs by ring-opening reactions between free carboxyl groups on the polymer, polymethacrylic acid (left and right sides), and molecules of the diglycidyl ether of bisphenol A (down the center).

linking combined with linear polymerization quickly converts small molecules to substances of enormous formula weights. With the extensive cross-linking in thermosetting polymers, one can almost say that an entire molded article is one vast molecule. Any effort to soften it for remolding is defeated by this size; the melting point is so high that thermal decomposition occurs long before the substance can soften and melt.

8.9 Polyurethanes

If the polyether made by the polymerization of ethylene oxide (previous section) is mixed with a diisocyanate, a polyurethane forms.

The urethane system is a mixed amide-ester that forms when an alcohol group reacts with the isocyanate group, —N=C=O. For example, (and we distort molecular geometries to make it easier to portray how bonds reorganize),

$$C_6H_5-N=C=O + H-O-R \longrightarrow C_6H_5-NH-C(=O)-O-R$$

Phenyl isocyanate | Alcohol | An alkyl phenyl urethane (amide bond; ester bond)

If we mix a diisocyanate and a diol, a polymer can form:

$$\text{etc.}-N=C=O + H-O(CH_2CH_2O)_nCH_2CH_2-O-H + O=C=N-C_6H_3(CH_3)-N=C=O + H-O-CH_2CH_2(OCH_2CH_2)_n$$

Toluene diisocyanate | Polyether | Toluene diisocyanate | Polyether

$$\Downarrow$$

$$\text{etc.}-NHCO(CH_2CH_2O)_nCH_2CH_2OCNH-C_6H_3(CH_3)-NHCOCH_2CH_2(OCH_2CH_2)_n\ \text{etc.}$$

Polyurethane (Spandex; Lycra)

A number of diols and diisocyanates may be used. Those illustrated result in some of the important elastomers in the world of synthetics. Common uses are in bathing suits and girdles. Medically, they have been studied as possible sutures, for replacement parts in the vascular system, diaphragms for blood pumps, and membranes for dialysis. If a triol is used instead of a diol, cross-linking occurs to give thermosetting polyurethanes; these have been investigated as bone glues, dentures, surgical sponges, and material for augmenting soft tissue in plastic surgery.

8.10 Condensation Polymers

Condensation polymers are usually copolymers. If "a" and "b" are parts of two functional groups that can react and in which a small molecule "a-b" splits, then condensation polymers may arise from difunctional compounds in the following general way:

$$\text{aXa} + \text{bYb} + \text{aXa} + \text{bYb} + \cdots + \text{etc.} \longrightarrow \text{XYXY(XY)}_n\text{XY-etc.} + m\ \text{a-b}$$

Often "a-b" is a water molecule.

Cross-linked, thermosetting condensation polymers form if at least one monomer is trifunctional. Polyamides, polyesters, and polysiloxanes are three important condensation polymers.

8.11 Polyamides. Nylon

The term **nylon** is a coined name that applies to any synthetic, long-chain, fiber-forming polymer with repeating amide linkages. One of the most common members of the nylon family, nylon-66, is made from hexamethylenediamine and adipic acid according to the following reaction in which water molecules split out:

$$\text{etc.}-\overset{\overset{\displaystyle O}{\|}}{C}-(OH + H)-\overset{\overset{\displaystyle H}{|}}{N}(CH_2)_6\overset{\overset{\displaystyle H}{|}}{N}-(H + H-O)-\overset{\overset{\displaystyle O}{\|}}{C}(CH_2)_4\overset{\overset{\displaystyle O}{\|}}{C}-(OH + H)-\overset{\overset{\displaystyle H}{|}}{N}-\text{etc.}$$

Hexamethylene diamine   Adipic acid

$$\downarrow$$

$$-\overset{\overset{\displaystyle O}{\|}}{C}-\overset{\overset{\displaystyle H}{|}}{N}(CH_2)_6\overset{\overset{\displaystyle H}{|}}{N}-\overset{\overset{\displaystyle O}{\|}}{C}(CH_2)_4\overset{\overset{\displaystyle O}{\|}}{C}-\overset{\overset{\displaystyle H}{|}}{N}-\text{etc.}$$

Nylon-66

(The "66" means that each monomer contains six carbons.) To be useful as a fiber-forming polymer, each nylon-66 molecule should contain from 50 to 90 of each of the monomer units. Shorter nylon molecules form weak or brittle fibers.

When molten nylon resin is being drawn into fibers, newly emerging strands are caught up on drums and stretched as they cool. Under tension, the long polymer molecules within the fiber line up side by side, overlapping each other to give a finished fiber of unusual strength. Nylon is more resistant to combustion than wool, rayon, cotton, or silk, and it is as immune to insect attack as Fiberglas. Molds and fungi do not attack nylon. In medicine, nylon is used in specialized tubing, and velour for blood contact surfaces. Nylon sutures were the first synthetic sutures and are still used.

8.12 Polyesters. Dacron

Dacron shares most of the desirable properties of nylon. Most distinctive is its ability to be set into permanent creases and pleats. **Dacron** is a polyester made by the copolymerization of ethylene glycol with terephthalic acid. Water molecules split out:

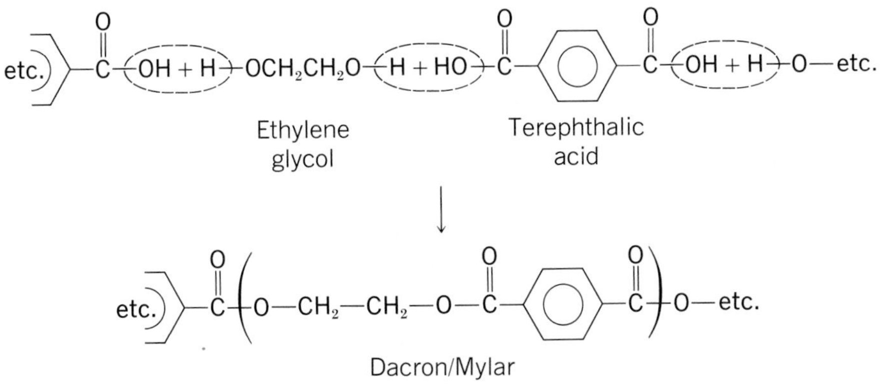

$$\text{etc.}\rangle\!-\overset{\overset{\displaystyle O}{\|}}{C}-(OH + H)-OCH_2CH_2O-(H + HO)-\overset{\overset{\displaystyle O}{\|}}{C}-C_6H_4-\overset{\overset{\displaystyle O}{\|}}{C}-(OH + H)-O-\text{etc.}$$

Ethylene glycol   Terephthalic acid

$$\downarrow$$

$$\text{etc.}\rangle\!-\overset{\overset{\displaystyle O}{\|}}{C}\!\left(O-CH_2-CH_2-O-\overset{\overset{\displaystyle O}{\|}}{C}-C_6H_4-\overset{\overset{\displaystyle O}{\|}}{C}\right)O-\text{etc.}$$

Dacron/Mylar

When Dacron is cast as a thin film it is called Mylar.

Fabrics of Dacron are used in medicine to repair or replace segments of blood vessels. The chapter opening figure pictures one such use.

Exercise 8.2 Write a structure that shows how ethylene glycol could be used to cross-link molecules of polyacrylic acid (from Exercise 8.1).

8.13 Polysiloxanes. Silicones

Silicones are polymers that have an inorganic backbone of alternating silicon and oxygen atoms and have alkyl or aryl side chains. The principal starting material is dichlorodimethylsilane. When mixed with water, the chlorines are replaced by hydroxyl groups, which then split out water between them to form chains (sometimes rings). When a little chlorotrimethylsilane is added at the start of the hydrolysis, it provides end groups that prevent the chains from becoming long enough to carry the polymerization to a solid product. In this manner silicone oils are prepared:

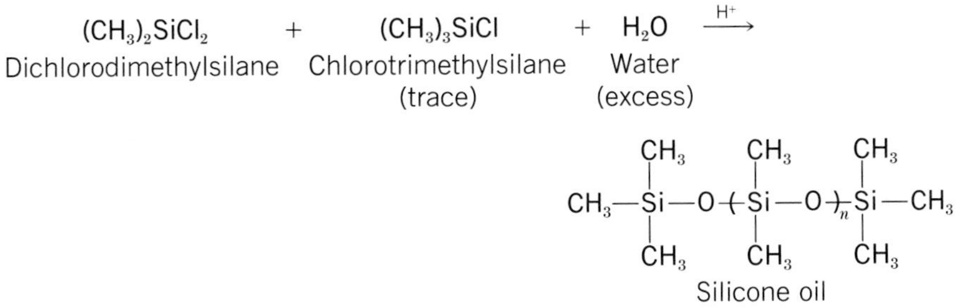

$$\underset{\text{Dichlorodimethylsilane}}{(CH_3)_2SiCl_2} + \underset{\substack{\text{Chlorotrimethylsilane}\\ \text{(trace)}}}{(CH_3)_3SiCl} + \underset{\substack{\text{Water}\\ \text{(excess)}}}{H_2O} \xrightarrow{H^+}$$

$$\underset{\text{Silicone oil}}{CH_3-\underset{\underset{\displaystyle CH_3}{|}}{\overset{\overset{\displaystyle CH_3}{|}}{Si}}-O\!\left(\underset{\underset{\displaystyle CH_3}{|}}{\overset{\overset{\displaystyle CH_3}{|}}{Si}}-O\right)_{n}\!\underset{\underset{\displaystyle CH_3}{|}}{\overset{\overset{\displaystyle CH_3}{|}}{Si}}-CH_3}$$

A solid polymer, silicone gum, is made by the hydrolysis and polymerization of dichlorodimethylsilane in the presence of a base.

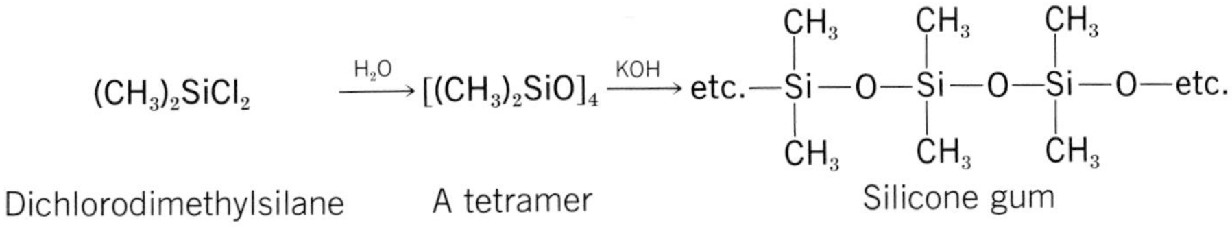

$$\underset{\text{Dichlorodimethylsilane}}{(CH_3)_2SiCl_2} \xrightarrow{H_2O} \underset{\text{A tetramer}}{[(CH_3)_2SiO]_4} \xrightarrow{KOH} \underset{\text{Silicone gum}}{\text{etc.}-\underset{\underset{\displaystyle CH_3}{|}}{\overset{\overset{\displaystyle CH_3}{|}}{Si}}-O-\underset{\underset{\displaystyle CH_3}{|}}{\overset{\overset{\displaystyle CH_3}{|}}{Si}}-O-\underset{\underset{\displaystyle CH_3}{|}}{\overset{\overset{\displaystyle CH_3}{|}}{Si}}-O-\text{etc.}}$$

Figure 8.6 A silicone board is all that separates the hand from the intense heat of the blowtorch. (Courtesy of the General Electric Corporation.)

When silicone gum is vulcanized, cross-links are introduced and the product is silicone rubber.

The silicones are unusually stable to heat and combustion. Silicon rubber can be heated with a blowtorch without melting or suffering serious change (Figure 8.6). The silicones not only do not react with water, they repel water. When a gauze or other fabric is sprayed with silicone, the pores and air spaces in the fabric are not closed; the fabric will "breathe." However, water beads so strongly on the silicone-treated surface that it is impossible for the water to go through the pores and holes. The silicone fluids are excellent in hydraulic pumps, as lubricants in machines run over a wide range of temperatures, as water-repellent films on concrete surfaces, and in protective hand creams. These fluids have been tried for soft tissue augmentation, as in breast surgery; as lubricants for hypodermic needles; as antiadhesion materials; as replacement for vitreous; and as a lubricant in the treatment of burns.

Solid silicones have also been used in tissue augmentation, as burr hole covers, and as implants in repair of the chin and nose. A number of implantable drainage tracts have been made of silicone rubber. Heart pacemakers are encapsulated in silicone rubber. Silicone rubber has also been used to make oxygenating membranes for heart lung machines, as coatings for sutures, and as diaphragms for blood pumps and other blood compatible surfaces.

Brief Summary

Polymers. Long molecules with repeating structural units form if simple or substituted alkenes or dienes add to each other or if the monomers each have two (or more) functional groups that can react together. Initiators cause polymerization via ionic or free radical intermediates. Cross-linking, which makes a polymer thermosetting, may be induced either by vulcanization or by the interaction of sidechains on polymer molecules. When one monomer is di- or polyfunctional and the other is at least difunctional, a cross-linked polymer generally forms.

Fibers from Polymers. Fibers can be drawn from those polymers whose molecules are very long relative to their cross section and are quite symmetrical, lacking bulky side chains. When tension is applied to the new fiber as it forms, its molecules line up better along the fiber axis and the fiber is stronger.

Useful Polymers. Of the simple polyolefins, polypropylene has developed into the most useful for making household articles, synthetic carpeting fiber, and medical implants where outstanding flexlife is needed. Ultrahigh-density polyethylene is useful in making sockets for ball-and-socket joint repair. Teflon, besides its familiar use in nonsticking pots and pans, makes an excellent self-lubricating material for hemodialysis shunts and some surgical drains and tubes. When a hard plastic is needed for dentures, dental filling agents, and for hard tissue augmentation, certain acrylates have been used. Cyanoacrylates make outstanding glues. Some epoxies also are excellent glues, but medically they are important in dental fillings and in hip joint replacements. Spandex-type polyurethanes are excellent elastomers. Three condensation polymers, nylons, Dacron/Mylar, and various silicones have all been used in medicine: nylon particularly as sutures, knitted Dacron mesh in replacing vascular segments, and silicones in protective hand creams as well as in soft-tissue augmentation. Solid silicones have seen use in corrective plastic surgery.

Selected References

Books

1. H. F. Mark. *Giant Molecules.* Life Science Library, 1966. Time, Inc., New York. Lavishly illustrated introduction to the history, development, chemistry, and many uses of polymers.
2. H. Lee and K. Neville. *Handbook of Biomedical Plastics.* Pasadena Technology Press, Pasadena, Calif., 1971. A review of the literature on plastic and restorative surgery and what plastics have been tried with what success.
3. H. Lee and J. Orlowski. *Handbook of Dental Composite Restoratives,* 3rd ed. Lee Pharmaceuticals, S. El Monte, Calif., 1974.
4. F. W. Billmeyer. *Textbook of Polymer Science,* 2nd ed. John Wiley & Sons, Inc., New York, 1971.

Articles

1. D. R. Uhlmann and A. G. Kolbeck. "The Microstructure of Polymeric Materials." *Scientific American,* December 1975, page 96.
2. M. W. Kelly. "Adhesive Bonding." *Chemistry.* July/August 1976, page 14.

Questions and Exercises

1. What is the difference between the members of each of the following pairs of terms?
   (a) A polymer and a synthetic resin
   (b) A synthetic resin and a plastic
   (c) An addition polymer and a condensation polymer
   (d) A catalyst and an initiator
2. Using the hypothetical symbols given, what would be the simplest type of polymer in each case?

(a) A linear addition polymer with "o" as the monomer
(b) A linear condensation polymer between cXc and dYd if the simpler compounds, Xc and Yd, react as follows:

$$\text{Xc} + \text{Yd} \longrightarrow \text{X—Y} + \text{c—d}$$

(c) A cross-linked condensation polymer between a X a (with a third a attached to X) and bYb where molecules of a-b split out. (Show enough structure to illustrate essential features.)
(d) A linear polymer in which a cyclic monomer reorganizes bonds into an open array, the monomer being represented by a three-membered ring of X, c, and d (X bonded to c and d, c bonded to d). (Note that "c" and "d" are the functional group parts.)

3. Examine the structure of each polymer and write the structure of its monomer.

(a) $\sim CH_2—CH(OCOCH_3)—CH_2—CH(OCOCH_3)—CH_2—CH(OCOCH_3)\sim$ (each CH bears —O—C(=O)—CH_3)

(b) $—C(CH_3)(C_6H_5)—CH_2—C(CH_3)(C_6H_5)—CH_2—C(CH_3)(C_6H_5)\sim$

(c) $\sim CClF—CF_2—CClF—CF_2—CClF—CF_2\sim$ (upper substituents Cl F Cl F Cl F; lower substituents F F F F F F)

Kel-F, fluorothene

(d) $\sim CH_2—C(Cl){=}CH—CH_2—CH_2—C(Cl){=}CH—CH_2—CH_2—C(Cl){=}CH—CH_2\sim$

Neoprene

(e) $\sim CH_2—CH(CN)—CH_2—CH(CN)—CH_2—CH(CN)\sim$

Orlon

4. In the manner used to write structures in Exercise 3, write structures illustrating the essential features of molecules in the polymers of each of the following monomers.
(a) $CH_2{=}CH—Cl$ (which gives polyvinyl chloride or PVC)
(b) $CH_2{=}CCl_2$ (which gives Saran)
(c) $CH_2{=}C(CO_2CH_3)—CN$

5. Nylon-6 is almost as important in the nylon family as nylon-66. The monomer for nylon-6 is the cyclic compound, ϵ-caprolactam. When ϵ-caprolactam polymerizes, a small molecule does not split out, and a linear polymer forms. Write enough of the structure of nylon-6 to illustrate its essential features. (Hint: Exercise 2-d gave the basic idea.)

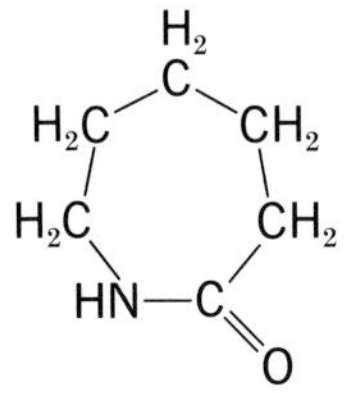

H_2
C
H_2C CH_2
H_2C CH_2
HN—C
O

ϵ-caprolactam

6. Why cannot a thermosetting polymer be remelted and reshaped?
7. If ingredients—monomer or partly polymerized monomer plus initiator—are put into a dental cavity to become a filling, what might happen if an inert filler is not also included?

chapter 9

Optical Isomerism

When two polarizing devices are "crossed" no light can get through as seen in the center of this photo. However, if an optically active substance is placed between them, then some light will be transmitted. We study this phenomenon in this chapter. Optically active molecular substances occur throughout living systems. (Courtesy Polaroid Corporation.)

9.1 Types of Isomerism

Compounds that share identical molecular formulas can be different in two general ways as illustrated in Figure 9.1. They may be structural isomers or stereoisomers.

Structural isomers generally have major differences in molecular frameworks, and three broad types exist. *Chain isomers* are structural isomers with variations in carbon chains or skeletons. *Position isomers* are structural isomers with the same chains and the same functional groups, but the groups are differently located. *Functional group isomers* are perhaps the most radically unlike; not only are their molecular chains usually different, but they also have different functional groups.

Stereoisomers, from *stereos,* Greek for "solid" (implying three-dimensional differences) are isomers differing only in some three-dimensional or geometric way. Stereoisomers generally have the same skeletons (not counting geometric variations) and identical functional groups located at identical sites on the chain, and yet subtle differences exist.

Two types of stereoisomers are known, the *cis-trans* (or geometric) isomers and a new type that we shall study in this chapter, **optical isomers.** Optical isomers are structurally dissimilar in the most minor ways, but biologically

Figure 9.1 Types of isomerism

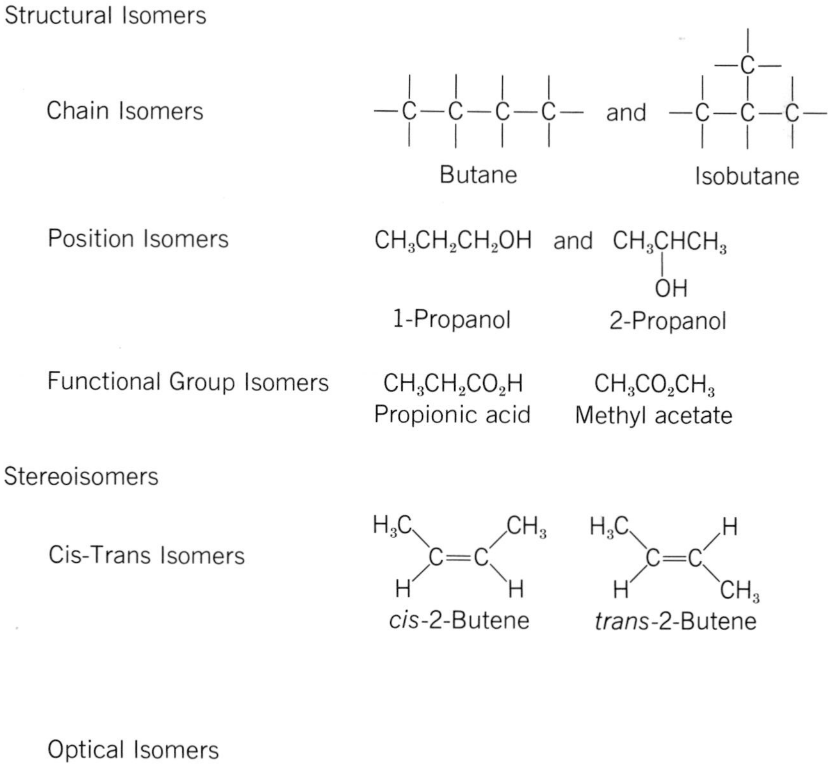

Structural Isomers

Chain Isomers — C—C—C—C and C—C(—C)—C

Butane Isobutane

Position Isomers $CH_3CH_2CH_2OH$ and $CH_3CH(OH)CH_3$

1-Propanol 2-Propanol

Functional Group Isomers $CH_3CH_2CO_2H$ $CH_3CO_2CH_3$

Propionic acid Methyl acetate

Stereoisomers

Cis-Trans Isomers $(H_3C)(H)C{=}C(CH_3)(H)$ $(H_3C)(H)C{=}C(H)(CH_3)$

cis-2-Butene *trans*-2-Butene

Optical Isomers

they display startling differences as illustrated by the bitter-sweet story of asparagine.

Exercise 9.1 Examine these structures and answer the questions that follow.

$$\underset{\text{(a)}}{CH_3CH(CH_3)C(=O)OH} \qquad \underset{\text{(b)}}{CH_3{-}O{-}CH_2CH_2CH{=}O} \qquad \underset{\text{(c)}}{CH_3CH_2CH_2C(=O)OH}$$

$$\underset{\text{(d)}}{CH_3CH_2{-}O{-}CH_2CH{=}O} \qquad \underset{\text{(e)}}{CH_3CH_2C(=O)OCH_3}$$

(1) Write the letters of any two that are chain isomers only.
(2) Write the letters of any two that are position isomers only.
(3) Write the letters of any two that are functional group isomers.

9.2 Asparagine—Two Forms

Asparagine (ăs-păr′-ȧ-jĭn) is a white solid first isolated from the juice of asparagus in 1806. When obtained from this source, asparagine has a bitter taste. Its molecular formula is $C_4H_8N_2O_3$, and its structure is now known to be

$$NH_2{-}\overset{\overset{\displaystyle O}{\|}}{C}{-}CH_2{-}\underset{\underset{\displaystyle NH_2}{|}}{CH}{-}\overset{\overset{\displaystyle O}{\|}}{C}{-}OH$$

1
Asparagine

In 1886, a chemist isolated from sprouting vetch[1] a substance of the same molecular formula and structure, but it had a sweet taste. To have names for these two, we call the one from asparagus "L-asparagine" and the one from vetch sprouts "D-asparagine". (The small capitals, D and L, although arbitrary here, acquire meaning in the next chapter.)

Taste is a chemical sense. The two samples of asparagine, therefore, do show one difference in a chemical property. Chemists have long worked with the principle "one substance, one structure." If two samples of matter have identical physical and chemical properties, they must be identical at the level of their individual molecules. If two samples differ in even one way in fundamental properties, their individual molecules must also differ. Under the one-substance–one-structure doctrine, the molecules of D- and L-asparagine must be structurally different in some way. The kind of isomerism they illustrate arises from a peculiar lack of symmetry in their molecules.

9.3 Molecular Symmetry and Chirality

Two partial ball-and-stick models of asparagine molecules are shown in Figure 9.2*a*. Examine each. Assure yourself that each is structure **1**. Now let us

[1] Vetch is a member of a genus of herbs, some of which are useful as fodder.

simplify the structures. Let us go to part *b* of Figure 9.2. In each structure, the same four groups are attached to the central carbon. Yet the two structures are not identical!

For two structures to be identical, it must be possible in your imagination (or with models) to superimpose them, as illustrated in Figure 9.2*c*. **Superimposition** is the fundamental criterion of identity for two structures. If two structures are identical, each must have a conformation (from twisting about single bonds) by which the two become superimposable. The two asparagine molecules are not identical, after all. Yet, they are related in a particularly interesting and significant way. They are related as an object to its mirror image. The fact that they cannot be superimposed raises a question. In exactly what way do they differ structurally?

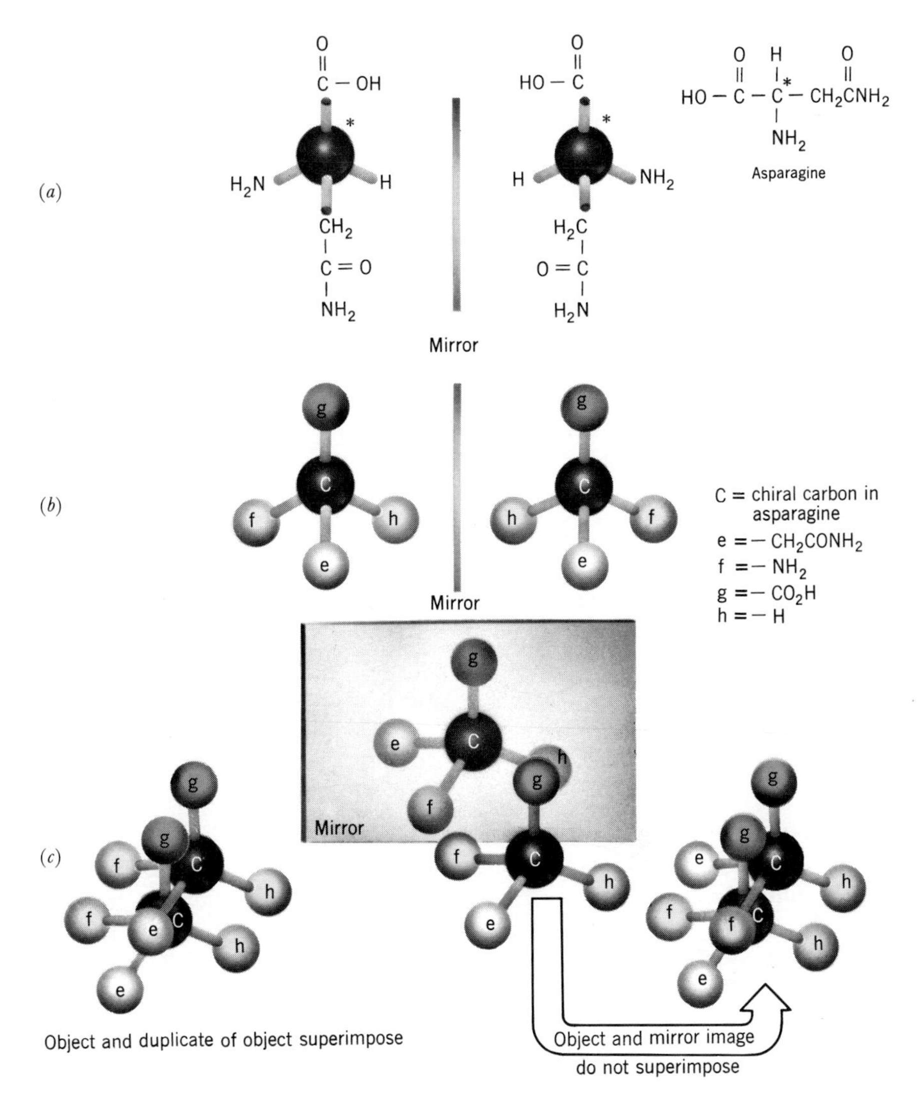

Figure 9.2 The enantiomers of asparagine (*a*). In (*b*) the groups joined to the central carbon, which do not further contribute to the isomerism, may be represented by simple model parts. (*c*) Attempts at superimposition using simplified models of asparagine.

The two asparagines differ only in the relative directions by which the four groups on the central carbon (Figure 9.2*b*) project into space. We say they have opposite configurations. We could call them configurational isomers. Instead, for historical reasons that will soon be apparent, we call the two optical isomers. Since there are several types of optical isomers, we need a name for the type illustrated by the asparagines.

9.4 Enantiomers

Isomers that, like the two asparagines, are related as object and mirror image that cannot be superimposed are called **enantiomers** (Greek, *enantios,* opposite, and *meros,* part).

To deal as swiftly as possible with one source of misunderstanding, any object will have a mirror image. Only when the object and mirror image cannot be superimposed are they said to be enantiomers. With this definition of enantiomers, optical isomers in general can be better defined. **Optical isomers** are members of a set of stereoisomers that includes at least two related as enantiomers.

Physical properties are identical for two enantiomers. They *must* be, because within two enantiomers intranuclear distances and bond angles are identical, and polarities must therefore be identical. Melting points, boiling points, densities, solubilities, and, of course, molecular weights have to be the same for two enantiomers. In thinking about intranuclear distances and angles in enantiomers, compare your two hands, disregarding differences in fingerprints and wrinkles. The left hand and the right hand illustrate enantiomers. They, too, are related as object and mirror image that cannot be superimposed. They differ only in that holding the palms facing us, we "read" clockwise from thumb to little finger in the left hand and counterclockwise in the right hand.

9.5 Chirality and Chiral Molecules

Enantiomers have the structural property of opposite "handedness." The Greek word for "hand" is *cheir,* and from this we have the words "chiral" and "chirality." We may call two enantiomers **chiral** or we may say they possess **chirality.**[2] In other words, they possess handedness. The opposite of chiral is **achiral;** it means symmetrical enough so that object and mirror image superimpose.

9.6 Enantiomers and Chiral Reagents

Chemical properties of enantiomers are also identical, provided that other reagents are symmetrical or achiral in the sense used here. If the reaction is with a chiral reagent, however, we observe striking differences in properties. A simple, widely used illustration involves a pair of gloves. Because two hands are related as enantiomers, so too are a pair of ordinary gloves:

[2] Teachers of organic chemistry will be keenly aware of the fact that very often objects having chirality have traditionally been called dissymmetric or (not as accurately) asymmetric. Both terms are widespread; both will be used for some time. Students may encounter them in other chemical or biochemical literature. They should be learned, and with almost no danger of serious inaccuracy the terms *chiral, asymmetric,* and *dissymmetric* may be considered to be synonyms. They are not, strictly speaking, but the distinctions are, for our purposes, unimportant.

right hand + right glove ⟶ gloved right hand

right hand + left glove ⟶ no fit

Taking the gloves as the chiral "reagents," we see that the ease and comfort with which each glove can "react" with each hand is certainly different. By contrast, if we use as the reagent system a pair of ski poles, for example, that are superimposable and achiral, no differences can be noted in the way these react with either hand.

9.7 Deciding Whether a Molecule is Chiral

Any structure that is identical with its mirror image has symmetry in some way. Three ways of having symmetry are recognized, but for our purposes we need only one, the plane of symmetry.

A **plane of symmetry** is an imaginary plane that divides an object into two halves that are mirror images of each other. Some simple planes of symmetry are shown in Figure 9.3. Some molecules have several planes, but we require only one. Just one guarantees that object and mirror image are superimposable and identical. On the other hand, if we cannot discover this (or any) symmetry in a structure, in at least one of its conformations, we are guaranteed that the object and mirror images are not superimposable and are optical isomers.

9.8 The Chiral Carbon—Another Shortcut

In nearly all examples of optical isomerism the molecules possess at least one carbon to which are attached four different atoms or groups. We saw this in asparagine (Figure 9.2). A carbon holding four different groups is defined as a **chiral carbon.** (In many references it will be called an asymmetric carbon.) A molecule with a chiral carbon will nearly always have no element of symmetry in any conformation.

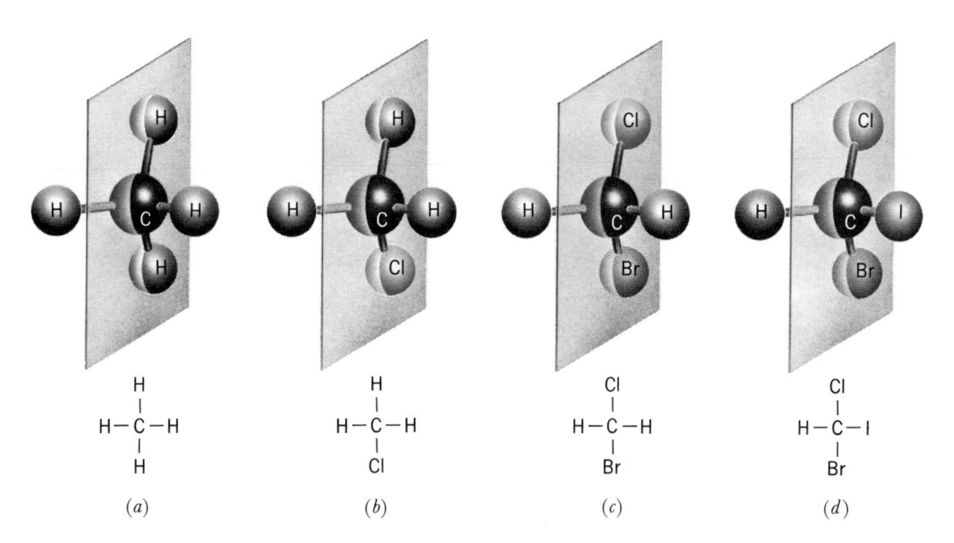

Figure 9.3 The plane of symmetry in progressively substituted methanes. Only in (*d*) is there no plane of symmetry, and (*d*) is chiral. It could not be superimposed on its mirror image.

The qualification "nearly always" is an important one. In a few substances the molecules contain chiral carbons but yet are symmetrical in some way (Figure 9.4), as we shall study in more detail later. In general, if a molecule has only one chiral carbon, it will be chiral as a whole, and it will belong to a set of optical isomers. We need not qualify that statement with "nearly always."

If the molecules of a substance each have n different chiral carbons, the number of possible optical isomers is 2^n. (Two chiral carbons are "different" if the sets of four groups attached to each are different in at least one way.)

When it works, the concept of a chiral carbon is extremely useful. It must be remembered that the shortcut is fallible, but with a little experience the exceptions are easily recognized. Moreover, they do not occur frequently. We turn next to the way we detect this kind of isomerism experimentally.

Exercise 9.2 Examine the given structures. Place an asterisk above each carbon to which four different groups are attached. If ball-and-stick sets or their equivalent are available to you, make molecular models of each structure and try to discover planes of symmetry. Make molecular models of the mirror images and check to see whether the objects and mirror images are superimposable.

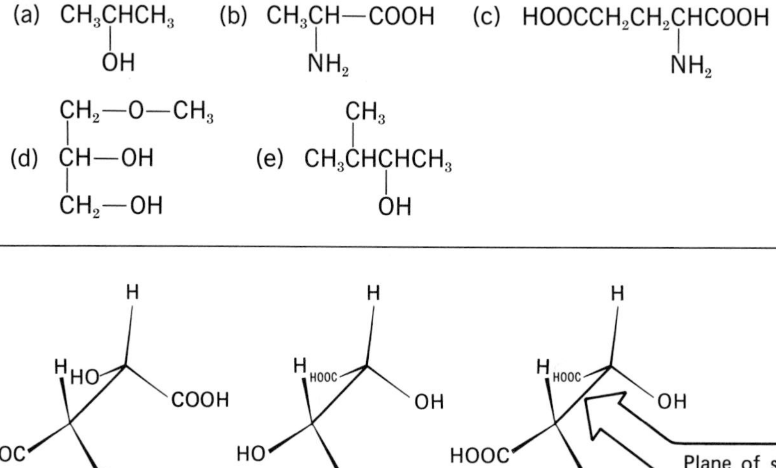

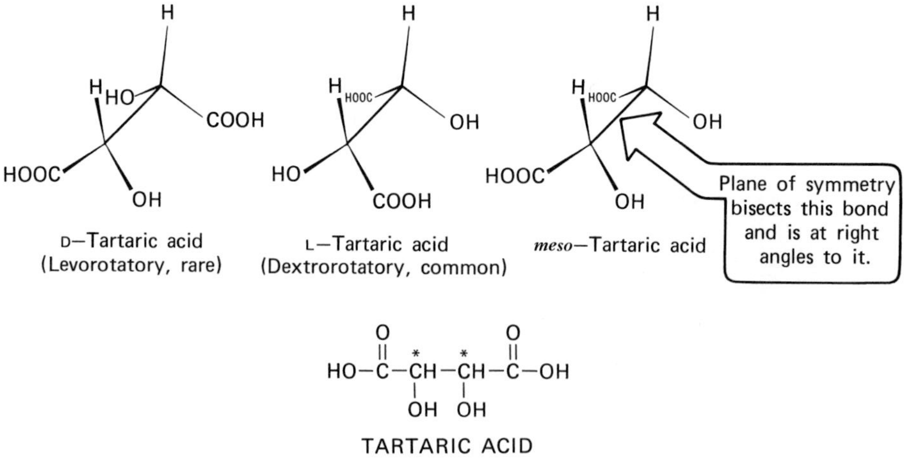

(a) $CH_3CH(OH)CH_3$ (b) $CH_3CH(NH_2)—COOH$ (c) $HOOCCH_2CH_2CH(NH_2)COOH$

(d) $CH_2(—O—CH_3)—CH(—OH)—CH_2—OH$ (e) $CH_3CH(CH_3)CH(OH)CH_3$

H HO COOH HOOC OH

D–Tartaric acid (Levorotatory, rare)

H HOOC OH HO COOH

L–Tartaric acid (Dextrorotatory, common)

H HOOC OH HOOC OH

meso–Tartaric acid

Plane of symmetry bisects this bond and is at right angles to it.

HO–C(=O)–C*H(OH)–C*H(OH)–C(=O)–OH

TARTARIC ACID

Figure 9.4 Chiral carbons do not guarantee chiral molecules. A tartaric acid molecule has two chiral carbons, but each has the same set of four groups. Each has one —H, one —OH, one $—CO_2H$, and one $—CHCO_2H$. Although two forms of tartaric acid are related as enantiomers, a third form has a plane of symmetry and is called *meso*-tartaric acid (after the Greek, *meso,* in the middle). Meso compounds occur when a molecule has two (or more) chiral carbons to each of which are joined the identical sets of four groups, each member of a set being different from every other.

Exercise 9.3 Examine the following structure and answer the questions about it.

$$\mathrm{HO{-}CH_2{-}\underset{\displaystyle OH}{\underset{|}{CH}}{-}\underset{\displaystyle OH}{\underset{|}{CH}}{-}\underset{\displaystyle OH}{\underset{|}{CH}}{-}\underset{\displaystyle OH}{\underset{|}{CH}}{-}\overset{\displaystyle O}{\overset{\|}{C}}{-}H}$$

(a) How many chiral carbons does it have?
(b) How many of these chiral carbons are different, as we have defined what it means to be different?
(c) How many optical isomers exist for this structure?

(Note: One of the isomers is D-glucose, perhaps the single most important sugar in metabolism. Only one other optical isomer, D-galactose, is nutritionally important. The rest cannot be used by the human body.)

Optical Activity

9.9 Polarized Light

Light is an electromagnetic radiation in which the strengths of electric and magnetic fields set up by the light source oscillate in a regular way. In ordinary light these oscillations occur equally in all directions about the line that defines the path of the light ray.

Certain materials affect ordinary light in a special manner. Polaroid film is an example. This material interacts with the oscillating electrical field of the incoming light to make this field oscillate in just one plane. The light is now **plane-polarized light** (Figure 9.5). If we look at some object through a Polaroid film and then place a second film in front of the first, we can rotate one film until the object no longer can be seen. If we now rotate one film by 90°, we see the object at maximum brightness again. The film seems to act as a lat-

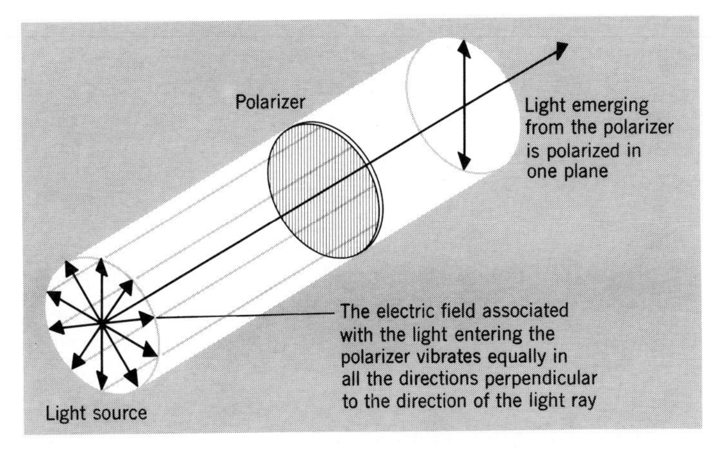

Figure 9.5 Light becomes polarized when it passes through certain materials called polarizers. Polaroid film is an example.

tice fence forcing any light going through to vibrate only in the direction allowed by the long spaces between the slats. This light then moves on to the molecular slats of the second film. If its slats are perpendicular to those of the first, the light has nowhere to oscillate and it cannot go any further. At intermediate angles fractional amounts of light can go through the second Polaroid film; when the slats of the two films are parallel, the light leaving the first most easily slips through the second.

9.10 Optical Rotation

If a solution of D-asparagine in water is placed in the path of plane-polarized light, the plane of polarization will be rotated. Any substance that will do this to plane-polarized light—rotate the plane—is said to be **optically active.** We have not explained how this phenomenon happens because we are unable to do so without a solid foundation in physics. We have only reported that it does happen.

9.11 The Polarimeter

The instrument that is used to detect and measure optical activity is called a **polarimeter.** See Figure 9.6. Its principal working parts consist of a polarizer, a tube for holding solutions in the light path, an analyzer (actually, just another polarizing device), and a scale for measuring degrees of rotation. As shown in Figure 9.6*a*, when the polarizer and analyzer are "parallel," and the tube contains no optically active material, the polarized light goes through at maximum intensity to the observer.

If a solution of an optically active material is placed in the light path, the plane-polarized light encounters molecules predominantly or wholly of just one chirality and its plane of oscillation is rotated. The polarized light that leaves the solution is no longer "parallel" with the analyzer (see Figures 9.6*a* and *c*), and the intensity of the observed light is reduced. To restore the original intensity, the analyzer is rotated to the right or to the left until it is again parallel with the light emerging from the tube. As the operator looks toward the light source, he will, if he rotates the analyzer to the right, record the degrees as positive, and the optically active substance will be called **dextrorotatory.** If the rotation is counterclockwise, the degrees will be recorded as negative, and the substance will be called **levorotatory.** In the example of Figure 9.6*c*, the record will read $\alpha = -40°$, where α stands for the **optical rotation,** the observed number of degrees of rotation.

Since the optical rotation varies with both the temperature of the solution and the frequency of the light, both data go into the record; for example, $\alpha_{D}^{20} = -40°$. The subscript D stands for the D-line of the spectrum of sodium, a yellow light and one particular frequency that electromagnetic radiation can have; and 20 means that the solution has a temperature of 20 °C.

When the sign of rotation of a compound is known, the name is written to include the sign. Naturally occurring glucose is dextrorotatory and it may be named "(+)-glucose." Naturally occurring fructose is levorotatory, and it may be named "(−)-fructose." Their enantiomers, which do not occur in nature, are respectively named (−)-glucose and (+)-fructose.

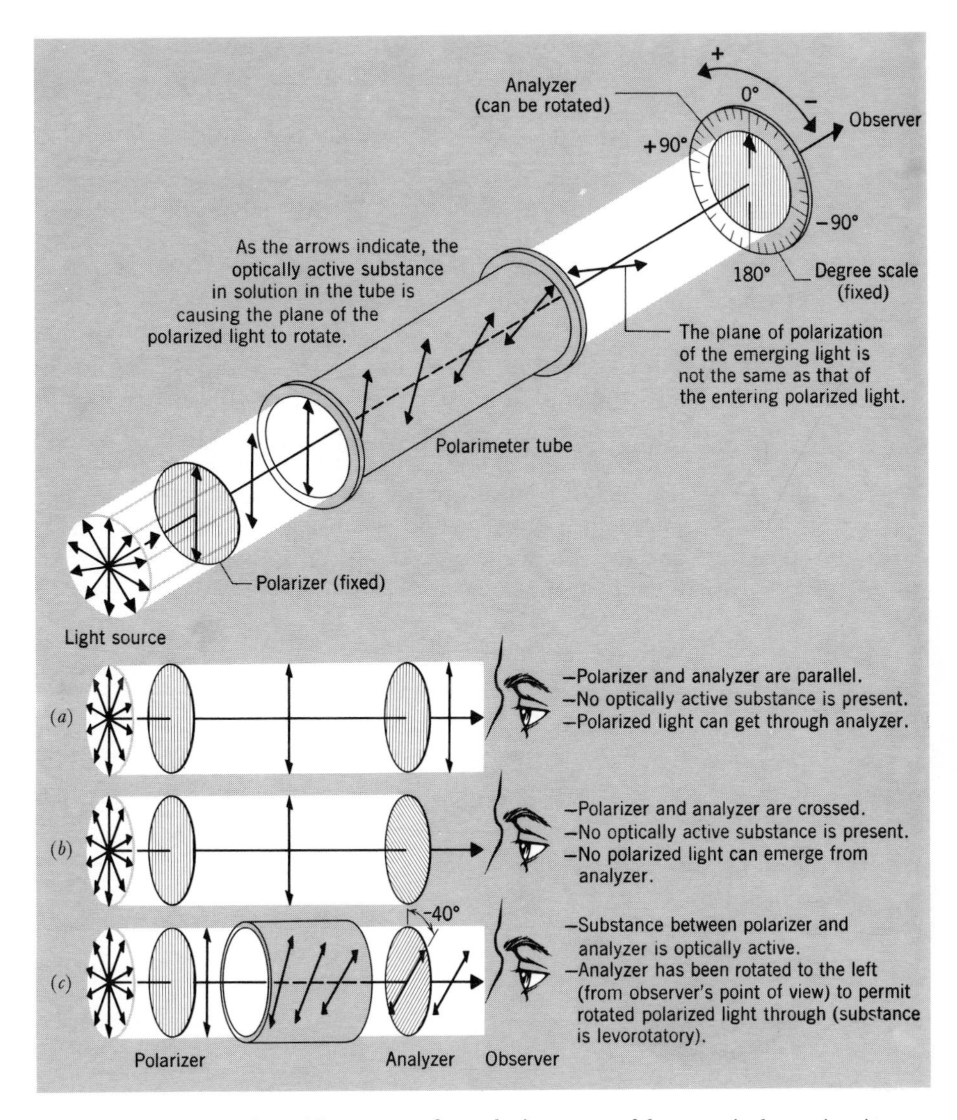

Figure 9.6 Principal working parts of a polarimeter and how optical rotation is measured.

9.12 Specific Rotation

The optical rotation of a given solution depends on the population of chiral molecules that the light beam encounters. The longer the tube, the more will the plane of the polarized light be rotated. Or, if the tube length is held constant, then the higher the concentration, the greater the observed rotation. To take these factors into account, we define a new quantity called the **specific rotation,** which is the number of degrees rotation per unit concentration per unit of path length. The symbol for specific rotations is $[\alpha]_{\lambda}^{t}$. By definition,

$$[\alpha]_{\lambda}^{t} = \frac{100\alpha}{cl}$$

where c = concentration in g/100 ml of solution
l = length of the light path in the solution, measured in decimeters (1 dm = 10 cm)
α = observed rotation in degrees (plus or minus)
λ = wavelength of light
t = temperature of the solution in degrees Celsius

The specific rotation of a compound is an important physical constant, comparable to its melting point or boiling point. It is one more physical characteristic that a chemist can use to identify a substance. Moreover, by measuring the actual rotation, α, of a solution of a substance of known specific rotation in a tube of fixed length, we can calculate the concentration of the solution. That is, with $[\alpha]$, α, and l known, the equation readily permits calculation for c. Hence optical rotation gives us a tool for the quantitative analysis of the concentrations of optically active compounds.

Table 9.1 contains specific rotations for several materials and Table 9.2 has physical constants for some sets of optical isomers. Pairs of enantiomers differ physically only in the sign or direction in which each rotates the plane of plane-polarized light. All other physical constants are the same, including even the number of degrees of rotation.

Exercise 9.4 An unknown compound was found to be one of those listed in Table 9.2. A solution of this unknown with a concentration of

TABLE 9.1
Specific Rotations of Some Substances

Physical State	Substance	Specific Rotation (sodium D light at 20 °C in solvent specified)
Solutions[a]	Asparagine	+5.41 (water)
	Albumin (a protein)	−25 to −38 (water)
	Cholesterol	−31.61 (chloroform)
	Glucose	+52.5 (aged solution in water)
	Sucrose (table sugar)	+66.4 (water)
	3-Methyl-2-butanol ("active amyl alcohol")	+5.34 (ethyl alcohol)
	Quinine sulfate	−214 (water, at 17 °C)
Pure liquids[a]	Turpentine	−37
	Cedar oil	−30 to −40
	Citron oil	+62 (at 15 °C)
	Nicotine	−162
Pure solids[b]	Quartz	+21.7
	Cinnabar (HgS)	+32.5
	Sodium chlorate	+3.13

[a] Specific rotation in degrees per decimeter.
[b] Specific rotation in degrees per millimeter of path length in the crystal.

TABLE 9.2
Physical Constants of Some Optical Isomers

Set	Members of Set	Mp (°C)	$[\alpha]_D^{20}$ (degrees)	Miscellaneous
$C_6H_5CH(OH)CO_2H$	(+) Mandelic acid	132.8	+155.5	Solubility: 8.54 g/100 ml water (at 20 °C)
Mandelic acids	(−) Mandelic acid	132.8	−155.4	Solubility: 8.64 g/100 ml water (at 20 °C)
$HO_2CCH(NH_2)CH_2CONH_2$	(+) Asparagine	234.5 (decomposes)	+5.41	d_4^{15} = 1.534 g/ml
Asparagines	(−) Asparagine	235 (decomposes)	−5.41	d_4^{15} = 1.54 g/ml
$HO_2CCH(OH)CH(OH)CO_2H$	(+) Tartaric acid	170	+11.98	d_4^{15} = 1.760 g/ml
	(−) Tartaric acid	170	−11.98	d_4^{15} = 1.760 g/ml
	meso-Tartaric acid	140	0	d_4^{15} = 1.666 g/ml
Tartaric acids	Racemic tartaric acid[a]	205	0	d_4^{15} = 1.687 g/ml

[a] This is a 50:50 "mixture" of the plus and minus forms of tartaric acid, which in this case happens to form what is known as a racemic compound.

2.50 g/100 ml in a tube 20 cm long gave an optical rotation of +7.77° at 20 °C. Which compound in Table 9.2 was the unknown?

Exercise 9.5 A solution of (−)-asparagine at 20 °C in a tube 20 cm long gave an optical rotation of −8.3°. What was its concentration in g/100 ml?

Other Kinds of Optical Isomers

9.13 Meso Compounds

In some sets of optical isomers, for example, the tartaric acids of Figure 9.4, one member is found to be optically inactive, even though it possesses chiral carbons. We are reminded that the fundamental criterion for chirality in a *molecule* is not the chiral carbon but rather the impossibility of superimposing object and mirror image. The isomer that belongs to a set of optical isomers but is optically inactive itself is called the **meso isomer.** The mirror image of *meso*-tartaric acid would be superimposable on the object. The acid also possesses a plane of symmetry within the molecule. Behind this plane is a reflection of what is in front. Hence, as one-half of the molecule tends to rotate polarized light to the left, the other half cancels this with an equal rightward rotation. The net effect is optical inactivity.

9.14 Diastereomers

Geometrical isomers that are not related as object to mirror image are called **diastereomers.** Figure 9.7 illustrates examples, and we shall encounter them again in carbohydrate chemistry. For two molecules to qualify as diastereomers, they must have identical molecular formulas, they must have identical nucleus-to-nucleus sequences (without regard to geometry), but they may

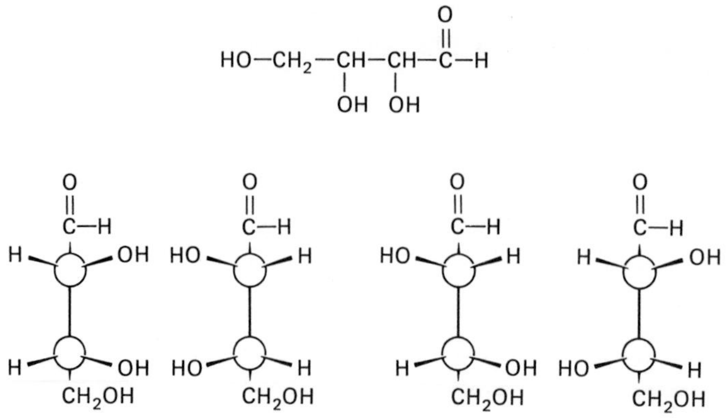

$HO{-}CH_2{-}CH(OH){-}CH(OH){-}C(=O){-}H$

D–Erythrose	L–Erythrose	D–Threose	L–Threose
$[\alpha]_0^{20}$ – 14.8°	+14.8°	$[\alpha]_0^{20}$ + 19.6°	–19.6°

Figure 9.7 Erythrose and threose, two simple sugars, share the basic structure given at the top, a molecule with two different chiral carbons and therefore $2^n = 2^2 = 4$ optical isomers. Two enantiomers of erythrose and two of threose exist. Any one of the enantiomers of erythrose is a diastereomer of any one of the threose forms. (Note the differences in specific rotations of enantiomers and diastereomers. Those of enantiomers differ only in sign; those of diastereomers differ in magnitude.)

not be related as object to mirror image. Generally, they will have different chemical and physical properties.

9.15 Racemic Mixtures

A substance that is composed of a 50 : 50 mixture of enantiomers is called a **racemic mixture.** One enantiomer cancels the effects of the other on polarized light, and a racemic mixture is optically inactive, emphasizing the point that the phenomenon of optical activity relates to a measurement that we can make on a substance. The explanation of the phenomenon relates to our theory about chiral molecules.

Exercise 9.6 Examine the following structures and answer the questions about them.

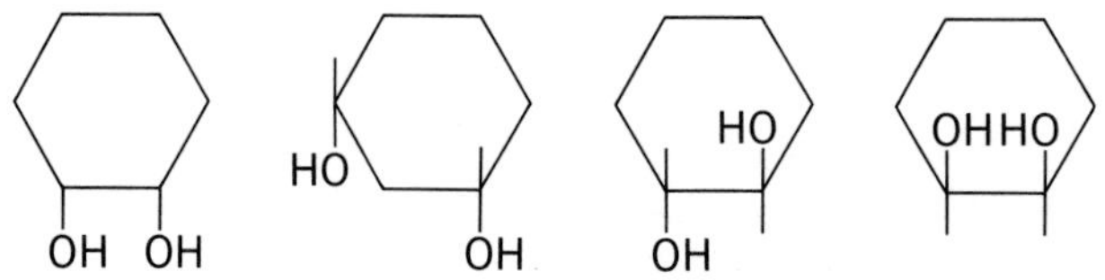

1 2 3 4

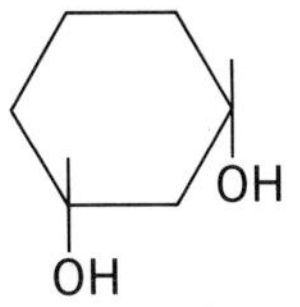

5 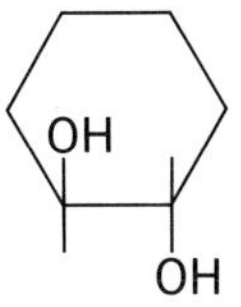 6 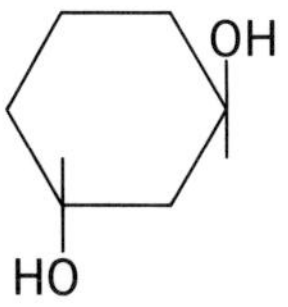 7

(1) Give the numbers of any two related as enantiomers. (More than two pairs may exist.)
(2) Give the numbers of any that are identical.
(3) Give the numbers of any meso compounds. For the meso compounds draw a dotted line connecting two corners of the six-membered ring through which the plane of symmetry would go perpendicularly.
(4) Give the numbers of any two related as diastereomers.
(5) Identify any two that in a 50:50 mixture would constitute a racemic mixture.

9.16 Configurational Changes in Chemical Reactions

A molecule of the compound *n*-butane has no chiral carbon, and it has a plane of symmetry. A molecule of bromine, Br_2, is also symmetrical. Yet, if bromine can be made to react with *n*-butane at the second carbon, this carbon becomes

$$CH_3CH_2CH_2CH_3 + Br_2 \longrightarrow CH_3CH_2\underset{\displaystyle Br}{\underset{|}{C}}HCH_3 + HBr$$

n-Butane　　　*sec*-Butyl bromide

chiral in the product. The question is whether the substance actually isolated as a product of this reaction and identified as *sec*-butyl bromide is optically active. The answer is *no*. The reaction produces a chiral carbon but in such a way that the product is a 50:50 mixture of enantiomers, a racemic, or an optically inactive mixture. We can understand this result by examining Figure 19.8. At the site where bromine will eventually become affixed, *n*-butane has two equivalent hydrogens. An equal chance exists that either hydrogen will be the one replaced. In a collection approaching Avogadro's number in size, the statistics make the production of a racemic mixture a certainty.

It is generally true that optically active substances cannot be synthesized from optically inactive reagents. Chiral *molecules* can easily be made, but the final *substance* will be a racemic mixture.

9.17 Absolute Configuration

The last major topic involving optical isomerism is about the absolute configuration of enantiomers. How do we specify what specific "handedness" a particular enantiomer has? We say "right hand" or "left hand" when we refer to our two hands. What is done with enantiomers?

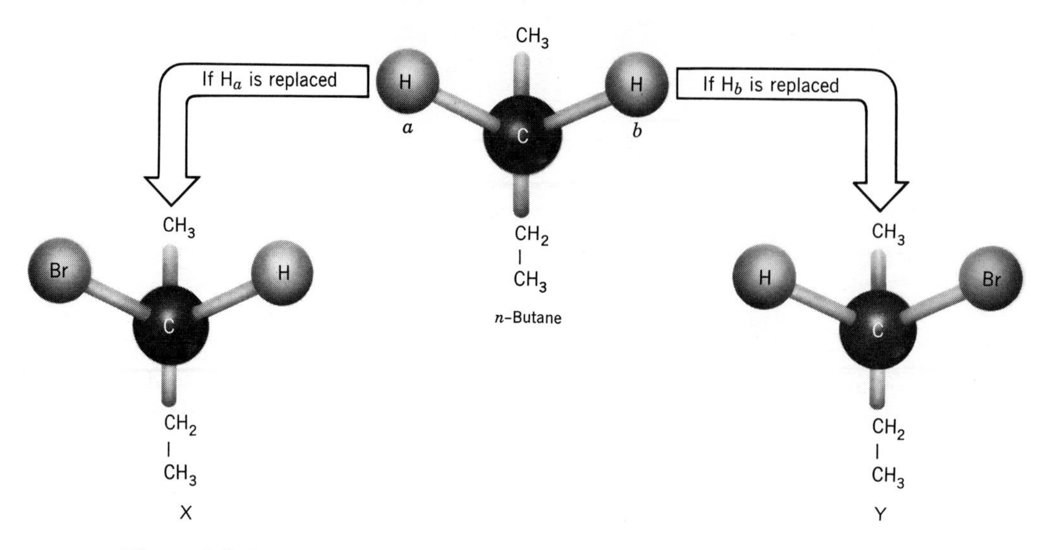

Figure 9.8 Conversion of *n*-butane into 2-bromobutane by direct bromination produces chiral carbons and chiral molecules, but the product is still optically inactive. The two chiral molecules, X and Y, differ only in configuration, are related as object to mirror image, and cannot be superimposed. They are formed in a 50:50 ratio—exactly. The product is therefore a racemic mixture.

In biochemistry, the literature is filled with the use of one system, the designations D or L as the two families of chiral molecules. We shall study that system in the next chapter.

Brief Summary

Optical Activity. Optical activity is a natural phenomenon detected by means of a polarimeter in which polarized light passing through the substance (or its solution) undergoes a rotation in the plane of polarization. Substances that will do this are called optically active.

Optical Isomerism. Optical isomerism is part of the explanation of optical activity. When optically active substances are examined at the molecular level, their molecules are found to be chiral. Molecules related as nonsuperimposable object and mirror images have chirality or handedness. Those of one chirality make up one isomer; those of the opposite chirality make up the other in a pair of substances called enantiomers. Each is an example of an optical isomer, and optical isomers belong to the broader family of stereoisomers. Besides one or more pairs of enantiomers, a set of optical isomers may include some related as diasteriomers and possibly one that happens to be optically inactive, despite having chiral carbons, called a meso compound.

Optical Inactivity. Substances whose molecules are all achiral are inactive. Substances consisting of a 50:50 mixture of enantiomers are inactive. Meso compounds have achiral molecules despite having chiral carbons and are optically inactive. Finally, some substances have such weak activity that it may go undetected.

Properties of Enantiomers. Enantiomers are identical in every physical respect except for the sign of rotation and for having structures related as object to mirror image without being superimposable. Enantiomers are identical in every chemical respect only if the reactant given them is itself an achiral substance, one whose atoms or molecules or ions are all symmetrical. However, if the reactant is chiral—for example, an enzyme, or almost any biological chemical—enantiomers often display unusual chemical differences.

Specific Rotation. The optical rotation given specifically by one unit of concentration (g/100 ml) in one unit of path length (1 dm), or the specific rotation, varies with both the temperature of the solution and the wavelength of the light used. Values of specific rotation aid in identifying an unknown compound or in determining its concentration in some solution.

Selected References

1. E. L. Eliel. "Stereochemistry Since LeBel and van't Hoff." *Chemistry,* January/February 1976, page 6.
2. G. B. Kauffman and R. D. Myers. "The Resolution of Racemic Acid." *Journal of Chemical Education,* December 1975, page 777. How Louis Pasteur carried out the first separation of a pair of enantiomers.

Questions and Exercises

1. What is the difference between optical activity and optical isomerism.
2. Examine the following structures and predict whether optical isomerism is a possibility.

(a) $CH_3C(CH_3)(OH)CH_2CH_3$ (b) $CH_3CH(CH_3)CH(OH)CH_3$ (c) $HOC(=O)CH(Br)CH_3$ (d) $HOC(=O)CH_2CH_2Br$

(e) *cis*-1,2-cyclopentanediol (f) *trans*-1,2-cyclopentanediol

3. Explain why enantiomers *should* have identical physical properties (except for the sign of rotation of plane-polarized light).
4. Explain why diastereomers *should not* be expected to have identical physical properties (except coincidentally).
5. A solution of sucrose in water at 25 °C in a tube 10-cm long gives an observed rotation of +2.0°. The specific rotation of sucrose in this solvent at this temperature is +66.4°. What is the concentration of the sucrose solution in grams per 100 ml?

6. A common name for glucose is "dextrose." Judging from that nickname, what is the likely sign for the specific rotation of glucose?
7. A common name of fructose, a particularly sweet sugar found in honey, is "levulose." What is the likely sign for the specific rotation of fructose.
8. Would molecules of the monomethyl ester of *meso*-tartaric acid be chiral? Explain.
9. The melting point of (−)-cholesterol is 148.5 °C. What is the melting point of (+)-cholesterol? (If you judge that insufficient evidence has been provided, state so.) In any case, explain your answer.

chapter 10

Carbohydrates

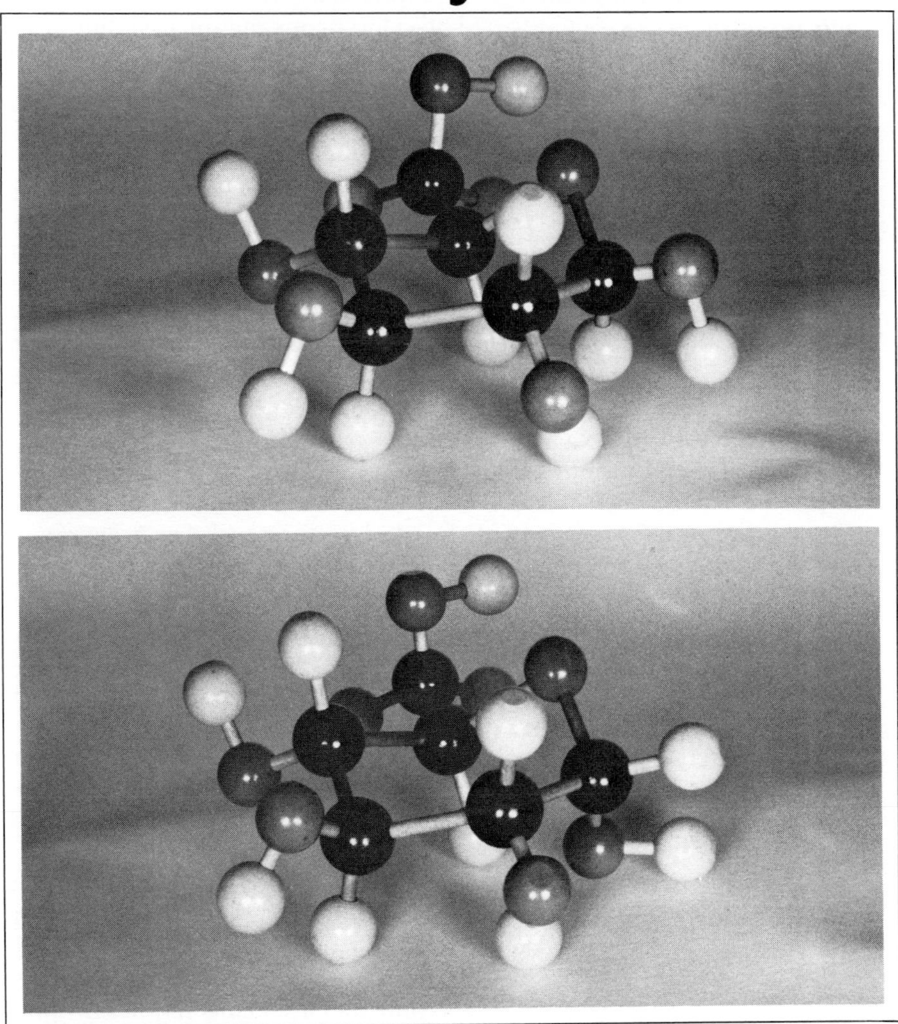

Glucose, probably the most widely occurring structural unit in the biosphere, exists chiefly in two forms, seen here in molecular models. Can you spot the difference between them? We study these in this chapter. (Photos by K. T. Bendo.)

10.1 Biochemistry—An Overview

Biochemistry is the systematic study of the chemicals of living systems, their organization, and the principles of their interaction as they participate in the processes of life.

The molecules of living systems are lifeless, yet life has a molecular basis. When isolated from their cells or when studied in their living environments, the chemicals at the foundation of life obey all the known laws of chemistry and physics. In isolation, however, not one single compound of a cell has life. The intricate cellular organization of interacting chemicals is as important to life as the chemicals themselves. The cell remains the smallest unit of matter that lives and can make a new cell exactly. A change in its environment—a change in temperature, pH, pressure, humidity, or availability of moisture, or the appearance of hostile chemicals—any of these changes can quickly render a cell nothing more than a lifeless bag of chemicals.

An ancient writer once said that we once were dust and to dust we shall return. That marvelous interlude between dust and dust called life endures only as long as the organism can maintain its dynamic organization of parts and defend it against natural tendencies to disintegrate and decay. An organism is a finely ordered and high-energy state of affairs compared to dust and ashes scattered to the four winds. We shall not speculate here on how life began, but as far back as we know, as well as in the present time, life begets life. One life shapes and fashions things into a new creature. One life, if a plant, takes ordinary substances such as water, carbon dioxide, and soil minerals, grows bigger, and then puts out seeds or other fruiting systems. As the seeds await their futures, the mature plant lives out the rest of its existence. After it dies, decay again returns the stuff of the plant to the much lower energy, much more disorganized forms of carbon dioxide, water, and soil minerals.

Animals must begin with substances more organized and more energy-rich than plant nutrients. Animals therefore feed on plants or on other animals that eat plants. Whether it be a plant or an animal, however, all life has three absolutely vital needs: materials, energy, and information. Our very limited purpose in the remainder of this book is to examine the molecular basis of these three needs and how one organism, the human body, makes use of the substances that supply them. As we look in a general way at life at the subcellular and molecular level, we shall also learn of several chemical dangers constantly faced (and nearly always met) by a living system.

We shall begin in the next three chapters to study the organic materials of life. We shall start with the three great classes of foods—carbohydrates, lipids, and proteins. From their molecules we build and run ourselves and try to stay in some viable state of repair. Proteins are particularly important in cell structure and function among both plants and animals. Plants rely heavily on a carbohydrate for cell walls, and animals obtain considerable energy from the carbohydrates made by plants. Lipids serve many purposes and are rich in energy.

Because of their central catalytic role in regulating chemical events in cells, a special family of proteins, the enzymes, will be studied immediately after proteins. Enzymes are actually a part of the molecular basis of information. Without plans or blueprints, materials and energy could combine to produce only rubble and rubbish. Monkeys swinging hammers would only reduce a stack of lumber to splinters. Expending no different amount of raw energy and using no different materials, carpenters can make the lumber into a dwelling because they have both a present plan and the past experience of applying knowledge and plans to materials and energy. Enzymes, however, do not originate the plans. They only help carry them out. The blueprint for an individual member of a species is in the organism's code on its cell's genetic material, a family of substances called nucleic acids. Nucleic acids direct the synthesis of a cell's enzymes, a set unique to the species. After we have learned what enzymes are and how they work, we shall study enzyme-makers. We shall examine how different species take essentially the same raw materials and energy and fashion unique enzyme systems.

To supply materials for any purpose—spare parts, energy, or information—each organism has basic nutritional needs, not only organic materials, including vitamins, but also minerals, water, and oxygen. All of the materials taken in the diet have to be processed by the digestive system and distributed by the bloodstream and the lymph if cells are to obtain what they need in forms suitable for that species. Therefore, following a chapter on the chemistry of heredity, we shall study the digestive needs of metabolism and then the fluid networks that process materials, distribute them, and carry wastes away.

The final broad topic, energy for living, will be studied in the last four chapters. How can gulps of air, a peanut butter sandwich, and a glass of milk be used by the body to make or repair, say, muscle cells? How do these materials make it possible for us to flex a muscle? As we study biochemical energetics, we shall have sufficient background to examine the molecular basis of a variety of metabolic disorders attributed to chemicals. With a solid background in the structures, properties, and metabolic uses of carbohydrates, lipids, proteins, nucleic acids, enzymes, hormones and vitamins we shall close our study of the molecular basis of life with a survey of basic nutritional requirements.

With this overview of where we are going, let us start back at the beginning, a new beginning in our study, with the carbohydrates.

10.2 Families of Carbohydrates

Carbohydrates are polyhydroxy aldehydes, polyhydroxy ketones, or substances that by simple hydrolysis yield these. Carbohydrates that cannot be hydrolyzed to simpler molecules are called **monosaccharides** or sometimes simple sugars, and their names end in -ose. Monosaccharides have the general formula $(CH_2O)_n$ and are classified in the following ways.

1. The number of carbons in one molecule or the value of n in $(CH_2O)_n$.

If the number of carbons is	the monosaccharide is a
3	triose
4	tetrose
5	pentose
6	hexose, and so forth

2. The nature of the carbonyl group present. If an aldehyde is present, the monosaccharide is an **aldose.** If a keto group is present, the monosaccharide is a **ketose.**

A hexose that has an aldehyde group is called an **aldohexose.** A **ketohexose** possesses a keto group and a total of six carbons. Thus word parts may be usefully combined.

Carbohydrates that can be hydrolyzed to two monosaccharides are called **disaccharides.** Sucrose, maltose, and lactose are common examples. (A broader family name, oligosaccharide, applies to those that can be hydrolyzed to 2 to 10 monosaccharides.) Starch and cellulose are common **polysaccharides.** Their molecules yield hundreds of monosaccharide units when they are hydrolyzed.

Carbohydrates are classified further as being either reducing carbohydrates or nonreducing carbohydrates. **Reducing carbohydrates** reduce (give positive tests with) Tollens', Benedict's, and Fehling's reagents. All monosaccharides and nearly all disaccharides are reducing sugars. Sucrose (table sugar) is not, and neither are the polysaccharides.

Monosaccharides

10.3 Aldohexoses

If relative abundance is our measure of importance, then the hexoses in general and the aldohexoses in particular are the most important saccharides. Two trioses are important, glyceraldehyde and dihydroxyacetone. One aldopentose, ribose, is essential to the structure of the ribonucleic acids (RNA), one of the chemicals of heredity. Among the hexoses of particular importance are **glucose** and **galactose**—both isomeric aldohexoses—and **fructose,** a ketohexose. All of the monosaccharides are white, crystalline substances. Most have a sweet taste, and all dissolve freely in water but do not dissolve in any nonpolar solvents.

Glucose

10.4 Occurrence

The glucose that occurs naturally is dextrorotatory and if we count all its combined forms, this glucose is perhaps the most abundant organic species on earth. **Cellulose,** a polysaccharide that yields nothing but glucose when it is hydrolyzed, is about 10% of all tree leaves of the world (dry weight basis), about 50% of the woody parts of plants, and nearly 100% of cotton. **Starch,** another polysaccharide, also yields nothing but glucose when it is hydrolyzed. Glucose and fructose are the major components of honey. Glucose is also commonly found in plant juices. Blood has roughly 100 mg glucose/100 ml, which explains why glucose is often called "blood sugar" in clinical and medical work.[1] Other names for glucose are "corn sugar" and **"dextrose"** (after its dextrorotatory property).

Glucose does not normally appear in the urine of healthy people for sustained periods of time. When it does appear, and persists in appearing, it is indicative of a malfunction in body chemistry (e.g., diabetes mellitus). The detection of glucose in the urine, therefore, is important in diagnosis. Benedict's test is most commonly used.

10.5 Structure of Glucose

The molecular formula of glucose is $C_6H_{12}O_6$. It forms a pentaacetate. Hence, five of the six oxygens are part of alcohol groups. Glucose can be oxidized by very mild oxidizing agents to a monocarboxylic acid having six carbons. Therefore, glucose has an aldehyde group. By vigorous means of reduction glucose can be changed to a derivative of *n*-hexane. The carbons of glucose are therefore in a straight chain. One carbon must bear the aldehyde group. Since 1,1-diols are unstable, the remaining five carbons carry one hydroxyl group each. The data are consistent with glucose being 2,3,4,5,6-pentahydroxyhexanal.

$$HO-CH_2-\underset{OH}{\underset{|}{CH}}-\underset{OH}{\underset{|}{CH}}-\underset{OH}{\underset{|}{CH}}-\underset{OH}{\underset{|}{CH}}-\overset{O}{\overset{\|}{C}}-H$$

2,3,4,5,6-Pentahydroxyhexanal

(basic system in glucose and all aldohexoses)

The glucose molecule has four different chiral carbons. Since the number of optical isomers = 2^n, where n is the number of different chiral carbons, $2^4 = 16$ optical isomers of 2,3,4,5,6-pentahydroxyhexanal are possible, or eight pairs of enantiomers. Dextrorotatory glucose is one of these.

10.6 Mutarotation

Complicating the structural picture of glucose is the fact that its optical rotation behaves strangely. When an aqueous solution of natural glucose is freshly made, the specific rotation is $[\alpha]_D^{20°} = +113°$, but this value gradually changes as the solution remains at room temperature to a value of $[\alpha]_D^{20°} = +52°$. We shall call a glucose solution with this specific rotation an "aged glucose solution." By a special method of recovery (which we shall not describe), crystalline glucose can be recovered from an aged solution, redissolved to make a fresh solution, and found to have, once again, $[\alpha]^{20°} = +113°$. We get the same glucose back. The new solution ages in exactly the same way to a value of $[\alpha]_D^{20°} = +52°$.

[1] Strictly speaking, "blood sugar" means the sum total of all types of carbohydrates in circulation in the bloodstream that are capable of giving a positive Benedict's test. Since glucose is by far the most abundant, the distinction is usually unimportant and "blood sugar" is often used as a synonym for "blood glucose."

By a different method of recovery a different crystalline compound can be isolated from the aged solution. When dissolved in water the fresh solution has a specific rotation of $[\alpha]_D^{20°} = +19°$, but that value also changes—to the same final value of $[\alpha]_D^{20°} = +52°$!

From the aged solution, to summarize, one method of recovering glucose gives back the form that, in a fresh solution, has $[\alpha]_D^{20°} = +113°$; the other method of recovery returns the second form of glucose with $[\alpha]_D^{20°} = +19°$ when freshly dissolved. Either substance in a fresh solution undergoes a change in optical rotation to a final value of $[\alpha]_D^{20°} = +52°$; either substance can again be recovered by a suitable choice of method. This change in optical rotation with time shown by an optically active substance, a substance that can be recovered from an aged solution without any other apparent change, is called **mutarotation.** All of the hexoses and most of the disaccharides exhibit it.

10.7 Haworth Forms of Glucose

Three forms of glucose exist, **1**, **2**, and **3**. Structure **2** is the open-chain form, 2,3,4,5,6-pentahydroxyhexanal, shown in a coiled conformation. The other two forms, one called α-glucose (**1**) and the other called β-glucose (**3**) are cyclic compounds—cyclic hemiacetals.

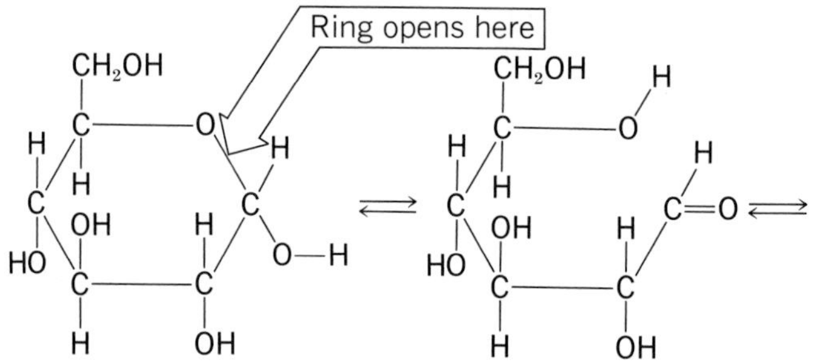

Ring opens here

α-Glucose

1

Open form
(polyhydroxyaldehyde)

2

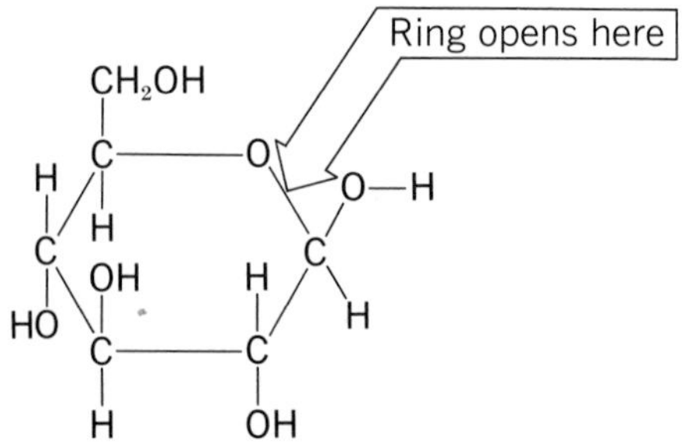

Ring opens here

β-Glucose

3

When α-glucose is dissolved in water, the solute molecules in the fresh solution are all in the cyclic form, **1**, with $[\alpha]_D^{20°} = +113°$. But hemiacetals are relatively unstable. First one molecule of **1** and then another opens up into the original aldehyde and original alcohol, but this time these two groups are on the same molecule, molecules of the open form, **2**. Molecules of **2** can close again into a ring. The —OH group at carbon-5 adds across the carbon-oxygen double bond to become the cyclic hemiacetal.

As ring closure occurs, the carbonyl group may be pivoted basically in one of two orientations as illustrated in Figure 10.1. When the carbonyl group is pointing one way (top of Figure 10.1) as it accepts the addition of the C_5 hydroxyl group, the new —OH group (at C-1) will point on the same side of the ring as the CH_2OH group. On the other hand, if the carbonyl group happens at the moment of ring closure to be pointing in the other direction (bottom of Figure 10.1), the new —OH group will point on the opposite side of the ring as the —CH_2OH group. Thus the two cyclic forms of glucose differ only in the orientation of the —OH group at carbon-1.

In the solution of freshly dissolved α-glucose ($[\alpha]_D^{20°} = +113°$) molecules open and close at random until eventually dynamic equilibrium is reached. The proportions at equilibrium are 36% α-glucose, 64% β-glucose, and 0.02% of the open-chain form. The same mixture eventually is established regardless of the starting material, whether it be pure α-glucose or pure β-glucose ($[\alpha]_D^{20°} = +19°$), and therefore from either the final specific rotation of $[\alpha]_D^{20°} = +52°$ develops. Under one method of recovery, only molecules in the α-form precipitate. Under the alternative method of recovery only the β-form precipitates.

Even though virtually no glucose in an aged solution is in the open-chain form, a solution of glucose gives nearly all the properties we expect of an aldehyde. Any reagent that specifically attacks an aldehyde (e.g., Tollens' reagent)

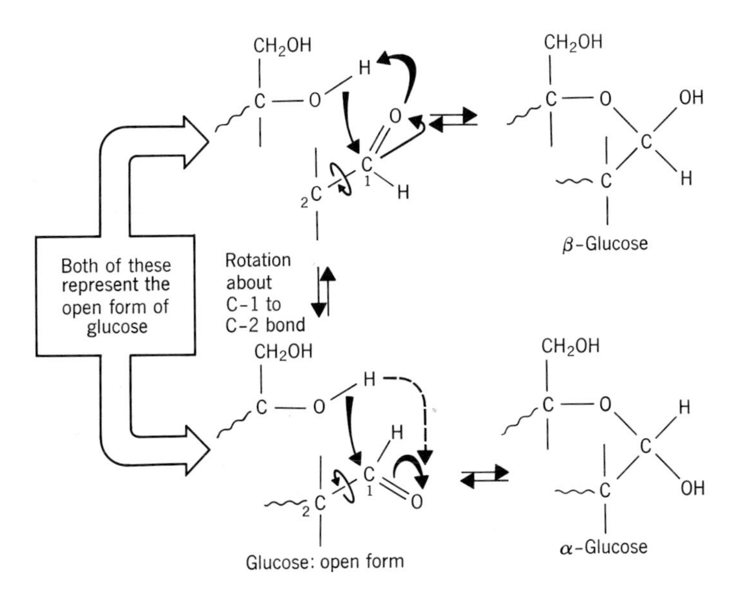

Figure 10.1 How the α- and β-forms of glucose arise from a common intermediate, the open form. (Only relevant parts of glucose molecules are shown.)

removes molecules of the open form. Because hemiacetals break down quite easily and without the need of a catalyst, closed forms of glucose open up to replace lost molecules of open form almost as rapidly as the latter disappear by some chemical change.

The cyclic forms of glucose are called **Haworth structures,** after Sir Walter Norman Haworth (1883–1950), a British chemist who shared the 1937 Nobel Prize in chemistry (with Paul Karrer of Switzerland) for work on carbohydrates and vitamin C. The Haworth structures we have given represent correctly most details of the structures of glucose. In particular they show the way the ring substituents project relative to each other. The differences between the optical isomers of glucose are in these projections. These differences are actually in relative configurations at chiral carbons, and we shall next study both the absolute and relative configurations of monosaccharides.

10.8 Absolute Configuration

To simplify, let us retreat from the complexities of glucose to the simplest monosaccharide with a chiral carbon, glyceraldehyde. The two enantiomers of glyceraldehyde are given in Figure 10.2, and both are known compounds. From experiments involving both chemical and spectrometric techniques, molecules in the dextrorotatory sample all have the absolute configuration given in Figure 10.2. **Absolute configuration** means the actual arrangement in space about chiral centers. Those in the levorotatory sample all have the mirror image of the dextrorotatory configuration, also shown in Figure 10.2. On the basis of these two absolute configurations, scientists have developed the concept of configurational families. Any compound that has a configuration like that of (+)-glyceraldehyde and can be related to (+)-glyceraldehyde by known reactions is said to be in the **D-family.** The mirror image of any compound in the D-family is in the **L-family.** Thus levorotatory glyceric acid is in the D-family because it can be made from D-(+)-glyceraldehyde by an oxidation that disturbs no bond to the chiral carbon, as this equation shows:

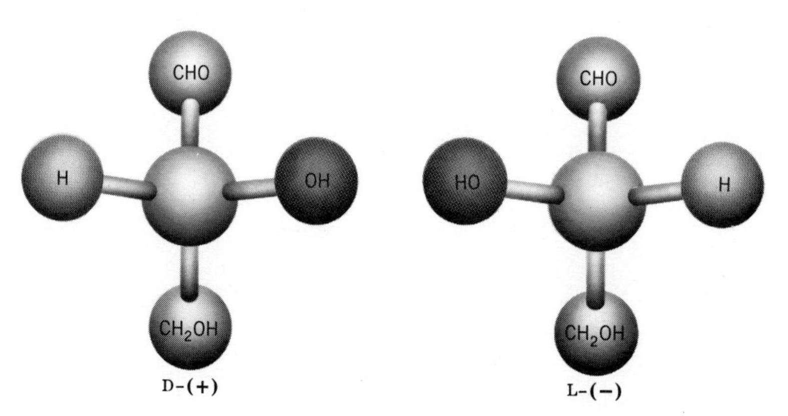

Figure 10.2 Absolute configurations of the glyceraldehyde enantiomers.

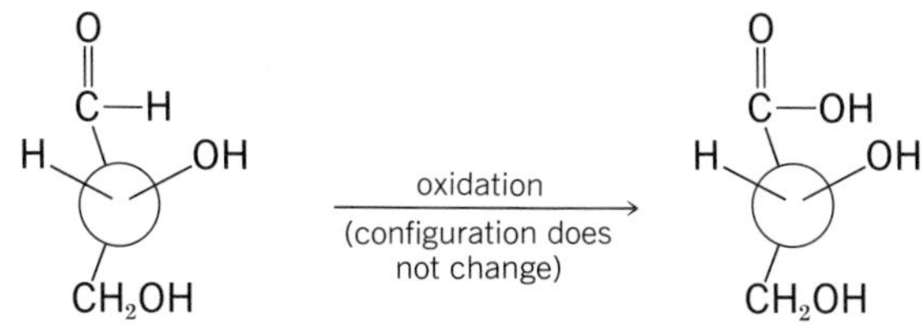

O
||
C—H
H OH
CH₂OH
D-(+)-Glyceraldehyde

oxidation
(configuration does not change)

O
||
C—OH
H OH
CH_2OH
D-(-)-Glyceric acid

The letters D and L are simply family names and have nothing to do with actual signs of rotation. They signify something about configurations only. These families were invented to handle matters of absolute and relative configurations of carbohydrates, but since they were invented many defects in the system became apparent and the D-L concept has been hard to use among noncarbohydrates. A vastly superior system has been devised, one not dependent on using chemical reactions that relate a substance to one of the glyceraldehyde enantiomers. The two families in the new system are the *R*-family (the family of D-(+)-glyceraldehyde) and the *S*-family, and details are given in an Appendix. Some time will elapse before the need for knowing the D-L system vanishes, and some current biochemical references and all the older literature use it. Because it is so common among both carbohydrates and proteins, we have elected to develop the D-L system here in the main body of the text and place the *R-S* system in an appendix for optional study or for your reference later, should you encounter the need.

10.9 Plane Projection Structures

Absolute configurations may be drawn with perspective figures, those giving a three-dimensional view, but perspectives are difficult to draw when a molecule has several chiral centers. A special convention using a few simple rules solves this problem by projecting onto a plane surface the three-dimensional model when it is oriented in a particular way—one defined by the rules. The result of this projection is called a plane projection diagram or a **plane projection structure.** The rules for writing plane projection structures are on the next page.

A plane projection diagram may have more than one intersection of lines, each intersection being a chiral carbon (see Figure 10.4). Just remember that at each intersection the horizontal lines are bonds coming toward you and each vertical line is a bond going back and away from you.

Once we have one plane projection structure, it is easy to draw the mirror image as illustrated in both Figures 10.3 and 10.4. We may easily check superimposition, too, provided that we heed one more important rule. We may never (in our imagination) lift a plane projection structure out of the plane of the paper. We may only slide and rotate it within the plane. The rule is required by the fact that if we turn a plane projection diagram out of the plane and over, we actually make groups projecting in one direction project oppositely, but the operation will not show this reversal.

Rules for Plane Projection Structures

Rule 1. We hold the molecule with the carbon chain vertical and we twist the structural members to make the chain go toward the rear at each chiral carbon. Carbon-1 is held at the top. In carbohydrates this means that the carbonyl group is at or near the top.

Rule 2. We imagine flattening the structure, chiral carbon by chiral carbon, onto a sheet of paper, as illustrated in Figures 20.3 and 20.4.

Rule 3. We represent each chiral carbon not by a "C" but by a simple intersection of two lines at right angles. (Sometimes you will see "C" used.)

Rule 4. The two sides of the horizontal line are bonds at the chiral center that actually project forward, out of the plane of the paper.

Rule 5. The lines going up and down are bonds at the chiral carbon that actually project rearward, behind the plane of the paper.

Exercise 10.1 Plane projection structures corresponding to various isomers of tartaric acid follow. Which are identical? Which are related as enantiomers? Which is a meso compound?

(a)	(b)	(c)	(d)	(e)
CO_2H	CO_2H	CO_2H	HO_2C	CO_2H
H—┼—OH	H—┼—OH	H—┼—OH	HO—┼—H	HO—┼—H
HO—┼—H	HO—┼—H	H—┼—OH	H—┼—OH	HO—┼—H
CO_2H	CO_2H	CO_2H	HO_2C	CO_2H

Exercise 10.2 Write plane projection structures for the enantiomers of glyceric acid, $HOCH_2CH(OH)CO_2H$ (OH on the central carbon).

10.10 D- and L-Families of Carbohydrates

Monosaccharides are assigned to the D-family or the L-family according to the projection of the —OH group in the chiral carbon farthest from the carbonyl group. By convention, if that —OH group projects to the right in a plane projection structure, the compound is in the D-family. If this —OH group projects to the left, the substance is in the L-family. D- and L-glyceraldehyde (Figure 10.3) illustrate these principles. D- and L-threose and D- and L-erythrose, Figure 10.4, are additional examples. It matters now how —OH groups at other chiral carbons project. Family membership is deter-

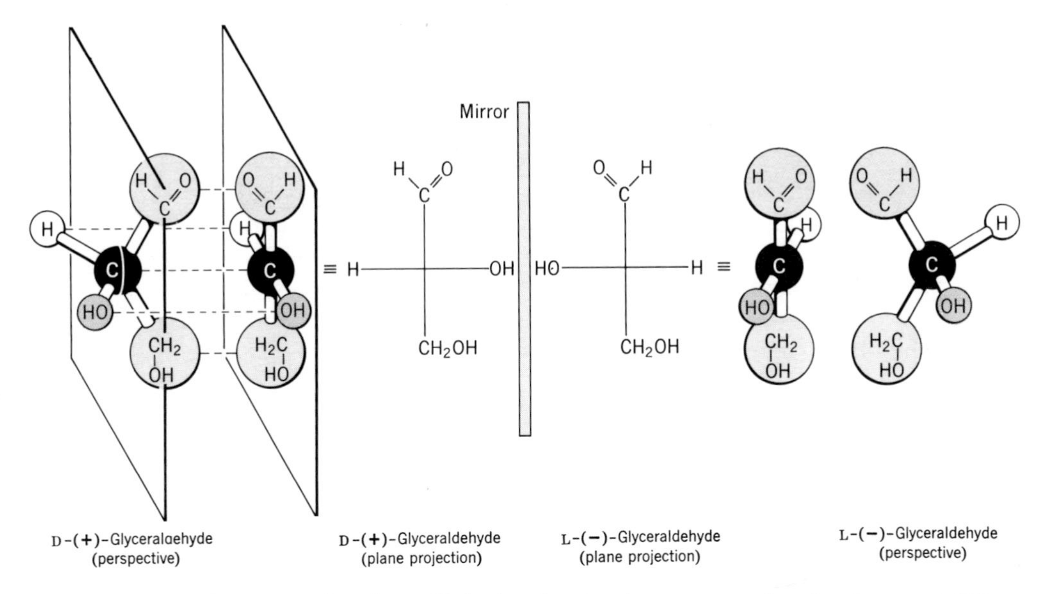

Figure 10.3 D- and L-glyceraldehyde, showing how perspective structures are projected onto plane surfaces to give plane projection structures.

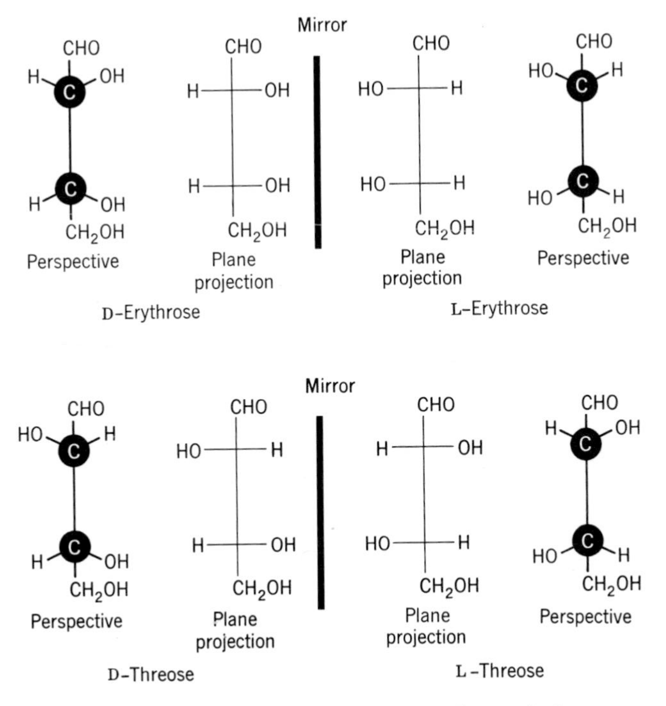

Figure 10.4 The aldotetroses in perspective and plane projection structures. Two chiral carbons in an aldotetrose mean four ($2^n = 4$) optical isomers, two pairs of enantiomers.

mined solely by the projection of the —OH on the chiral carbon farthest from the carbonyl group.

The nutritionally important carbohydrates are all in the D-family. In one of the "tremendous trifles" of nature, we cannot use their L-enantiomers. We lack enzymes with the proper "handedness" and therefore the proper "fit" to act on saccharides of the L-family. Hence, for the remainder of the book, if family membership for a carbohydrate is not given, the D-family is meant. Figure 10.5 gives plane projection structures for all aldoses in the D-family from D-glyceraldehyde through the eight aldohexoses. The other 8 aldohexoses, to make up the full complement of 16 optical isomers of glucose, are the mirror images of those shown in the figure.

10.11 Glucose in Chair Forms

How one gets from the plane projection to Haworth structures is illustrated in Figure 10.6. Six-membered rings, of course, may be either in a boat

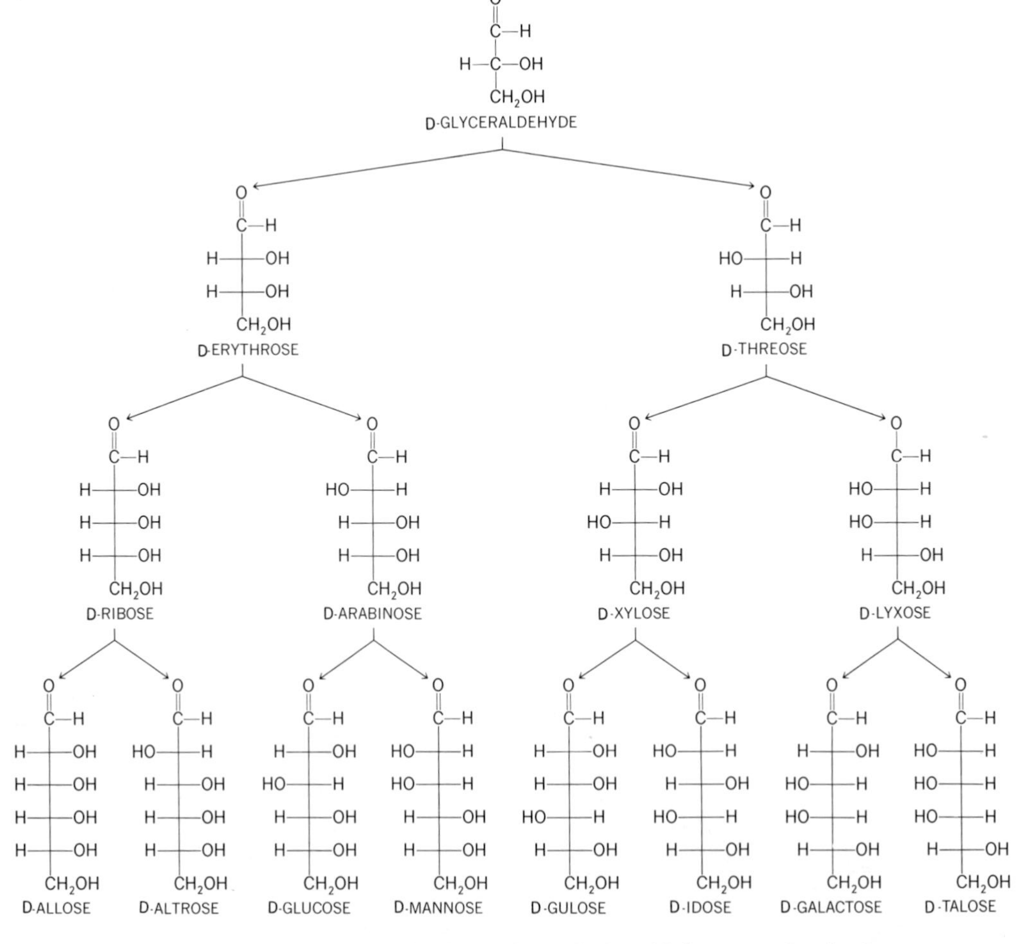

Figure 10.5 The D-family of aldoses through the aldohexoses. On the bottom row are eight of the sixteen optical isomers of the aldohexoses. The other eight are the mirror images of these. In all structures the —OH on the chiral carbon farthest from the carbonyl group projects to the right.

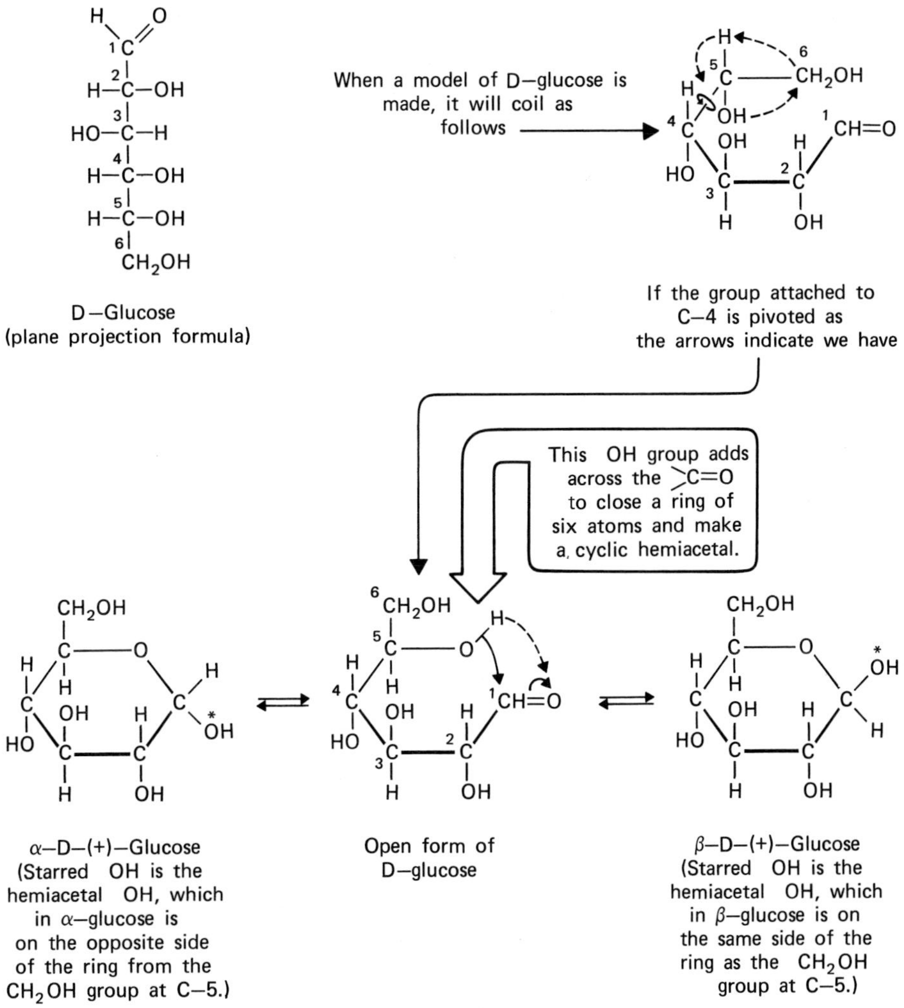

D–Glucose
(plane projection formula)

When a model of D–glucose is made, it will coil as follows

If the group attached to C–4 is pivoted as the arrows indicate we have

This OH group adds across the >C=O to close a ring of six atoms and make a, cyclic hemiacetal.

α–D–(+)–Glucose
(Starred OH is the hemiacetal OH, which in α–glucose is on the opposite side of the ring from the CH_2OH group at C–5.)

Open form of D–glucose

β–D–(+)–Glucose
(Starred OH is the hemiacetal OH, which in β–glucose is on the same side of the ring as the CH_2OH group at C–5.)

Figure 10.6 Relating the plane projection structure of D-glucose to its Haworth structures.

or one of two chair forms, but the chair forms are more stable. The flat rings in the Haworth structures, therefore, represent a simplification, one that we shall find convenient to continue, however. The final refinement is shown in Figure 10.7. Both α- and β-glucose exist in chair forms, and are in those particular chair forms that get the maximum number of bulky substituents (—OH and CH_2OH groups) sticking outward and away from the ring as much as possible. In this manner they crowd each other the least. An interesting feature of β-glucose is the fact that it is the only aldohexose in which all bulky substitutents can stick as far in an outward uncrowded configuration as possible. Perhaps—and we only repeat a speculation—this explains why β-glucose is the single most abundant aldohexose in nature.

We shall not normally use chair-type structures to represent cyclic forms. Flat-ring Haworth representations will serve nicely, and Figure 10.8 demonstrates how the condensation we shall use of the Haworth structure of glucose may be written.

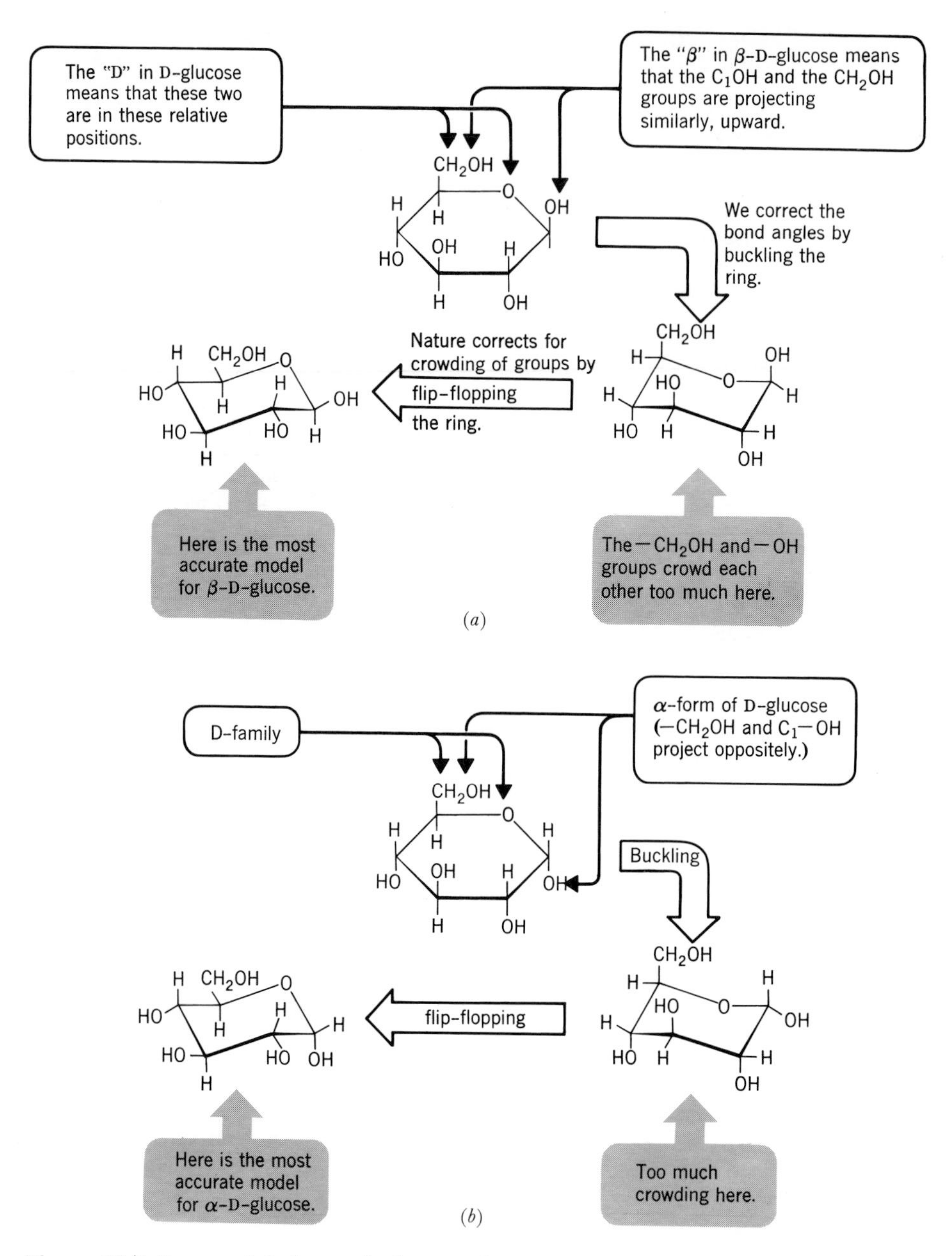

Figure 10.7 Best models for cyclic forms of D-glucose. (*a*) β-D-Glucose. (*b*) α-D-Glucose.

10.12 Other Properties of Glucose

The monosaccharides, in general, and glucose, in particular, are insoluble in nonpolar solvents, but they dissolve in water. This behavior may be understood in terms of the poly-hydroxy feature of their molecules. The —OH groups hydrogen-bond to each other in the presence of nonpolar solvents and will not let their molecules dissolve in these solvents.

The hemiacetal system in cyclic forms of monosaccharides may easily be changed to an acetal by reaction with an alcohol in the presence of a mineral acid. These acetals are called, in general, **glycosides.** Those from glucose are specifically called **glucosides,** α-glucosides and β-glucosides depending on the projection of the new group. For example, methyl alcohol reacts as follows:

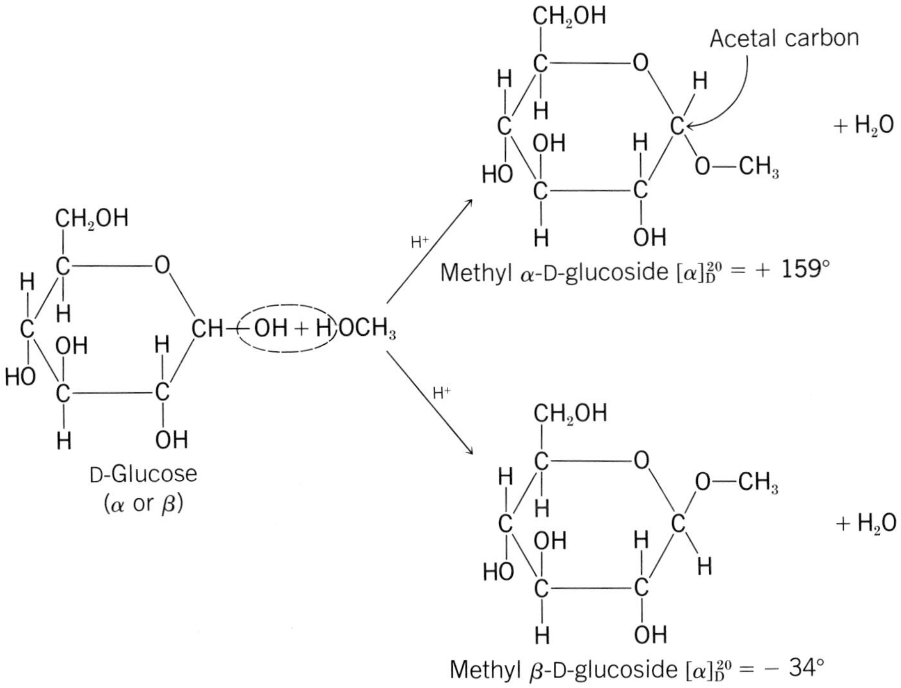

D-Glucose (α or β)

Acetal carbon

Methyl α-D-glucoside $[\alpha]_D^{20} = +159°$ $+ H_2O$

Methyl β-D-glucoside $[\alpha]_D^{20} = -34°$ $+ H_2O$

Since acetal systems do not spontaneously open and close, like hemiacetals, the glycosides do not mutarotate. Like acetals, the glycosides are stable in base but are hydrolyzed by dilute acid (or special enzymes). They give negative tests with Tollens' and Benedict's reagents.

Even though glucose circulates freely in the bloodstream, only traces are present inside cells. Glucose does not exist inside cells in its free form but as negative ions of one or another phosphoric acid ester, usually ions either of glucose 1-phosphoric acid or of glucose 6-phosphoric acid.

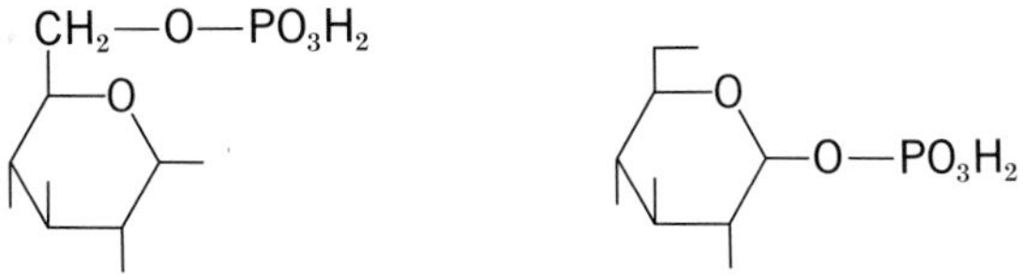

$CH_2—O—PO_3H_2$ $—O—PO_3H_2$

Glucose 6-phosphoric acid   Glucose 1-phosphoric acid

The ions of these acids are generally called phosphates, as in "glucose 1-phosphate" (G-1-P) and "glucose 6-phosphate" (G-6-P), without specifying the state of ionization. Ions of phosphoric acid esters of other monosaccharides are also known.

1. First write a six-membered ring with an oxygen in the upper right-hand corner.
2. Next "anchor" the —CH_2OH on the carbon to the left of the oxygen. (Let all the —Hs attached to ring carbons be "understood.")
3. Continue in a counterclockwise way around the ring, placing the —OHs first down, then up, then down.
4. Finally, at the last site on the trip, how the last —OH is positioned depends on whether the alpha or the beta form is to be written. The alpha is "down," the beta "up."

If this detail is immaterial, or if the equilibrium mixture is intended, the structure may be written as.

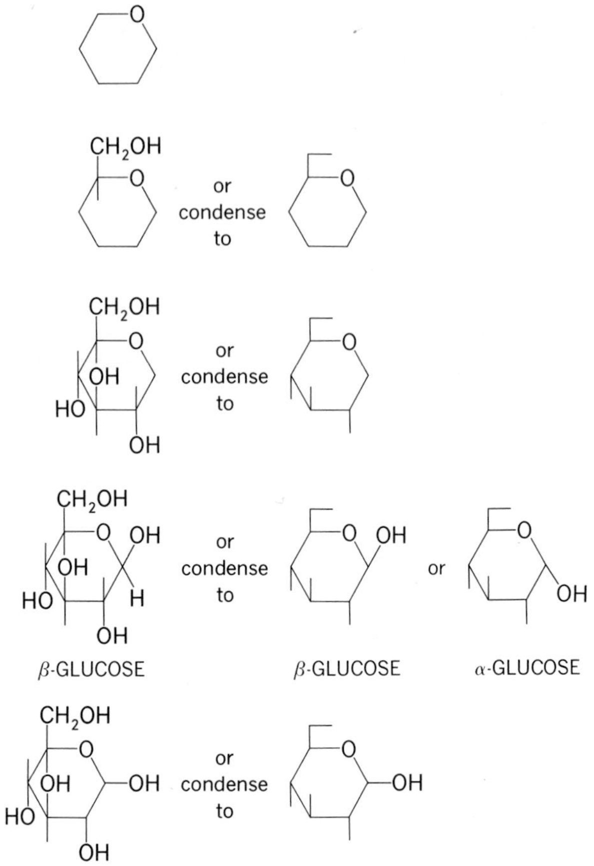

Figure 10.8 Condensing Haworth structures of D-glucose.

10.13 Galactose

Most galactose occurs naturally in combined forms, especially in the disaccharide lactose (milk sugar). Galactose differs from glucose only in the orientation of the C-4 hydroxyl. Like glucose, it is a reducing sugar; it mutarotates; and it exists in three forms, alpha, beta, and open.

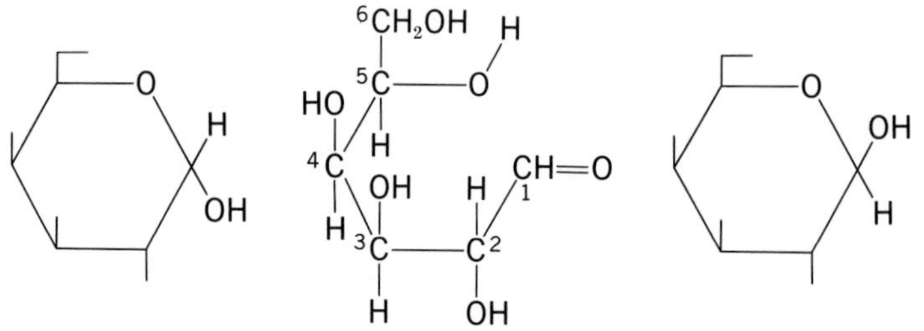

α-D-Galactose

D-Galactose (open form)

β-D-Galactose

10.14 Fructose, A Ketohexose

Fructose, the most important ketohexose, is found together with glucose and sucrose in honey and fruit juices. It can exist in more than one form, and as a building unit in sucrose it exists in a cyclic, five-membered hemiketal form.

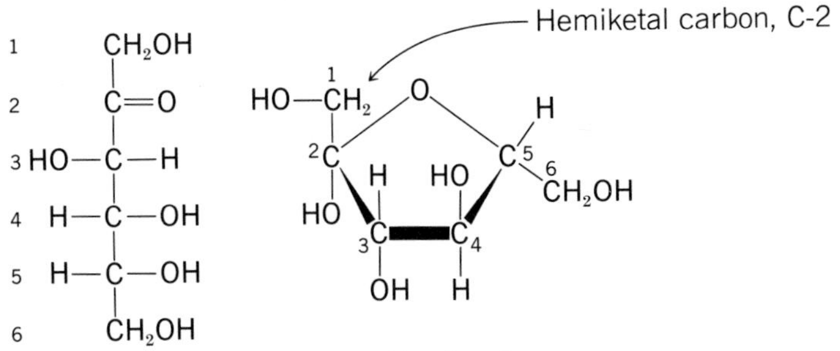

1 CH_2OH
2 C=O
3 HO—C—H
4 H—C—OH
5 H—C—OH
6 CH_2OH

D-Fructose

Hemiketal carbon, C-2

α-D-Fructose[2]

Because fructose is strongly levorotatory, we sometimes call it levulose; the specific rotation is $[\alpha]_D^{20°} = -92.4°$. Because fructose has the α-hydroxyketone system, it is reducing sugar. Its cyclic form, shown above, has a hemiketal system. Therefore, it can react with alcohols to give ketals called fructosides—table sugar (sucrose) is one example. Mono- and diphosphoric acid esters of fructose are important compounds in the metabolism of carbohydrates.

10.15 Amino Sugars

Amino derivatives of glucose and galactose occur widely in nature as building units for some important polymeric substances. Glucosamine is a structural unit in chitin, the principal component of the shells of lobsters, crabs, and certain insects and in heparin, a powerful blood anticoagulant. Galactosamine is a structural unit of chondroitin, a polymeric substance that is important to the structure of cartilage, skin, tendons, adult bone, the cornea, and heart valves. Amino sugars combined with proteins form walls of certain cells such as red blood cells.

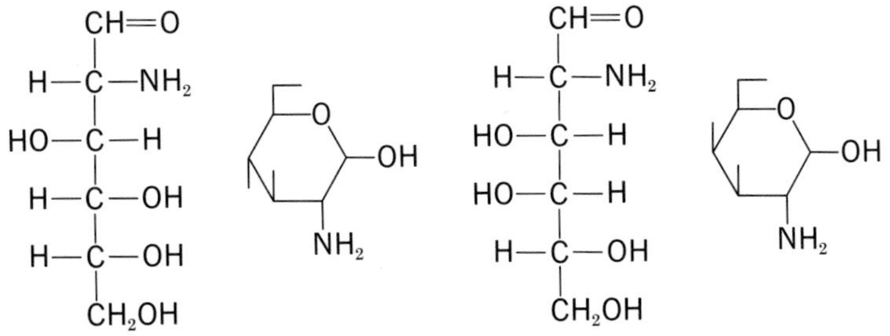

D-Glucosamine: Open form, Closed form

D-Galactosamine: Open form, Closed form

[2] This is the alpha-form because the hemiketal —OH group at C-2 is on the *same* side of the ring as the group that includes carbon number 6.

Disaccharides

10.16 The Important Disaccharides

Nutritionally, the three important disaccharides are **maltose, lactose,** and **sucrose.** All are in the D-family. How these are related to the monosaccharides may be seen in the following equations:

$$\text{maltose} + H_2O \longrightarrow \text{glucose} + \text{glucose}$$

$$\text{lactose} + H_2O \longrightarrow \text{glucose} + \text{galactose}$$

$$\text{sucrose} + H_2O \longrightarrow \text{glucose} + \text{fructose}$$

All these disaccharides are glycosides and are formed by one monosaccharide acting as a hemiacetal and the other as an alcohol to make an acetal.

10.17 Maltose

Although present in germinating grain, maltose or malt sugar does not occur widely in the free state in nature. It can be prepared by the partial hydrolysis of starch and it occurs in corn syrup. As with glucose, two forms of maltose are known and each mutarotates.

Because the bridging oxygen leaves the first ring in an alpha orientation, the glycosidic or acetal "bridge" is an alpha bridge. Therefore, maltose is an α-glucoside. The bridge goes from C-1 of one glucose unit to C-4 of the second unit. The written symbol for such a bridge is "$\alpha(1 \rightarrow 4)$." Maltose is a reducing sugar and hydrolyzes to two molecules of glucose. The enzyme maltase catalyzes this reaction, and it is known to act only on α-glucosides, not on β-glucosides.

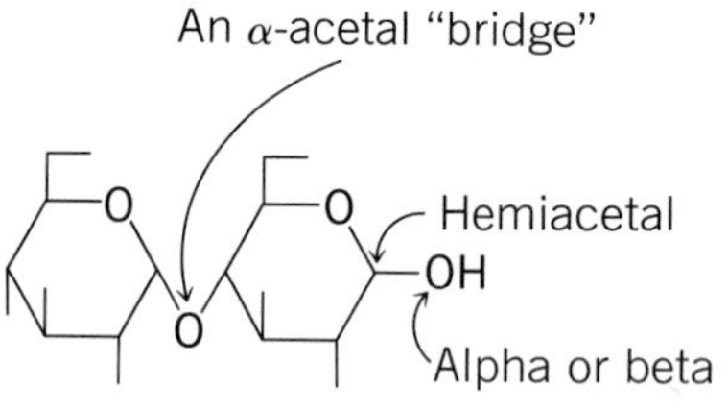

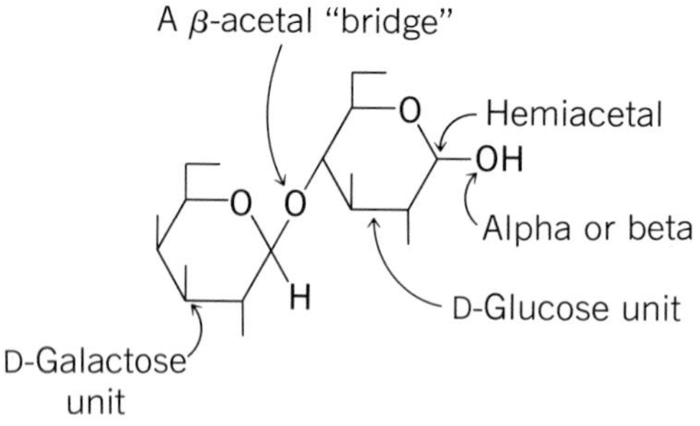

10.18 Lactose

Lactose or milk sugar occurs in the milk of mammals, 5 to 8% in human milk, 4 to 6% in cow's milk. It is obtained commercially as a by-product in the manufacture of cheese. Like maltose, it exists in alpha and beta forms; it is a reducing sugar; and it mutarotates. Its hydrolysis yields galactose and glucose, and these are joined $\beta(1 \rightarrow 4)$ from galactose to glucose.

10.19 Sucrose

The juice of sugar cane, which contains about 14% sucrose, is obtained by crushing and pressing the canes. It is then freed of proteinlike substances by the precipitating action of lime. Evaporation of the clear liquid leaves a semi-solid mass from which raw sugar is isolated by centrifugation. (The liquid that is removed is black-strap molasses.) Raw sugar (95% sucrose) is processed to remove odoriferous and colored contaminants. The resulting white sucrose, our table sugar, is probably the largest volume pure organic chemical produced. Much of our supply of sucrose now comes from sugar beets; so-called beet sugar and cane sugar are chemically identical. Structurally, a molecule of sucrose is derived from one glucose unit and one fructose unit:

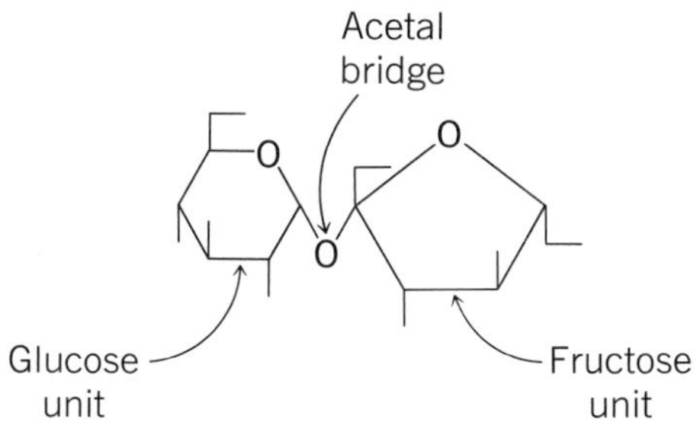

Acetal bridge

Glucose unit

Fructose unit

Sucrose

An acetal oxygen bridge links the two units. No hemiacetal group is present in the sucrose molecule. Sucrose is not a reducing sugar.

The 50:50 mixture of glucose and fructose that forms when sucrose is hydrolyzed is often called **invert sugar.** Honey, for example, consists of this sugar for the most part. The term "invert" comes from the change or inversion in sign of optical rotation that occurs when sucrose, $[\alpha]_D^{20} = +66.5°$, is converted into the 50:50 mixture, $[\alpha]_D^{20} = -19.9°$.

Polysaccharides

10.20 General Features

Much of the glucose that a plant makes by photosynthesis goes to its cell walls and its rigid fibers. Much is also stored for the food needs of the plant. Instead of being stored as free glucose, which is too soluble in water, most is converted to a much less soluble form, **starch.** This polymer of glucose is particularly abundant in plant seeds.

We use starch for food. During digestion we break it down to glucose, and what we do not need right away for energy we store. We do not normally excrete excess glucose but instead convert it to a starchlike polymer, **glycogen,** or to fat. In this section we concentrate on the three types of glucose polymers: starch, glycogen, and **cellulose.**

10.21 Starch

The skeletal outline of what is believed to constitute the basic structure of starch is shown in Figure 10.9. Starch is actually a mixture of polyglucose molecules, some linear and some branched. One type, **amylose,** consists of long, unbranched polymers of α-glucose with bridges just as in maltose, $\alpha(1 \rightarrow 4)$. The long polymer molecules coil into a helix (Figure 10.10). The other polymer of α-glucose present in starch is called **amylopectin.** In addition to the $\alpha(1 \rightarrow 4)$ glycosidic bridges, like those in amylose, amylopectin molecules also have $\alpha(1 \rightarrow 6)$ glycosidic links, bridges from carbon-1 at the very tip of one long amyloselike chain to carbon-6 at some point along another chain. Many such links occur to give numerous branches. Since amylopectin is branched, not coiled, its —OH groups are more exposed; therefore, amylopectin is somewhat more soluble in water than amylose. Natural starches are about 10 to 20% amylose and 80 to 90% amylopectin.

The acetal oxygen bridges linking glucose units together in starch are easily hydrolyzed, especially in the presence of acids or certain enzymes. We have

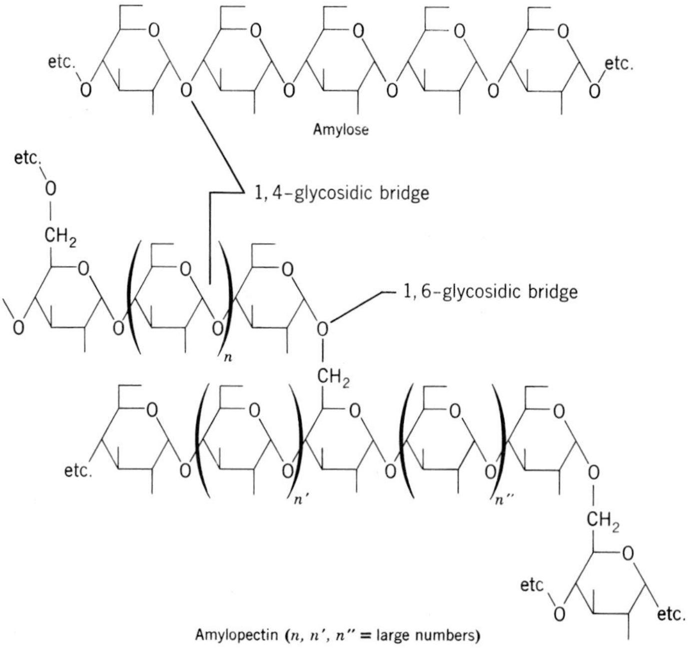

Figure 10.9 The glucose polymers in starch. Formula weights of various kinds of starch range from 50 000 to several million. (A formula weight of 1 million corresponds to about 6000 glucose units per molecule.)

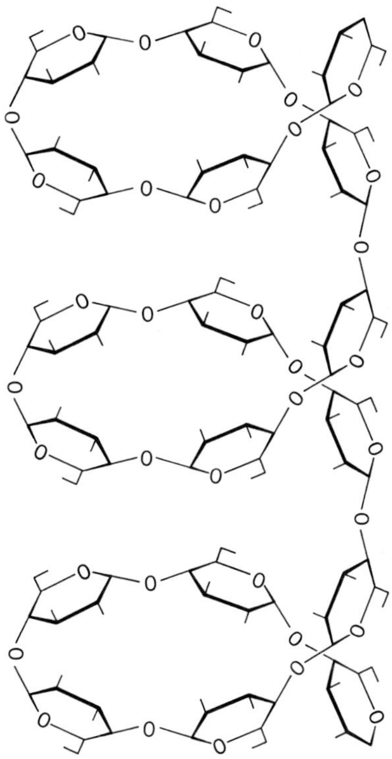

Figure 10.10 Amylose, coiled state.

learned to expect this behavior of acetals and, when we digest starch, digestive juices do nothing more than hydrolyze it, step by step as illustrated in Figure 10.11.

The partial breakdown products of amylopectin are still large molecules called the **dextrins.** They are used to prepare mucilage, pastes, and fabric sizes.

Starch is not a reducing carbohydrate. The potential aldehyde groups would be at the ends of chains only in percentages too low for detection by Benedict's reagent. Starch does, however, give an intense, brilliant blue-black color with iodine.[3] This **iodine test** for starch can detect extremely minute traces of starch in solution. The chemistry of the test is not definitely known, but iodine molecules are believed to become trapped within the vast network of starch molecules. Should this network disintegrate, as it does during the hydrolysis of starch, the test will fail.

10.22 Glycogen

Liver and muscle tissue are the main sites of glycogen production and storage in the body. Under the control of enzymes, some of which are in turn con-

[3] The starch-iodine reagent is iodine, I_2, dissolved in an aqueous solution of sodium iodide, NaI. Iodine, by itself, is only slightly soluble in water. When iodide ion is present, however, it combines with iodine molecules to form a triiodide ion, I_3^-, which liberates iodine easily on demand.

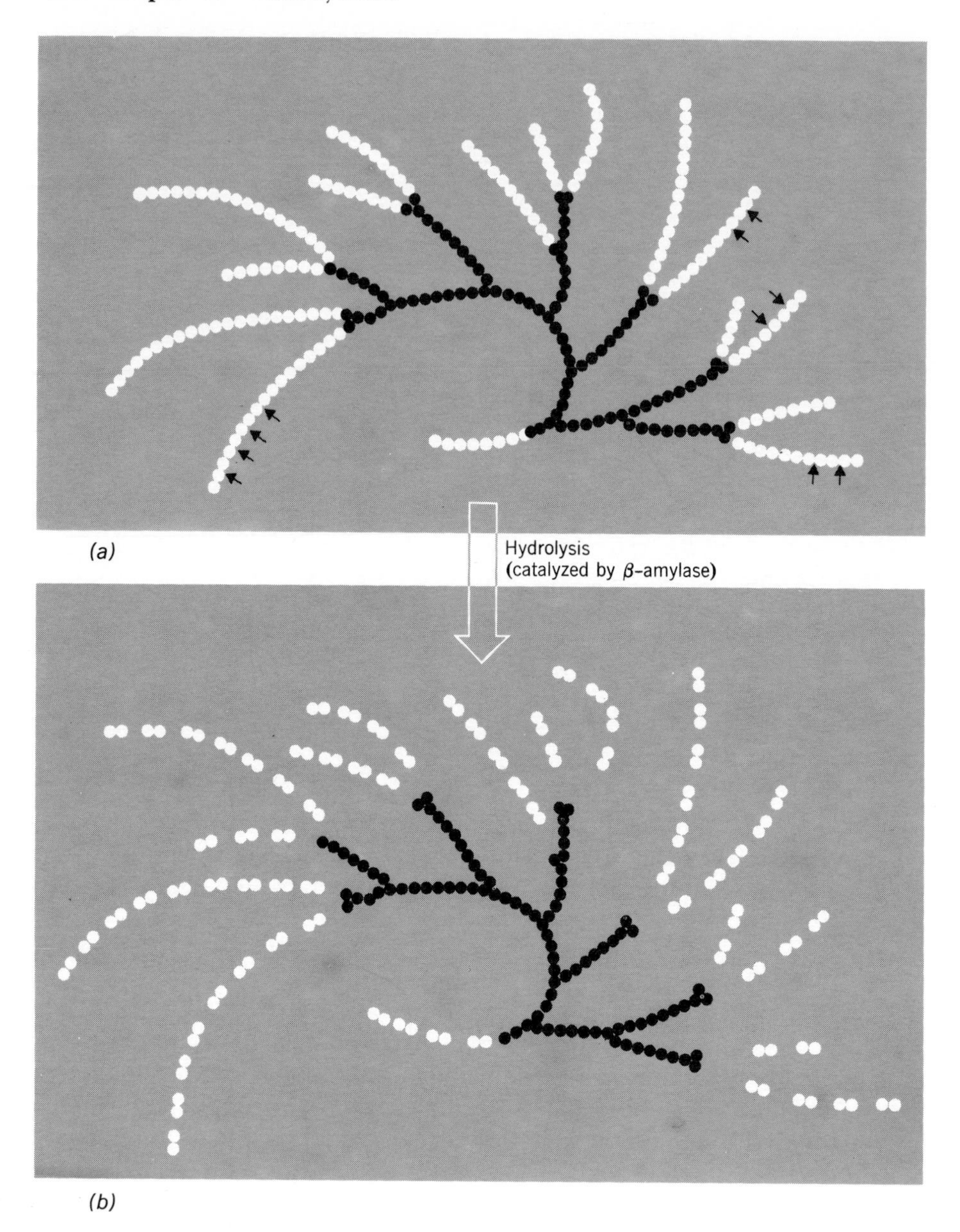

Figure 10.11 The hydrolysis of amylopectin. (*a*) Model of amylopectin. (*b*) Dextrins of medium formula weight (dark color) give purple to red colors with iodine. (After P. Bernfeld. *Advances in Enzymology*. Vol. 12 (1951), page 390.)

trolled by hormones, glucose molecules can be built into or mobilized from these glycogen reserves to store or supply chemical energy for the body.

Glycogen differs from starch by the apparent absence of any molecules of the unbranched, amylose type. It is branched very much like amylopectin, perhaps even more so. Molecular weights of various glycogen preparations have been reported over a range of 300 000 to 100 000 000, corresponding roughly to 1800 to 60 000 glucose units. During the digestion and absorption

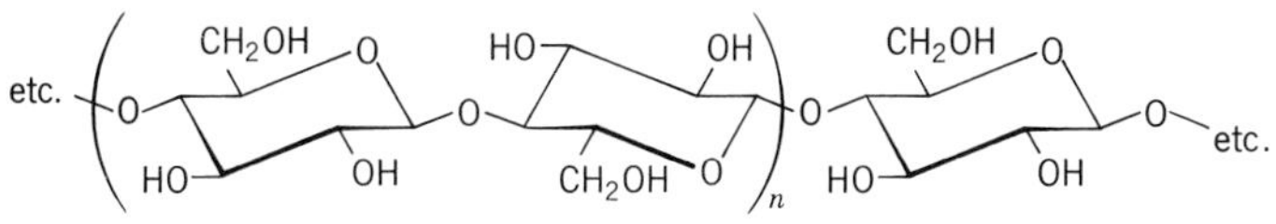

Figure 10.12 Cellulose, a linear polymer of β-glucose. Correct chair forms of the glucose units are used here to indicate that cellulose molecules are almost like long, flat ribbons. (Every other glucose unit is flipped over from our usual manner of drawing them.) In cotton the value of n ranges from 1000 to 4500 (or 2000 to 9000 glucose units). The cotton fiber gets its strength from the hydrogen bonds that can form as these relatively flat molecules nestle close together.

of a meal containing carbohydrates, the body builds up its glycogen deposits. Between meals or during fasts the deposits are made to deliver glucose.

10.23 Cellulose

Starch and glycogen are polymers of the alpha form of glucose. Cellulose is a polymer of the beta form. Its molecules are unbranched and resemble amylose; the bridges are all $\beta(1 \rightarrow 4)$. A portion of the structure of cellulose is shown in Figure 10.12.

The structural difference between amylose and cellulose is another of the "tremendous trifles" in nature. In humans amylose is digestible, but cellulose is not. Yet the only difference between them structurally is the orientation of the oxygen bridges. Starch has alpha bridges; cellulose, beta.

Brief Summary

Biochemistry. The study of the chemical inventory of living systems, how the parts are organized into the living cell and the principles of their interaction reveal that the dynamic organization is as important to the cell as its parts. Materials, energy, and information are the three rock-bottom essentials for life, and we obtain these from food, water, air, and hereditary materials passed from life to life. The chief organic substances in cells are carbohydrates, lipids, proteins, and nucleic acids.

Carbohydrates. Molecules of this major family of food and fiber chemicals are polyhydroxy aldehydes or ketones, acetals or ketals. None dissolves in nonpolar solvents. Those that cannot be hydrolyzed, the monosaccharides, give positive tests with Tollens' and Benedict's reagents. Monosaccharides of four or more carbons exist in cyclic hemiacetal (or hemiketal) forms (Haworth structures), mutarotate, and form sugar acetals called glycosides. Disaccharides are glycosides between two monosaccharide units, and polysaccharides are polyglycosides. The most nutritionally important carbohydrates are three monosaccharides, glucose, galactose, and fructose; three disaccharides, maltose, lactose, and sucrose; and one polysaccharide, starch, a mixture of amylose and amylopectin. The amylopectinlike glycogen is a storage form of glucose units in animals, whereas starch serves this function in plants. Cellulose, a polymer of β-glucose, is an important plant fiber.

Glucose. The single most abundant molecular unit on earth and the single most important in human metabolism is one of 16 optically isomeric aldohexoses. Natural glucose is dextrorotatory in all of its three forms, and is in the D-family. The α- and β-forms are cyclic hemiacetals with six-membered heterocyclic rings. Glucose exists in the free state in the bloodstream (and is often called "blood sugar"), but inside cells it exists as ions of one or another ester of phosphoric acid.

Galactose. The molecules of galactose differ from those of glucose only in the configuration of the —OH group at carbon-4. We obtain this aldohexose from the digestion (hydrolysis) of milk sugar.

Fructose. This reducing ketohexose differs from glucose by the location of the carbonyl group—being at C-2 in fructose and at C-1 in glucose. We obtain fructose in our diet from the digestion of sucrose as well as directly from certain fruits and honey.

Disaccharides. These are glycosides or sugar acetals. Maltose is made of two glucose units linked by an $\alpha(1 \rightarrow 4)$ glycosidic link. Lactose, a galactoside, joins a galactose ring by a $\beta(1 \rightarrow 4)$ bridge to a glucose unit. Sucrose joins C-1 of a glucose ring by an α-bridge to C-2 of fructose (in its five-membered ring form). The digestion of these disaccharides occurs by the following reactions, each catalyzed by a special enzyme.

$$\text{maltose} + H_2O \longrightarrow 2\ \text{glucose}$$

$$\text{lactose} + H_2O \longrightarrow \text{glucose} + \text{galactose}$$

$$\text{sucrose} + H_2O \longrightarrow \text{glucose} + \text{fructose}$$

Maltose and lactose retain hemiacetal systems in one ring. Therefore, each exists in α-, β-, and open forms; mutarotates; and is a reducing sugar. Sucrose has no potential aldehyde group—no hemiacetal—and therefore neither mutarotates nor is a reducing sugar.

Polysaccharides. Three important polysaccharides of glucose are starch (a plant product), glycogen (an animal product), and cellulose (a plant fiber). In molecules of each $(1 \rightarrow 4)$-glycosidic links occur: alpha bridges in starch and glycogen, and beta-bridges in cellulose. Polysaccharides are not reducing materials. Starch may be detected by a special reaction, the iodine test. Most starch molecules, those of the amylopectin fraction, as well as glycogen molecules, have additional 1,6-glycosidic links making the molecules highly branched. Humans lack enzymes for catalyzing the hydrolysis of beta-links in cellulose and cannot digest this material. Intermediate breakdown products from starch are called dextrins.

Absolute Configuration. In carbohydrate chemistry the D/L system is a commonly used convention for specifying absolute configuration. Plane projection structures of open-chain forms are made according to a set of rules. If the —OH group on the chiral carbon farthest away from the carbonyl group projects to the right, the substance is in the D-family; if to the left, the L-family. With very few exceptions the naturally occurring carbohydrates are in the D-family, and humans lack enzyme systems for doing anything with the enantiomeric L-family.

Questions and Exercises

1. Examine these structures and identify by letter(s) which fit(s) each label. If a particular label is not illustrated by the structures given, state so.

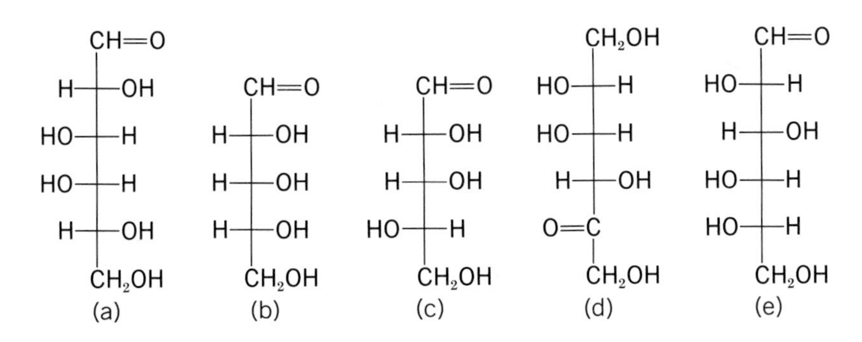

(1) Aldopentose(s)
(2) Aldohexose(s)
(3) Ketose(s)
(4) Any compound(s) in the D-family
(5) Any compound(s) in the L-family
(6) L-glucose
(7) D-Galactose
(8) D-Fructose

2. Practice writing the following structures (all in the D-family) until you can write them without any notes: (a) α-Glucose, (b) β-Glucose, (c) Open form of glucose (either the plane projection structure or the coiled form), (d) α-Maltose, (e) β-Maltose, and (f) Amylose.

3. Name the three nutritionally most important monosaccharides and give a source of each.

4. Name the three nutritionally most important disaccharides and give a source of each.

5. Name the polysaccharides that give only glucose when completely hydrolyzed.

6. How do amylose and cellulose differ, structurally?

7. What is the difference between amylopectin and dextrin?

8. Examine the following structure and answer the questions about it.

(a) Does it have a hemiacetal system? Where? (Draw an arrow to it or circle it.)
(b) Does it have an acetal system? Where? (Circle it.)
(c) Would this substance give a positive Benedict's test? Explain.
(d) In what specific structural way(s) does this compound differ from maltose?
(e) Write the structure(s) and name(s) of the products of the hydrolysis of this compound.

9. Trehalose is a disaccharide found in young mushrooms and yeast and is the chief carbohydrate in the hemolymph of certain insects. On the basis of its structural features answer the questions.

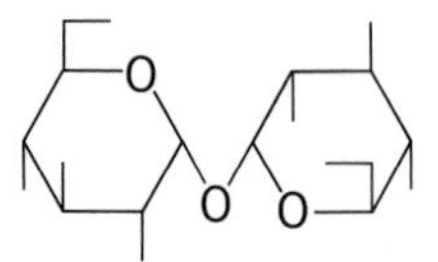

Trehalose

   (a) Is trehalose a reducing sugar? Explain.
   (b) Can trehalose mutarotate? Explain.
   (c) Identify by name only the product(s) of the hydrolysis of trehalose.

10. Maltose has a potential aldehyde group. Write the structure of maltose in which this group has changed to the open, free aldehyde system.

11. Repeat Exercise 10 for lactose.

12. Are D-glucose and D-galactose related as enantiomers or as diastereomers; or are they related in some other way?

13. What is the name of the enantiomer of D-ribose?

14. Write the plane projection structure of L-arabinose. (Obtain the plane projection structure of D-arabinose from Figure 10.5.)

chapter 11
Lipids

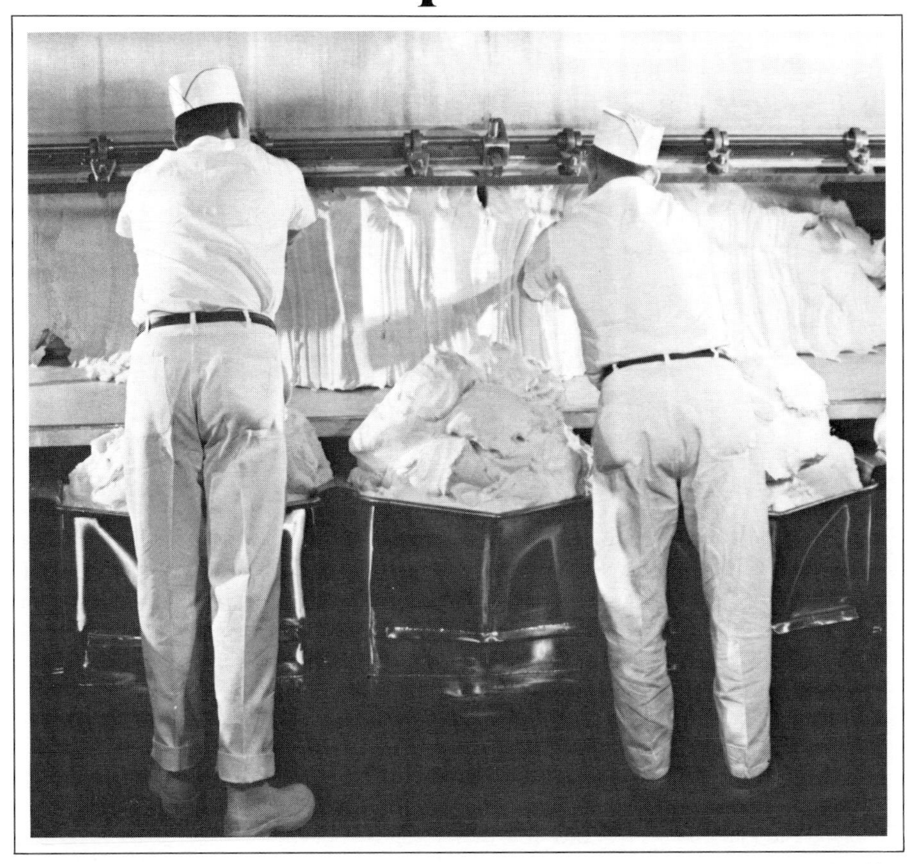

The carbon-hydrogen bonds in fats and oils are rich in chemical energy. When we take this energy in quantities greater than our needs we get chubby. Butter makers, seen here pulling new butter from a giant churn, make their contribution by making butter irresistibly flavorful. (USDA Photo.)

11.1 What Lipids Are

Lipids are defined by an experimental operation, not by molecular structure. The operation is extraction with a nonpolar solvent. When plant or animal material is crushed and ground with nonpolar solvents such as ether, benzene, chloroform, or carbon tetrachloride (fat solvents), whatever dissolves in the solvent is classified as **lipid.** Carbohydrates, proteins, water, ions, and very polar compounds do not dissolve. Depending on the plant or animal, the extracted material includes such a variety of compounds that we cannot define lipids in simple structural terms. Lipids include neutral fats containing only carbon, hydrogen, and oxygen; phosphorus-containing compounds called phospholipids; aliphatic alcohols; waxes; steroids; terpenes; and "derived lipids," substances resulting from partial or complete hydrolysis of some of the foregoing. Lipids, therefore, are plant and animal products that are soluble in ether or similar nonpolar solvents.

Although there are many different types, all lipid molecules are substantially hydrocarbonlike. One major class consists of those that can be saponified. Their molecules have one or more ester linkages, and the chief subclasses of the **saponifiable lipids** are waxes, neutral fats (triacylglycerols), phospholipids, and glycolipids. The principal **nonsaponifiable lipids** are the steroids and the terpenes, and we shall study the steroids.

11.2 Waxes

Waxes, esters of long-chain monohydric alcohols with long-chain carboxylic acids, are protective coatings on feathers, fur, skin, leaves, and fruits. In the esters of the glossy wax layer on tree leaves, alcohol portions and acid portions each have 26 to 34 carbons, making these esters almost totally hydrocarbonlike.

$$\mathrm{R{-}O{-}\overset{\displaystyle O}{\overset{\|}{C}}{-}R'}$$

Alcohol unit — Fatty acyl unit

Components of waxes

Beeswax is similar, with palmitic acid being the predominant acid portion. The lanolin (wool fat) prized for skin lotions is a mixture of esters involving steroid alcohols (discussed later) and long-chain carboxylic acids. A secretion of human skin, sebum, contains waxes as well as other lipids and serves to keep the skin supple.

Exercise 11.1 One particular ester in beeswax was hydrolyzed and the alcohol was found to have 26 carbons and the acyl unit had 28 carbons. Write the structure of this component of beeswax.

11.3 Triacylglycerols

The most abundant group of lipids are the **neutral fats,** called **triacylglycerols** or **triglycerides.**[1] Included in this group are lard, tallow and butterfat, all **animal fats;** and olive oil, cottonseed oil, corn coil, peanut oil, linseed oil, coconut oil and soybean oil, all **vegetable oils.** Their molecules consist of esters between glycerol and long-chain **fatty acids.** The three acid units in a typical triacylglycerol are not identical. They usually are from three different fatty acids.

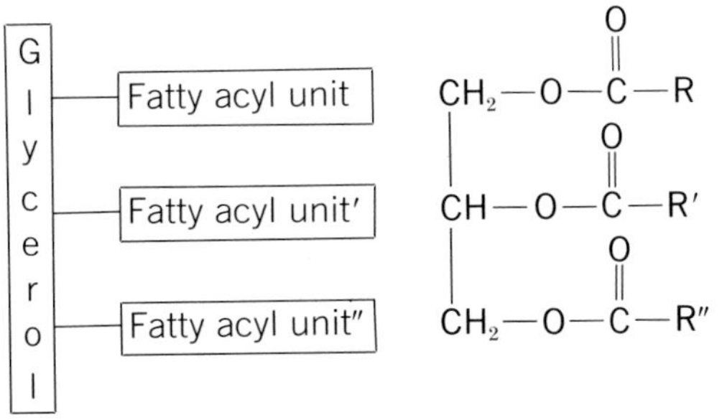

Glycerol		
	Fatty acyl unit	$CH_2-O-\overset{O}{\overset{\|\|}{C}}-R$
	Fatty acyl unit′	$CH-O-\overset{O}{\overset{\|\|}{C}}-R'$
	Fatty acyl unit″	$CH_2-O-\overset{O}{\overset{\|\|}{C}}-R''$

Components of neutral fats (triacylglycerols)

Fats and oils of whatever source are mixtures of different molecules. The differences are in specific fatty acids incorporated, not in the general structural features. Thus we cannot write the structure of cottonseed oil, for example. We are limited to describing a typical molecule such as the one shown in structure **1**. In a particular fat or oil, certain fatty acids tend to predominate, certain others are either absent or are present in trace amounts, and virtually all the molecules are triacylglycerols. Data for several fats and oils are in Table 11.1.

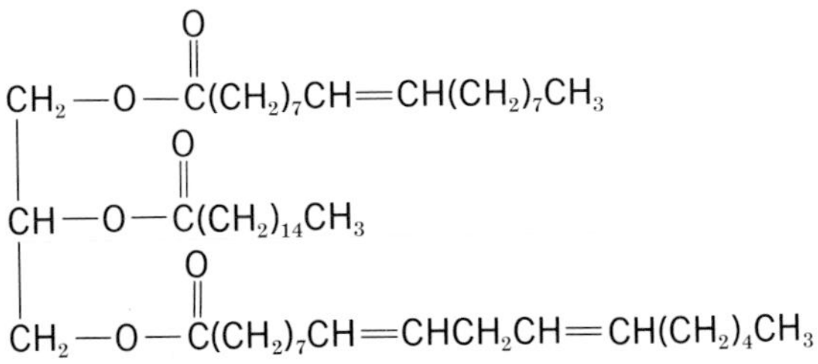

$$CH_2-O-\overset{O}{\overset{||}{C}}(CH_2)_7CH=CH(CH_2)_7CH_3$$
$$CH-O-\overset{O}{\overset{||}{C}}(CH_2)_{14}CH_3$$
$$CH_2-O-\overset{O}{\overset{||}{C}}(CH_2)_7CH=CHCH_2CH=CH(CH_2)_4CH_3$$

1

11.4 Fatty Acids

Fatty acids obtained from the lipids of most plants and animals tend to share the following characteristics.

[1] Until recently the most commonly used name was "triglyceride," but an international scientific commission has recommended that this chemically inaccurate term no longer be used. Nonetheless, its use no doubt will continue for some time; therefore, it should be learned.

TABLE 11.1

Composition of the Fatty Acids Obtained by Hydrolysis of Common Neutral Fats and Oils

				Average Composition of Fatty Acids (%)					
	Fat or Oil	Iodine Number	Saponification Value	Myristic Acid	Palmitic Acid	Stearic Acid	Oleic Acid	Linoleic Acid	Others
Animal fats	Butter	25–40	215–235	8–15	25–29	9–12	18–33	2–4	[a]
	Lard	45–70	193–200	1–2	25–30	12–18	48–60	6–12	[b]
	Beef tallow	30–45	190–200	2–5	24–34	15–30	35–45	1–3	[b]
Vegetable oils	Olive	75–95	185–200	0–1	5–15	1–4	67–84	8–12	
	Peanut	85–100	186–195	—	7–12	2–6	30–60	20–38	
	Corn	115–130	188–194	1–2	7–11	3–4	25–35	50–60	
	Cottonseed	100–117	191–196	1–2	18–25	1–2	17–38	45–55	
	Soybean	125–140	190–194	1–2	6–10	2–4	20–30	50–58	[c]
	Linseed	175–205	190–196	—	4–7	2–4	14–30	14–25	[d]
Marine oils	Whale	110–150	188–194	5–10	10–20	2–5	33–40		[e]
	Fish	120–180	185–195	6–8	10–25	1–3			[e]

[a] Three to four percent butyric acid, 1 to 2% caprylic acid, 2 to 3% capric acid, 2 to 5% lauric acid.
[b] One percent linolenic acid.
[c] Five to ten percent linolenic acid.
[d] Forty-five to sixty percent linolenic acid.
[e] Large amounts of other highly unsaturated fatty acids.

1. They are usually monocarboxylic acids, R-CO_2H.
2. The R- group is usually an unbranched chain.
3. The number of carbon atoms is almost always even.
4. The R-group may be saturated, or it may have one or more double bonds, which are *cis*.

The most abundant saturated fatty acids are palmitic acid, $CH_3(CH_2)_{14}CO_2H$, and stearic acid, $CH_3(CH_2)_{16}CO_2H$, having 16 and 18 carbons, respectively. Others are included in Table 7.1, the acids above acetic with an even number of carbons, but they are present in only small amounts.

The most frequently occurring unsaturated fatty acids are listed in Table 11.2. The 18-carbon skeleton of stearic acid is duplicated in oleic, linoleic, and linolenic acids. Oleic acid is the most abundant and most widely distributed fatty acid in nature.

Exercise 11.2 Write the structure of linoleic acid in a way that correctly shows the geometry at the alkene group—a *cis*-relation for the carbons joined to the double bonds.

The properties of the fatty acids are those to be expected of compounds having a carboxyl group, double bonds (in some), and long hydrocarbon chains. They are insoluble in water and soluble in nonpolar solvents. They can form salts, be esterified, and be reduced to the corresponding long-chain alcohols. Where present, alkene linkages react with bromine and take up hydrogen in the presence of a catalyst.

TABLE 11.2
Common Unsaturated Fatty Acids

Name	Number of Double Bonds	Total Number of Carbons	Structure	Mp (°C)
Palmitoleic acid	1	16	$CH_3(CH_2)_5CH{=}CH(CH_2)_7CO_2H$	32
Oleic acid	1	18	$CH_3(CH_2)_7CH{=}CH(CH_2)_7CO_2H$	4
Linoleic acid	2	18	$CH_3(CH_2)_4CH{=}CHCH_2CH{=}CH(CH_2)_7CO_2H$	−5
Linolenic acid	3	18	$CH_3CH_2CH{=}CHCH_2CH{=}CHCH_2CH{=}CH(CH_2)_7CO_2H$	−11
Arachidonic acid	4	20	$CH_3(CH_2)_4CH{=}CHCH_2CH{=}CHCH_2CH{=}CHCH_2CH{=}CH(CH_2)_3CO_2H$	−50

Chemical Properties of Triacylglycerols

11.5 Iodine Number

The degree of unsaturation in a lipid is measured by its iodine number. The **iodine number** is the number of grams of iodine that would add to the double bonds present in 100 g of the lipid if iodine itself could add to alkenes. The reagent used is I-Br, iodine bromide. The data are calculated as if iodine had been used. Saturated fatty acids, with no alkene linkages, have zero iodine numbers. Oleic acid has an iodine number of 90, linoleic acid 181, and linolenic acid 274. Animal fats have low iodine numbers while vegetable oils, the polyunsaturated oils of countless advertisements, have higher values as the data of Table 11.1 show. The iodine number of structure **1** is calculated to be 89, in the range of olive oil or peanut oil.

11.6 Hydrolysis in the Presence of Enzymes

Enzymes in the digestive tracts of human beings and animals act as efficient catalysts for the hydrolysis of the ester links of triacylglycerols.

In general,

$$
\begin{array}{l}
RC(=O){-}OCH_2 \\
\quad | \\
R'C(=O){-}OCH \\
\quad | \\
R''C(=O){-}OCH_2 \\
\text{Triacylglycerol}
\end{array}
+ 3H_2O \xrightarrow{\text{enzyme}}
\begin{array}{l}
RC(=O){-}OH \\
\quad + \\
R'C(=O){-}OH \\
\quad + \\
R''C(=O){-}OH \\
\text{Fatty Acids}
\end{array}
+
\begin{array}{l}
HOCH_2 \\
\quad | \\
HOCH \\
\quad | \\
HOCH_2 \\
\text{Glycerol}
\end{array}
$$

A specific example is

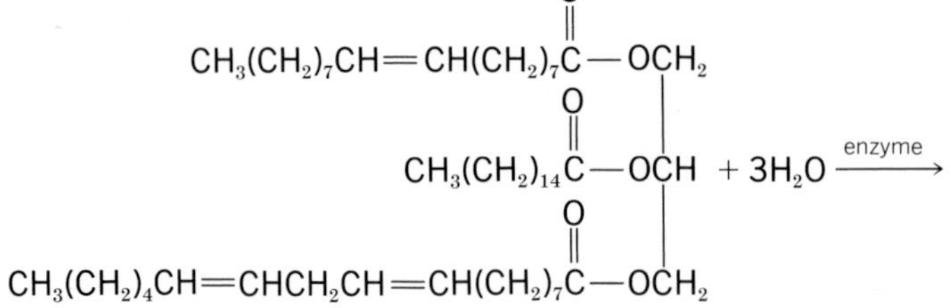

$$
\begin{array}{l}
CH_3(CH_2)_7CH{=}CH(CH_2)_7\overset{\overset{\displaystyle O}{\|}}{C}{-}OCH_2 \\
\qquad\qquad\qquad CH_3(CH_2)_{14}\overset{\overset{\displaystyle O}{\|}}{C}{-}OCH \; + 3H_2O \xrightarrow{\text{enzyme}} \\
CH_3(CH_2)_4CH{=}CHCH_2CH{=}CH(CH_2)_7\overset{\overset{\displaystyle O}{\|}}{C}{-}OCH_2 \\
\qquad\qquad\qquad \mathbf{1}
\end{array}
$$

$$
\begin{array}{r}
CH_3(CH_2)_7CH{=}CH(CH_2)_7\overset{\overset{\displaystyle O}{\|}}{C}OH \\
\text{Oleic acid} \\
+ \\
\text{Glycerol} + CH_3(CH_2)_{14}\overset{\overset{\displaystyle O}{\|}}{C}OH \\
\text{Palmitic acid} \\
+ \\
CH_3(CH_2)_4CH{=}CHCH_2CH{=}CH(CH_2)_7\overset{\overset{\displaystyle O}{\|}}{C}OH \\
\text{Linoleic acid}
\end{array}
$$

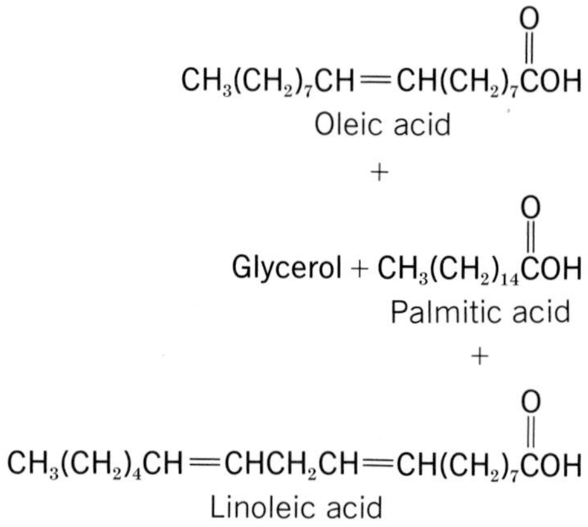

The enzyme-catalyzed hydrolysis is the only reaction neutral fats undergo during digestion.

11.7 Saponification

When the ester links in neutral fats are saponified by the action of a strong base (e.g., NaOH, KOH), glycerol plus the salts of the fatty acids are produced. These salts are soaps, and how they exert detergent action will be described later in this chapter.

In general,

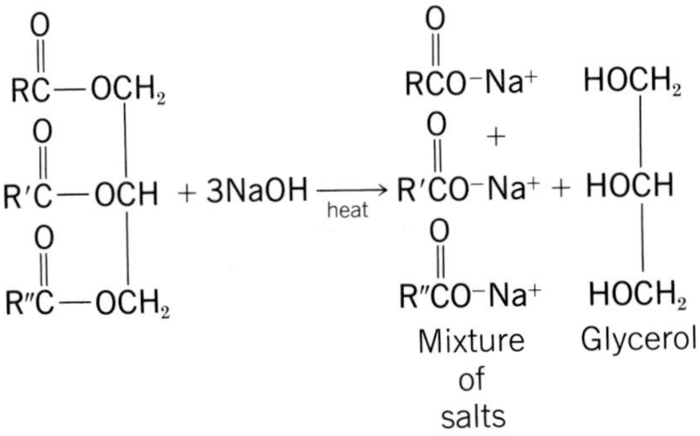

$$
\begin{array}{l}
R\overset{\overset{\displaystyle O}{\|}}{C}{-}OCH_2 \\
R'\overset{\overset{\displaystyle O}{\|}}{C}{-}OCH \; + 3NaOH \\
R''\overset{\overset{\displaystyle O}{\|}}{C}{-}OCH_2
\end{array}
\xrightarrow[\text{heat}]{}
\begin{array}{l}
R\overset{\overset{\displaystyle O}{\|}}{C}O^-Na^+ \\
\qquad + \\
R'\overset{\overset{\displaystyle O}{\|}}{C}O^-Na^+ \\
R''\overset{\overset{\displaystyle O}{\|}}{C}O^-Na^+ \\
\text{Mixture of salts}
\end{array}
+
\begin{array}{l}
HOCH_2 \\
\\
HOCH \\
HOCH_2 \\
\text{Glycerol}
\end{array}
$$

Exercise 11.3 Write a balanced equation for the saponification of **1** with sodium hydroxide.

11.8 Hydrogenation

If some of the double bonds in vegetable oils were hydrogenated, the oils would become like animal fats and would be solids at room temperature. Many people prefer solid to liquid shortening in baking, and the hydrogenation of vegetable oils increases the supply of lardlike materials. Complete hydrogenation to an iodine value of zero is not desirable, for the product would be as brittle and unpalatable as tallow. Manufacturers of commercial hydrogenated vegetable oils such as Crisco, Fluffo, Mixo, and Spry limit the degree of hydrogenation to leave some double bonds. The product has a lower iodine number (e.g., about 50 to 60), a higher degree of saturation, and a melting point that makes it a creamy solid at room temperature, similar to lard or butterfat. The oils of soybean and cottonseed, much more abundant than animal fats, are raw materials for hydrogenated products. If just one molecule of hydrogen were added to the molecule of structure **1**, the iodine number would drop from 89 to 59, from the olive or peanut-oil range to that of lard. The peanut oil in popular brands of peanut butter has been partially hydrogenated.

Oleomargarine is made from hydrogenated, carefully selected, and highly refined oils and fats. One goal is to produce a final product that will melt readily on the tongue, a feature that makes butter so desirable.

Exercise 11.4 Write the balanced equation for the complete hydrogenation of all alkene links in **1**.

11.9 Rancidity

When fats and oils are left exposed to warm, moist air for any length of time, they become **rancid,** meaning that they develop disagreeable flavors and odors. Two kinds of reactions are chiefly responsible: the hydrolysis of ester links and the oxidation of double bonds.

The hydrolysis of butterfat would produce a variety of relatively volatile and odorous fatty acids, as the data in footnote *a* of Table 11.1 indicate. Water for such hydrolysis is present in butter, and airborne bacteria furnish enzymes.

If the reaction producing rancidity is oxidation, attack by atmospheric oxygen occurs at unsaturated side chains in triacylglycerols. Eventually, short-chain and volatile carboxylic acids and aldehydes form. Both types of substances have extremely disagreeable odors and flavors.

11.10 Phospholipids

Phospholipids are esters either of glycerol or of sphingosine, a long-chain dihydric amino alcohol with one double bond. Molecules of the glycerol-based

$$\underset{\text{Sphingosine}}{CH_3(CH_2)_{12}CH{=}CH\underset{\substack{|\\ OH}}{C}H{-}\underset{\substack{|\\ NH_2}}{C}H{-}CH_2{-}OH}$$

phospholipids have two ester bonds from glycerol to fatty acids and one ester bond to phosphoric acid, which in turn is joined as a phosphate ester to some small alcohol molecule. Without this small alcohol the substance is phosphatidic acid and with the alcohol, a phosphoglyceride.

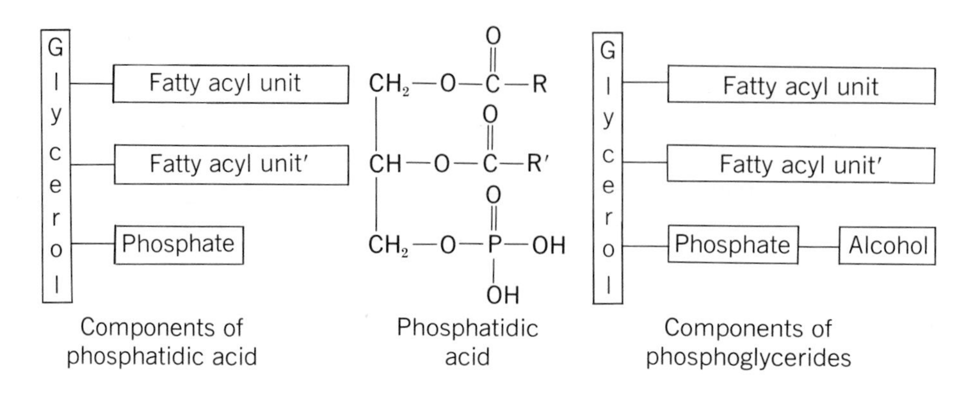

Components of phosphatidic acid — Phosphatidic acid — Components of phosphoglycerides

11.11 Phosphoglycerides

The three principal **phosphoglycerides** incorporate into their molecules units of choline, ethanolamine, or serine to give phosphatidylcholine (lecithin),

$HOCH_2CH_2\overset{+}{N}(CH_3)_3$ — Choline (a cation)

$HOCH_2CH_2NH_2$ — Ethanolamine

$HOCH_2CH(NH_2)CO_2H$ — Serine (an amino acid)

2, phosphatidylethanolamine, **3**, and phosphatidylserine, **4**. As the structures of **2** – **4** show, one part of each phosphoglyceride molecule is very polar, having electrically charged sites. The remainder is nonpolar and hydrocarbonlike. These characteristics have important implications in understanding how phosphoglycerides are used to make cell membranes (discussed later).

$$\begin{array}{l} CH_2OC(=O)R \\ | \\ CHOC(=O)R' \\ | \\ CH_2OP(=O)(O^-)OCH_2CH_2\overset{+}{N}(CH_3)_3 \end{array}$$

2
Phosphatidylcholine (lecithin)

$$\begin{array}{l} CH_2OC(=O)R \\ | \\ CHOC(=O)R' \\ | \\ CH_2OP(=O)(O^-)OCH_2CH_2\overset{+}{N}H_3 \end{array}$$

3
Phosphatidylethanolamine

$$\begin{array}{l} CH_2OC(=O)R \\ | \\ CHOC(=O)R' \\ | \\ CH_2OP(=O)(O^-)OCH_2CH(NH_3^+)CO_2H \end{array}$$

4
Phosphatidylserine

11.12 Plasmalogens

Plasmalogens, another family of glycerol-based phospholipids, are widely distributed in the membranes of nerve cells and muscle cells. They differ from the other phosphoglycerides in having an unsaturated ether link instead of an acyl unit at one end of the glycerol unit.

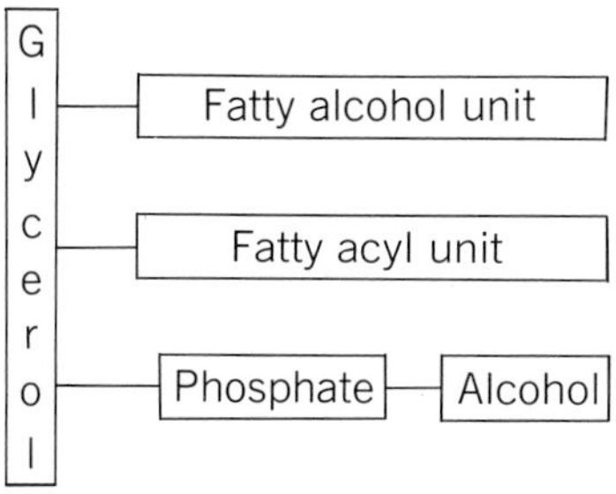

Glycerol — Fatty alcohol unit; Fatty acyl unit; Phosphate — Alcohol

Components of plasmalogens

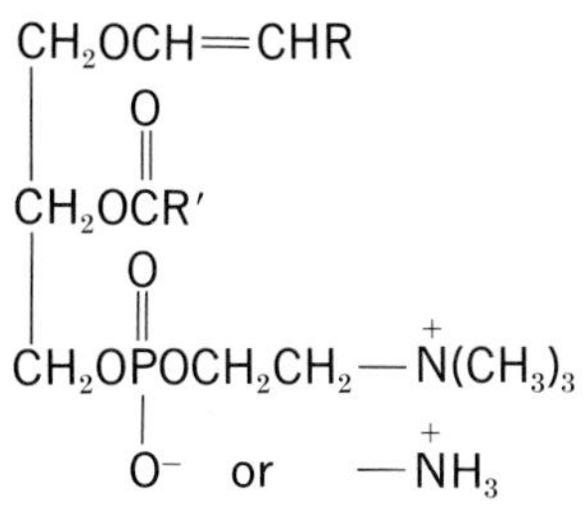

$$\begin{array}{l} CH_2OCH{=}CHR \\ | \\ CH_2OC(=O)R' \\ | \\ CH_2OP(=O)(O^-)OCH_2CH_2{-}\overset{+}{N}(CH_3)_3 \ \text{or} \ {-}\overset{+}{N}H_3 \end{array}$$

Plasmalogens

11.13 Sphingolipids

The two types of lipids based on sphingosine, the sphingomyelins and the cerebrosides, are also important constituents of cell membranes. The cerebrosides are not actually phospholipids but instead are **glycolipids,** lipids incorporating a sugar ("glycose") unit and having no phosphate system. The sugar units in glycolipids are usually D-galactose or D-glucose or derivatives of these (e.g., amino sugars).

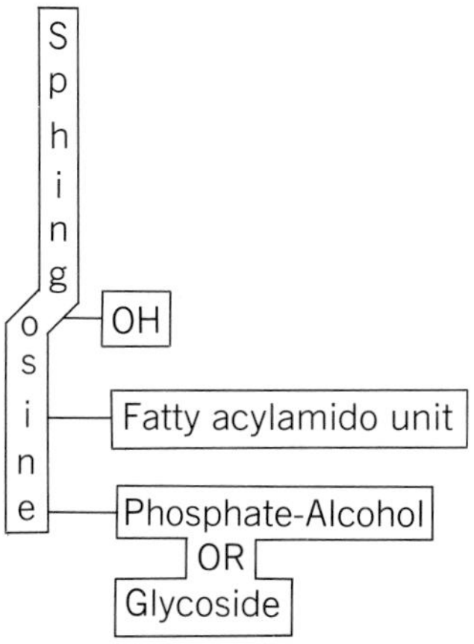

Sphingosine — OH; Fatty acylamido unit; Phosphate-Alcohol OR Glycoside

Components of sphingolipids

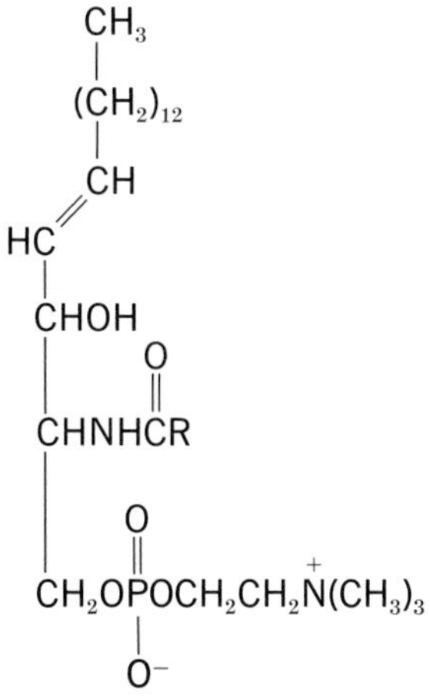

$$\begin{array}{l} CH_3 \\ | \\ (CH_2)_{12} \\ | \\ CH{=}CH \\ | \\ CHOH \\ | \\ CHNHC(=O)R \\ | \\ CH_2OP(=O)(O^-)OCH_2CH_2\overset{+}{N}(CH_3)_3 \end{array}$$

Sphingomyelin

$$\begin{array}{l} CH_3 \\ | \\ (CH_2)_{12} \\ | \\ CH{=}CH \\ | \\ CHOH \\ | \\ CHNHC(=O)R \\ | \\ CH_2O{-}\text{(sugar ring: OH, OH, OH, HO, O)} \end{array}$$

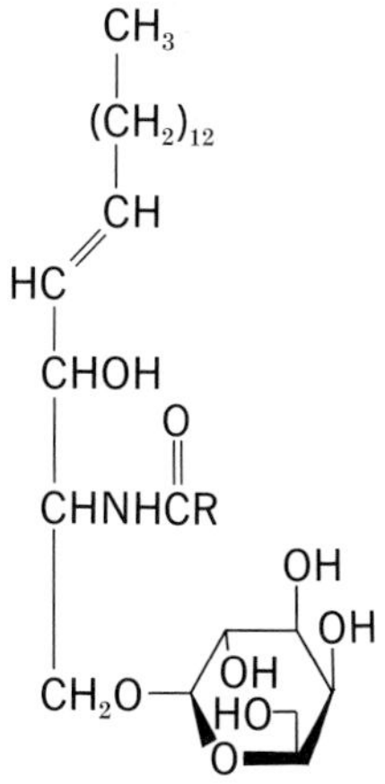

Cerebrosides

The acyl units in the acylamido parts of the **sphingolipids** are not the usual fatty acids found in neutral fats. A 24-carbon acid, lignoceric acid, is incorporated into sphingomyelin and into one of the cerebrosides, kerasin.

Other cerebrosides contain 24-carbon acyl units. Phrenosin has cerebronic acid; nervon, nervonic acid; and oxynervon, oxynervonic acid.

$CH_3(CH_2)_{22}CO_2H$ — Lignoceric acid

$CH_3(CH_2)_{21}CH(OH)CO_2H$ — Cerebronic acid

$CH_3(CH_2)_7CH{=}CH(CH_2)_{13}CO_2H$ — Nervonic acid

$CH_3(CH_2)_7CH{=}CH(CH_2)_{12}CH(OH)CO_2H$ — Oxynervonic acid

11.14 Steroids

Steroids are high-formula-weight aliphatic compounds whose molecules include the characteristic steroid nucleus, four rings fused: three 6-membered

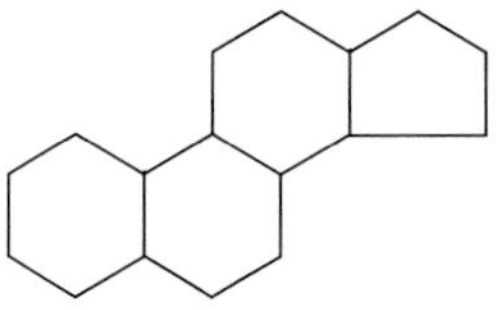

Steroid nucleus **5**

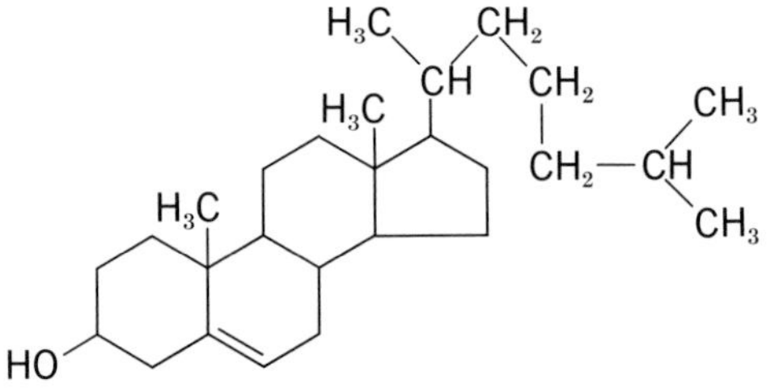

Cholesterol (Greek: *chole,* bile; *stereos,* solid; -ol, alcohol)

rings and one 5-membered ring as seen in structure **5**. Several steroids are very active, physiologically, and their effects vary widely, ranging from vitamin activity to the action of sex hormones. One stimulates the heart; another ruptures red blood cells. Table 11.3 lists several and their functions.

11.15 Cholesterol

Cholesterol is an unsaturated steroid alcohol or sterol that makes up part of the membranes of certain cells and is the body's raw material for making the bile salts and several steroid hormones, including the sex hormones in Table 11.3. Cholesterol is also the chief constituent of gallstones. It is routinely made in the liver from acetate units by a number of steps. Cholesterol in the diet leads to some suppression of this synthesis. Cholesterol biosynthesis is also inhibited when you go without food, but it is accelerated when your diet includes relatively large amounts of animal fat, such as butterfat or fat from juicy, well-marbled, premium grade roasts and steaks. A number of mechanisms work to control how much cholesterol is made and how it is transported, but few are yet well understood. When cholesterol deposits in blood vessels, atherosclerosis exists. As these deposits slowly accumulate, the heart must work harder to keep sufficient blood flow going, an extra strain that can

TABLE 11.3
Important Individual Steroids

Vitamin D_2 precursor

Irradiation of this hormone, the commonest of all plant hormones, by ultraviolet light opens one of the rings to produce vitamin D_2.

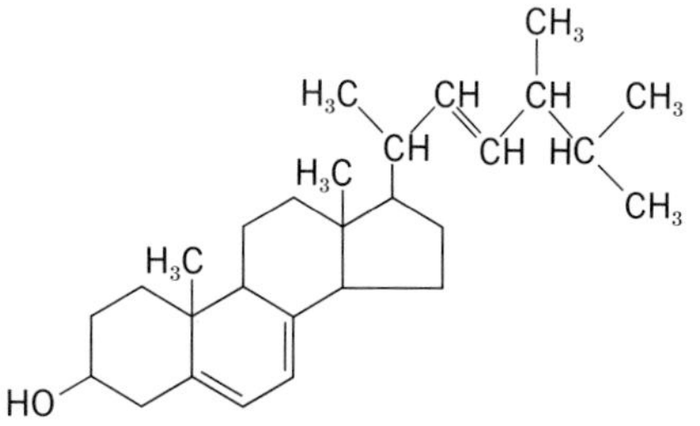

Ergosterol

Ultraviolet light

A deficiency of this antirachitic factor causes rickets, an infant and childhood disease characterized by faulty deposition of calcium phosphate and poor bone growth.

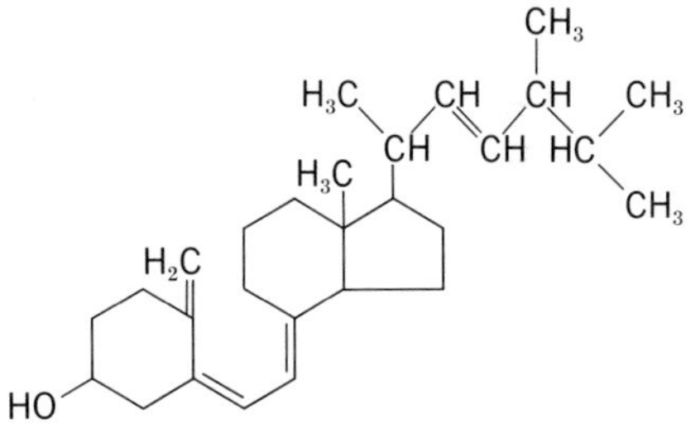

Vitamin D_2

Bile acid

Cholic acid is found in bile in the form of its sodium salt. This and closely related salts are the bile salts that act as powerful emulsifiers of lipid material awaiting digestion in the upper intestinal tract. The sodium salt of cholic acid is soaplike because it has a very polar head and a large hydrocarbon tail.

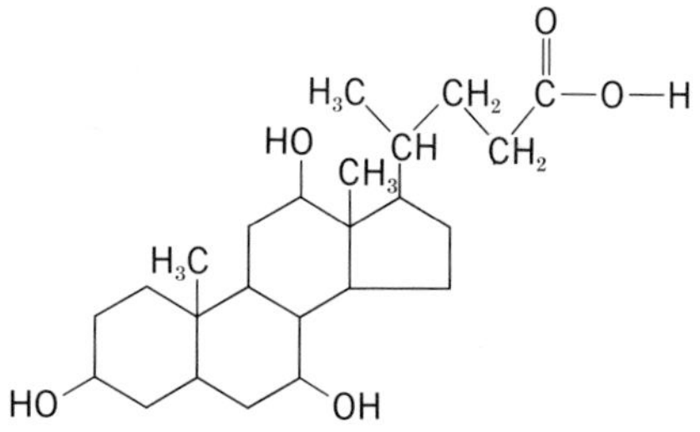

Cholic acid

Antiarthritic compound

One of the 28 adrenal cortical hormones, cortisone is not only important in the control of carbohydrate metabolism but also effective in relieving the symptoms of rheumatoid arthritis.

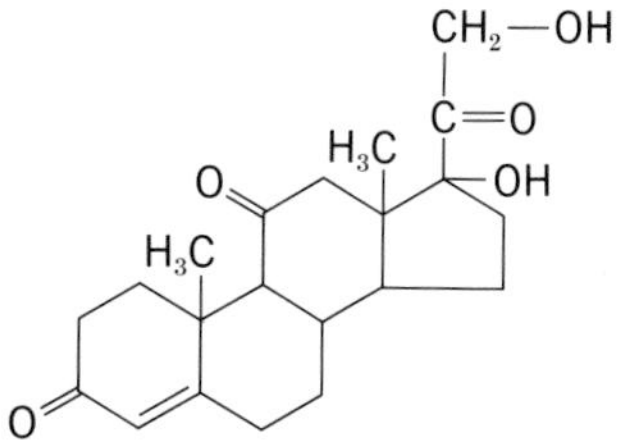

Cortisone

TABLE 11.3 (continued)

Cardiac aglycone	
Digitoxigenin is found in many poisonous plants, notably digitalis, as a complex glycoside. In small doses it stimulates the vagus mechanism and increases heart tone. In larger doses it is a potent poison.	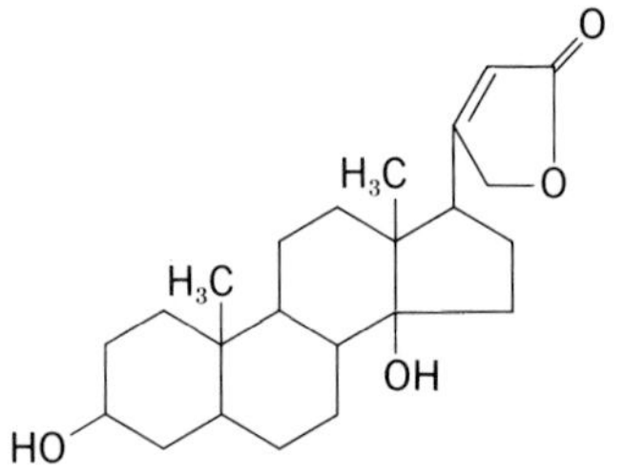 H_3C H_3C OH HO O O Digitoxigenin
Sex hormones	
This is one human estrogenic hormone.	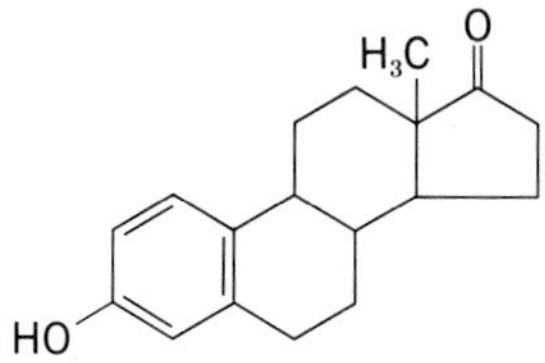 H_3C O HO Estrone
This human pregnancy hormone is secreted by the corpus luteum.	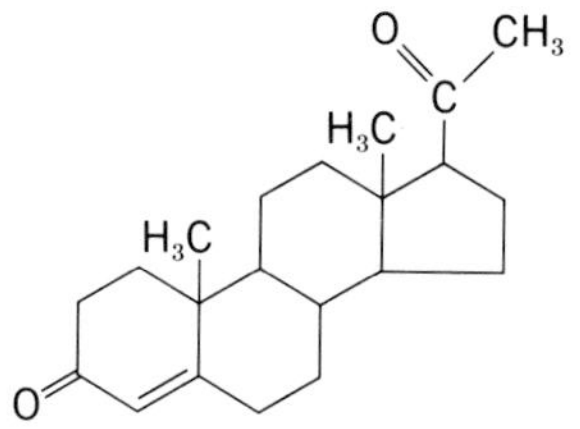 O CH_3 C H_3C H_3C O Progesterone
This male sex hormone regulates the development of reproductive organs and secondary sex characteristics.	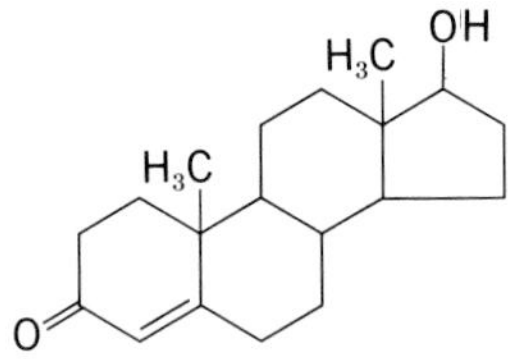 H_3C OH H_3C O Testosterone
Androsterone, a second male sex hormone, is less potent than testosterone.	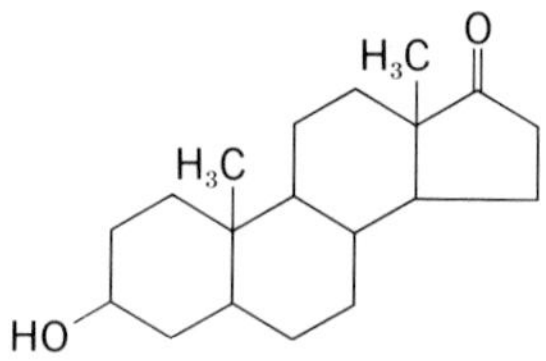 H_3C O H_3C HO Androsterone

TABLE **11.3 (continued)**

Synthetic hormones in fertility control	
Most oral contraceptive pills contain one or two synthetic, hormonelike compounds. (Synthetics must be used because the real hormones are broken down in the body.)	Synthetic Estrogens 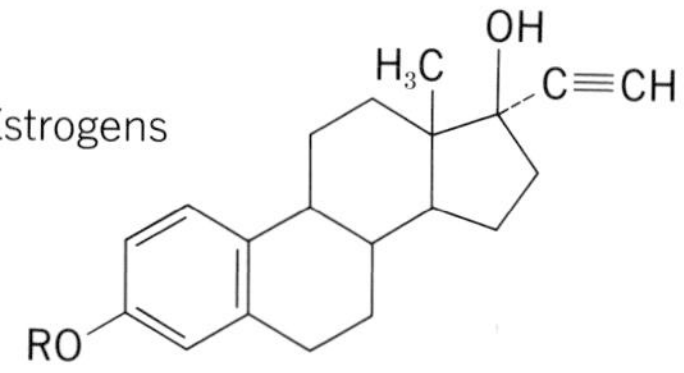 If R = H, Ethynylestradiol R = CH_3, Mestranol
The most widely used pill is a combination of an estrogen (not more than 30 micrograms) and a progestin (up to 4 milligrams).[a]	Synthetic Progestins 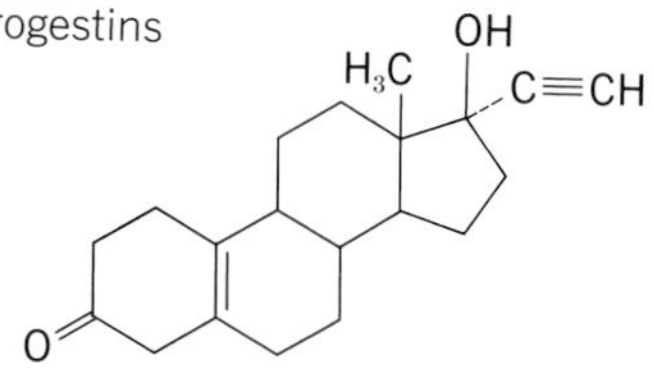 Norethynodrel 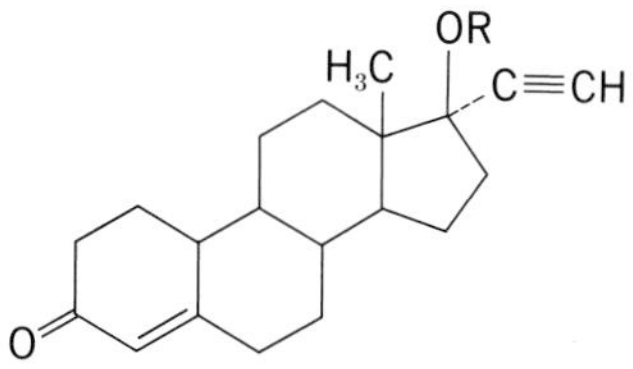 If R = H, Norethindrone $R = \overset{O}{\overset{\|}{C}}-CH_3$, Norethindrone acetate 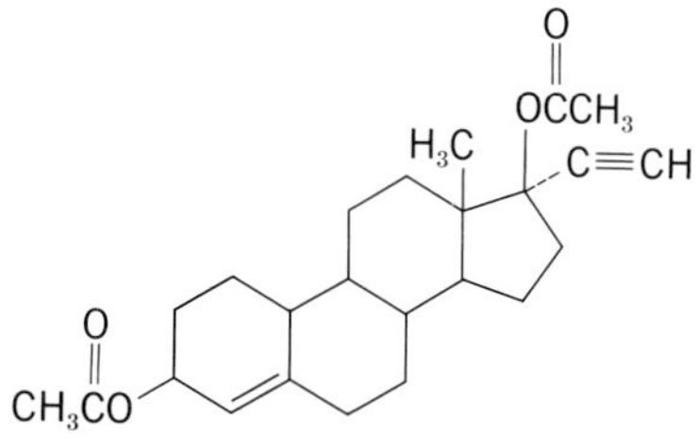 Ethynodiol Diacetate

[a] In October 1976 the U.S. Food and Drug Administration issued a warning that progestins must not be taken during pregnancy because they can cause severe birth defects. Prior to this the estrogens had caused concern that their use might endanger the health of users.

lead to a heart attack. Some evidence implicates a diet high in animal fats to heart disease.

11.16 Biological Membranes

Biological membranes are structures that hold individual cells together or that hold together organelles, small bodies inside cells such as the nucleus. These membranes are not simply passive bags for holding chemicals. They are actively involved in the life of the cell. The membrane has sites that recognize particular chemicals such as hormones, and the arrival of a chemical stimulus induces the cell membrane to engage in an important chemical activity. That activity may be changes that let particular nutrient chemicals get inside the cell, or let a chemical made by the cell and needed elsewhere get out and into circulation. Membranes, in other words, control the flow of both information and materials to and from the cell. Some membranes are the actual sites of syntheses of chemicals needed by the organism.

Cell membranes are made of both lipids and proteins. The principal lipids are the phospholipids, glycolipids, and cholesterol although all three are not present in all membranes. Molecules of each of these lipids have parts that are either polar or electrically charged and therefore are attracted to water molecules and attract water molecules in turn. These molecular parts are called **hydrophilic groups** (*hydro,* water; *philos,* loving—"water loving"). In phospholipids the hydrophilic group is the phosphate ester, partly ionized, and the attached alcohol, which often includes a substituted ammonium ion. In glycolipids, the hydrophilic group is a sugar unit such as a glucose or galactose system. In cholesterol, the hydrophilic unit—the –OH group—is only a tiny part of the molecule.

Membrane lipids all have relatively large **hydrophobic** parts (*hydro,* water; *phobic,* hating—"water hating"), the hydrocarbon systems. These parts are not attracted to water and cannot attract water molecules or break into the hydrogen-bonding network of water. In a sense, the best they can do in an aqueous medium is avoid water as much as they can by "dissolving" in each other, intermingling almost as if they formed a hydrocarbon solution.

11.17 The Lipid Bilayer of Membranes

When phospholipids or glycolipids are mixed with water, they spontaneously form a **lipid bilayer,** a sheetlike array consisting of two layers of lipid molecules aligned side by side as illustrated in Figure 11.1. Considerable evidence exists that all cellular membranes include lipid bilayers. The hydrophobic "tails" of lipid molecules intermingle in the center of the bilayer as much away from the water phase as possible. The hydrophilic "heads" stick out into the aqueous phase, where they are solvated.

If a pin were stuck through the bilayer and removed, the layer would close back spontaneously. No covalent bonds hold neighboring lipid molecules to each other. Only the net forces of attraction that we imply when we use the terms *hydrophobic* and *hydrophilic* are at work. Yet the bilayer is strong enough to be a membrane; it is highly impermeable to ions and polar molecules, with

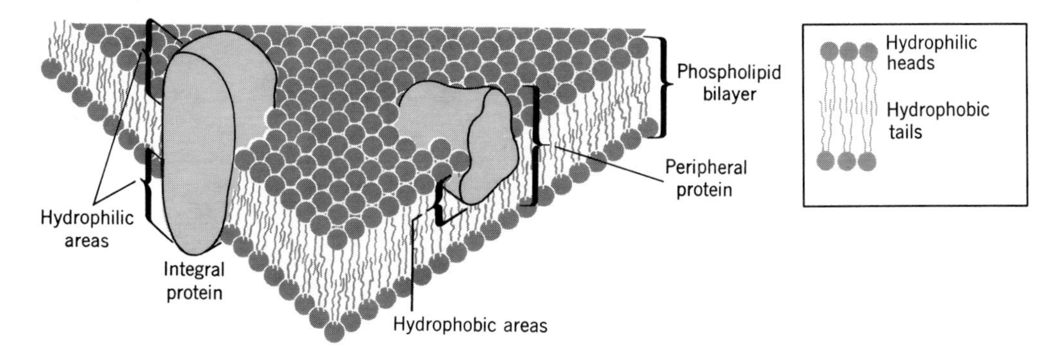

Figure 11.1 Cell membrane—fluid mosaic model.

the enormously important exception of water. Water molecules move back and forth easily.

The proteins of cell membranes control what gets in and out, and they are the sites of all active operations of the membrane: gates, pumps, enzymes, energy converters, and receptors for stimuli. Proteins are polymers whose molecules are polyamides of high formula weight. Details beyond that, to be studied in the next chapter, are not essential here. Protein units lie on the surfaces of some membranes but are integral parts of others as seen in Figure 11.1.

The two surfaces of a membrane, one on the outside of the cell and the other inside, are not identical. When glycolipids are used to make the bilayer, the sugar unit evidently is always on the outer surface of the cell. Moreover, the proteins of the membrane are themselves oriented in special ways peculiar to the cell.

11.18 Active Transport and the Bilayer

Membrane proteins involved in moving polar molecules and ions through the bilayer are evidently oriented. One kind of protein actively moves sodium ions to the outside of a cell and potassium ions to the inside of the cell. These movements run counter to natural ionic **concentration gradients;** that is, counter to the changes in concentration of each ion along the path from the inside to the outside of the cell.

Ions will spontaneously diffuse from a region where their concentration is high to one where their concentration is low until the concentration is the same all over. Once that happens, equilibrium is reached and one will never see the original concentration gradient spontaneously restored. To recreate that original gradient, you would have to do work—use energy to evaporate the solution and carefully redissolve the solute to the specifications of the gradient. That kind of work is done by proteins, energized by different reactions, in cell membranes. Proteins maintain concentration gradients.

Outside cells, the sodium ion concentration is relatively high, and between the outside and the inside is a concentration gradient. Sodium ions that leak along the gradient into the cell are moved back out again by energy-consuming chemical events involving some of the proteins in and on the cell membrane. Coupled to this is the movement of potassium ions from outside the cell to the inside. The movement of any solute against a concentration gradient is called **active transport;** the molecular system responsible for that counter-gradient movement is often called simply a "pump." These pumps are integral parts of cell membranes, and their operation requires energy from metabolism.

11.19 Detergents

Detergents are surface-active agents, substances that lower the surface tension of water. Water alone is a poor cleansing agent. Its molecules are so polar that they stick to each other (via hydrogen bonds) rather than penetrate into a nonpolar region such as a film of grease on soiled hands or clothing. The best detergents have molecules with strongly hydrophilic "heads" and hydrophobic "tails," such as illustrated in Figure 11.2. Ordinary **soap** consists of a mixture of sodium (or potassium) salts of long-chain fatty acids obtained by the saponification of neutral fats. Because the negative ions of soap form precipitates with metallic ions in hard water (Ca^{2+}, Mg^{2+} or Fe^{2+}/Fe^{3+}), synthetic detergents (**"syndets"**) were developed that were fully soluble in such water. How detergents work to loosen greasy films is illustrated in Figure 11.3. Bile salts (see cholic acid, Table 11.3) are important detergents we make and release into the digestive tract to wash dietary lipids off other food particles and help digest the lipids themselves.

11.20 Biodegradability

Because even traces of detergents can produce suds, we do not want them in our drinking water supply. The negative ions in ordinary soap are destroyed by bacteria in the soil or in sewage treatment plants. These bacteria metabolize such ions. We say that soap is **biodegradable,** because biological organisms degrade it. Syndets originally developed to replace soap were not biodegradable. The structure of one, given in Table 11.4, has several alkyl

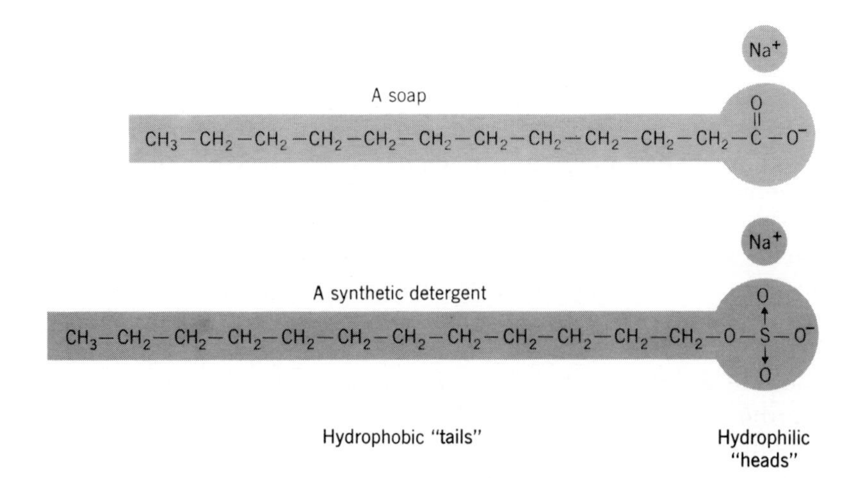

Figure 11.2 Two organic salts with detergent properties.

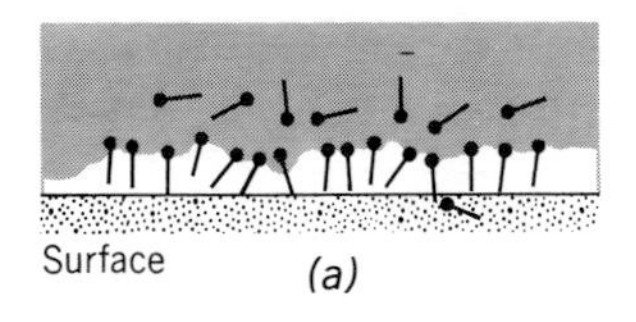

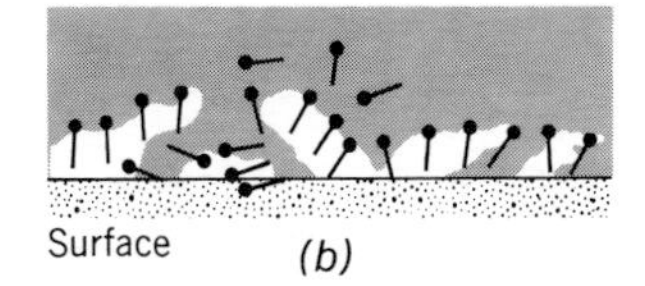

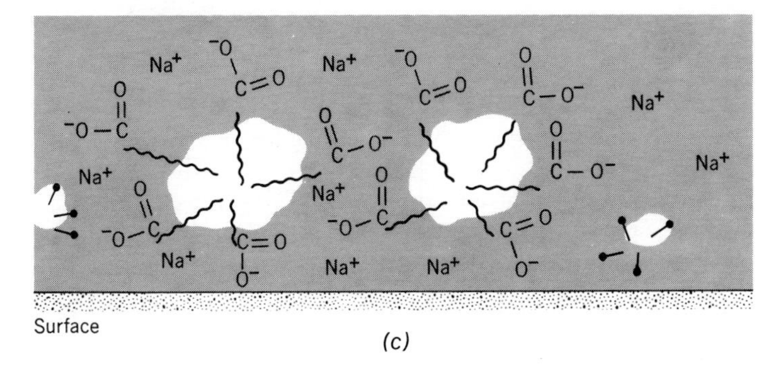

Figure 11.3 How detergents work. (*a*) Hydrophobic tails of detergent molecules work into the hydrophobic environment of the greasy layer. (*b*) Hydrophilic heads of the detergent molecules, remaining in contact with the water phase, help work the greasy layer loose. This process goes faster if the water phase is hot and mechanical agitation is used (as in washing machines). (*c*) Globules of greasy material form, pincushioned by detergent molecules. The "head" of the pins all bear electrical charges, which helps keep the globules from rejoining and also keeps them in solution.

TABLE **11.4**
Biodegradability of Some Synthetic Detergents

Types of Synthetic Detergents	Examples to Illustrate Structural Features	Biodegradable?
Sodium alkyl sulfate	$CH_3(CH_2)_{10}CH_2OSO_2O^-Na^+$	Yes
Sodium alkylbenzene sulfonate, tetrapropylene-based	$CH_3CH(CH_3)CH_2CH(CH_3)CH_2CH(CH_3)CH_2—CH(CH_3)—C_6H_4—S(\rightarrow O)_2—O^-Na^+$	No
Sodium alkylbenzene sulfonate, *n*-paraffin or α-olefin-based	$CH_3(CH_2)_nCH(CH_3)—C_6H_4—S(\rightarrow O)_2—O^-Na^+$ (n = 7 − 11)	Largely so
Sodium alkane sulfonate	$C_nH_{2n+1}SO_3^-Na^+$ (Chain is largely straight; n is 15 to 18.)	Almost completely

branches. Bacteria evidently cannot work their way around these branches. The most biodegradable detergents have straight-chain hydrophobic "tails."

Brief Summary

Lipids. Ether-extractable substances in animals and plants include saponifiable esters and nonsaponifiable compounds. The esters are generally of glycerol or sphingosine with acyl portions contributed by long-chain carboxylic acids. Because molecules of all lipids are mostly hydrocarbonlike, they are soluble in nonpolar solvents but not in water.

Fatty Acids. The carboxylic acids released when saponifiable lipids are hydrolyzed generally have long chains of even numbers of carbons; seldom are branched; and often have one or more alkene groups. The latter are generally *cis*. The sodium and potassium salts of these acids are surface-active agents and are used as soap.

Waxes. Molecules of waxy coatings on leaves and fruit, or in beeswax or sebum, are simple esters between long-chain monohydric alcohols and fatty acids.

Triacylglycerols. Molecules of neutral fats, those without electrically charged sites or those that are similarly polar, are esters of glycerol and a variety of fatty acids. The degree of unsaturation may be determined by measuring the iodine number of the substance. Vegetable oils generally have higher iodine numbers than animal fats and are more unsaturated (have more alkene double bonds per molecule). The triacylglycerols (or triglycerides) may be hydrogenated, hydrolyzed (digested), and saponified.

Phosphoglycerides. Molecules of these phospholipids are also esters of glycerol, but instead of three fatty acyl units they have two, which provide hydrophobic parts to the molecules. The third ester is to phosphoric acid, which on its other side is engaged in a phosphate ester link to a small alcohol. The alcohol may have a positively charged aminelike group as well, making this part of the phospholipid particularly hydrophilic. In plasmalogens only one fatty acyl unit occurs; the second is an unsaturated alkyl system joined to glycerol by an ether link.

Sphingomyelins. These phospholipids are esters of sphingosine, a dihydric amino alcohol, not glycerol. Their molecules also have the strongly hydrophilic phosphate system.

Glycolipids. Also sphingosine-based, these use a monosaccharide instead of the phosphate-small alcohol unit to provide a hydrophilic part. Otherwise, they resemble the sphingomyelins.

Steroids. These nonsaponifiable lipids all have the unique steroid nucleus of four fused rings. Several steroids are sex hormones, and oral fertility control drugs mimic their structure and functions. Cholesterol, the raw material used by the body to make other steroids, is also made by the body. If its synthesis or delivery is upset, cholesterol can be a problem in the bloodstream.

Membranes. A double layer of lipid molecules—phospholipids or glycolipids with cholesterol sometimes also incorporated—comprise the lipid bilayer part of a membrane. The hydrophobic tails of these lipids are lined up side by side within the bilayer away from the aqueous phase of the tissue. The hydrophilic heads poke out into the aqueous medium. On or in the bilayer are the proteins that do all the active work of the membrane and control what gets in or out of the cell.

Selected References

Book

1. L. Stryer. *Biochemistry.* W. H. Freeman and Company, Publishers, San Francisco, 1975. Chapter 10, "Introduction to Biological Membranes," is a beautifully illustrated and lucid discussion of membranes.

Articles

1. S. E. Luria. "Colicins and the Energetics of Cell Membranes." *Scientific American,* December 1975, page 30. The first part of this article describes what membranes are and how they are structured. Colicins, antibiotics made by bacteria, can inhibit active transport.
2. J. L. Marx. "Birth Control: Current Technology, Future Prospects." *Science,* March 23, 1973, page 1222. A survey of oral contraceptives, IUDs, prostaglandins, the luteinizing hormone-releasing hormone, "morning after" pills, and oral contraceptives for males.
3. J. E. Rothman and J. Lenard. "Membrane Asymmetry." *Science,* 25 February 1977, page 743. A discussion of the composition and organization of biological membranes.
4. E. P. Benditt. "The Origin of Atherosclerosis." *Scientific American,* February 1977, page 74.
5. C. E. Meloan. "Detergents—Soaps and Syndets." *Chemistry,* September 1976, page 6.

Questions and Exercises

1. Write the structure of a mixed glyceride between glycerol and the following three acids: linolenic, oleic, and myristic.
   (a) Write an equation illustrating the hydrogenation of this mixed glyceride to an iodine number of zero.
   (b) Calculate the iodine number of this mixed glyceride. (This is a problem in calculating moles. First, calculate the molecular formula of the lipid, and then the formula weight. Note how many moles of iodine (I_2) are needed per mole of lipid. Calculate how many moles of lipid there are in 100 grams; then determine how many moles of iodine are needed for that amount of lipid. Finally, convert moles iodine to grams, the iodine number.)
   (c) Would this glyceride most likely be found in a vegetable oil or an animal fat? Why?
   (d) What would be the iodine number of the glyceride resulting from the addition of hydrogen to just one of the double bonds in the mixed glyceride?
2. Write the structures of all the products that would form from the complete digestion of the following lipid.

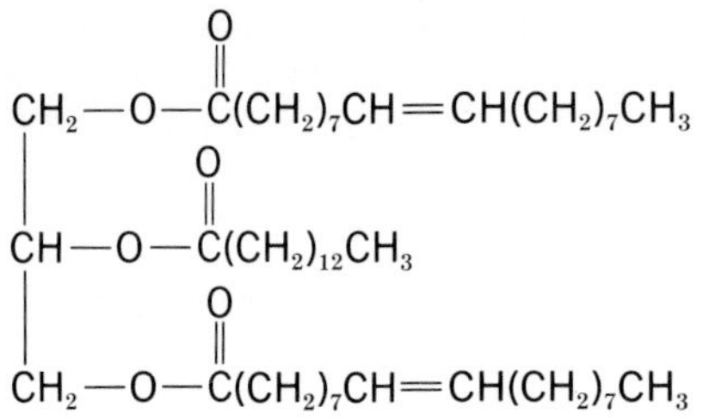

$$\begin{array}{l} O \\ \| \\ CH_2-O-C(CH_2)_7CH{=}CH(CH_2)_7CH_3 \\ |O \\ |\| \\ CH-O-C(CH_2)_{12}CH_3 \\ |O \\ |\| \\ CH_2-O-C(CH_2)_7CH{=}CH(CH_2)_7CH_3 \end{array}$$

3. Write the structures of all the products that would form from the saponification of the lipid with the structure shown in Exercise 2.
4. The hydrolysis of a lipid produced glycerol, lauric acid, linoleic acid, and oleic acid. Write a structure of the lipid consistent with these results.
5. When hydrochloric acid is added to a solution of the soap,

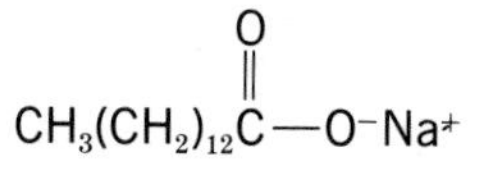

$$CH_3(CH_2)_{12}\overset{\overset{\displaystyle O}{\|}}{C}-O^-Na^+$$

a precipitate forms that is soluble in ether. What is its structure and common name?
6. Some low-sudsing detergents for automatic washers consist of molecules of the following type:

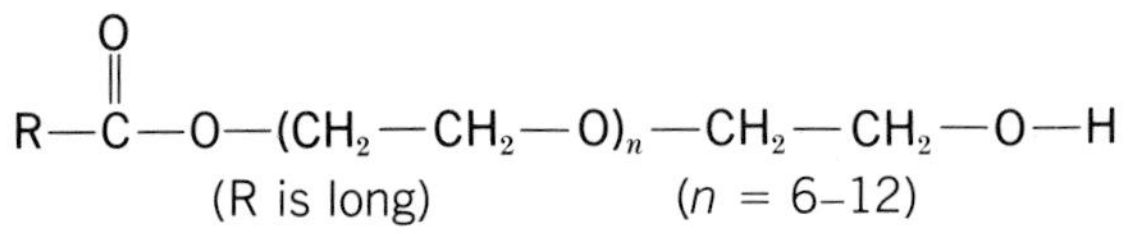

$$\underset{\text{(R is long)}}{R-\overset{\overset{\displaystyle O}{\|}}{C}-O-}(CH_2-CH_2-O)_n-CH_2-CH_2-O-H \qquad (n = 6\text{–}12)$$

Note that the structure is not ionic; yet it has detergent properties. Explain.
7. Write the formula for the calcium salt of stearic acid, a salt that would be part of soap scum if soap were used in hard water containing calcium ions.
8. What is the structural difference between the triacylglycerols of animal fats and vegetable oils?
9. A company might truthfully advertise that its brand of shortening is more "polyunsaturated" than a rival brand. The two brands were found to have iodine numbers of 85 and 98. Which value is of the rival brand? Explain your answer and state what difference it might make.
10. What are the two chief kinds of phospholipids? (Name them only.)
11. How do the phosphoglycerides and plasmalogens differ, structurally?
12. How are sphingomyelins and cerebrosides alike, structurally?
13. What structural unit provides the hydrophilic "head" on a glycolipid molecule? (Name it.)
14. Phospholipids are not classifed as neutral fats. Explain?
15. Phospholipids are particularly common in what part of a cell?
16. How do hydrophobic "tails" of lipid molecules avoid exposure to an aqueous medium in a lipid bilayer?
17. What kind of substance is present on or in a lipid bilayer to perform the active functions of a cell membrane? (Name it.)

18. Explain why a cellular membrane is described as being almost in the fluid state.
19. What are two types (by name) of sphingosine-based lipids?
20. Why are steroids classified as lipids?
21. What steroid occurs as a detergent in our systems?
22. What vitamin is made in our systems from a steroid by action of sunlight on the skin?
23. What steroid is a potent antiarthritic compound?
24. Name three steroidal sex hormones.
25. What steroid is part of cell membranes in many tissues?

chapter 12
Proteins

A tiny molecular alteration in hemoglobin changes the shape of a red blood cell from the normal dented sphere form (top) to the sickle-shaped form (bottom) in sickle cell anemia. Protein function is unusually sensitive to protein structure and shape—topics that we study in this chapter. (Top, courtesy François Morel. From *J. Cell Biol.*, **48,** 91–100, 1971; bottom, courtesy Springer-Verlag, Publishers, from *Corpuscles* by Marcel Bessis.)

12.1 Occurrence

Proteins, found in all cells and in virtually all parts of cells, constitute about half of the body's dry weight. Proteins hold an animal together and run it. As skin they give it a shell. As muscles and tendons they provide levers. As one substance inside bones, like steel in reinforced concrete, they constitute a reinforcing network. Some proteins—in antibodies, hemoglobin, and serum albumin—circulate as protectors and long-distance haulers. Others form the communication network or nerves. Certain proteins, such as enzymes, hormones, and gene regulators direct and control all forms of body repair, construction, and energy conversion. No other class of compounds is involved in such a variety of functions, all essential to life. Proteins deserve their name, taken from *proteios,* Greek for "of the first rank."

Proteins are polymers, and their monomers are **α-amino acids.** Twenty different amino acids have been identified, the same set of 20 regardless of the form or level of life, from bacteria to humans. These 20 amino acids are given in Table 12.1. In very rare instances a few others have been found, and some of the 20 amino acids occur as simple derivatives. Basic to any study of proteins, therefore, is a study of their monomers.

Amino Acids

12.2 Common Structural Features

In each molecule of an amino acid one carbon holds an amino group, a carboxyl group, another group (*G*) called a **side chain,** and in all but one, a hydrogen. Since the amine function is alpha to the carboxyl group, these are α-amino acids.

The amino group can accept a proton and the carboxyl group can donate one. Therefore, the exact molecular-ionic condition of an amino acid in water varies with the pH, being largely in form **1** in acid (at pH 1 or lower) and form **3** in alkali (pH 11 or higher) and a mixture of **1**, **2**, and **3** between pH 1 and 11. The proportions depend on the exact pH and on the specific amino acid. In the solid state as well as in solutions at or near neutrality amino acids exist as **dipolar ions, 2**, sometimes called **zwitterions.**

$$\underset{\mathbf{1}}{\overset{+}{\mathrm{N}}\mathrm{H_3}-\underset{\underset{G}{|}}{\overset{\overset{\mathrm{H}}{|}}{\mathrm{C}}}-\overset{\overset{\mathrm{O}}{\|}}{\mathrm{C}}-\mathrm{OH}} \underset{\mathrm{OH^-}}{\overset{\mathrm{H^+}}{\rightleftharpoons}} \underset{\mathbf{2}}{\overset{+}{\mathrm{N}}\mathrm{H_3}-\underset{\underset{G}{|}}{\overset{\overset{\mathrm{H}}{|}}{\mathrm{C}}}-\overset{\overset{\mathrm{O}}{\|}}{\mathrm{C}}-\mathrm{O^-}} \underset{\mathrm{H^+}}{\overset{\mathrm{OH^-}}{\rightleftharpoons}} \underset{\mathbf{3}}{\mathrm{NH_2}-\underset{\underset{G}{|}}{\overset{\overset{\mathrm{H}}{|}}{\mathrm{C}}}-\overset{\overset{\mathrm{O}}{\|}}{\mathrm{C}}-\mathrm{O^-}}$$

12.3 Common Properties of Amino Acids

As dipolar ions, amino acids are saltlike. Like salts, they are all solids and have high melting points, so high that they generally char before they change to the molten state. They are insoluble in nonpolar solvents and soluble in water.

TABLE 12.1
Common Amino Acids $\overset{+}{N}H_3-CH(G)-C(=O)-O^-$

	G (Side Chain)	Name	Three-Letter Symbol[a]	pI
Side chain is nonpolar	$-H$	Glycine	Gly	5.97
	$-CH_3$	Alanine	Ala	6.00
	$-CH(CH_3)_2$	Valine	Val	5.96
	$-CH_2CH(CH_3)_2$	Leucine	Leu	5.98
	$-CH(CH_3)CH_2CH_3$	Isoleucine	Ile	6.02
	$-CH_2-C_6H_5$ (phenyl ring)	Phenylalanine	Phe	5.48
	$-CH_2-$ (3-indolyl, N–H)	Tryptophan	Trp	5.89
	$HOC(=O)-CH-CH_2-CH_2-CH_2-NH$ (ring) (complete structure)	Proline	Pro	6.30
Side chain has a hydroxyl group	$-CH_2OH$	Serine	Ser	5.68
	$-CHOH(CH_3)$	Threonine	Thr	5.64
	$-CH_2-C_6H_4-OH$	Tyrosine	Tyr	5.66
Side chain has a carboxyl group (or an amide)	$-CH_2CO_2H$	Asparatic acid	Asp	2.77
	$-CH_2CH_2CO_2H$	Glutamic acid	Glu	3.22
	$-CH_2CONH_2$	Asparagine	Asn	5.41
	$-CH_2CH_2CONH_2$	Glutamine	Gln	5.65
Side chain has a basic amino group	$-CH_2CH_2CH_2CH_2NH_2$	Lysine	Lys	9.74
	$-CH_2CH_2CH_2NH-C(=NH)-NH_2$	Arginine	Arg	10.76
	$-CH_2-C=CH-N=CH-NH$ (imidazole ring)	Histidine	His	7.59
Side chain contains sulfur	$-CH_2S-H$	Cysteine	Cys	5.07
	$-CH_2CH_2SCH_3$	Methionine	Met	5.74

[a] These three-letter symbols are recommended by the Joint Commission on Biochemical Nomenclature of IUPAC IUB.

Exercise 12.1 Alanine begins to char at 290 °C. Its ethyl ester melts at 87 °C. Account for this enormous difference in property.

In dipolar forms, **2**, most amino acids are poor proton donors and very poor proton acceptors. They have neutralized themselves internally.

When electrodes are immersed in an aqueous solution of an amino acid, molecules that happen to be in form **1** migrate to the cathode, those in forms **3** to the anode, and those in the dipolar form, **2**, do not migrate at all. Molecules in form **2** are isoelectric; the number of (+) charges equals the number of (−) charges and the molecule is electrically neutral. At a particular pH, the system will behave as if all ions were isoelectric and no migration in an electric field could occur. The value of pH at which no net migration occurs is called the **isoelectric point** of the amino acid and is symbolized by **pI.** Table 12.1 gives these values. In a solution at its pI value, any amino acid molecule in form **1** that starts to move to the cathode soon flips a proton to something else and becomes form **2** (and therefore stops moving) or becomes form **3** (and reverses its direction). At the pI value, the rates of proton exchange are exactly adjusted to insure that each unit that is not **2** spends an equal amount of time as **1** and as **3** with the result that no net migration to either electrode can occur. We introduce the concept of pI among amino acids because it applies also to proteins. The ability of a protein to function physiologically is sensitive to the pH of the medium, because a change in pH may change the pattern of (+) and (−) charges on a protein and may affect its solubility.

The collection of side chains brought to the polymer by its monomer units also affects the properties of a protein. Let us next survey the types of side chains and how they affect properties. (You should then memorize a set of five amino acids that illustrate these types of side chains, for example, glycine, alanine, cysteine, lysine, and glutamic acid.)

12.4 Amino Acids with Nonpolar Side Chains

The first amino acids in Table 12.1 have essentially nonpolar, hydrophobic side chains.[1] When a huge protein molecule folds into its distinctive shape, hydrophobic groups tend to be folded next to each other rather than next to highly polar groups or to water molecules in a solution.

Exercise 12.2 Write the structures of glycine, alanine, leucine, and phenylalanine in the manner of structure **2**.

12.5 Amino Acids with Hydroxyl-Containing Side Chains

The second set of amino acids in Table 12.1 consists of those whose side chains carry alcohol or phenol groups. In cellular environments these side chains are neither basic nor acidic, but they are polar and hydrophilic. They

[1] The 2° amino group (and therefore a polar group) in tryptophan is unable to coordinate with a proton. It is not basic. The nitrogen's unshared pair is delocalized into the aromatic system. Hence tryptophan is put in this group because, for all practical purposes, its side chain is nonbasic and only slightly polar.

can be either hydrogen-bond donors or acceptors. As a long protein chain folds into its final shape, these side chains will tend to stick out into the surrounding aqueous phase.

12.6 Amino Acids with Acidic Side Chains

Molecules of aspartic acid and glutamic acid have extra carboxyl groups on their side chains, but in proteins these extra groups exist largely as carboxylate ions, $—CO_2^-$, in media of roughly neutral or slightly basic pH, as in body fluids.

Aspartic acid and glutamic acid often occur as asparagine and glutamine, forms in which the side-chain carboxyl groups are amides instead. These are also good polar hydrophilic groups.

Exercise 12.3 Write the structures of aspartic acid and glutamic acid in the manner of structure **2**:
(a) With the side-chain carboxyl in its carboxylate ion form
(b) With the side-chain carboxyl in its simple amide form

12.7 Amino Acids with Basic Side Chains

A lysine molecule has an extra amino group that makes its side chain basic, hydrophilic, and a hydrogen-bond donor or acceptor. The side chain in arginine has the guanidine group, $—NH—\overset{\overset{NH}{\|}}{C}—NH_2$. One of the most powerful proton-accepting groups found in organisms, it exists almost exclusively in its protonated form, $—NH—\overset{\overset{NH_2^+}{\|}}{C}—NH_2$ in a medium of even slightly basic pH. Lysine and histidine side chains are likewise usually in protonated forms.

Exercise 12.4 Write the structure of lysine in the manner of **2** with its side chain amino group in its protonated form.

12.8 Amino Acids with Sulfur-Containing Side Chains

The side chain in cysteine has an —SH group, called the mercaptan group or the sulfhydryl group. Mercaptans are easily oxidized to disulfides and disulfides are easily reduced to mercaptans:

$$2\ RSH \underset{(H)}{\overset{(O)}{\rightleftharpoons}} R—S—S—R + H_2O$$

Because the sulfhydryl group makes cysteine especially reactive toward mild oxidizing agents, it is a particularly important amino acid. Cysteine and its oxidized form, cystine, are interconvertible by oxidation and reduction, a property of far-reaching importance in some proteins.

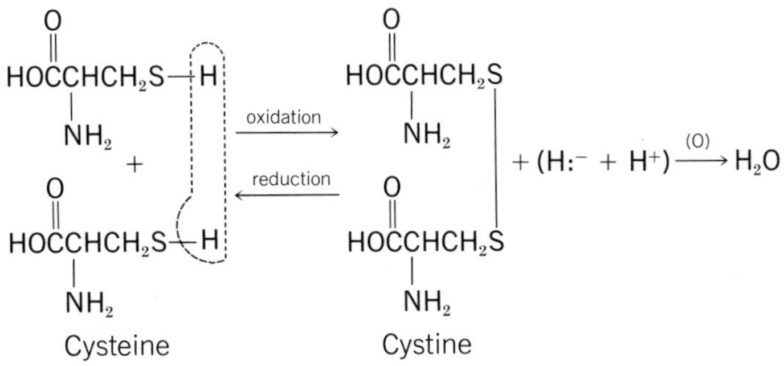

$$2\ HOCCH(NH_2)CH_2S{-}H \underset{\text{reduction}}{\overset{\text{oxidation}}{\rightleftharpoons}} HOCCH(NH_2)CH_2S{-}SCH_2CH(NH_2)COH + (H:^- + H^+) \xrightarrow{(O)} H_2O$$

(each HOCC group carries C=O)

Cysteine Cystine

The **disulfide link** in cystine is common to several proteins, as we shall see. It is especially prevalent in proteins having a protective function, such as those forming hair, fingernails, and shells.

12.9 Optical Activity of Amino Acids

The α-carbon is chiral in all naturally occurring amino acids except glycine. The chiral amino acids, therefore, can exist as enantiomers which belong to one or the other of the two optical families, D or L. Members of only one family occur naturally (with very rare exceptions).

The amino acid serine is the reference compound in amino acid chemistry and is related configurationally to glyceraldehyde, the reference in carbohydrate chemistry. Naturally occurring serine is in the L-family if we let the —NH_2 group of serine correspond to the C-2-OH group of glyceraldehyde and the CO_2H of serine correspond to the aldehyde group of glyceraldehyde. (In the *R*/*S* family L-serine is *S*-serine. See Appendix.)

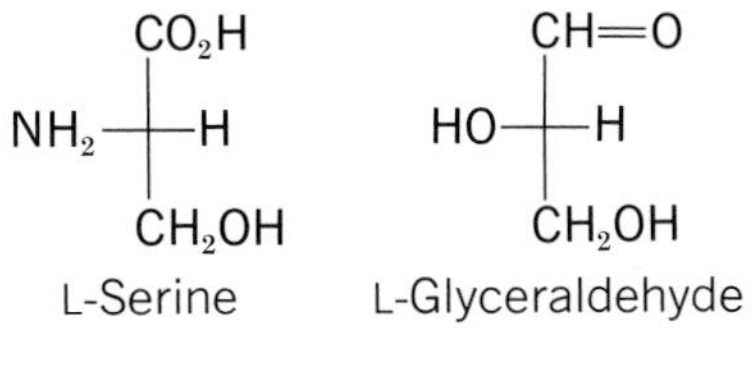

$$\begin{array}{ccc} & CO_2H & \\ NH_2 & - & H \\ & CH_2OH & \end{array} \qquad \begin{array}{ccc} & CH{=}O & \\ HO & - & H \\ & CH_2OH & \end{array}$$

L-Serine L-Glyceraldehyde

All other optically active amino acids in nature belong to the L-family. The proteins in our bodies are made only from L-amino acids. If we were to ingest the D-enantiomers, which can be made in the laboratory, we could not use them. Our enzymes, themselves made from L-amino acids, are of one handedness or chirality and can work only with L-amino acids or proteins made from them.

Exercise 12.5 Write the plane projection structure of L-alanine.

12.10 Protein Structure

Proteins are mostly or entirely polyamides called **polypeptides,** but much more of importance must be added to that definition of structure. Four levels of complexity exist in the structural features of proteins, and if structure is undone at any level, the protein generally can no longer perform its biological function. The first and most basic level is called the **primary structure** of a protein, the network of covalent bonds that hold monomer units together.

The long molecule described by the primary structure seldom exists as a randomly flexed, stringlike system. Once the long molecule forms, it assumes a particular flexed conformation. This level of protein structure is the **secondary structure.** Noncovalent forces associated with the hydrogen bond force the long molecule generally into one or another of three secondary structures, the alpha helix, the beta-pleated sheet, or the elongated three-stranded cable in collagen.

Just as metal springs may be twisted and curved, folded and bent, so the secondary structures of proteins are usually twisted and folded into forms called the **tertiary structures** of proteins. Noncovalent forces, generally hydrogen bonds, stabilize these structures.

Finally, many native proteins, those that occur in a living system, are assemblages of two or more polymer molecules, sometimes with a nonpolypeptide group. Hemoglobin, the oxygen-carrying molecule in the blood, has four polypeptide molecules, two of one kind and two of another, and each bears a molecule of heme, not a polypeptide. This final level of complexity, the particular collection of polypeptides making up a protein (each having its own primary, secondary, and tertiary features), together with any other group, is the **quaternary structure** of the protein. Let us now look at each of these four features in greater detail.

Primary Structures of Protein

12.11 The Peptide Bond

The amide bond, carbonyl-to-nitrogen, is the chief covalent bond that forms when proteins are put together. In protein chemistry this bond is called the **peptide bond,** and the protein polymer is called a polypeptide. To illustrate this bond as well as show how polypeptides acquire their primary structural features, we simplify the structures of amino acids by showing free amino groups, —NH_2 (not —NH_3^+) and free carboxyl groups —CO_2H (not —CO_2^-). Suppose that glycine acts as a carboxylic acid, and alanine acts as an amine in amide formation:

$$\underset{\substack{\text{Glycine}\\ \text{(acid)}}}{NH_2CH_2\overset{O}{\overset{\|}{C}}\text{—}OH} + \underset{\substack{\text{Alanine}\\ \text{(amine)}}}{H\text{—}NH\underset{CH_3}{\underset{|}{C}}H\overset{O}{\overset{\|}{C}}\text{—}OH} \longrightarrow \underset{\substack{\mathbf{4}\\ \text{Glycylalanine, Gly}\cdot\text{Ala}\\ \text{(a dipeptide)}}}{NH_2CH_2\overset{O}{\overset{\|}{C}}\text{—}NH\underset{CH_3}{\underset{|}{C}}H\overset{O}{\overset{\|}{C}}\text{—}OH} + H_2O$$

(Peptide bond: the C—N bond between the glycine carbonyl and the alanine nitrogen.)

Of course, there is no reason why we could not picture the roles reversed, that alanine acts as the carboxylic acid and glycine acts as the amine:

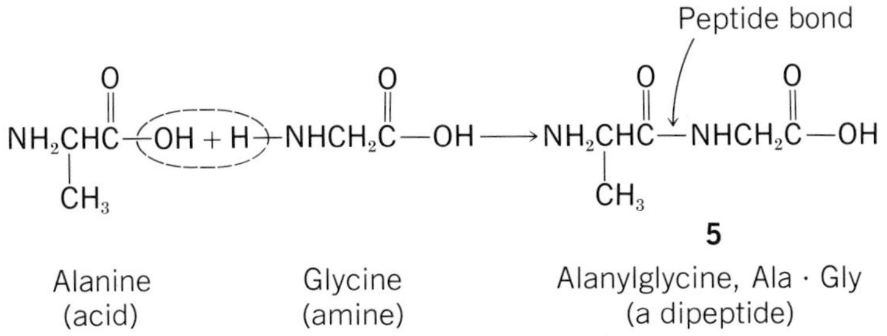

$$\underset{\substack{\text{Alanine}\\\text{(acid)}}}{NH_2\underset{\substack{|\\CH_3}}{CH}\overset{\substack{O\\\|}}{C}-OH} + \underset{\substack{\text{Glycine}\\\text{(amine)}}}{H-NHCH_2\overset{\substack{O\\\|}}{C}-OH} \longrightarrow \underset{\substack{\mathbf{5}\\\text{Alanylglycine, Ala}\cdot\text{Gly}\\\text{(a dipeptide)}}}{NH_2\underset{\substack{|\\CH_3}}{CH}\overset{\substack{O\\\|}}{C}-NHCH_2\overset{\substack{O\\\|}}{C}-OH}$$

Peptide bond

The union of two amino acids by a peptide bond produces a **dipeptide.** Three amino acid units linked by peptide bonds make a tripeptide, and so forth. Each unit in the polypeptide contributed by an amino acid is called a residue.

Exercise 12.6 Write the structures of two dipeptides that could be made from alanine and phenylalanine.

To make it easier to write polypeptide structures, protein chemists use the three-letter symbols of amino acids (Table 12.1) in accordance with specific conventions. A series of three-letter symbols, each separated by a dot raised slightly above the line, represents a polypeptide structure, provided that the first three-letter symbol (reading left to right) is the free amino end of the polypeptide, called the **N-terminal end,** and the last symbol is the free carboxyl end, called the **C-terminal end.** Thus the structure of glycylalanine, **4**, is written simply as Gly·Ala; the isomer, alanylglycine, **5**, is simply Ala·Gly. In both dipeptides the "backbones" are the same:

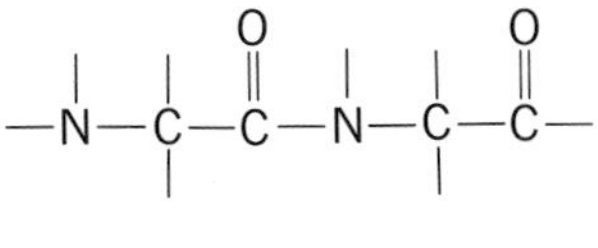

$$-\overset{|}{N}-\underset{|}{\overset{|}{C}}-\overset{\substack{O\\\|}}{C}-\overset{|}{N}-\underset{|}{\overset{|}{C}}-\overset{\substack{O\\\|}}{C}-$$

The two differ in the sequence in which the side chains, —H and $-CH_3$, occur on α-carbons.

Exercise 12.7 Write the structures of the two dipeptides of Exercise 12.6 using three-letter symbols.

Exercise 12.8 Write out the structures of each.
(a) Lys·Cys (b) Glu·Ala (c) Gly·Gly

Amino acids are often called the letters of the protein alphabet and proteins are the long—very long—words, chapters, and books. The protein "book" of one species is not identical with that of another, but the same alphabet is used for both. Just as the two letters "N" and "O" can be arranged to give two different words, "NO" and "ON," so two different amino acids can be joined to give two isomeric dipeptides that have identical "backbones" and differ only in the sequence of side chains.

Dipeptides, of course, are still amino acids although they are no longer α-amino acids. A third α-amino acid can react at one end or the other.

In general,

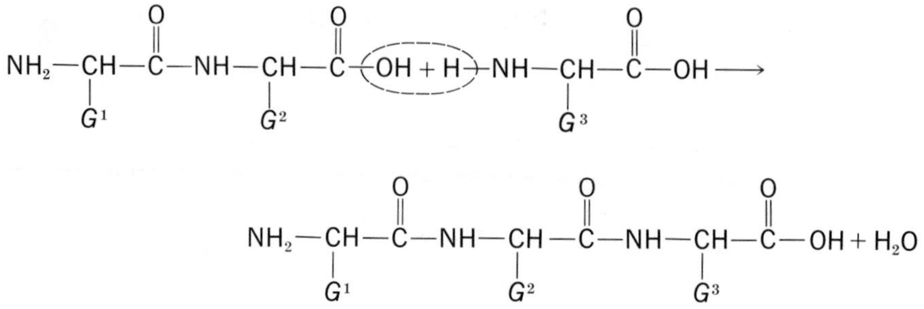

$$NH_2-\underset{G^1}{\underset{|}{CH}}-\overset{O}{\overset{\|}{C}}-NH-\underset{G^2}{\underset{|}{CH}}-\overset{O}{\overset{\|}{C}}-(OH+H)-NH-\underset{G^3}{\underset{|}{CH}}-\overset{O}{\overset{\|}{C}}-OH \longrightarrow$$

$$NH_2-\underset{G^1}{\underset{|}{CH}}-\overset{O}{\overset{\|}{C}}-NH-\underset{G^2}{\underset{|}{CH}}-\overset{O}{\overset{\|}{C}}-NH-\underset{G^3}{\underset{|}{CH}}-\overset{O}{\overset{\|}{C}}-OH+H_2O$$

A tripeptide

A specific example is

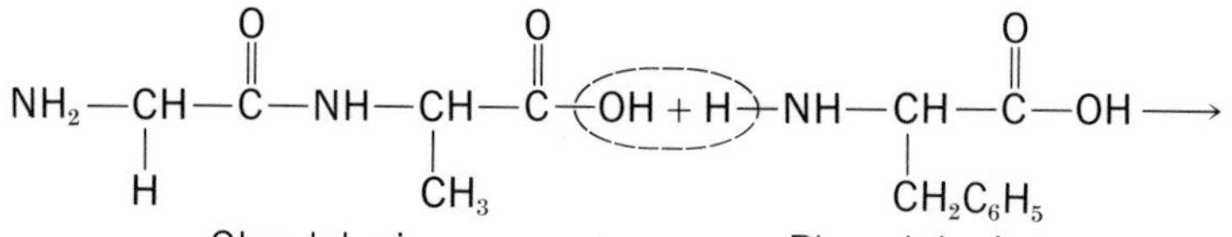

$$\underbrace{NH_2-\underset{H}{\underset{|}{CH}}-\overset{O}{\overset{\|}{C}}-NH-\underset{CH_3}{\underset{|}{CH}}-\overset{O}{\overset{\|}{C}}-OH}_{\text{Glycylalanine (Gly · Ala)}}+\underbrace{H-NH-\underset{CH_2C_6H_5}{\underset{|}{CH}}-\overset{O}{\overset{\|}{C}}-OH}_{\text{Phenylalanine (Phe)}} \longrightarrow$$

$$NH_2-\underset{H}{\underset{|}{CH}}-\overset{O}{\overset{\|}{C}}-NH-\underset{CH_3}{\underset{|}{CH}}-\overset{O}{\overset{\|}{C}}-NH-\underset{CH_2C_6H_5}{\underset{|}{CH}}-\overset{O}{\overset{\|}{C}}-OH+H_2O$$

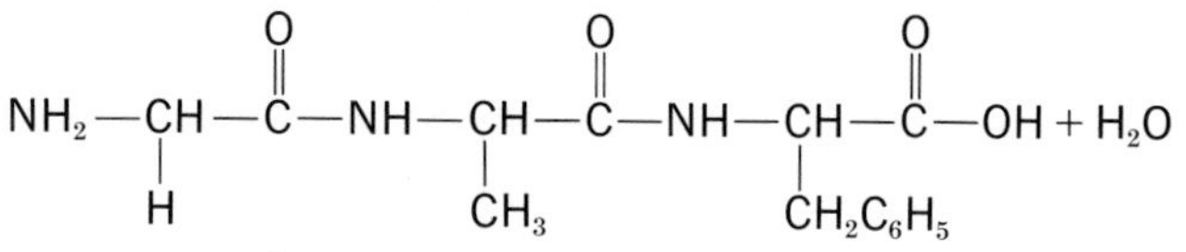

Glycylalanylphenylalanine
(Gly · Ala · Phe)

The tripeptide shown in the last example is only one of six possible isomers involving these three different amino acids. The set of all possible sequences for a tripeptide made from glycine, alanine, and phenylalanine is as follows.

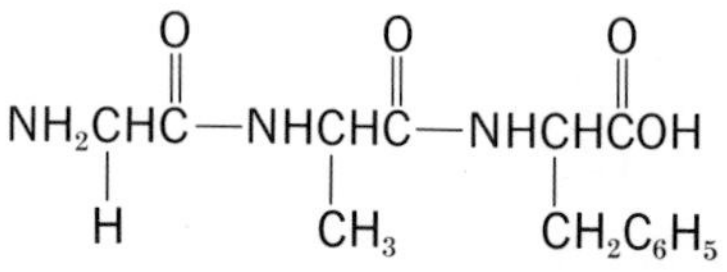

$$NH_2\underset{H}{\underset{|}{C}}H\overset{O}{\overset{\|}{C}}-NH\underset{CH_3}{\underset{|}{C}}H\overset{O}{\overset{\|}{C}}-NH\underset{CH_2C_6H_5}{\underset{|}{C}}H\overset{O}{\overset{\|}{C}}OH \qquad \text{Gly · Ala · Phe}$$

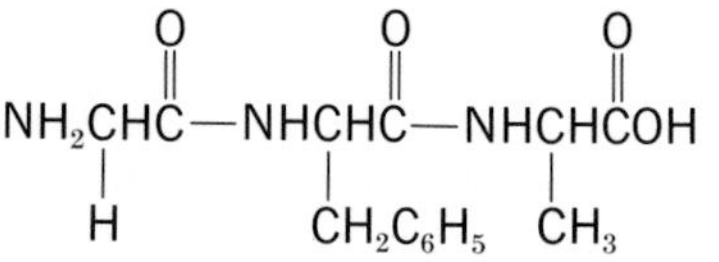

$$NH_2\underset{H}{\underset{|}{C}}H\overset{O}{\overset{\|}{C}}-NH\underset{CH_2C_6H_5}{\underset{|}{C}}H\overset{O}{\overset{\|}{C}}-NH\underset{CH_3}{\underset{|}{C}}H\overset{O}{\overset{\|}{C}}OH \qquad \text{Gly · Phe · Ala}$$

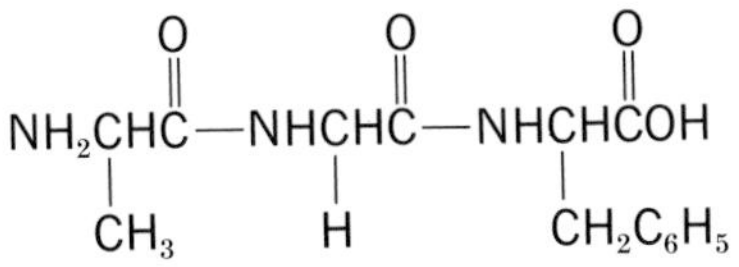

$$\underset{\underset{CH_3}{|}}{NH_2CH}\overset{O}{\overset{\|}{C}}-\underset{\underset{H}{|}}{NHCH}\overset{O}{\overset{\|}{C}}-\underset{\underset{CH_2C_6H_5}{|}}{NHCH}\overset{O}{\overset{\|}{C}}OH \qquad \text{Ala} \cdot \text{Gly} \cdot \text{Phe}$$

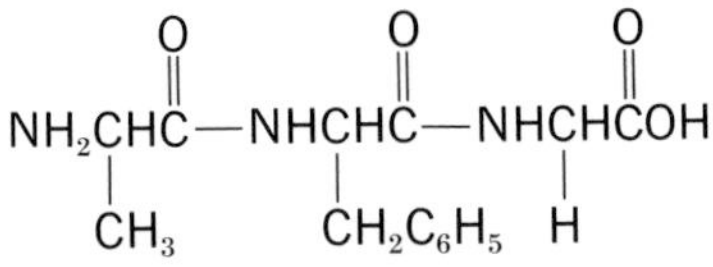

$$\underset{\underset{CH_3}{|}}{NH_2CH}\overset{O}{\overset{\|}{C}}-\underset{\underset{CH_2C_6H_5}{|}}{NHCH}\overset{O}{\overset{\|}{C}}-\underset{\underset{H}{|}}{NHCH}\overset{O}{\overset{\|}{C}}OH \qquad \text{Ala} \cdot \text{Phe} \cdot \text{Gly}$$

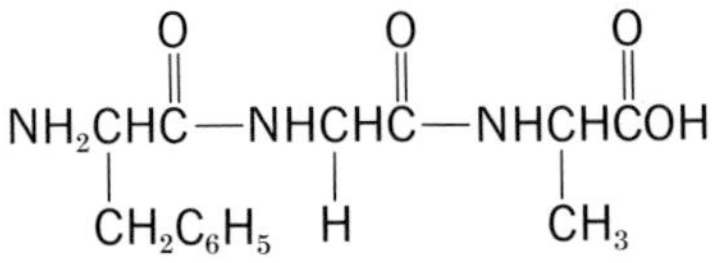

$$\underset{\underset{CH_2C_6H_5}{|}}{NH_2CH}\overset{O}{\overset{\|}{C}}-\underset{\underset{H}{|}}{NHCH}\overset{O}{\overset{\|}{C}}-\underset{\underset{CH_3}{|}}{NHCH}\overset{O}{\overset{\|}{C}}OH \qquad \text{Phe} \cdot \text{Gly} \cdot \text{Ala}$$

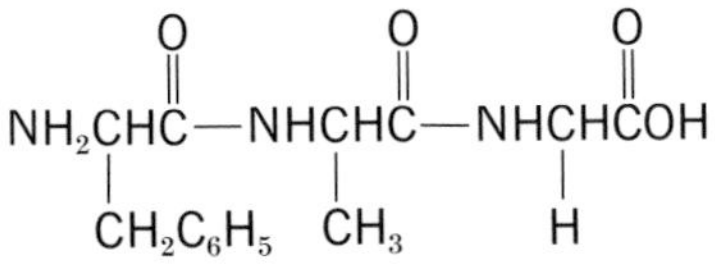

$$\underset{\underset{CH_2C_6H_5}{|}}{NH_2CH}\overset{O}{\overset{\|}{C}}-\underset{\underset{CH_3}{|}}{NHCH}\overset{O}{\overset{\|}{C}}-\underset{\underset{H}{|}}{NHCH}\overset{O}{\overset{\|}{C}}OH \qquad \text{Phe} \cdot \text{Ala} \cdot \text{Gly}$$

Each of these structures represents a different compound with its own unique set of physical properties. The chemical properties are quite similar, however, because the same functional groups are present.

Each tripeptide is also an amino acid, although none is an α-amino acid. Each can combine with still another amino acid at one end or the other. By repeating this process, we can see how a protein molecule is put together. All protein molecules have this repeating sequence in the backbone:

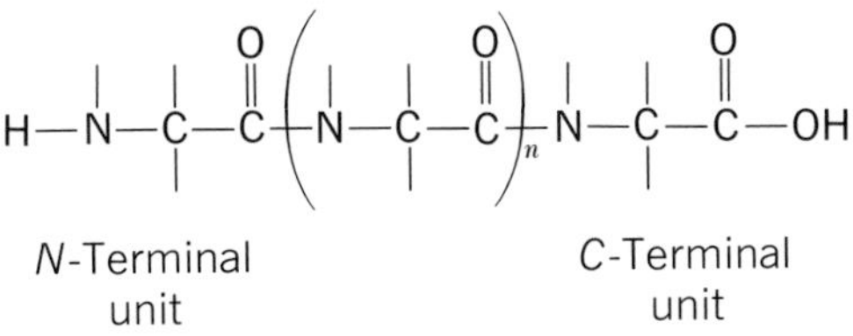

$$H-\overset{|}{N}-\underset{|}{\overset{|}{C}}-\overset{O}{\overset{\|}{C}}\left(-\overset{|}{N}-\underset{|}{\overset{|}{C}}-\overset{O}{\overset{\|}{C}}-\right)_n\overset{|}{N}-\underset{|}{\overset{|}{C}}-\overset{O}{\overset{\|}{C}}-OH$$

N-Terminal unit

C-Terminal unit

(*n* may vary from dozens to thousands.)

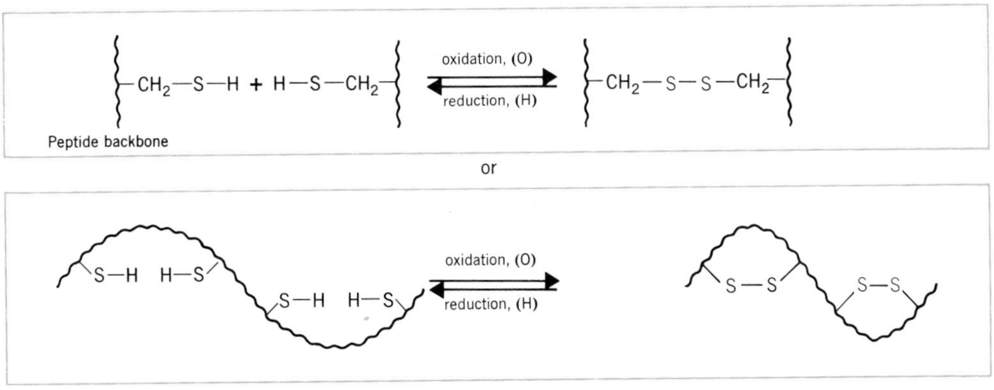

$$\sim CH_2-S-H + H-S-CH_2 \sim \; \underset{\text{reduction, (H)}}{\overset{\text{oxidation, (O)}}{\rightleftharpoons}} \; \sim CH_2-S-S-CH_2 \sim$$

Peptide backbone

or

S—H H—S S—H H—S oxidation, (O) reduction, (H) S—S S—S

Figure 12.1 The disulfide link. Sulfhydryl groups on cysteine residues can furnish disulfide links to hold one chain to another (top) or force a single strand into a particular shape (bottom).

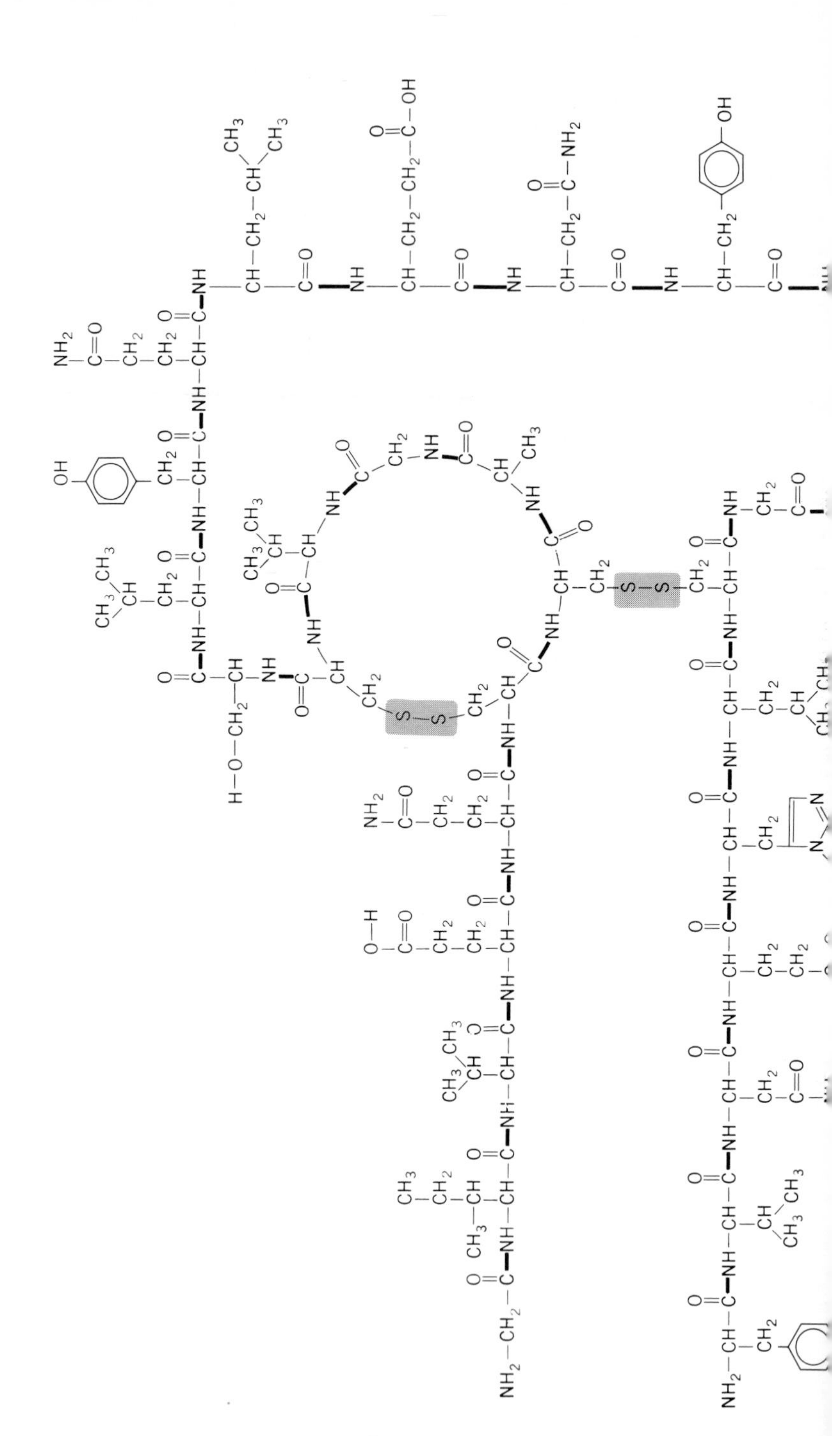

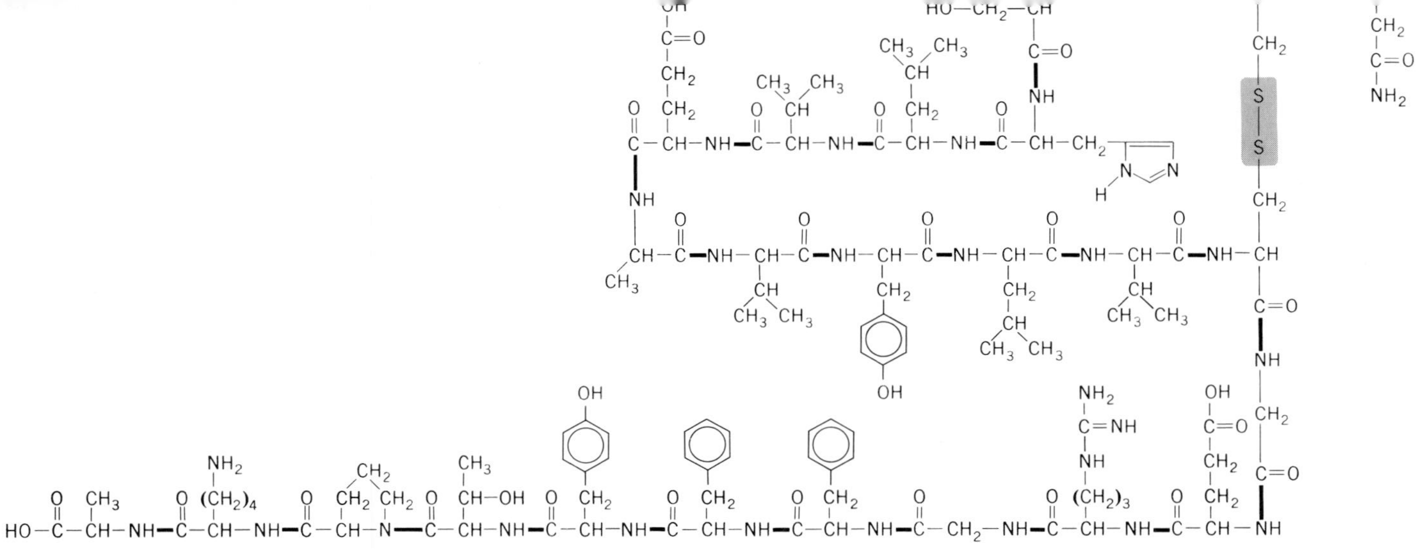

Figure 12.2 Sheep insulin. Boldface lines are peptide bonds. Disulfide links are highlighted by rectangles.

This repeating sequence, together with the sequence of side chains, constitutes the primary structure of all proteins. The peptide bond is thus the chief covalent linkage holding amino acid units together in proteins. The only other covalent bond at this level of protein structure is the disulfide linkage.

As the number of amino acid units per polypeptide molecule increases, some used several times, the number of isomeric polypeptides rises rapidly. For example, if 17 different amino acids were incorporated, 3.56×10^{14} isomeric polypeptides would be possible. A protein of formula weight 34 000, in which 12 different amino acids were used and 288 peptide bonds were formed, could exist in an estimated 10^{300} isomeric structures. Each molecule would weigh about 10^{-20} g, and a collection of one molecule of each isomer would weigh 10^{280} g. But the actual mass of the earth is only about 10^{28} g. Clearly, the versatile protein alphabet makes possible many more protein "books" than needed to supply every living species with its unique set of proteins.

12.12 Disulfide Bonds[2]

If the sulfhydryl-bearing side chain of cysteine appears on two neighboring protein molecules, mild oxidation will link the two molecules through a disulfide bond, as illustrated in Figure 12.1. The cross-linking may occur between segments of the same protein strand, in which case a closed loop results.

The structure of what is regarded as a relatively simple protein, the hormone insulin, is shown in Figure 12.2. Three disulfide bonds are important to this structure.

Secondary Structures of Proteins

12.13 Alpha Helix

Once a cell puts together a long protein molecule, noncovalent forces of attraction between parts of the structure make it twist into a particular shape. The principal noncovalent force is the **hydrogen bond.** Using data obtained

$$-\overset{|}{N}-\overset{\delta+}{H}\cdots\overset{\delta-}{O}=C\langle$$

with X rays, Linus Pauling (Nobel Prize, 1954) and R. B. Corey first showed that a coiled configuration is one geometric form. They called it the ***α*-helix** ("alpha," because it was the first such form to be verified), and it is illustrated in Figure 12.3. In the *α*-helix the backbone coils as a right-handed screw with all the side chains sticking outside. Hydrogen bonds extend from oxygens of carbonyl groups to hydrogens of —NH— groups in peptide bonds down the spiral.

[2] Although many biochemists consider the disulfide bond as a tertiary structural feature, others prefer to view it as a primary feature because, like the peptide bond, it is a covalent bond and is not easily broken (except by reducing agents).

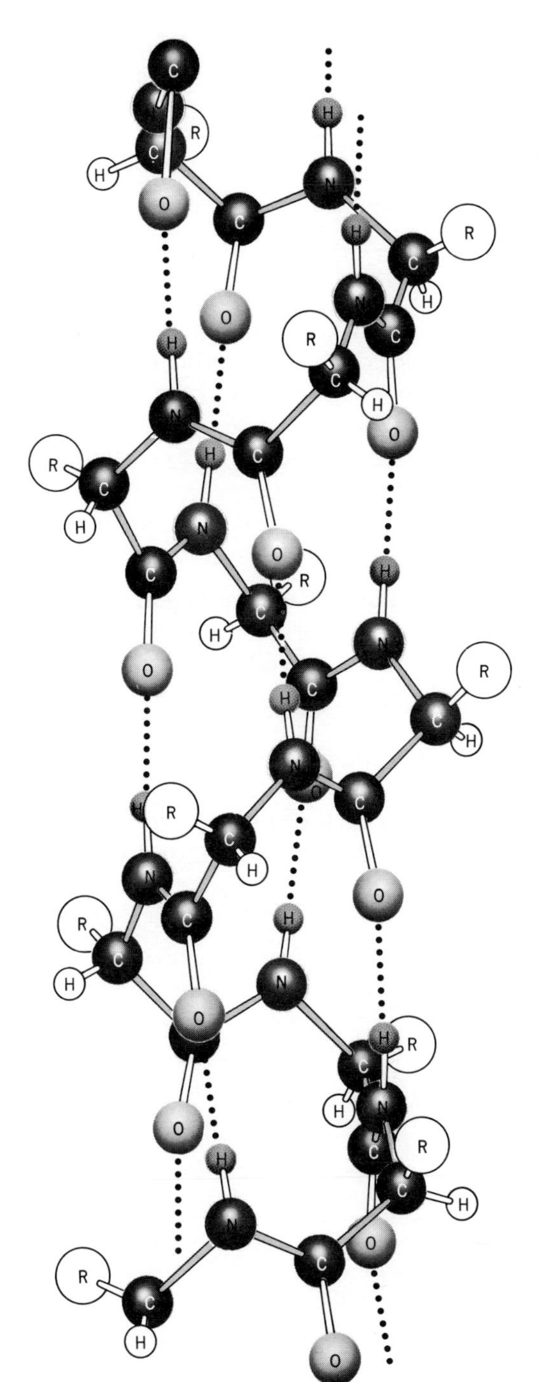

Figure 12.3 The α-helix. Dotted lines are hydrogen bonds. Side chains (labeled R) extend away from the spiral. (Adapted from B. Low and J. T. Edsall. *Currents in Biochemical Research.* 1956. Wiley-Interscience, New York.)

Individually, these hydrogen bonds are weak forces of attraction. Collectively, however, they add up in much the same way that forces holding shut a zipper add up. In fact, the hydrogen bond is often called the zipper of protein molecules.

Among proteins studied thus far, only segments of molecules, not entire lengths, are in the α-helix configuration.

12.14 Beta-Pleated Sheet

Pauling and Corey also discovered that molecules in some proteins line up side by side, becoming regularly pleated as they do, to form a sheetlike array. This ***β*-pleated sheet** configuration, illustrated in Figure 12.4, is the dominant feature in fibroin, the protein found in silk. In other proteins these sheets are a feature, if at all, of only small portions. Hydrogen bonds between parallel strands hold them together.

12.15 Triple Helix

The third important secondary structure is featured in **collagen,** the protein that gives enormous strength to bone, teeth, cartilage, tendons, and skin (Figure 12.5). The polypeptide in collagen, called tropocollagen, is unusual in being dominated by three amino acid residues, those of glycine, proline, and hydroxyproline. The latter is not listed in Table 12.1 because it is not used directly to make proteins. Instead, its alcohol group is put on after proline residues have been incorporated into the polypeptide. Carbon-14 labeling studies show that if hydroxyproline containing carbon-14 is fed to rats, it is not used to make new collagen. But if proline with carbon-14 is used, the new

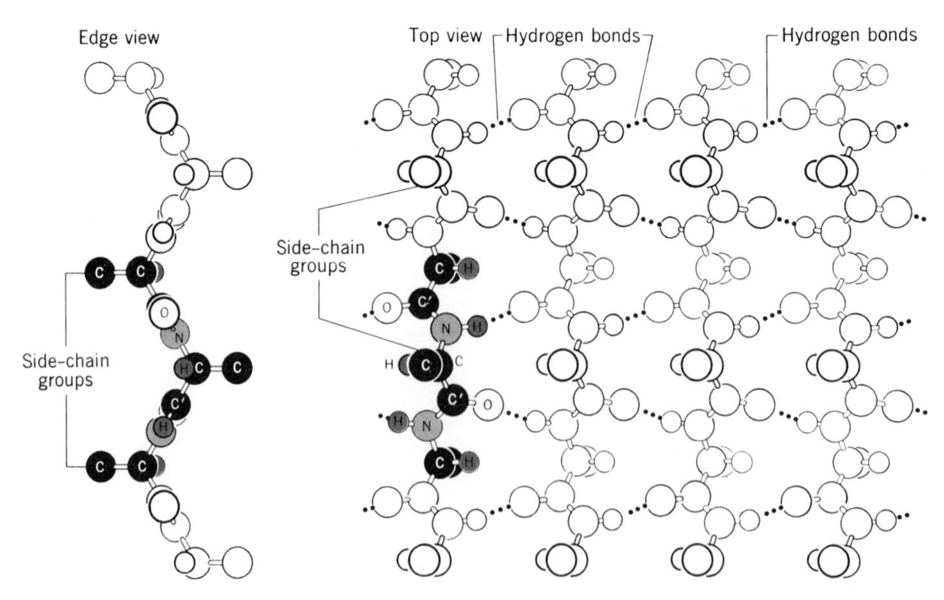

Figure 12.4 The β-pleated sheet. Hydrogen bonds (dotted lines) provide the forces keeping polypeptide strands aligned side by side. The edge view shows what is meant by the "pleats." (By permission of L. Pauling and R. B. Corey.)

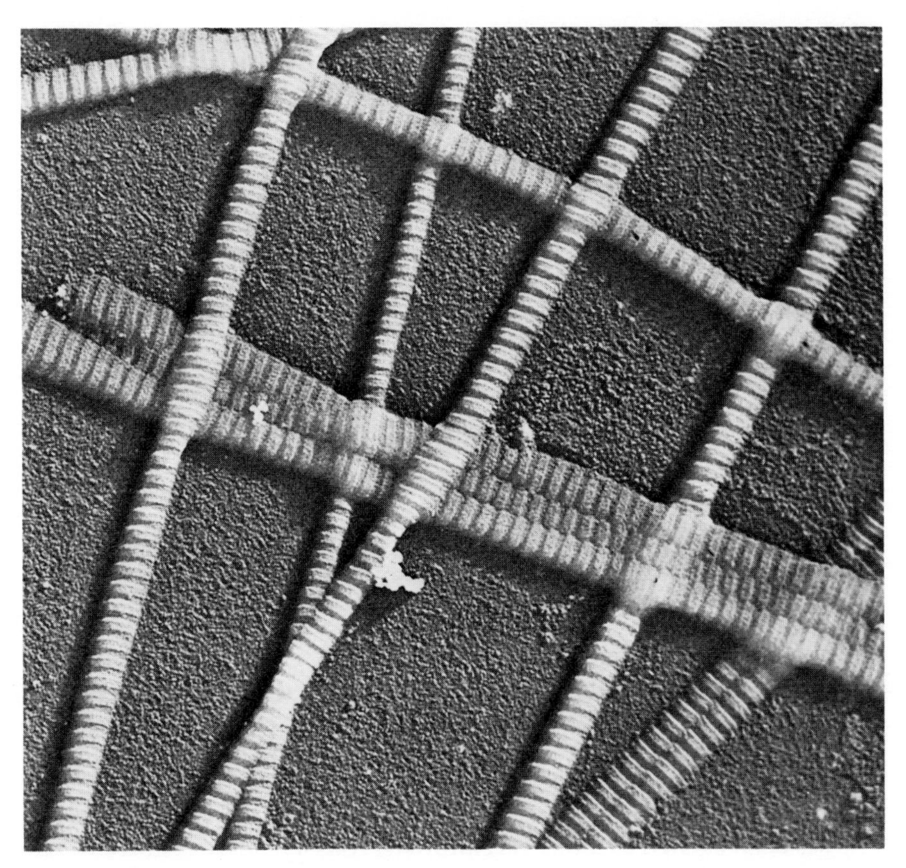

Figure 12.5 Collagen fibrils. Electron micrograph of intact fibrils from skin. (Courtesy of Dr. Jerome Gross.)

collagen contains carbon-14 in *both* proline and hydroxyproline residues. This proves that proline is used to make polypeptide, which becomes collagen after proline residues are hydroxylated.

Some lysine residues are also hydroxylated after the polypeptide forms. Hydroxylation requires the presence of vitamin C (ascorbic acid), and the absence of this vitamin means formation of defective collagen. The result is poor bone tissue, and the disease is known as scurvy.

Each tropocollagen molecule consists of three polypeptide chains, each with about 1000 amino acid residues, twisted in a **triple helix.** Very nearly every third residue is from glycine, the smallest amino acid. Another unusual feature of tropocollagen is the occurrence of sugar residues. An uncommon disaccharide, glucose joined to galactose, is bound to some of the hydroxylysine units.

The individual chains in tropocollagen strands exist in a very open helix not stabilized by hydrogen bonds, but these open helices wrap around each other into a right-handed helical cable. Within the cable hydrogen bonds help hold the system together, but in addition a few complex changes occur to form covalent cross-links between particular residues. (We shall not study how the covalent links form.) The collagen fibril (small fiber) forms when individ-

OH, HC—CH_2, H_2C, CH—CO_2^-, N^+, H, H

4-Hydroxyproline (Hyp)

OH
$NH_2CH_2CHCH_2CH_2CHCO_2^-$
NH_3^+

5-Hydroxylysine (Hyl)

α-D-galactose unit

O O—CH 5-Hydroxylysine unit

O

α(1 → 2) bridge

O

α-D-glucose unit

Disaccharide unit in collagen

ual tropocollagen molecular cables overlap lengthwise. See Figure 12.6. Gaps from the head of one tropocollagen molecule to the tail of another occur, and these may be sites where, in bone tissue, mineral deposits become tied into the protein.

The tensile strength of collagen fiber is exceptional. A fiber only 1 mm in diameter will hold a load as heavy as 10 kg (22 lb). Such strength is obviously important in the tissues served by collagen, a protein that in bone and teeth has been likened to the steel reinforcing rods in concrete-steel structures. The molecular bases of this strength are simple covalent forces in the primary structure as well as the cross-links, hydrogen bonds in the higher structures, and the mechanical advantage of being a twisted cable.

12.16 Tertiary Structures of Proteins

When α-helices take shape, side chains stick out. In a typical protein as many as 40% of the side chains are hydrophobic, even in water-soluble proteins. To

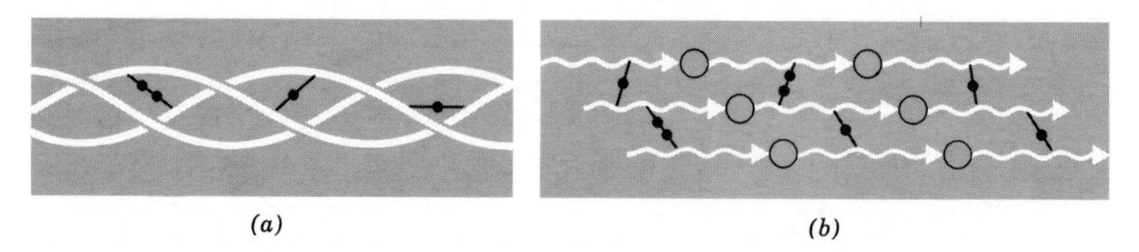

(a) *(b)*

Figure 12.6 Collagen. (*a*) Segment of tropocollagen showing how three polypeptide strands wrap together to form the tropocollagen cable. (Not shown is the open-helix form of each individual strand.) (*b*) Tropocollagen molecules, represented here by wavy arrows, line up side by side, overlapping as shown and leaving gaps (shown as circles) between individual ends to give collagen fibrils. Lines that include solid dots represent covalent cross-links both within a collagen molecule (*a*) and between collagen molecules in a fibril (*b*).

minimize contact between hydrophobic groups and an aqueous medium, helices usually fold and twist into arrangements that tuck hydrophobic groups as much as possible to the inside, away from water, and leave hydrophilic side chains exposed to water. The final shape of the polypeptide, its tertiary structure, thus emerges in response to simple molecular forces, to water-avoiding and water-attracting properties of the side chains. These features are particularly common in nonfibrous proteins such as globular proteins, which are usually in an aqueous environment.

The tertiary structure of **myoglobin** is given in Figure 12.7. About 75% of the polypeptide is in an α-helix form. By the folding shown in the figure, the polypeptide has managed to place virtually all hydrophobic groups inside and has left the hydrophilic groups outside.

12.17 Prosthetic Groups

Myoglobin is an oxygen-holding protein in muscle tissue, where it maintains an oxygen reserve and helps move oxygen in the tissue. Myoglobin is not

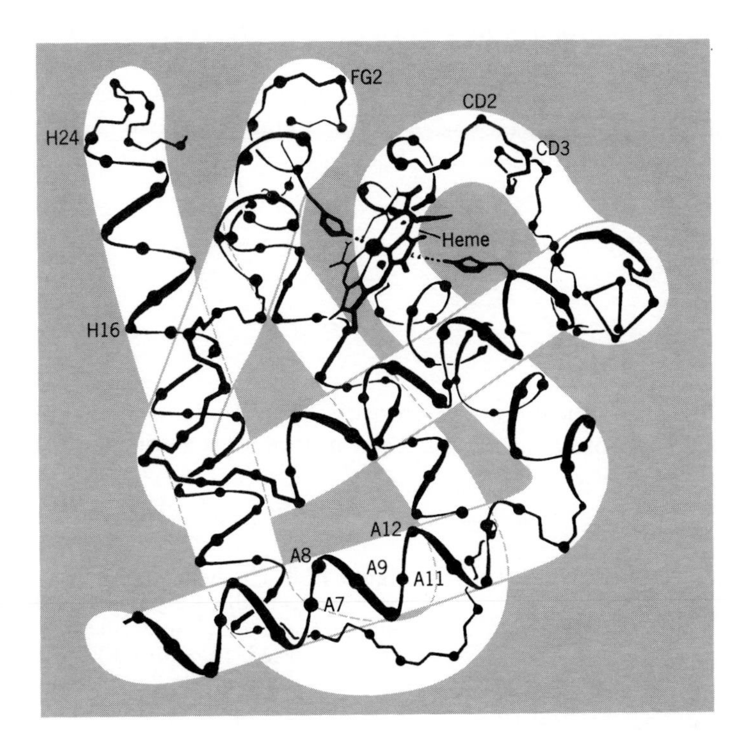

Figure 12.7 Myoglobin, the oxygen carrier in muscles. The black dots are individual alpha carbons of amino acid residues. Long tubelike forms outline segments in the α-helix configuration. The identities of all residues are known. A few are lettered (to indicate a segment) and numbered (to indicate location in the segment.) Hydrophilic groups are at FG2 and H16 (lysine), H24 (glutamic acid), CD2 (aspartic acid), and CD3 (arginine), to name a few. Their side chains are on exposed sites, where they can attract water molecules of the medium. Hydrophobic groups are at A7 and A9 (leucine), A8 and A11 (valine), and A12 (tryptophan), among others. Their side chains tend to be tucked inside away from the medium. (Reproduced by permission, from R. E. Dickerson, "X-Ray Analysis of Proteins," in *The Proteins,* H. Neurath, ed. Academic Press, New York; © 1964, all rights reserved.)

just a polypeptide. A nonprotein group called **heme** (Figure 12.8) is bound together with it, and this group actually holds the oxygen molecule. Heme occurs also in **hemoglobin,** the oxygen carrier of red blood cells. Nonpolypeptide molecules associated with proteins are called **prosthetic groups,** and many proteins have them. Proteins with such groups are called conjugated proteins.

12.18 Salt Bridges

Another force that can affect tertiary structure is the attraction between a full positive and a full negative charge, each occurring on particular side chains. When side chains of aspartic and glutamic acid residues carry carboxylate ions, $—CO_2^-$, and side chains of arginine and lysine residues exist in protonated forms, for example, $—NH_3^+$, these oppositely charged groups will naturally attract each other, just as sodium and chloride ions attract each other in a salt crystal. The resulting force of attraction in a protein is called a **salt bridge;** see Figure 12.9.

Once the covalent bonds in a polypeptide have been formed and the unique sequence of amino acid residues has been fixed, the molecule automatically twists itself in response to forces of potential hydrogen bonds, salt bridges, and water-avoidance or water-attraction properties into the secondary and tertiary structural features. Some proteins also have a quaternary structure.

12.19 Quaternary Structures of Proteins

The enzyme called phosphorylase consists of two identical polypeptides. Neither alone functions as a catalyst. They must exist together in the enzyme

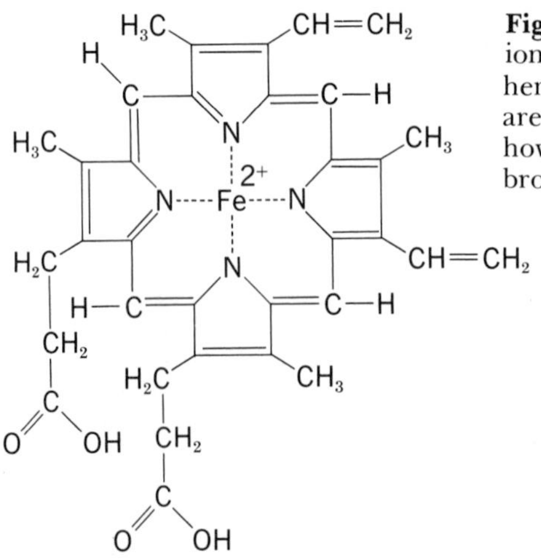

Figure 12.8 The heme molecule. The Fe^{2+} ion remains in the +2 state in oxygenated hemoglobin in which molecules of oxygen are transported. In oxidized hemoglobin, however, iron is in the Fe^{3+} state and has a brown color (as in cooked meat).

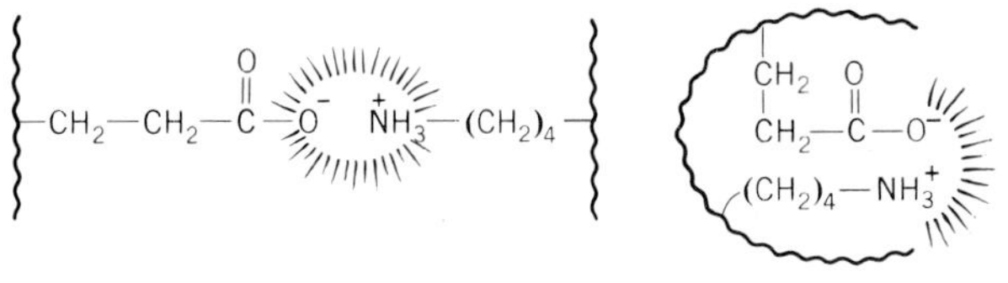

(*a*) (*b*)

Figure 12.9 The salt bridge in protein structure. The attraction between unlike charges may help hold one polypeptide to another (*a*) or affect the coiling or folding of an individual strand (*b*).

to have catalytic activity. Many proteins are aggregations of two or more subunits, which may be identical or different; this feature of protein structure is called the **quaternary structure.** Hemoglobin has four subunits, two of one kind and two of another, and each subunit holds one molecule of heme. See Figure 12.10. Forces between oppositely charged sites, hydrogen bonds, and charged sites on prosthetic groups are involved in holding subunits together to stabilize the quaternary structure.

12.20 Sickle-Cell Anemia and Altered Hemoglobin

The decisive importance of the primary structure to all other features is illustrated by the grim story of **sickle-cell anemia.** This inherited disease is widespread in central and western Africa and among descendants of those who came from there. In its mild form, where only one parent carried the genetic trait, symptoms are seldom noticed except when the environment is low in oxygen (e.g., high altitude; the depressurized cabin of an airplane). In the severe form, where both parents carried the trait, the infant usually dies by the age of two. The problem is an impairment in blood circulation traceable to the altered shape of hemoglobin in sickle-cell anemia, particularly after the hemoglobin has delivered oxygen and is on its way back to the heart and lungs for more.

Each of the four subunits in hemoglobin is made of about 300 amino acid residues. In the hemoglobin of those with sickle-cell anemia, only one of these residues is changed. Instead of a glutamic acid unit (with a carboxylate side chain, $—CO_2^-$) a valine unit (with an isopropyl side chain) occurs sixth in from the N terminus of the β-chain. Normal hemoglobin, symbolized as HHb, and sickle-cell anemia hemoglobin, HbS, therefore, have different patterns of electrical charges. Both have about the same solubility in well-oxygenated blood, but HbS without its oxygen molecules tends to precipitate inside red cells, distorting these cells into a characteristic curved, sickle shape. Distorted red cells are harder to pump; the heart bears a greater strain; and sometimes sickle cells clump together to plug a capillary. Sometimes they split open. Thus a subtle change in one side chain affects all other structural features. One amino acid unit out of 300 is different at the molecular level, but at the human level life is gravely threatened.

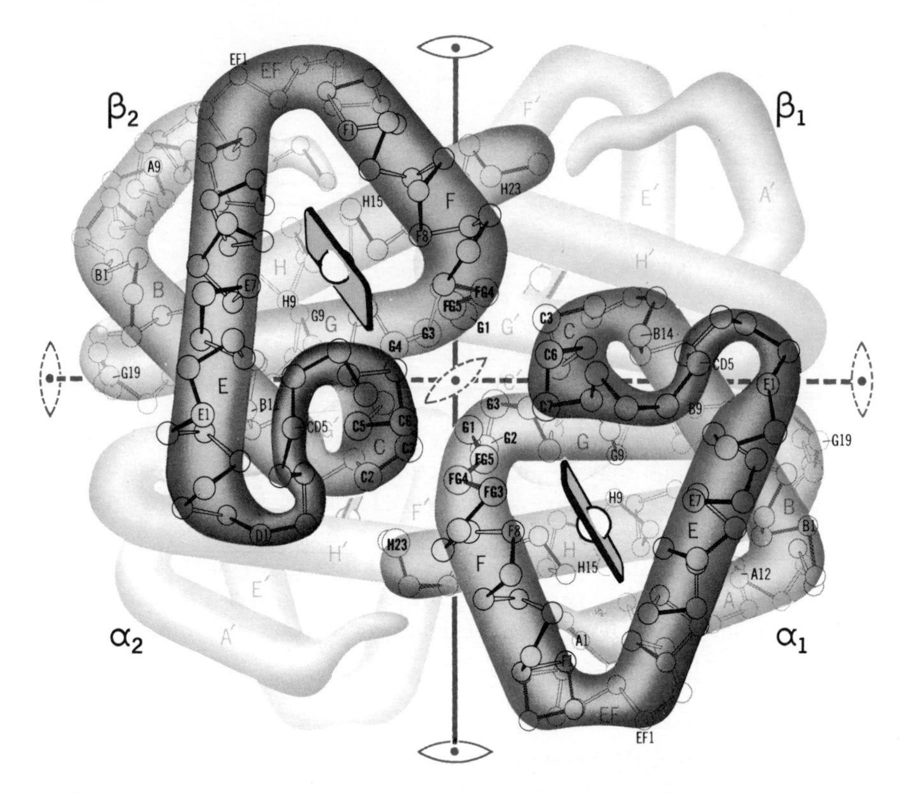

Figure 12.10 Hemoglobin. Four chains are folded and packed together, labelled α_1, α_2, β_1 and β_2. The straight segments are in α-helix configurations. Heme units are seen as flat disks. (From R. E. Dickerson and I. Geis, *The Structure and Action of Proteins,* W. A. Benjamin, Inc., Menlo Park, California. © 1969. All rights reserved. Used by permission.)

Common Properties of Proteins

12.21 Hydrolysis and Digestion

The digestion of proteins is nothing more than the hydrolysis of peptide bonds (amide linkages) as illustrated in Figure 12.11 with a hypothetical polypeptide. Huge protein molecules are hydrolyzed during the digestive process to their several amino acid units. Special digestive enzymes, themselves proteins, aid in the changes. The amino acids, being very soluble in aqueous media, are rapidly absorbed from the small intestine into the bloodstream and transported to sites where they are needed.

12.22 Denaturation

A wide variety of reagents and conditions do not hydrolyze peptide bonds but still destroy the biological nature and activity of a protein. When this has happened, the protein is denatured. After denaturation coagulation usually happens. Several of the more common chemicals or conditions that denature a protein are listed in Table 12.2.

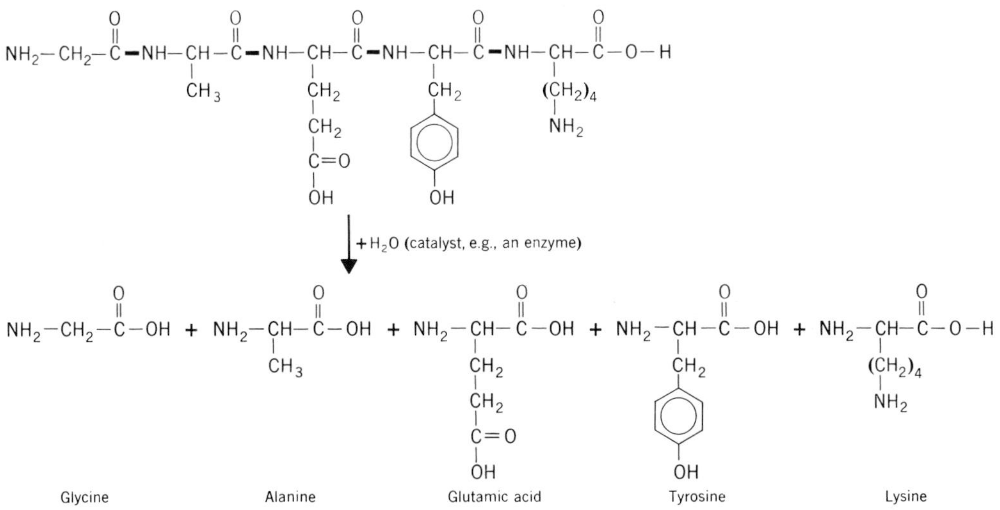

$+H_2O$ (catalyst, e.g., an enzyme)

Glycine Alanine Glutamic acid Tyrosine Lysine

Figure 12.11 The hydrolysis (or digestion) of a protein, illustrated here by the hydrolysis of a pentapeptide. Only peptide bonds break.

TABLE 12.2
Chemicals and Conditions That Cause Protein Denaturation

Denaturing Agent	How the Agent May Operate
Heat	Disrupts hydrogen bonds by making molecules vibrate too violently. Produces coagulation as in the frying of an egg
Solutions of urea $(NH_2{-}\overset{O}{\overset{\|}{C}}{-}NH_2)$	Disrupt hydrogen bonds. Since it is amidelike, urea can form hydrogen bonds of its own
Ultraviolet radiation	Appears to operate the same way that heat operates (e.g., sunburning)
Organic solvents (e.g., ethyl alcohol, acetone, isopropyl alcohol)	May interfere with hydrogen bonds in protein, since alcohol molecules are themselves capable of hydrogen bonding. Quickly denatures the proteins of bacteria, thus killing them (e.g., disinfectant action of ethyl alcohol, 70% solution)
Strong acids or bases	Can disrupt hydrogen bonds Prolonged action of aqueous acids or bases leads to actual hydrolysis of proteins
Detergent	May affect hydrogen bonds
Salts of heavy metals (e.g., salts of the ions Hg^{2+}, Ag^{+}, Pb^{2+})	Ions combine with SH groups. These ions usually precipitate proteins
Violent whipping or shaking	May form surface films of denatured proteins from protein solutions (e.g., beating egg white into meringue)

Structurally, **denaturation** is a disorganization of the molecular configuration of a protein. It can occur as an unfolding or uncoiling of a pleated or coiled structure or as the separation of the proteins into subunits, which may then unfold or uncoil (see Figure 12.12).

The fact that certain heavy-metal salts denature and coagulate proteins is the basis of an important method of poison treatment. Mercury(II), silver, and lead salts are poisons because their metallic ions, combining with sulfhydryl groups, wreak havoc among important body proteins once they enter general circulation. However, they can be kept in the stomach from general circulation by giving the poison victim the handiest protein available, for example, raw egg albumin (egg white). Of course, if the coagulated poison-albumin mixture remains too long in the stomach, digestion will eventually "dissolve" the protein and liberate the heavy-metal ions. An emetic must also be given following the raw egg white. (Also, call a doctor at once.)

The so-called "alkaloidal reagents"—tannic acid, picric acid, and phosphomolybdic acid—have been used in studying structures of alkaloids, hence, their group name. They also denature proteins.

Heat and other denaturing agents can be used to detect protein in urine. (Urine normally does not contain protein.) The urine specimen becomes cloudy when heated if protein is present, because the protein denatures and coagulates. Many clinical analyses of blood samples can be performed only in the absence of protein. Hence the analyst will add a "deproteinizing" agent such as an acid (e.g., trichloroacetic acid or sulfosalicylic acid) or some metal ion in alkaline solution to precipitate protein.

Denaturing agents differ widely in their action, depending largely on the protein. Some proteins (e.g., skin and hair) strongly resist most denaturing actions, because they are rich in disulfide linkages.

12.23 Effect of pH on Protein Solubility

Because some side chains, as well as the end groups on polypeptides, bear electrical charges, (+) and (−), the entire molecule may carry a net charge.

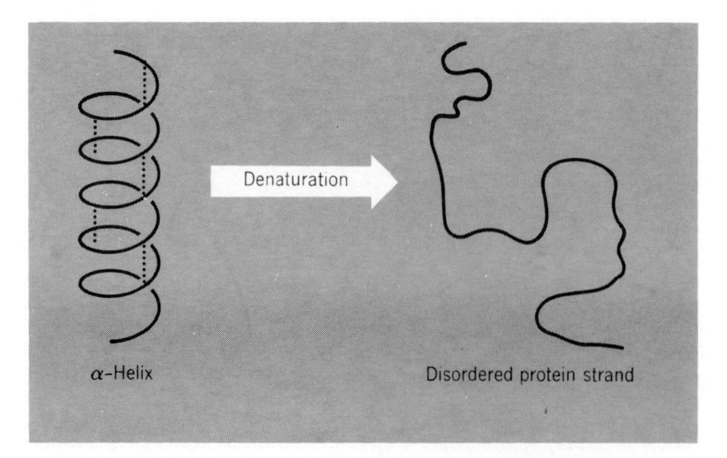

Figure 12.12 Denaturation of a protein—fundamentally a disorganization of secondary, tertiary, or quaternary structures without rupture of peptide bonds.

Suppose that the net charge is −1 and an "extra" —CO_2^- group is present. By adding mineral acid (by trying to lower the pH), that group will acquire a proton and become electrically neutral.

NH_3^+ CO_2^- CO_2^- + H_3O^+ ⟶ NH_3^+ CO_2^- CO_2H + H_2O

Polypeptide, net charge = −1 | Added acid | Polypeptide, net charge = 0

Now the polypeptide has an equal number of positive and negative charges and is isoelectric. On the other hand, a polypeptide might have a net charge of +1, and therefore an extra —NH_3^+ group. By adding hydroxide ion (by trying to raise the pH), we could pull a proton from one —NH_3^+ group; make it into an amino group, —NH_2; and erase the net +1 charge. This polypeptide would then also be isoelectric.

NH_3^+ CO_2^- NH_3^+ + OH^- ⟶ NH_3^+ CO_2^- NH_2 + H_2O

Polypeptide, net charge = +1 | Added base | Polypeptide, net charge = 0

Each protein, like each amino acid, has a characteristic pH, called its isoelectric point, pI, at which its net charge is zero and at which it cannot migrate in an electrical field. The significance of this is that polypeptide molecules can aggregate and precipitate, as seen in Figure 12.13, when they are isoelectric. Because protein is least soluble in water where the pH equals the pI for that protein, the pH of an aqueous fluid in which proteins must remain in solution to work (as with some enzymes) has to be buffered by the organism.

Casein, milk protein, precipitates if the pH of milk changes to 4.7, the pI for casein. Ordinarily, the pH of milk is 6.3 to 6.6. When bacteria grow in milk, however, they release acids that lower the pH and make the milk sour. The milk than "curdles"; that is, the casein separates into curds. As long as the pH of milk is something other than the isoelectric pH for casein, the protein remains colloidally dispersed. At or near its pI value, however, it has its lowest solubility.

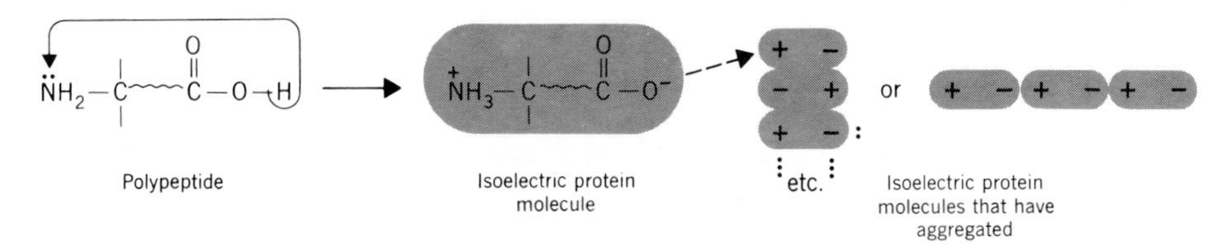

Figure 12.13 Aggregation of isoelectric protein molecules, which reduces their solubility in water. Proteins are least soluble at their isoelectric points.

Classes of Proteins

12.24 Classification According to Solubility

When proteins are classified according to their solubilities, two major families are the fibrous and globular proteins. Fibrous proteins that are insoluble in water include the following.

1. ***Collagens***—in bone, teeth, tendons, skin, and soft connective tissue. When such connective tissue is boiled with water, its collagen changes to gelatin, a much more soluble form. The change of this collagen to gelatin when meat is cooked is an important preliminary step to later digestion.
2. ***Elastins***—in many places where collagen also occurs, but particularly in ligaments, the walls of blood vessels, and the necks of grazing animals. Elastin has little hydroxyproline, no hydroxylysine, and is rich in hydrophobic side chains. Cross-links between elastin strands are important to its recovery after stretching. Elastin is not changed to gelatin by hot water.
3. ***Keratins***—in hair, wool, animal hooves, nails, and porcupine quills. The keratins are exceptionally rich in disulfide links.
4. ***Myosins***—proteins in contractile muscle.
5. ***Fibrin***—the protein of a blood clot, formed from a more soluble form, fibrinogen, when clotting must occur.

Globular proteins that are soluble in water or water containing certain salts include the following.

1. ***Albumins***—in egg white; in circulation in the blood where they perform various duties—buffers, carriers of lipids, carbohydrates, metal ions, and other small things that otherwise would not be soluble in blood. Albumins are soluble (or easily dispersed colloidally) in pure water.
2. ***Globulins***—include the gamma (γ)-globulins, part of the body's defense against infectious diseases. Globulins need the presence of dissolved salts to dissolve.

12.25 Classification According to Composition

The nature of the prosthetic group in conjugated proteins provides another way of classifying proteins.

1. *Glycoproteins*—proteins with sugar units attached; γ-globulin is an example. The sugar unit is a complex system.
2. *Hemoproteins*—proteins with heme units. Hemoglobin, myoglobin and an enzyme for using oxygen, cytochrome *c*.
3. *Lipoproteins*—proteins carrying lipid molecules.
4. *Metalloproteins*—proteins incorporating a metallic ion; several enzymes have this feature.
5. *Nucleoproteins*—proteins bound to nucleic acids, such as ribosomes and some viruses.
6. *Phosphoproteins*—proteins with an ester between the side chain of a serine residue and a phosphate unit. Milk casein is an example.

12.26 Classification According to Biological Function

Perhaps no other classification quite so clearly dramatizes the enormous importance of proteins.

1. *Enzymes*—All known enzymes are proteins, simple or conjugated.
2. *Contractile proteins*—In muscle both the stationary filaments, myosin, and the moving filaments, actin, are examples.
3. *Hormones*—Growth hormone, insulin and many other hormones are proteins.
4. *Storage proteins*—These store nutrients needed by the organism or used by another species feeding on it. Examples are the seed proteins in grains, casein in milk, ovalbumin in egg white and ferritin, the protein that stores iron for us in the spleen.
5. *Transport proteins*—These carry things from one place in the body to another. Examples are hemoglobin and myoglobin that carry oxygen, serum albumin that carries fatty acids in the blood, ceruloplasmin that carries copper ions, and a globulin that carries iron ions in blood.
6. *Structural proteins*—Proteins that hold structure together. Examples are collagen, elastin, keratin, and proteins in cell membranes.
7. *Protective proteins*—These participate in the body's defensive mechanisms. Examples are antibodies, fibrinogen, and thrombin (both important in the formation of a blood clot), and complement (a material that sometimes participates with antibodies in handling foreign proteins or antigens).
8. *Toxins*—Proteins that are poisons. Examples are snake venom, diphtheria toxin, and the toxic element in bacterial food poisoning, clostridium botulinum toxin.

Brief Summary

Amino Acids. Twenty α-amino acids are the monomers for proteins. (A few others occur rarely, and some are made after a polypeptide chain has been put together.) All but one of these 20 is optically active and in the L-family. In the solid state, as well as in aqueous solution at or near a pH of 7, most amino acids exist in a dipolar ionic form making them saltlike in solubility and melting points. Each has a characteristic pI, or isoelectric point, at which it will not migrate in an electrical field. Several have hydrophobic side chains: alkyl or aryl groups. The rest have hydrophilic side chains with —OH, $—CO_2^-$, $—NH_3^+$, or similar groups. The sulfydryl group in cysteine is easily oxidized to a disulfide.

Polypeptides. Covalent forces in the peptide bond and the disulfide link hold amino acid residues together in these polyamides. The repeating unit is nitrogen-carbon-carbonyl: $—N—\overset{|}{\underset{|}{C}}—\underset{|}{C}{=}O$, each carbon (the alpha-carbon) bearing a particular side chain. The primary structure of a protein is the sequence of side chains in amino acid residues held by these covalent forces. Once the primary structure is put together, it spontaneously takes up higher levels of structural features in response largely to forces acting in hydrogen bonds, but also reflecting salt bridges and the avoidance or attraction of water by side chains. Three secondary structures place neighboring side chains in particular positions. These are the α-helix, a tightly coiled, right-handed helix; the β-pleated sheet, a rippled sheet of polypeptide molecules lying side by side; and the triple helix, three relatively open, left-handed helices coiled together in a right-handed fashion in the tropocollagen cable. Helices often fold and twist giving the polypeptide its tertiary structure in which hydrophobic groups are turned to the inside, away from water, as much as possible and hydrophilic groups are on the outside. Prosthetic groups may be folded in, also. When two or more polypeptides form a complex, we have the final level of protein structure, the quaternary structure.

Proteins. Simple proteins may be made of single polypeptide chains, or of two or more such chains in a quaternary structure. Conjugated proteins may be of single chains bound to some nonprotein group (a prosthetic group), or they may have two or more chains holding one or more prosthetic groups. (Often, however, "protein" and "polypeptide" are used as synonyms.) A protein, therefore, is a molecular species consisting mostly or entirely of amino acid residues and held together in a particular overall shape by both covalent and noncovalent forces. Proteins may be classified according to solubility, composition, or function. Because of acidic and basic side chains, the solubility of a protein is sensitive to the pH of the medium and is least at the pI or isoelectric point of the protein. Because of the peptide (amide) bonds, proteins can be hydrolyzed (digested). Because noncovalent forces are individually weak, proteins can be denatured, suffering the loss of 2°, 3°, and 4° structural features.

Selected References

Books

1. L. Stryer. *Biochemistry.* W. H. Freeman and Company, San Francisco, 1975. Part I, "Conformation," discusses the relation between all structural features of proteins and their biological functions; beautifully illustrated and very lucid.
2. A. L. Lehninger. *Biochemistry.* Worth Publishers, Inc., New York, 1970. Chapter 6, "Proteins: Conformation," is both well written and illustrated.
3. R. E. Dickerson and I. Geis. *The Structure and Action of Proteins.* Harper & Row, Publishers, New York, 1969. Another outstandingly well written and illustrated description of protein structure and function.

Articles

(Many of these are old, but they are classics still being cited as references in the most recent textbooks of biochemistry.)

1. J. C. Kendrew. "The Three-Dimensional Structure of a Protein Molecule." *Scientific American,* December 1961, page 96. Myoglobin is considered to be a small protein—only 150 amino acid units in one polypeptide chain plus a heme group. Kendrew reports on its three-dimensional structure. A striking model of myoglobin, in color, is pictured.
2. E. Frieden. "Non-Covalent Interactions." *Journal of Chemical Education,* December 1975, page 754. A survey of all the kinds of attractions and repulsions of a noncovalent nature in biological compounds.
3. J. Gross. "Collagen." *Scientific American,* May 1961, page 121. Magnificent electron photomicrographs of collagen fibrils illustrate this article about the principal component of connective tissue.

4. R. D. B. Fraser. "Keratins." *Scientific American,* August 1969, page 87. The details of the molecular structure of keratins are described.
5. M. F. Perutz. "The Hemoglobin Molecule." *Scientific American,* November 1964, page 64. Perutz, sharer of the 1962 Nobel Prize in Chemistry with J. C. Kendrew, describes his work on determining the structure of hemoglobin.
6. Choh Hao Li. "The ACTH Molecule." *Scientific American,* July 1963, page 46. How the structure of adrenocorticotropic hormone (ACTH), a protein, relates to its function.
7. N. Sharon. "Glycoproteins." *Scientific American,* May 1974, page 78. Some of the rare sugar units in glycoproteins and how they might affect function.
8. J. M. Murray and A. Weber. "The Cooperative Action of Muscle Proteins." *Scientific American,* February 1974, page 59. How muscle action might correlate with the interactions of four major proteins, calcium ions, and sources of chemcial energy.
9. C. Cohen. "The Protein Switch of Muscle Contraction." *Scientific American,* November 1975, page 36. How changes in the conformations of two proteins, tropomyosin and troponin, triggered by calcium ions, lead to the contraction of muscle.
10. D. L. Martin and J. E. Huheey. "Sickle-Cell Anemia, Hemoglobin, Solubility, and Resistance to Malaria." *Journal of Chemical Education,* March 1972 (Vol. 49), page 177. A nice introduction to this disease with thoughts on sickling as a contribution to resistance to malaria in central and west Africa.
11. N. Isenberg and M. Grdinic. "Cyclic Disulfides. Their Functions in Health and Disease." *Journal of Chemical Education,* June 1972, page 392. Disulfide groups occur not only in proteins but in other biological systems, where their ease of reduction to sulfhydryl groups contributes in many ways to biological functions.

Questions and Exercises

1. Draw the structure of the polypeptide that would form from the following amino acids if they became joined in the following sequence. (Let the first function as a carboxylic acid leaving its amino group intact. Refer to Table 12.1 for the structure of amino acids.)

   alanine-valine-phenylalanine-glycine-leucine

2. Repeat Exercise 1 with the sequence:

   aspartic acid-threonine-lysine-glutamic acid-tryrosine

3. Study the two structures you have written for Exercises 1 and 2.
   (a) Which one is the more hydrocarbonlike?
   (b) Which one would tend to be more soluble in a nonaqueous, fatlike medium?
   (c) Which one would tend to be more soluble in water? Why?
   (d) Which one would be more capable of participating in salt bridges? Why?
4. If the tripeptide, Gly·Cys·Ala, were subjected to mild oxidizing conditions, what would form? Write its structure using three-letter symbols.
5. At which level of protein structure are covalent forces exclusively at work? What are these covalent forces?

6. By coiling into a helix or aligning in a pleated sheet, which force(s) of attraction is used by a polypeptide?
7. Vitamin C is necessary to what molecular task in the formation of strong bones?
8. In what way does hemoglobin illustrate the following proteins?
   (a) A conjugated protein (b) A protein with quaternary structure
9. Give a brief description of the organization of collagen.
10. Why are not "polypeptide" and "protein" exact synonyms?
11. How do the side chains—hydrophilic and hydrophobic—influence the tertiary structure of a protein?
12. What would be the products of the digestion of the following polypeptide?

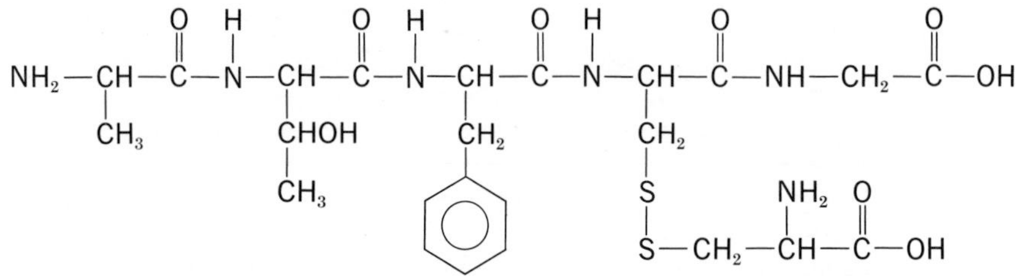

```
             O  H         O  H         O  H        O                  O
             ‖  |         ‖  |         ‖  |        ‖                  ‖
NH2—CH—C—N—CH—C—N—CH—C—N—CH—C—NH—CH2—C—OH
     |           |           |           |
    CH3        CHOH         CH2         CH2
                 |           |           |
                CH3      (benzene)       S          NH2  O
                                         |           |   ‖
                                         S—CH2—CH—C—OH
```

13. If the polypeptide of Question 12 were subjected to mild reducing conditions, what products would form? (Show their structures.)
14. Why is a protein least soluble in a medium at its isoelectric point?
15. What is the difference between digestion and denaturation of a protein?
16. What is the relation between collagen and gelatin?
17. How are collagens and elastins alike? How are they different?
18. Explain how raw egg may be used as first aid in the case of poisoning by a copper insecticide or a lead-based paint? Why must "second aid" be given, and of what should this consist?

chapter 13

Metabolism-Its Regulation and Defense Enzymes, Vitamins, Hormones, Chemotherapy

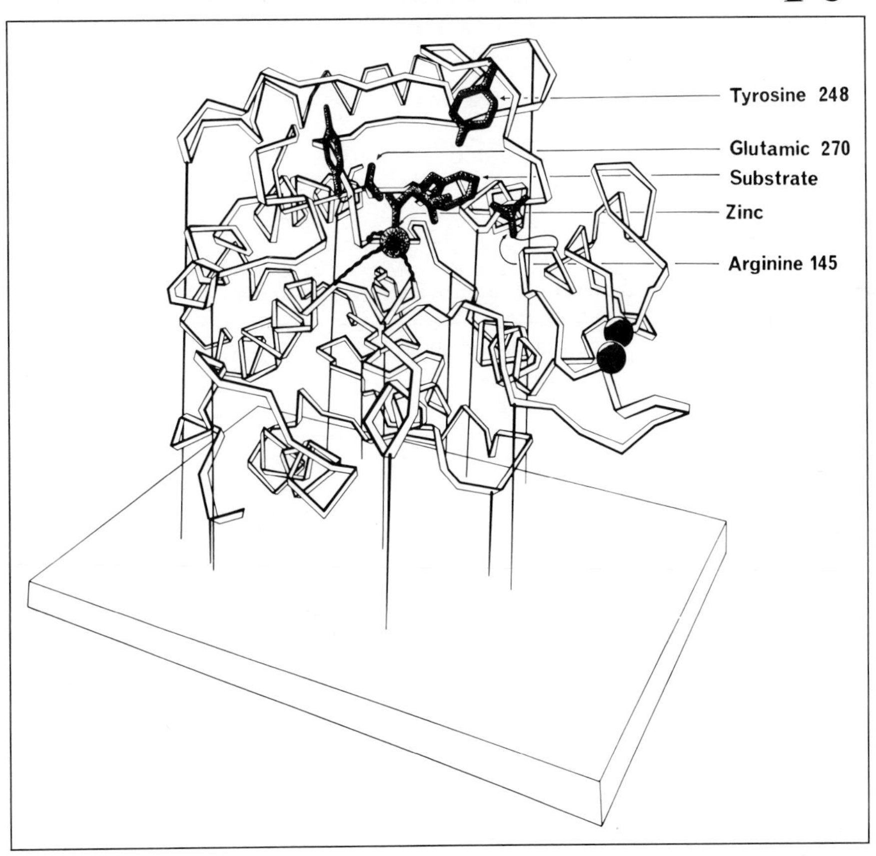

The molecules of an enzyme, carboxypeptidase A, have this shape, at least when they are in the solid state. The substrate, the molecule whose reaction is being catalyzed, nestles in the folds of the enzyme, and particular sidechains participate in the catalysis. Substrates fit to enzymes like locks to keys, and we learn more about this in this chapter. (Courtesy William Lipscomb, Harvard University.)

13.1 Introduction to Important Terms

Thousands of chemical events are necessary to life and health, but they may not all go at the same time. Metabolism running wild brings death just as surely as metabolism turned off. Regulation is essential, and the regulation of metabolism is the heart of our study in this chapter. Involved are enzymes, some of the vitamins, the hormones, and the chemical defenses of the body. While we introduce vitamins and hormones and describe how they are used, we become more detailed about them in later chapters. This chapter is mostly about enzymes and how they are regulated.

Enzymes are catalysts essential to chemical reactions in living systems. Without enzymes reactions in cells and fluids would be much too slow to serve any form of life that we know. The absence or insufficiency of an enzyme can have very serious consequences. Albinism and phenylketonuria (PKU) are two of hundreds of examples.

Vitamins are nonprotein organic substances that must be in the diet in trace amounts. The body cannot make these compounds. Specific diseases are associated with the absence of vitamins in the diet, vitamin deficiency diseases such as xerophthalmia (eye disease), rickets, beriberi, pellagra, pernicious anemia, and scurvy.

Minerals are inorganic ions necessary for growth, repair, and operation, and which also must be included in the diet. Many are needed in relatively large amounts: Ca^{2+}, Mg^{2+}, Na^{+}, K^{+}, Fe^{2+}/Fe^{3+}, $H_2PO_4^{-}/HPO_4^{2-}$, and Cl^{-}.

TABLE 13.1
Trace Elements of Nutritional Significance for Humans[a]

Element	Approximate Amount in a 154-lb (70 kg) Man, in mg	Some Uses
Chromium	<6	Probably an essential cofactor to insulin (glucose metabolism)
Cobalt	Up to 3	Part of vitamin B_{12}
Copper	100	Cofactor for respiratory enzymes and other enzymes; in all body tissue
Fluorine	Trace	Bone and teeth formation
Iodine	30	Needed to make the hormone thyroxine; if deficient—goiter
Manganese	20	In several enzymes; peptidases, decarboxylases, kinases
Molybdenum	Up to 5	Enzymes for metabolizing nucleic acids; enzymes for oxidizing aldehydes
Zinc	1600	Many dehydrogenating enzymes; carbonic anhydrase

[a] In addition to the minerals on this list there are several others that may be important to man but for which the picture is incomplete. These include tin, selenium, nickel, silicon, and vanadium. (Data from R. L. Pike and M. L. Brown. *Nutrition: An Integrated Approach.* John Wiley & Sons, Inc., New York, 1967, page 98.)

Sulfur as cysteine and methionine is also needed. Others, called the **trace elements,** are required in very small amounts (Table 13.1).

Hormones are chemical messengers in specialized glands, secreted into the bloodstream as needed and taken by "target cells," where the chemical "message" is delivered. The overabundance of hormones on the one hand or their insufficiency on the other is related to some dramatic disfunctions: cretinism, goiter, diabetes, dwarfism, and gigantism, for examples.

Enzymes

13.2 Chemical Nature of Enzymes

All known enzymes are proteins. A cell may contain as many as 3000 different enzymes. Over 1000 enzymes have been identified; several dozen have been isolated, purified, and crystallized; and the amino acid sequences of several have been determined.

Many enzymes are conjugated proteins in which the prosthetic group has the actual catalytic activity. Sometimes that group is made from a vitamin. Many enzymes have quaternary structure. They consist of two or more polypeptide chains, each with secondary and tertiary structure. Sometimes the chains are identical, and yet an individual chain, alone, will not function as a catalyst.

Because enzymes are proteins, they are sensitive to any or all of the denaturing agents, including changes in the pH of the medium. The most dangerous poisons we know act by denaturing enzymes.

13.3 Cofactors

The nonprotein group in an enzyme is called a **cofactor.** Enzymes requiring cofactors lack catalytic activity without them. Where a cofactor is needed, the wholly protein portion of the enzyme is called the **apoenzyme.** The cofactor may be an organic species called a **coenzyme,** or it may be a metal ion—often one of the trace elements—and then is called a **metal-ion activator.**

We may express the relations of these substances by word equations:

coenzyme + apoenzyme ⟶ enzyme (sometimes, holoenzyme)
metal ion + apoenzyme ⟶ enzyme
activator

13.4 Vitamins and Coenzymes

Some vitamins (but not all) are used to make molecules of coenzymes, many of which are merely mono- or diphosphate esters of these vitamins. Structures **1** to **4** that follow are not to be memorized but note that the shaded portions are from molecules of one of the vitamins. The coenzyme thiamine pyrophosphate, **1**, is a diphosphate ester of thiamine (vitamin B_1) used in the metabolism of glucose.

Pyridoxal phosphate, **2**, the phosphate ester of pyridoxal, is a coenzyme needed in those reactions in which amino groups are transferred. Pyridoxal, an aldehyde, is made from vitamin B_6 (pyridoxine), by the oxidation of a $—CH_2OH$ group.

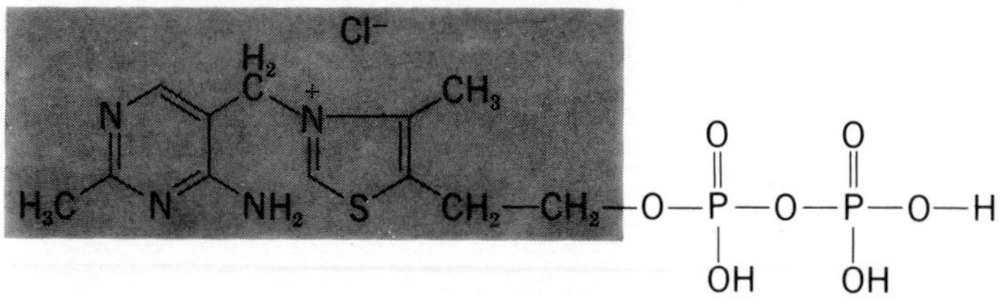

1
Thiamine pyrophosphate
(Cocarboxylase)

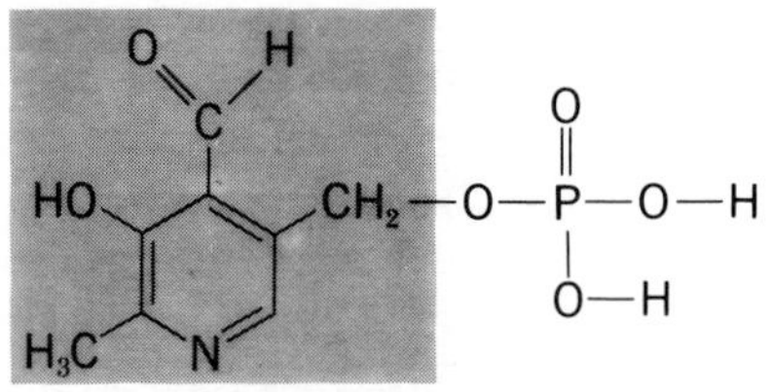

2
Pyridoxal phosphate

Nicotinamide adenine dinucleotide, **3-a**, is usually called NAD^+ (or even more simply, but less descriptively, NAD). A phosphate ester of NAD^+, another important coenzyme, also has a name to match its size—nicotinamide adenine dinucleotide phosphate—but is more easily known as $NADP^+$ (the "P" signifying the extra phosphate group), **3-b**. NAD^+ and $NADP^+$ are coenzymes for a large number of oxidations and reductions. The nicotinamide unit (shaded) with its (+) charge is the electron-accepting site.[1]

Flavin adenine dinucleotide (FAD), **4**, which requires riboflavin (vitamin B_2), is also involved in biological oxidation-reductions. Another flavin coenzyme, flavin mononucleotide or FMN, is simply the monophosphate ester of riboflavin (i.e., FAD, **4**, minus the top network of adenine, ribose, and the upper phosphate unit).

13.5 How Enzymes Are Named and Classified

Many enzymes are named by attaching the suffix -ase to the name of the compound, called the **substrate,** whose reaction the enzyme catalyzes. Thus a lipase is an enzyme for a reaction of a lipid. A peptidase is an enzyme for some reaction of a peptide.

[1] The positive charge of the nicotinamide unit is intimately involved in the catalytic activity of the corresponding enzymes; that is why the (+) signs are included in NAD^+ or $NADP^+$. The negative charges on the phosphate units vary with the pH of the medium and are not considered in writing the very abbreviated symbols for the coenzymes.

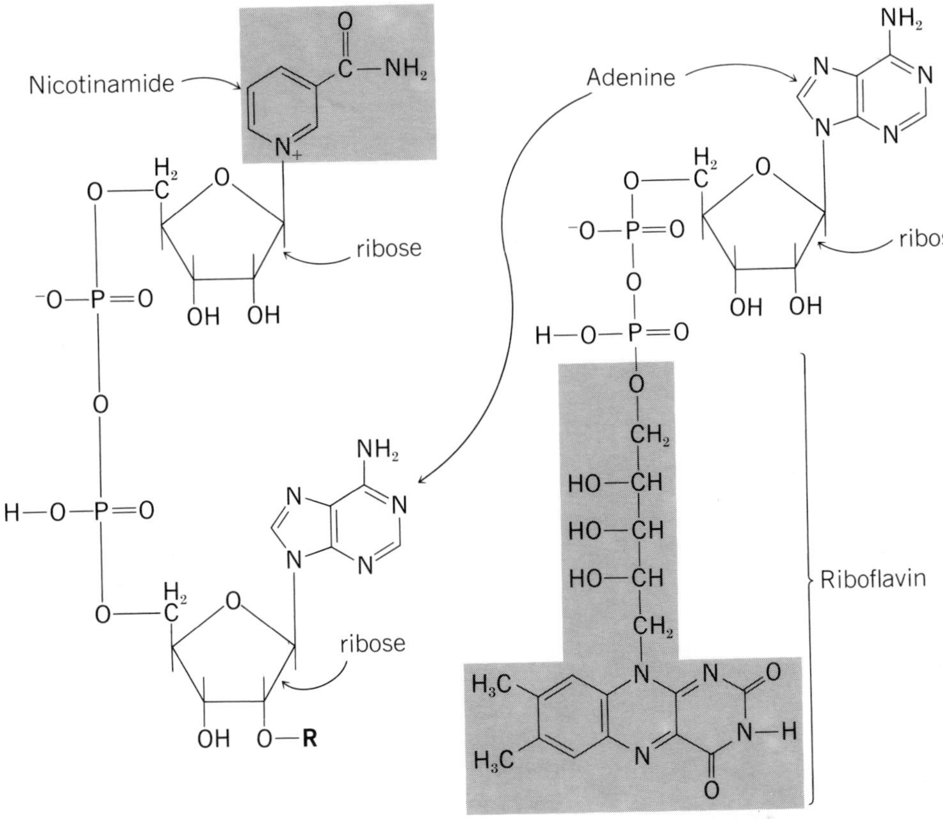

3

a NAD^+ **R** = H

b $NADP^+$ **R** = $-OPO_3^{2-}$

4

FAD

Exercise 13.1 What is the likely kind of substrate for each enzyme?
(a) Sucrase (b) Glucosidase (c) Protease
(d) Phosphatase (e) Esterase

Other enzymes are named by affixing -ase to the name of the reaction being catalyzed. An **oxidase** is an enzyme catalyzing an oxidation. **Reductase** and **hydrolase** are other examples of enzymes named after reactions not substrates. **Transferases** are enzymes to aid the transfer of a group from one molecule to another. Transaminase, for example, is an enzyme for the transfer of an amino group. Transacylases (for acyl group transfers) and transglucosidase (for glucose transfers) are other examples of transferases. Transphosphatases, often called **kinases,** help transfer a phosphate unit from one molecule to another.

Lyases aid in the addition of something (e.g., water) to a double bond. Isomerases catalyze the isomerization of one molecule into its isomer. Phosphohexose isomerase, for example, helps glucose 6-phosphate change into fructose 6-phosphate in the metabolism of glucose. A ligase is an enzyme that "knits" new bonds to carbon while the chemical energy of adenosine triphosphate (ATP) is consumed.

A few enzymes, known the longest, have names without the -ase ending, for example, pepsin and trypsin, two digestive enzymes. The least informative, chemically, these names are nonetheless widely used.

13.6 Systematic Names for Enzymes

The confusion in enzyme names led to a recommendation by an International Enzyme Commission for a new system. Our needs require only an awareness of its existence. We shall do no more than give one illustration of how systematic, chemically informative names for enzymes are devised. The name(s) of the principal reactant(s) and the type of reaction are included in the name. For example, in the following reaction an amino group is transferred from the glutamate ion to the pyruvate system.

$$\underset{\text{Glutamate}}{^{-}OOCCH_2CH_2CH(NH_3^+)COO^-} + \underset{\text{Pyruvate}}{CH_3COCOO^-} \longrightarrow \underset{\alpha\text{-Ketoglutarate}}{^{-}OOCCH_2CH_2COCOO^-} + \underset{\text{Alanine}}{CH_3CH(NH_3^+)COO^-}$$

The systematic name for the enzyme is "glutamate:pyruvate aminotransferase." The common name is simply transaminase (and you perhaps can appreciate why in routine conversation biochemists use common names). We shall use common names almost exclusively.

Catalytic Properties of Enzymes

13.7 Rates of Enzyme-Catalyzed Reactions

The rates of some enzyme-catalyzed reactions are spectacularly high. The enzyme for the interconversion of carbon dioxide, water, and carbonic acid is carbonic anhydrase.

$$CO_2 + H_2O \xrightleftharpoons{\text{carbonic anhydrase}} \underset{\text{Carbonic acid}}{H_2CO_3}$$

When carbon dioxide is plentiful, each molecule of this enzyme can convert 600 000 molecules of carbon dioxide each second into carbonic acid, the fastest known rate of reaction for an enzyme, and 10 million times faster than the reaction uncatalyzed.

Beta-amylase, which catalyzes the hydrolysis of oxygen bridges between glucose units in amylose, can handle 4000 bridges/sec for each molecule of enzyme. Acetylcholinesterase, an enzyme essential to transmitting nerve impulses between the end of one nerve cell to the next cell, handles 25 000 acetyl groups/sec for each molecule of enzyme. Most enzymes catalyze events at the rate of 1 to 10 000/sec for each enzyme molecule.

Each enzyme works best at a particular pH (or pH range). The optimum pH 1.5, for pepsin, a protease acting in the stomach, is quite acidic like the stomach contents. Trypsin, another protease, has an optimum pH of 7.0 to 7.7, slightly alkaline, like the medium in which it works, the juices in the upper intestinal tract where pepsin could not work.

Exercise 13.2 Each enzyme works best at a particular temperature. Offer an explanation why rates of enzyme-catalyzed reactions do not keep increasing with temperature as is normal for most uncatalyzed reactions.

13.8 Specificity of Enzymes

Enzymes have a remarkable ability to "pick and choose" not only the types of reactions they catalyze, but also the substrates on which they work. In a test tube, any mineral acid will catalyze the hydrolysis of esters, acetals, or amides, and the hydrolysis essentially of any member of any of these families. In contrast, an enzyme that catalyzes the hydrolysis of an amide will work only poorly or not at all on an ester or an acetal. Trypsin, a digestive enzyme, acts far better on peptide bonds than ester bonds, for example, and trypsin works best of all on a peptide bond when the carbon alpha to the carbonyl group at that bond has the side chain of either lysine or arginine. In other words, trypsin, although a peptidase, works best not on any peptide bond but just on those having particular neighboring groups. Urease is an enzyme that approaches absolute specificity. It catalyzes the hydrolysis of the amide bonds in urea, but not in such closely similar compounds as biuret or amides in general.

$$\underset{\text{Urea}}{NH_2-\overset{\overset{\displaystyle O}{\|}}{C}-NH_2} \qquad \underset{\text{Biuret}}{NH_2-\overset{\overset{\displaystyle O}{\|}}{C}-NH-\overset{\overset{\displaystyle O}{\|}}{C}-NH_2} \qquad \underset{\text{Amides}}{R-\overset{\overset{\displaystyle O}{\|}}{C}-NH_2}$$

Since substrate molecules are usually much smaller than those of the enzyme, only a relatively small site on the enzyme molecule can be responsible for catalytic properties. Such a site is called the **active site,** and the functional groups that do the catalytic work are called the **catalytic groups.**

13.9 Control of Enzyme Activity

The enzymes in a cell cannot all be active simultaneously or the cell would probably explode in a furious chemical storm. What cells must do, chemically, varies with time. When the brain signals a muscle to do some work, for example, enzymes must quickly go into action. When the signal comes to shut down, these same enzymes should quietly retire. The body's enzymes must therefore be under tight control. Activation is required at times, inhibition at others. Several strategies handle these needs. As we survey theories of how enzymes work, we shall study nine control mechanisms.

How Enzymes Work

13.10 Enzyme-Substrate Complex

A molecule of enzyme and a molecule of substrate combine by noncovalent forces to form an **enzyme-substrate complex.** Some complexes have been isolated and studied, and their structures are known in great detail. In equation form we may write:

$$E + S \rightleftharpoons E—S \rightleftharpoons E—S^* \rightleftharpoons$$

E = Enzyme; S = Substrate; E—S = Enzyme substrate complex; E—S* = * Means that the substrate has become activated

$$E—P \rightleftharpoons E + P$$

E—P = Complex of enzyme and product; E = Enzyme (recovered); P = Product(s)

13.11 Binding Sites

The specificity of an enzyme is explained by the need for the shape of at least part of the enzyme molecule to be complementary and matching to the shape of the substrate, as illustrated in Figure 23.1. The enzyme has binding sites as well as an active site. The **binding sites** are a pattern of electrical charges and hydrogen-bond donating or accepting groups arrayed on or sticking out from the protein part of the enzyme. These attract groups with complementary patterns on substrate molecules. As a molecule of substrate nestles to the binding sites of the enzyme, the enzyme's active catalytic site is brought to that part of the substrate about to react chemically. The fact that each enzyme has an optimum pH is explained by the tendency of patterns of (+) and (−) charges to change with pH.

13.12 Lock and Key Theory

With some enzymes, binding sites "match" substrates much as a lock will accept only one or maybe a few keys. In this **lock and key theory** (or enzyme-substrate theory), as the substrate settles onto these sites, the forces of attraction drawing it down set up strains in the bonds of the substrate, strains that are the early stages in the reorganization of chemical bonds leading to products.

13.13 Induced Fit Theory

With other enzymes, binding sites are not initially such exact matches for substrates. Several keys will work. These enzymes can draw substrates of similar shapes, not necessarily exactly matching shapes. As a substrate starts to form weak bonds to the enzyme, strains occur in the enzyme's noncovalent bonds of its secondary and higher structural features. These features change.

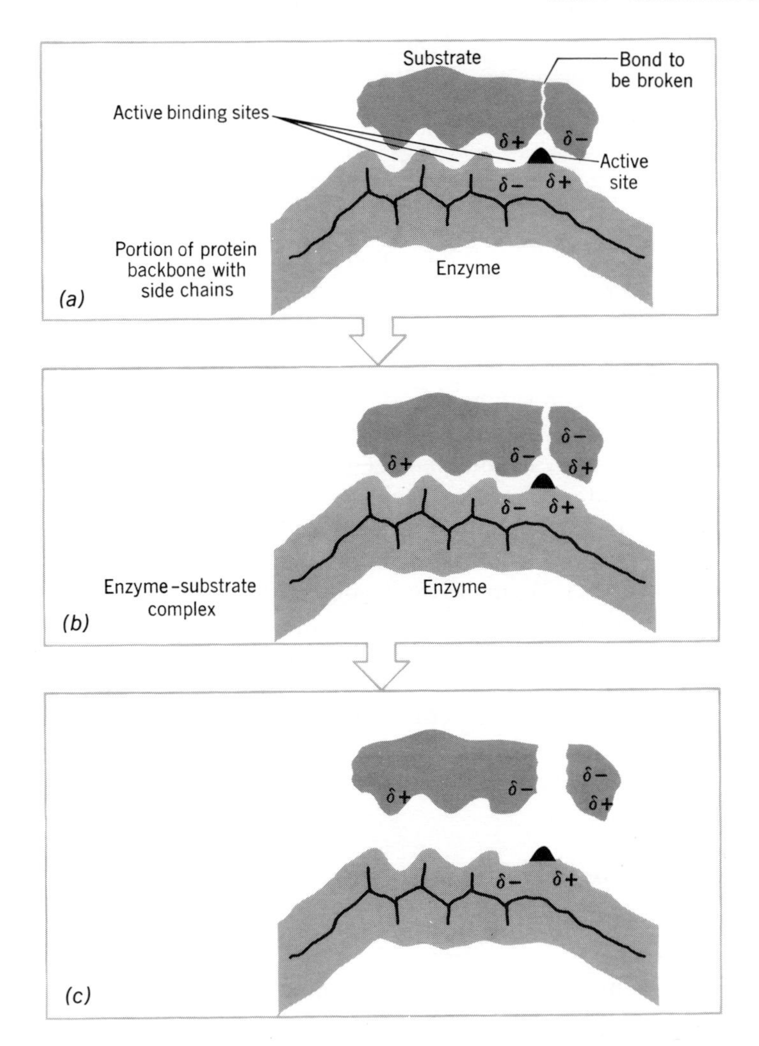

Figure 13.1 Lock and key model of enzyme action. (*a*) Enzyme and substrate have complementary shapes and electrical binding forces bring the two together. In the induced fit model, the substrate induces the enzyme to shift into the correct shape. (*b*) In the complex, bonds in the substrate may already be strained. In any case, the reaction proceeds. (*c*) Product molecules, having different patterns of electrical charge, separate from the surface of the enzyme. The active site is often supplied by a molecule of a B-vitamin.

The shape of the enzyme is induced by the substrate to alter itself until the fit of the substrate over the active site is achieved.

13.14 Saturation of Enzymes

For a given molar concentration of the enzyme, the overall rate at which the product forms usually rises sharply with an increase in the molar concentration of the substrate. See Figure 13.2. Eventually, however, the rate levels off. Still higher concentrations of substrate have little or no effect on the rate.

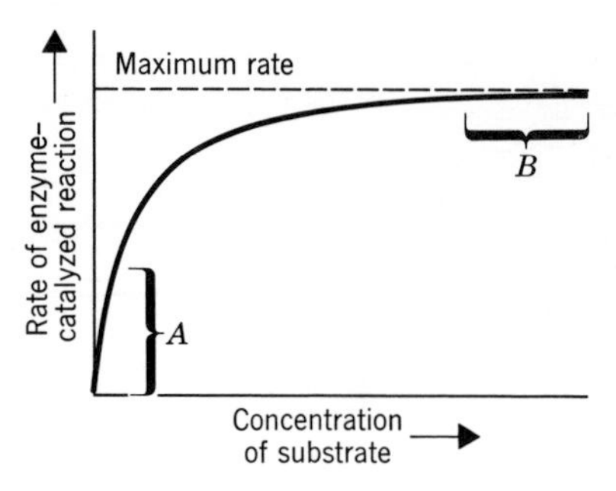

Figure 13.2 Effect of changes in concentration of substrate on the rate of a reaction catalyzed by a fixed concentration of enzyme. In region *A* a small change in substrate concentration causes a large rise in rate because substrate molecules can still find a large number of enzyme molecules not yet working. In region *B* enzyme molecules are all working and any additional substrate molecules have to wait. Now the rate has leveled off.

Normally, the rate of a reaction keeps increasing as the concentrations of reactants increase, but not with reactions catalyzed by enzymes. Enzyme molecules must be treated as reactants that are in effect removed as they become saturated with substrate units. More substrate molecules cannot be handled until earlier ones have been changed and moved out.

With some enzymes the sharp rise in rate at low substrate concentration is delayed. See Figure 13.3. A small rise in substrate concentration has a small effect on overall rate. Then, a small additional increase in substrate concentration accelerates the rate dramatically. Finally, the rate levels off. The graph of rate versus concentration for these enzymes has a "lazy S" shape called a sigmoid shape. Either kind of rate curve demonstrates that enzyme-catalyzed reactions increase in rate with increasing substrate concentration only up to a point, up to a certain maximum velocity.

Enzyme Activation and Inhibition

13.15 Allosteric Effects

Enzymes displaying sigmoid rate curves may have two (or more) active sites. To simplify, we assume an enzyme with just two polypeptide chains, each one bearing an active site. Molecules of substrate at very low concentration (region A in Figure 13.3) bind to one of the two active sites, but they encounter some resistance. See Figure 13.4. The shapes do not initially "fit." They must be forced. The rate of the reaction, therefore, does not in-

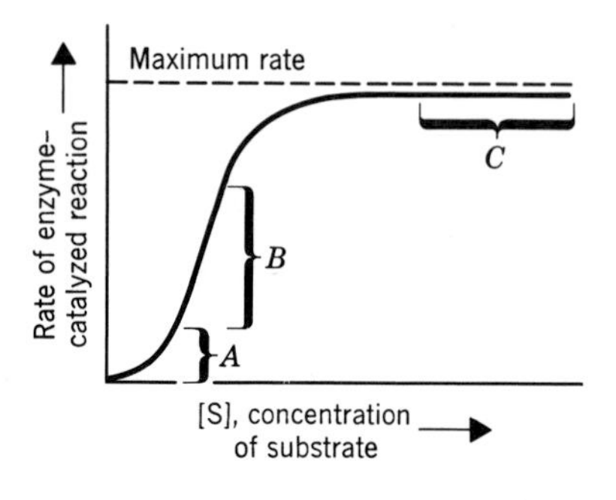

Figure 13.3 Effect of changes in substrate concentration on rate when the fixed concentration of enzyme is of an enzyme with two (or more) active sites. At low [S] in region *A*, small changes in [S] give only small changes in rate. But substrate molecules are doing more than binding to their own active sites, doing more than reacting to give products. They are activating still other active sites on the enzymes. Therefore, at medium values of [S], region *B*, additional substrate encounters much more active enzyme and the rate rises sharply with small increases in [S]. Eventually, in region *C* the enzyme is saturated.

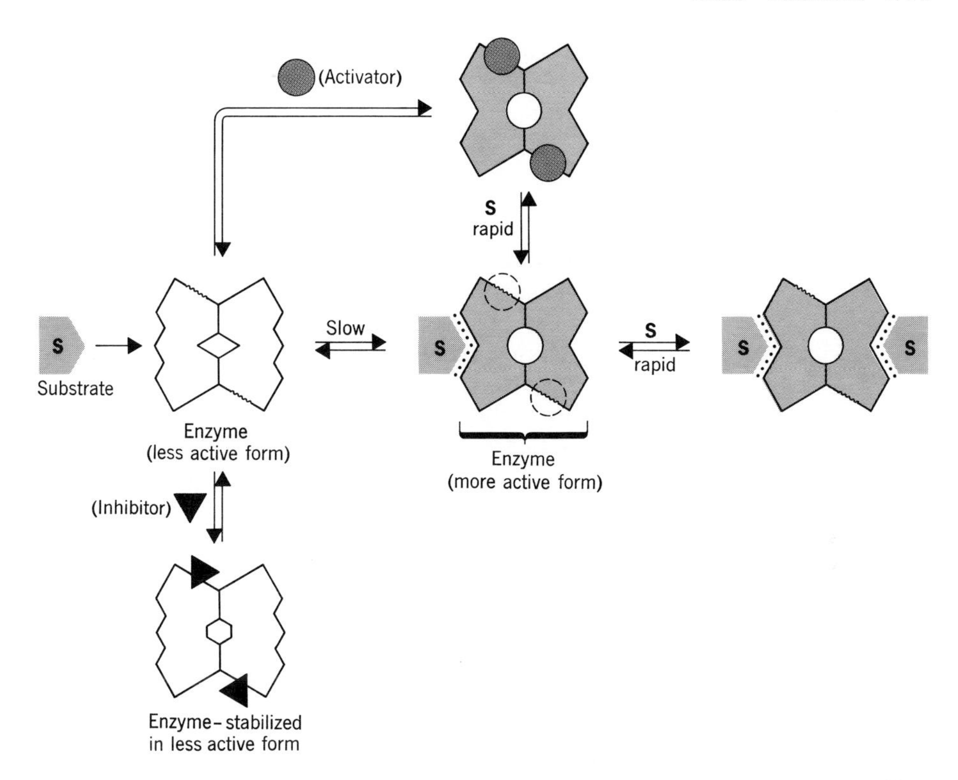

Figure 13.4 Allosteric effects by effectors or modulators (molecules represented by circles) or by inhibitors (molecules represented by triangles). The enzyme can exist in a less active and a more active form. In the absence of either effector or inhibitor, the first substrate molecule to combine (represented by a pentagon) forces the "fit," and that action pops the remaining active site into the proper shape. This makes the remaining site much more active toward the remaining substrate. An activator, by combining elsewhere, forces both active sites into their right shapes. An inhibitor stabilizes the less active form.

crease markedly at low substrate concentration. However, this first binding alters the shape of the entire two-stranded enzyme, including the other active site. The second site now much more easily accepts the second substrate molecule. Thus the first substrate molecules actually activate the remaining sites, and the overall rate can now increase very markedly with increasing concentration of the substrate (region *B*, Figure 13.3). Eventually, of course, all active sites become saturated and the rate tapers off. This phenomenon, called **allosteric activation** (from *allo-*, other; *steric* space, other space or other site), is the activation of one active site by a change in shape at a location somewhere else—at another site—in the enzyme. In a later chapter we shall see allosteric activation at work to permit the nonenzymic protein, hemoglobin, to transport oxygen efficiently, showing that allosterism is not limited to enzymes.

13.16 Effectors

The allosteric activation just described was by an actual substrate. In some enzymes, an entirely different, nonsubstrate molecule can cause the allosteric change. A substance that can do this is called an **effector,** and whenever the cell has some mechanism for controlling the presence of an effector it controls an enzyme. The circles in Figure 13.4 represent molecules of effector acting on the two-stranded enzyme to change the shape of both active sites; such action would change the rate curve for the enzyme to that of Figure 13.2.

13.17 Competitive Inhibition of Enzymes

Sometimes the cell has a substance whose molecules resemble those of a particular substrate and can be accepted and held by the substrate's enzyme without any further reaction occurring. When such a substance is present along with the substrate, the two compete for active sites with the "look-alike" acting as an inhibitor. The higher the concentration of the competitive inhibitor, the more the enzyme is inhibited and the slower the remaining enzyme will catalyze a change in the true substrate. See Figure 13.5.

Sometimes competitive inhibitor is a product made by the enzyme directly or it may be made later farther down a sequence of reactions. Then competitive inhibition is called **feedback inhibition.** For example, the amino acid isoleucine is made from threonine by a series of steps, each with its own enzyme. Molecules of isoleucine inhibit the first enzyme, E_1, in the sequence. As the

$$\underset{\text{Threonine}}{CH_3\underset{\underset{OH}{|}}{C}H\overset{\overset{NH_3^+}{|}}{C}HCO_2^-} \xrightarrow{E_1} \xrightarrow{E_2} \xrightarrow{E_3} \xrightarrow{E_4} \xrightarrow{E_5} \underset{\text{Isoleucine}}{CH_3CH_2\underset{\underset{CH_3}{|}}{C}H\overset{\overset{NH_3^+}{|}}{C}HCO_2^-}$$

inhibition of E_1 by molecules of isoleucine

concentration of isoleucine rises, the concentration of effective, uninhibited E_1 declines and the rate at which threonine can be changed to isoleucine also falls off. When the system uses up its isoleucine, those of its molecules inhibiting E_1 one by one leave, letting the concentration of active enzyme, E_1, rise. The rate of the whole sequence starts to rise, too. Thus the system makes isoleucine only when needed and at a rate matching that need. It operates like a thermostat shutting down whatever operation (e.g., running of a furnace) that produces a sufficient supply (e.g., heat) to hold the system for awhile.

A competitive inhibitor need not be a product of the enzyme's own work. It may be quite a different molecule that the cell makes or it might be a mole-

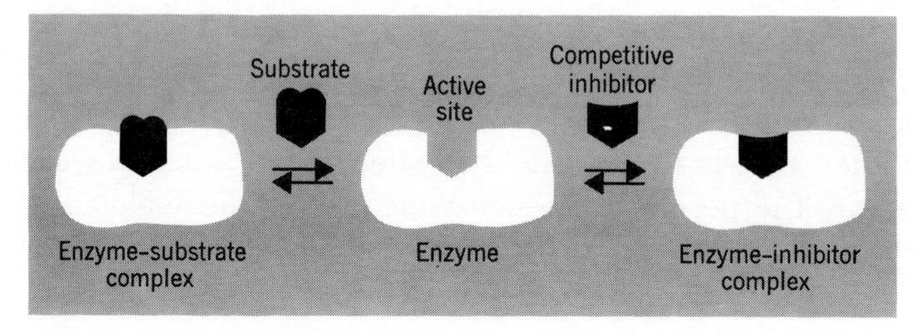

Figure 13.5 Competitive inhibition of an enzyme by a molecule, the inhibitor, that resembles the substrate enough to be accepted and held by the enzyme.

cule of some drug (of which more later). It does, however, resemble the enzyme's actual substrate molecule.

13.18 Noncompetitive Inhibition

In **allosteric inhibition,** also illustrated in Figure 13.5, the inhibitor binds to an enzyme at a site that differs from the active site. The shape of the enzyme is altered, and the substrate molecule either does not fit or fits less well to it. On the other hand, the active site may be disturbed, reducing the ability of the enzyme to catalyze the reaction in the complex itself.

13.19 Other Methods of Enzyme Control

Thus far, we have studied the following means of controlling enzymes.

1. Allosteric activation by substrate molecules.
2. Allosteric activation by nonsubstrate molecules called effectors.
3. Competitive inhibition by a product, called feedback inhibition.
4. Competitive inhibition by a nonproduct (which might be a drug).
5. Noncompetitive inhibition, or allosteric inhibition.

Other means of control include the following.

6. Conversion of a proenzyme or zymogen to the enzyme. The body makes some enzymes in an inactive form where a polypeptide chain, longer than in the real enzyme, is folded in some way that covers up the active site. These forms are called **proenzymes** or **zymogens.** Several digestive enzymes are made first as zymogens and are changed to active enzymes only as food arrives needing digestion. The zymogen-to-enzyme conversion is done by breaking one or two peptide bonds, which splits off a small fragment of the polypeptide, evidently exposing or creating a region responsible for catalytic activity.

7. Attachment of coenzyme or metal-ion activator. If the cell can control the flux of coenzymes or metal-ion activators to their apoenzymes, it thereby controls the availability of finished enzymes.

8. Hormones activate some enzymes. The arrival of a hormone at a target cell activates an enzyme within the cell membrane itself, initiating events inside the cell.

9. Genes control enzyme activity by the simple means of directing the syntheses of enzymes. Anything that controls a gene controls the enzyme made by the gene. (This will be discussed further in the next chapter.)

We thus see that a number of avenues are available for the activation and inhibition of enzymes. Except for the zymogen-enzyme change, which breaks covalent bonds, these avenues are reversible. The reason certain poisons are so dangerous is that they pose an irreversible stress.

13.20 Irreversible Inhibition by Poisons and Toxins

The extremely dangerous poisons are effective in trace concentrations because they are powerful inhibitors of key enzymes. They stick to the enzyme so strongly they cannot be removed or they come off too slowly to help. Thus they cause irreversible inhibition.

The cyanide ion, CN^-, forms strong bonds to metal-ion activators, particularly those needed by enzymes directly involved in our use of oxygen, and both hydrogen cyanide gas and metal cyanides are powerful poisons.

Heavy metal ions, for example, Hg^{2+} or R—Hg^+, Pb^{2+}, and Cu^{2+}, combine with sulfhydryl groups (—SH) on side chains of cysteine residues in polypeptides. If that happens to an enzyme, it is denatured. Methyl-mercury (the cation, CH_3—Hg^+) is particularly dangerous because its hydrophobic methyl group helps it migrate into hydrophobic areas, such as cell membranes; this poison causes permanent, crippling damage to the central nervous system. In soil and lake bottoms, mercury in the environment is changed partly to CH_3—Hg^+, and that is why mercury-based pesticides are now banned in the United States.

Arsenic poisons such as sodium arsenate, Na_3AsO_4, provide the arsenate ion, AsO_4^{3-}, which closely resembles the phosphate ion, PO_4^{3-}. An enzyme made with arsenate instead of phosphate evidently cannot work but cannot be unmade or replaced fast enough to prevent illness or death.

Nerve poisons inhibit or deactivate enzymes needed to send nerve signals. One chemical event necessary for these signals in the cholinergic nerves is the hydrolysis of acetylcholine:

$$(CH_3)_3\overset{+}{N}CH_2CH_2O\text{—}\overset{O}{\overset{\|}{C}}CH_3 \xrightarrow[H_2O]{\text{cholinesterase}} (CH_3)_3\overset{+}{N}CH_2CH_2OH + CH_3CO_2^-$$

Acetylcholine — Choline — Acetate

Nerve gases, both the powerful agents of potential gas warfare as well as weaker poisons used as pesticides—organophosphate and carbamate insecticides—deactivate cholinesterase, the enzyme, and are called anticholinesterases. Other poisons that affect various operations of the nervous system are given in Figure 13.6.

Toxins are poisons made and secreted by certain bacteria. In botulinus food poisoning, the botulinus bacillus secretes the most powerful poison yet discovered, one that blocks the synthesis of acetylcholine (Figure 13.6). Death is caused by respiratory failure in 12 to 36 hours after exposure.

13.21 Chemotherapy and Enzyme Inhibition

Chemotherapy is the use of chemicals—drugs—that destroy infectious organisms without seriously harming human protoplasm. **Antibiotics** ("against life") are chemotherapeutic agents extracted from microorganisms such as fungi and bacteria. These agents inhibit the growth of, or even destroy, other microorganisms. Examples include Terramycin, Aureomycin, streptomycin, and chloramphenicol. The antibiotics belong to a large family of chemicals called **antimetabolites,** whose antibacterial action stems specifically from their ability to inhibit enzymes that the bacteria need for their own metabolism.

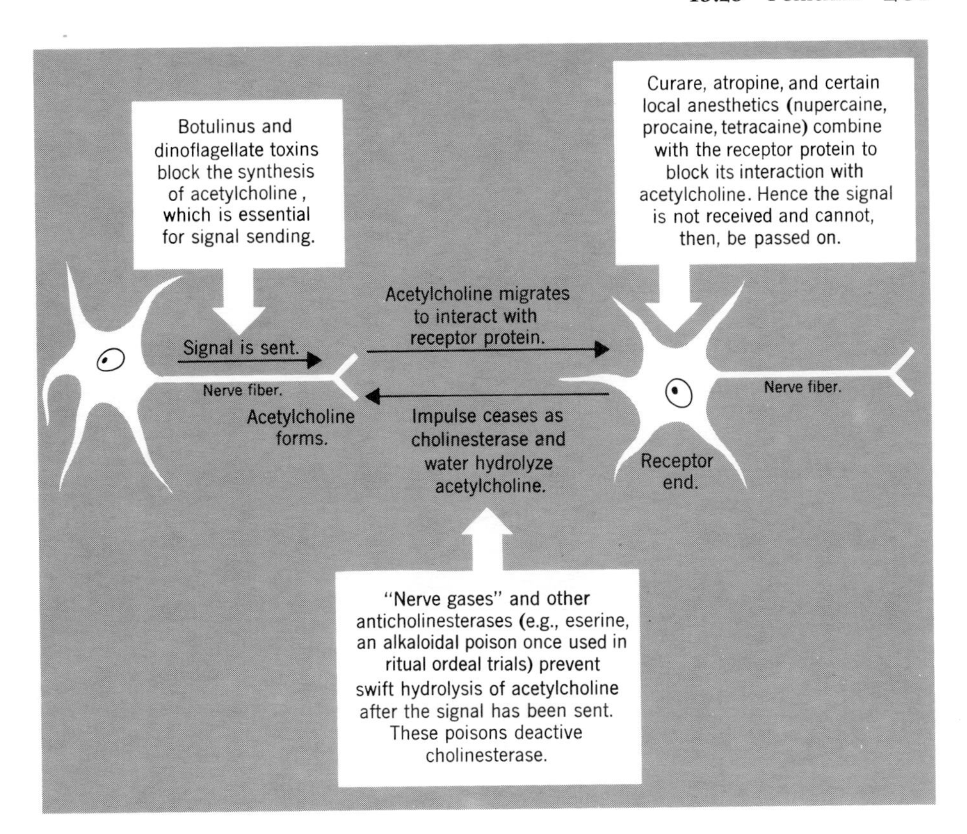

Figure 13.6 Acetylcholine cycle.

13.22 Sulfa Drugs as Antimetabolites

The sulfa drugs function by interfering with an important enzyme in several pathogenic microbes. These, unlike human beings, require para-aminobenzoic acid, a B-vitamin, to complete the formation of folic acid, an important vitamin itself. (Human beings must apparently obtain folic acid intact in the diet.) Structures of several important sulfa drugs are indicated in Figure 13.7. To affected bacteria these molecules apparently look enough like para-aminobenzoic acid to "fool" them into using sulfa drug molecules to make a coenzyme. See Figure 13.7. However, the altered coenzyme will not work. This means that an important metabolic process in the microbe is inhibited, and the microbe dies.

13.23 Penicillin

The wall of a bacterial cell is essentially one huge, enveloping molecule, a gigantic polymer put together from carbohydrate derivatives and several amino acids. Figure 13.8 shows the basic features. In the closing stages of the cell's work to make its cell wall—a wall outside its cellular membrane—it must close the cross-links. It does this at the sites to which arrows point in Figure 13.8. Glycine units (Gly) form peptide bonds to side-chain groups of lysine residues (Lys). The cell has worked hard, spent a great deal of chemical energy, and used considerable material—only to have molecules of penicillin

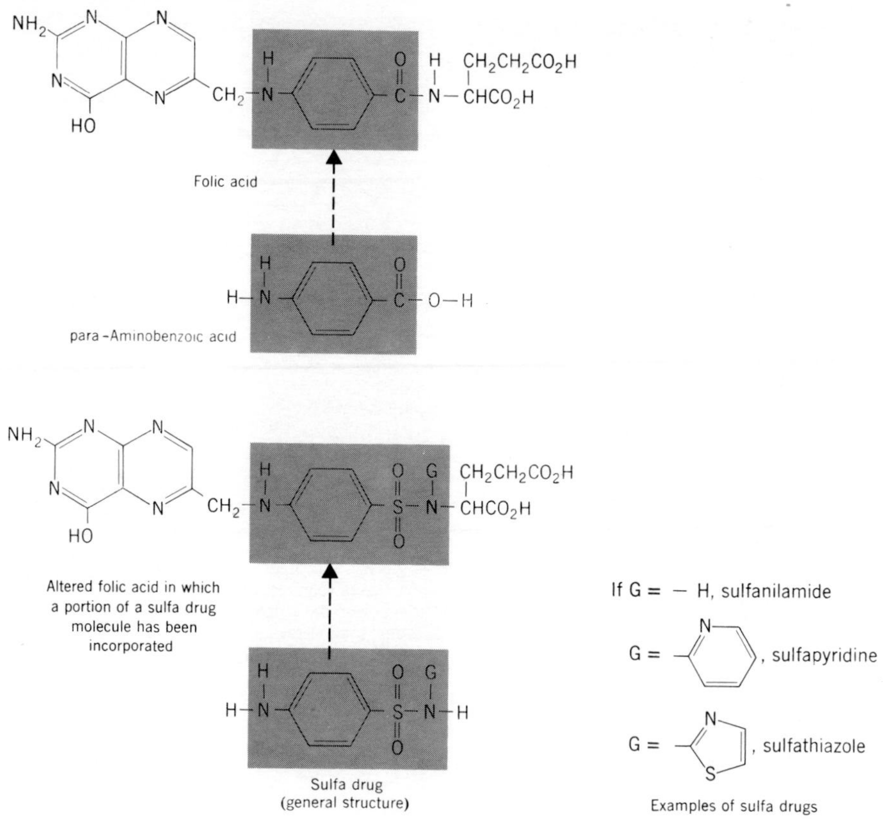

Folic acid

para-Aminobenzoic acid

Altered folic acid in which a portion of a sulfa drug molecule has been incorporated

Sulfa drug (general structure)

If G = —H, sulfanilamide

G = 2-pyridyl, sulfapyridine

G = 2-thiazolyl, sulfathiazole

Examples of sulfa drugs

Figure 13.7 Sulfa drugs as antimetabolites. If a bacterium uses sulfa drug to make folic acid (bottom) instead of *p*-aminobenzoic (top), it makes a coenzyme that cannot be used and it dies.

inhibit this final, cross-linking step. The cell, not properly formed, cannot last, and the spread of the bacterial infection is halted.

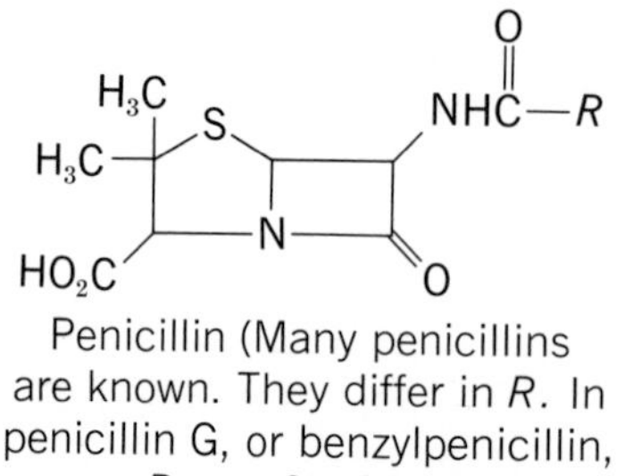

Penicillin (Many penicillins are known. They differ in R. In penicillin G, or benzylpenicillin, $R = —CH_2C_6H_5$.)

13.24 Resistance-Transfer Factor

The number of drug-resistant species of disease bacteria is increasing throughout the world. When an antibiotic fails to kill all of the target bacteria, the survivors are likely to include individuals that happen to be able to resist

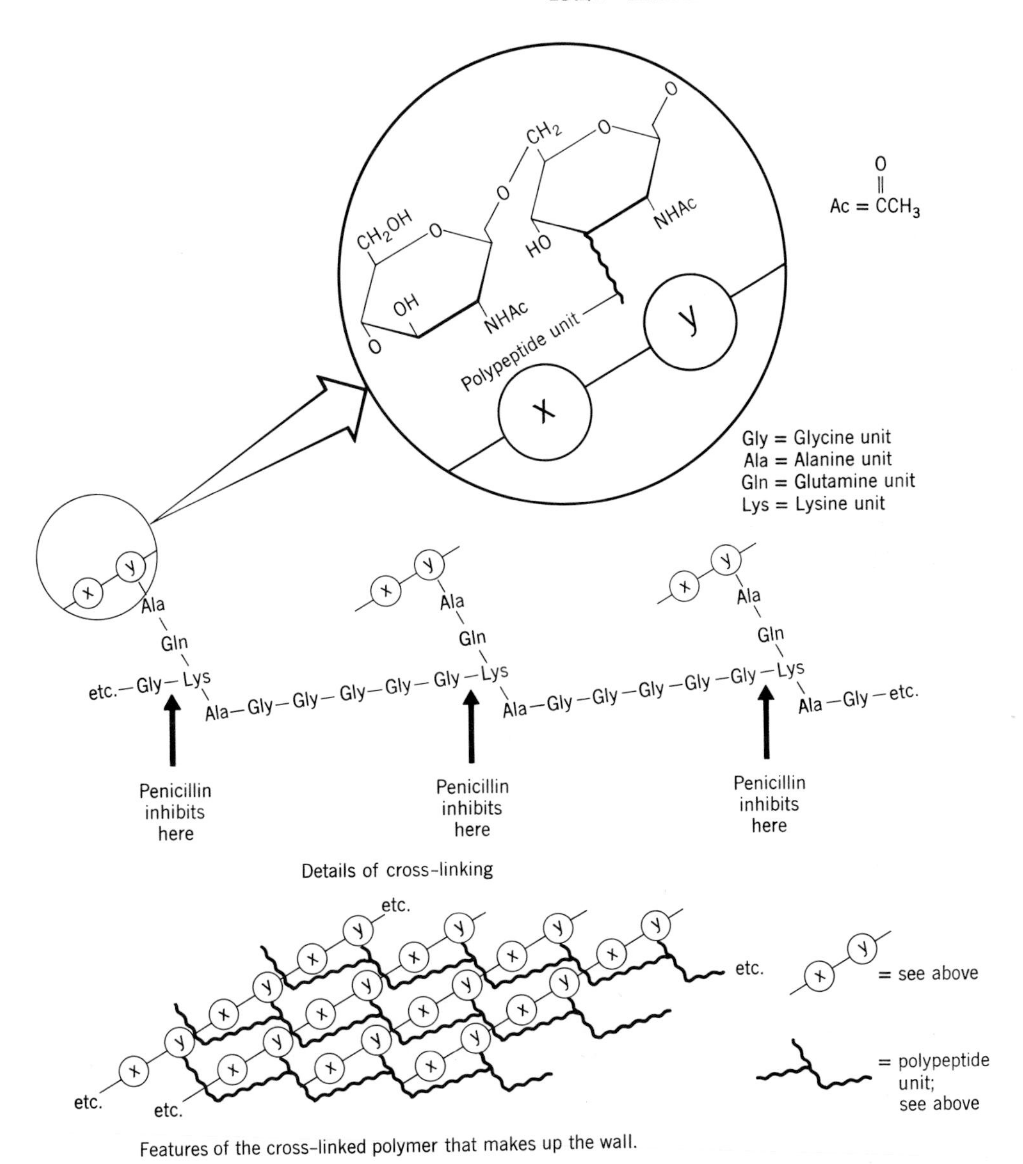

Figure 13.8 Penicillin and the bacterial cell wall. Arrows point to the sites of final cross-linking, where penicillin inhibits the needed enzyme. Note the occurrence of amino sugars (Section 10.15) in X and Y.

the antibiotic. The offspring of the survivors have a good chance to inherit this resistance, and in this way drug-resistant species develop. (The same kind of problem occurs as insects become resistant to various insecticides.) What is particularly serious is that the drug-resistant bacteria apparently manufacture a factor, called *R*, a resistance-transfer factor, which they can pass to bacteria never exposed to an antibiotic. Drug-resistant bacteria are a particular problem in hospitals, where both drugs and bacteria meet in spite of stringent methods of control.

13.25 Antigen-Antibody Reaction

A "lock and key" type of interaction evidently is at work in one of the body's own defensive mechanisms, the immune response. **Immunity** is a built-in capacity to resist infectious diseases, and the **immune response** is the set of events that build immunity and use it as needed.

If "bodies" foreign to the bloodstream enter it, specific, serum-soluble proteins called **antibodies** are usually generated. Any foreign macromolecule that can cause the generation of an antibody is called an **antigen** (antibody generator) or an immunogen, and all pathogenic microbes are in this category. Toxins are also antigens. Small molecules generally are not antigens, unless they are carried aboard attached to a macromolecule. Small molecules that stir a system in this way to make antibodies are called **haptens.**

The purpose of an antibody is to combine with its antigen and render it harmless. This interaction, called the **antigen-antibody reaction,** is an important part of the defensive mechanism of a body. The antigen-antibody reaction is at least as specific as the action of an enzyme. Antibodies that protect a person against diphtheria give no protection against streptococcal infections, for example. Once formed, antibodies can "recognize" differences between various antigens. The specificity of this reaction is explained by a "lock and key" theory very similar to that used to explain enzyme activity (see Figure 13.9).

According to the **selective theory** of antigen action, an antigen influences cells that make antibodies to select for accelerated manufacture just that one antibody that will combine with the antigen.

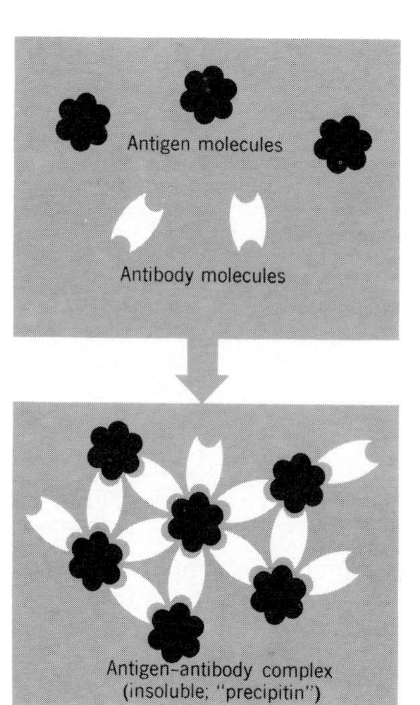

Figure 13.9 Antigen-antibody reaction. Shapes and binding sites on molecules of antigen (seen here as multivalent) take up complementary shapes and sites on molecules of an antibody.

13.26 Clonal Selection Theory of Antibody Synthesis

MacFarlane Burnet (Nobel Prize in Physiology and Medicine, 1960) and Niels Jerne, who advanced the selective theory of antigen action, were also active in proposing the **clonal selection theory** of the immune response. The principal points of this theory are the following.

1. An antibody's specificity is set by its amino acid sequence.
2. A single antibody-producing cell is genetically programmed to make just one amino acid sequence for one antibody. (The antigen has no influence on this sequence; genes already in the cell control this.)
3. As each antibody-producing cell matures in the fetus, it makes a small amount of its unique antibody, and some reaches the immature cell's surface where it is bound.
4. When, before it matures, an antibody-producing cell is hit by the antigen unique to it, the cell is killed. In this way the organism's own macromolecules—circulating proteins, carbohydrates, and nucleic acids—kill off all antibody-producing cells in the fetus that would make antibodies against the fetus, against its own life. (The organism could not survive if it were allergic to itself!)
5. After antibody-producing cells mature, they can no longer be killed by their antigens. Instead, they are stimulated by the antigen to make antibody. They are also stimulated to divide, and divide again and again, and so forth, until the descendents of the original cell make up a cluster of identical cells called a **clone.** All cells of that clone can now be stimulated by the same antigen to make one antibody whenever the antigen invades.
6. Clones can sometimes survive for the life of the individual. Those that survive confer lifetime immunity to their particular antigen.

Vaccines are substances that immunize against particular infections by mimicking the antigens for those infections. Without seriously endangering life, vaccines provoke the organism to be prepared to fight the real antigen should it invade.

In a later chapter we shall study a different line of inner defense, the mechanism by which a blood clot forms.

13.27 Hormones

We shall do no more than introduce hormones here and save for later chapters discussion of how certain specific hormones work.

Specialized tissues and organs in multicelled forms of life communicate with each other and the environment basically in two ways. One is by external stimuli—sight, sound, smell, taste, and touch—which activate various parts of the nervous system, and messages are sent electrically. The other is by the circulation of blood. The chemical messengers carried by the blood are called **hormones** (from the Greek, *hormon,* arousing, exciting). The principal human hormones are given in Table 13.2. Organs that manufacture hormones are called **endocrine glands** (from the Greek, *endon,* within; *krinein,* to separate; that is, "to separate or to secrete within"); see Figure 13.10. Hormones are

TABLE **13.2**
Principal Endocrine Glands and Hormones in Humans

Gland or Tissue	Hormone	Major Function of Hormone
Thyroid	Thyroxine	Stimulates rate of oxidative metabolism and regulates general growth and development.
	Thyrocalcitonin	Lowers level of calcium in blood.
Parathyroid	Parathormone	Regulates the levels of calcium and phosphorus in the blood.
Pancreas (Islets of Langerhans)	Insulin	Decreases blood glucose level.
	Glucagon	Elevates blood glucose level.
Adrenal medulla	Epinephrine (adrenalin)	Various "emergency" effects on blood, muscle, temperature.
Adrenal cortex	Cortisone and related hormones	Control carbohydrate, protein, mineral, salt, and water metabolism.
Anterior pituitary	1. Thyrotropic	1. Stimulates thyroid gland functions.
	2. Adenocorticotropic	2. Stimulates development and secretion of adrenal cortex.
	3. Growth hormone	3. Stimulates body weight and rate of growth of skeleton.
	4. Gonadotropic (2 hormones)	4. Stimulate gonads.
	5. Prolactin	5. Stimulates lactation.
Posterior pituitary	1. Oxytocin	1. Causes contraction of some smooth muscle.
	2. Vasopressin	2. Inhibits excretion of water from the body by way of urine.
Ovary (follicle)	Estrogen	Influences development of sex organs and female characteristics.
Ovary (corpus luteum)	Progesterone	Influences menstrual cycle, prepares uterus for pregnancy; maintains pregnancy.
Uterus (placenta)	Estrogen and progesterone	Function in maintenance of pregnancy.
Testis	Androgens (testosterone)	Responsible for development and maintenance of sex organs and secondary male characteristics.
Digestive system	Several gastrointestinal	Integration of digestive processes.

From G. E. Nelson, G. G. Robinson, and R. A. Boolootian. *Fundamental Concepts of Biology*, 2nd ed., 1970. John Wiley & Sons, New York, page 114. Used by permission.

made and secreted generally in response either to a change in the concentration of some component in blood or the digestive tract or to some signal sent via the nervous system.

Each hormone affects a particular organ or tissue called the **target organ** or target tissue. This fine selectivity is generally understood in terms of some "lock and key" type of chemical recognition much as in enzyme-substrate or antigen-antibody interactions.

Three principal means of hormonal activation have been identified. A

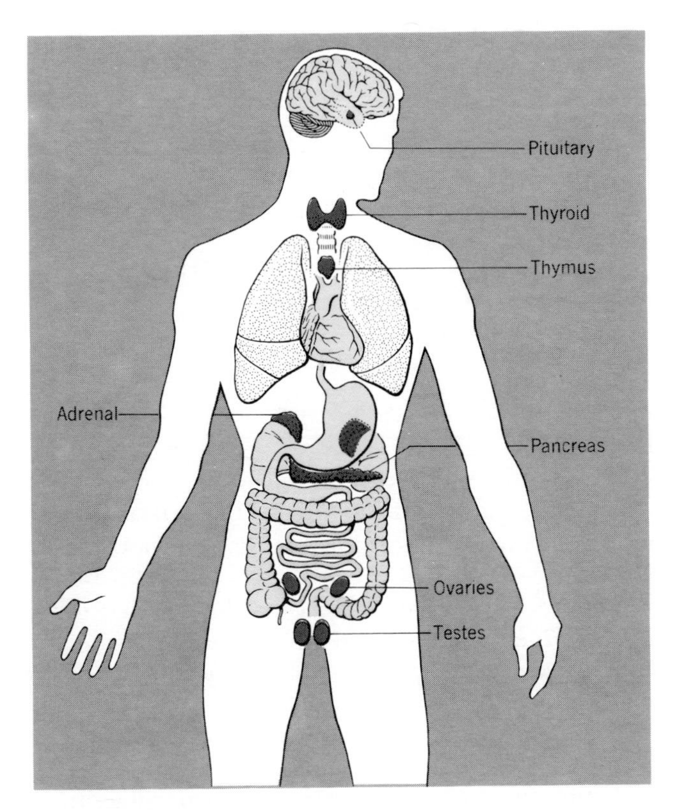

Figure 13.10 The endocrine glands.

hormone might activate a gene; it might activate an enzyme; or it might change the ability of a cell wall to let something in or out.

Sex hormones appear to function as gene activators. Their molecules, all steroids (Table 11.3), are fairly small and they can migrate through the membranes of their target cells, into the cell nuclei and there interact with chemicals of heredity, nucleic acids (discussed in the next chapter).

Adrenaline, some thyroid hormones, and a few others work as enzyme activators. When their molecules arrive at target cells, they activate an enzyme within the cell membrane. This enzyme, in turn, activates changes inside the cell. Thus the "message" of the hormone gets through the cell membrane even though the messenger stays outside. Because nonsteroid hormones are polypeptides or proteins, and therefore have very large molecules, this ability to send the message inside without going in is clearly necessary.

Hormones that work by affecting the permeability of cell walls include several involved in stimulating the secretion of digestive juices, as well as insulin and human growth hormone. The latter changes the ability of cell walls to let amino acids in. If this hormone is deficient, proteins needed for growth are not made at the necessary rate and physical stunting results.

13.28 Prostaglandins

Prostaglandins are a family of oxygenated, 20-carbon fatty acids with five-membered rings. More than a dozen have been identified. The structures of two follow. Prostaglandins were named after the organ, the prostate gland,

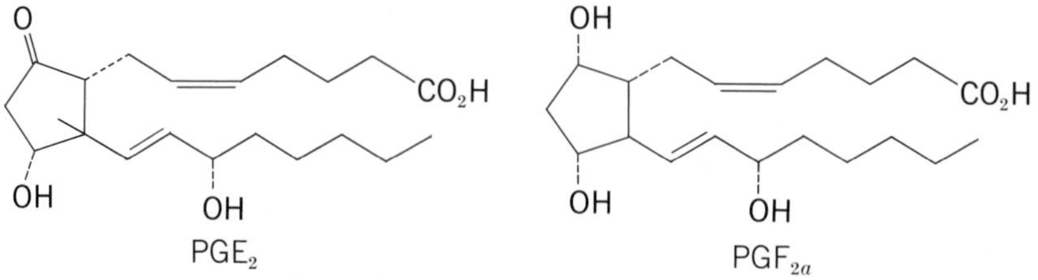

PGE_2 $PGF_{2\alpha}$

Two of the many prostaglandins

where their presence was first discovered in 1935 by Ulf von Euler, a Swedish scientist (Nobel Prize, 1970).

Prostaglandins work inside cells. Just how they work is not yet clear. They are not hormones, but they appear to modulate the effects of hormones. Some prostaglandins will enhance a hormonal message; others tone the hormonal message down.

Prostaglandins have been implicated in an astonishing variety of metabolic activities, and therefore they must be used very carefully. Some have been used with enormous success as drugs. One will induce labor at the end of a pregnancy. Another will stop the flow of gastric juice and thereby let the body heal a stomach ulcer. One prostaglandin was injected directly into the hearts or aortas of newborns with cyanotic, congenital heart disease ("blue babies"). By arresting the condition successful heart surgery was possible. In some patients with peripheral arteriosclerosis so developed that the only way to relieve the pain appeared to be leg amputation, the injection of a prostaglandin relieved the pain and seemed to increase the flow of blood to the legs. Other uses of prostaglandins being investigated are to relieve nasal congestion, and treat asthma, or high blood pressure, or rheumatoid arthritis.

13.29 Enzymes in Medical Diagnosis and Therapy

Enzymes in the body normally are confined to intracellular fluids. Only minute traces of enzymes are present in such extracellular fluids as urine, plasma, cerebrospinal fluid, and bile. In certain diseases, however, the concentration of an enzyme in one or another of these extracellular fluids is known to increase greatly.

In human infectious hepatitis the concentration in blood of the enzyme glutamic: pyruvic aminotransferase increases tenfold even before the symptoms appear. Assay of blood serum for the enzyme is a much simpler way of detecting the disease or its carriers than microscopic examination of a sample of liver tissue.

In experimental rabbits the level in the blood of the enzyme proline hydroxylase is high when plaque deposits are being formed in arteries. Such deposits are associated with hardening of the arteries or arteriosclerosis. Tests for this enzyme might give forewarning of heart disease.

The blood serum levels of four enzymes increase following heart attacks, the increases being as high as six times normal in severe cases. The amount of increase is closely related to various complications associated with heart attacks. Since enzymes in blood are easy to monitor, this clinical analysis gives physicians a powerful tool in assessing how much disability and how much recovery to expect from a heart attack patient. (The four enzymes are serum glutamic: oxaloacetic aminotransferase, lactic dehydrogenase, creatine phosphokinase, and hydroxybutyrate dehydrogenase.)

Brief Summary

Regulation of Metabolism. When basic nutritional needs are met, metabolism is controlled principally by the regulation of genes (which direct the synthesis of enzymes), by the regulation of enzyme activity, or by the operation of hormones. For internal self-defense the system has the immune response and the blood clotting mechanism.

Enzymes. Some enzymes are made wholly of one or more polypeptide strands and others have a cofactor. The names of enzymes nearly always end in "-ase" and also indicate either the substrate or the type of reaction. Enzymes work best under an optimum pH and temperature. The reaction rates they make possible are extremely high. Enzymes are very specific in both the type of reaction and the specific substrate they affect. At low substrate concentrations some enzymes produce dramatic increases in rates with even slight increases in substrate concentration. Other enzymes are more sluggish first, being activated by substrate at low concentrations and then showing dramatic rate increases. Rates for both types of enzymes eventually level off at some maximum value when the substrate concentration is high enough to saturate the enzyme. In contrast, uncatalyzed reaction rates tend to keep on increasing as reactant concentrations increase. Conditions that denature proteins denature enzymes. The analysis of enzyme concentrations in body fluids is sometimes useful in diagnosing or treating a disease.

Cofactors. Most enzymes need either a coenzyme or a metal-ion activator to be activated. Some coenzymes are made from B-vitamins and often these nonprotein substances provide the completed enzyme's active site.

Theories of Enzyme Action. An enzyme-substrate complex must form to bring the active site to that part of the substrate that is to react. Binding sites on the enzyme guide the substrate to the correct position. The lock and key model explains the specificity of some enzymes' actions as the fitting of substrate to a complementary shape on the enzyme. The induced fit model explains other enzymes by postulating that the substrate induces a conformational change in the substrate.

Regulation of Enzymes. Some enzymes are made in a proenzyme or zymogen form which is activated by the splitting of one (or more) peptide bonds to expose active or binding sites. Because other enzymes become activated only when they affix their cofactors, they are regulated through control of the cofactors. Some enzymes are regulated by competitive feedback inhibition caused by a product of the enzyme's work; others, by competitive inhibition caused by some nonproduct. Some enzymes are affected by noncompetitive inhibition brought about by a substance acting allosterically at some site on the enzyme other than the active or binding sites. Allosteric interactions in other enzymes can activate them, not inhibit them. This activation might occur at one or two (or more) active sites, done by the first substrate molecule that combines with the enzyme molecule. Allosteric activation can also be caused by something other than the substrate. Another control over enzymes is exerted by factors that control genes, because genes direct the synthesis of enzymes. Finally, some hormones are enzyme activators; others are gene activators and through genes regulate enzymes.

Poisons. Irreversible inhibition of enzymes is accomplished by poisons that bind to the active site or the binding sites or otherwise distort the shape of the enzyme.

Chemotherapy. Drugs that act against microbes include antibiotics in particular and antimetabolites in general. These act by inhibiting enzymes in bacteria. Overuse of these drugs contributes to the development of drug-resistant bacteria.

Immune Response. In the growing fetus only those antibody-producing cells survive into maturity that can make antibodies for foreign macromolecules, antigens. None survive that could make antibodies against the organism's own macromolecules. Later, when some antigen invades, it selects and activates a unique antibody-producing cell both to make antibody and to grow and generate a clone. The clone's ability to make considerable antibody later in life gives the system immunity to that antigen. Special vaccines can induce the building of clones, too.

Hormones. In response to external or internal stimuli endocrine glands secrete hormones—chemical messengers that can instruct a gene in a target cell nucleus, or an enzyme in that cell's membrane, or affect the permeability of that membrane. A lock and key interaction between the hormone and something on the cell membrane permits the hormone to identify just its own target cells. The prostaglandins seem to regulate how fast or slow the hormone's action at the cell will be.

Selected References

Books

1. L. Stryer. *Biochemistry.* W. H. Freeman and Company, San Francisco, 1975. Chapter 6, "Introduction to Enzymes," and Chapter 34, "Hormone Action," are clearly written and illustrated.
2. E. E. Conn and P. K. Stumpf. *Outlines of Biochemistry,* 3rd ed. John Wiley & Sons, New York, 1972. Chapter 9, "Vitamins and Coenzymes," provides a nice survey.

Articles

1. E. Frieden. "The Chemical Elements of Life." *Scientific American,* July 1972, page 52. The emphasis is on the trace elements needed for life.
2. S. M. Mellinkoff. "Chemical Intervention." *Scientific American,* September 1973, page 103. Included are lists of vitamins required by humans, the major hormones, 19 diseases against which vaccines exist, major antimicrobial drugs, and drugs that affect the nervous system.
3. D. E. Koshland, Jr. "Protein Shape and Biological Control." *Scientific American,* October 1973, page 52. The ability of protein molecules to bend flexibly is essential in enzyme-substrate interaction.
4. N. K. Jerne. "The Immune System." *Scientific American,* July 1973, page 52. Only a small patch, called an epitope, on the surface of a foreign protein molecule signals the immune response that is able to respond to millions of foreign epitopes without being confused by the proteins that are normal to the organism.

5. A. R. Magliulo. "The Chemistry of Immunity. Part I. General Introduction." *Chemistry,* November 1975, page 10. A nice introduction to the terms.
6. R. C. Clowes. "The Molecule of Infectious Drug Resistance." *Scientific American,* April 1973, page 19. How the resistance factors of bacteria that have survived an antibiotic are transferred to other bacterial strains.
7. B. W. O'Malley and W. T. Schrader. "The Receptors of Steroid Hormones." *Scientific American,* February 1976, page 32. Included is a discussion of the steroid hormones.
8. J. D. Capra and A. B. Edmundson. "The Antibody Combining Site." *Scientific American,* January 1977, page 50.
9. N. G. Anderson. "Technology vs. Disease." *Chemtech,* March 1977, page 174. In a wide-ranging discussion of the impact technology has had on the diagnosis of diseases, the inventor of some of the most sophisticated diagnostic analyzers includes a discussion of the analysis of amniotic fluid.

Questions and Exercises

1. In general terms what is the difference between an apoenzyme and a coenzyme?
2. What relation exists between certain B-vitamins and some coenzymes?
3. Briefly describe nine mechanisms by which various types of enzymes are regulated—either activated or inhibited.
4. What is the difference between lactose and lactase?
5. What is the difference between hydrolysis and hydrolase?
6. Why do enzymes have optimum pHs and are not as effective at other pHs?
7. The following reaction occurs in a long series of steps that degrade glucose:

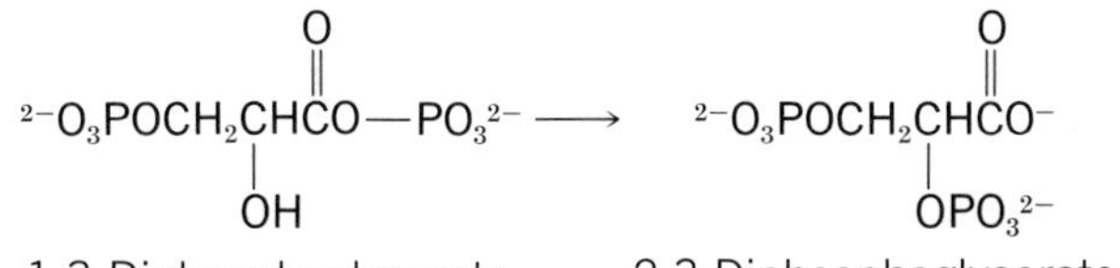

$$^{2-}O_3POCH_2CH(OH)C(=O)O{-}PO_3^{2-} \longrightarrow {}^{2-}O_3POCH_2CH(OPO_3^{2-})C(=O)O^-$$

1,3-Diphosphoglycerate (1,3-DPG) → 2,3-Diphosphoglycerate (2,3-DPG)

   The enzyme for this change is inhibited by 2,3-DPG. What kind of control is exerted by 2,3-DPG? (Name it.)
8. If enough methanol is drunk, the individual will become blind or will die. One strategy to counteract methanol poisoning is to give a nearly intoxicating drink of ethanol. Methanol will then be slowly and harmlessly excreted. Otherwise, it will be oxidized to formaldehyde, which is actually the poison.

$$CH_3OH \xrightarrow{\text{dehydrogenase}} CH_2{=}O$$

Methanol → Formaldehyde

   What kind of inhibition might ethanol be achieving here? (Name it.)
9. In general terms how do the lock and key mechanism and the induced fit mechanism of enzyme action differ?

10. What probably converts a zymogen to the active enzyme?
11. How can a huge protein molecule act as a hormone and get its "message" inside the target cell when it cannot go through the cell membrane?
12. Why do the increases in the rates of enzyme-catalyzed reactions eventually level off as the concentration of substrate increases?
13. When an effector is controlled, the rate of some enzyme-catalyzed reaction is controlled. Explain how that works, in general terms.
14. Enzyme deficiency diseases cannot be treated by supplying the enzyme in the patient's food or drink, but vitamin deficiency diseases can be treated by supplying vitamins in food or drink. Explain.
15. How does cyanide ion work as a poison?
16. In general terms, how do some of the most dangerous poisons work?
17. When an antigen invades, one that the body has not experienced before, what does it do to an antibody-producing cell? How does this differ from what the antigen does if, some time later, it invades another time?
18. Explain in general terms how sulfa drugs work. (You need not write any structures; use the names of substances.)
19. Explain in general terms how penicillin works.
20. To get data on mercury pollution in the environment decades ago, scientists have analyzed feathers of museum bird specimens for their mercury content. Why would keratin, the protein of feathers, tend to accumulate mercury?
21. In general terms discuss three ways by which different hormones exert their effects.

chapter 14

Nucleic Acids and Heredity

A blur of dots that made scientific history was obtained in the laboratories of M. H. F. Wilkins in England. This X-ray diffraction photograph of DNA convinced many scientists, including James Watson and F. H. C. Crick, that DNA has a helical shape. The Crick-Watson theory, studied in this chapter, revolutionized biology. (M. H. F. Wilkins. *Science,* **140,** 141–950, 1963. Copyright 1963 by the American Association for the Advancement of Science.)

14.1 Heredity and Enzymes

We have learned that nearly every reaction in a living organism requires its special enzyme. With the aid of its set of enzymes the organism makes all other cellular substances. The set of enzymes in one organism is not identical with the set of another. Each has a unique set, although many are similar. Each can transmit the capacity to possess that set to new daughter cells made by cell division or to new offspring begun at conception. Prior to reproduction the organism does not duplicate its enzymes to pass them on directly. Instead, it duplicates another set of chemicals, those in a family of polymers called nucleic acids, which then direct the syntheses of enzymes. Before continuing with their study, we shall review certain features of cells and the physical basis of heredity.

14.2 The Cell

The cell is the structural unit of life and chemicals associated with living things are, by and large, organized in these units. An artist's rendering of a typical cell is given in Figure 14.1. Cells of different tissue differ widely in shape and size; the figure simply displays what most cells typically include regardless of tissue.

The boundary of a cell is the cell membrane, which we studied earlier. Everything inside has the general name of protoplasm. Within it are a number of discrete "bodies" called organelles ("little organs"). Prominent among them are the mitochondria, which make adenosine triphosphate (ATP) for the cell's immediate needs of chemical energy. The endoplasmic reticulum (ER) consists of membrane-lined tubules with many convoluted branches in which fluids and chemicals move and interact. The endoplasmic reticulum generally extends from the cell membrane to the wall of the cell nucleus. The nucleus has its own membrane with pores called annuli. All the cell content outside the nucleus is called the cytoplasm. Pinocytotic vesicles are a part of the way a cell can move materials to the outside of the cell. The Golgi complex or Golgi apparatus makes lipids, collagen, and packages of enzymes. The centriole is involved in cell division. Polysomes (polyribosomes) are strands, like strung beads, consisting mostly of nucleic acid; they help make enzymes under the direction of nucleic acid found in the nucleus as part of the chromatin. Some of these cell parts will be described further later.

14.3 Chromosomes, Genes, and Heredity

The twisted, intertwined filaments inside a cell nucleus are a substance called **chromatin.** Each chromatin strand consists of one kind of nucleic acid, called **DNA,** associated with a mixture of polypeptides. Individual sections of DNA molecules are individual **genes,** the basic units of heredity. When cell division, called mitosis, begins, each chromatin strand may be seen to thicken. As they do, rodlike bodies become quite visible under the microscope if a staining

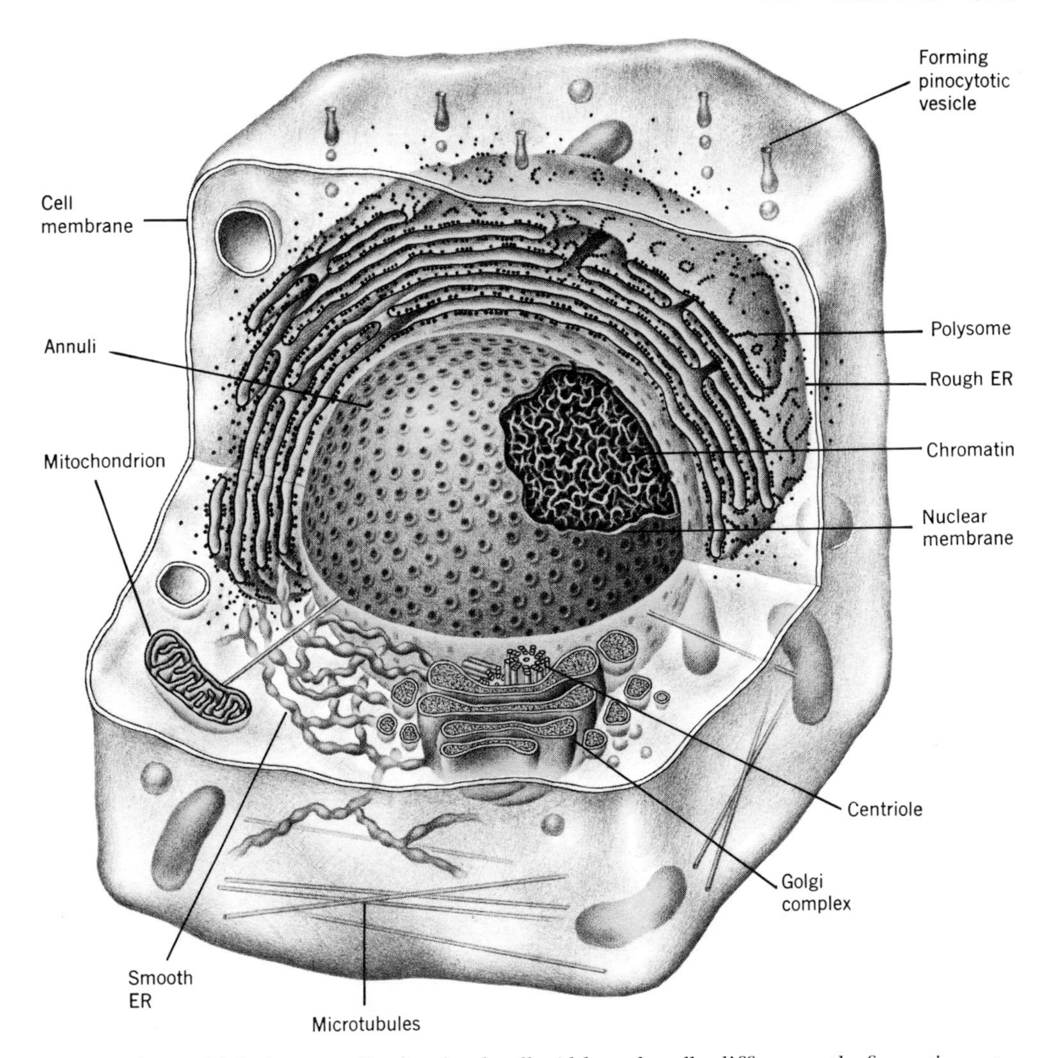

Figure 14.1 A generalized animal cell. Although cells differ greatly from tissue to tissue, most have the features shown here. (From A. Nason and R. L. Dehan. *The Biological World,* 1973. John Wiley & Sons, New York. Used by permission.)

agent is used. These discrete bodies are the **chromosomes,** and they are made of chromatin. The thickening of chromatin is caused by the synthesis of new DNA (and associated proteins). The new DNA is an exact copy of the old—if all goes well, and it nearly always does. The cycle of cell division illustrated in Figure 14.2 has now reached the prophase stage. Each gene has been replicated—reproduced in duplicate. Division then continues until the new daughter cells are complete. Thus through DNA **replication** the genetic message of the first cell is passed to each of the two new cells.

14.4 Germ Cells

In a typical animal, the union of a sperm cell from the male with an egg cell from the female produces a new cell called a zygote. From subsequent cell

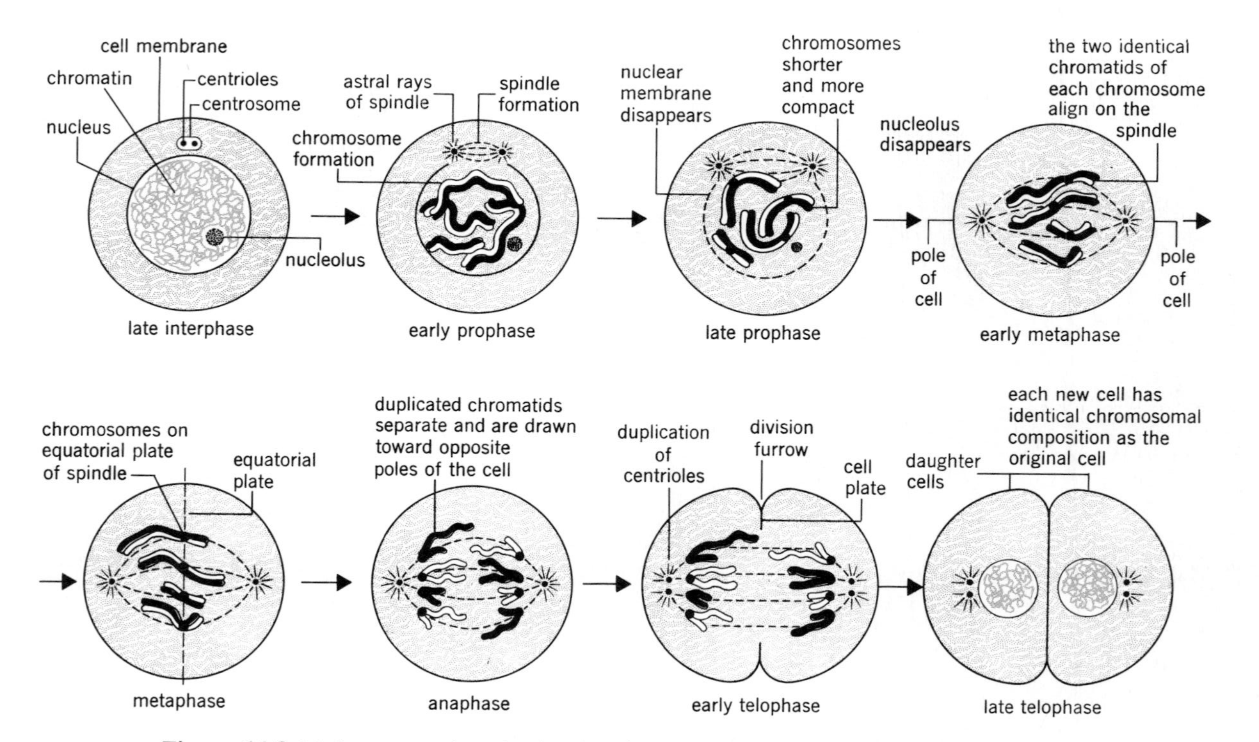

Figure 14.2 Major events in mitosis, showing a cell having five chromosomes (From J. R. McClintic, *Basic Anatomy and Physiology of the Human Body*, John Wiley & Sons, Inc., New York, 1975. Used by permission.)

divisions of the zygote an embryo gradually takes form. Early in this stage of the development, two fundamentally different kinds of cells appear: germ cells, which will give rise either to sperm or eggs; and somatic cells, from which will form all the myriad tissues and organs unique to the body of the species. Just how cells that come from the original zygote undergo differentiation and become cells of different tissue we shall not study. Little is yet known of this question.

The germ cells are cells set apart, protected from change and unaffected by the tremendous variations taking place among the somatic cells. When the somatic cells have proceeded in their development enough so that the gonads are fully elaborated and sexual maturity is reached, the germ cells become active. They develop sperm or eggs, depending on the sex of the individual. If the sperm and the egg of the parents contained the essentials to produce a unique enzyme system, then the germ cells of the children must possess these essentials, too, in order that they, in turn, may pass them on to the next generation. How germ cells divide and how hereditary traits from each parent have chances of being expressed in offspring are other topics we shall leave to other courses. At the foundation of such questions, however, are the chemical composition and properties of the units of heredity, the genes.

Chemistry of Hereditary Units

14.5 Nucleic Acids

Deoxyribonucleic acid, DNA and ribonucleic acid, RNA, belong to the family of polymers called **nucleic acids.** Their monomer units are called **nucleotides.** Unlike the monomers of other polymers we have studied, the nucleotides can be hydrolyzed. As illustrated in Figure 14.3, the hydrolysis of a mixture of nucleotides produces three kinds of products: inorganic phosphate, a pentose sugar, and a group of heterocyclic amines.

Five important heterocyclic amines or bases are obtainable from nucleic acids. Three are common to both DNA and RNA: adenine, guanine, and cytosine. Of the remaining two, one is found only in DNA, thymine; the other is found only in RNA, uracil. See Figure 14.3. The abbreviated symbols for these bases are A = adenine, T = thymine, U = uracil, G = guanine, and C = cytosine. Two, A and G, are related to purine and are often called the purine bases; three, T, U and C, are related to pyrimidine and are called the pyrimidine bases.

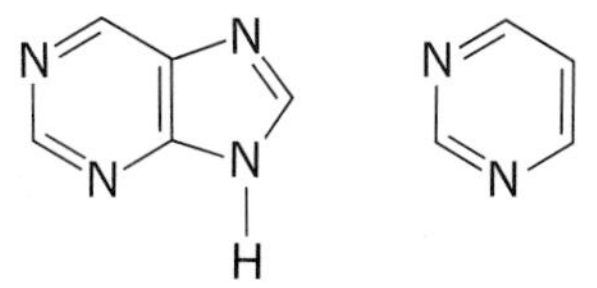

N N N N H

Purine

N N

Pyrimidine

The hydrolysis of any given nucleotide may also produce one of two aldopentoses, ribose and deoxyribose ("de-" means lacking; "deoxy-" means lacking in an oxygen atom found in a close structural relative). Nucleic acids

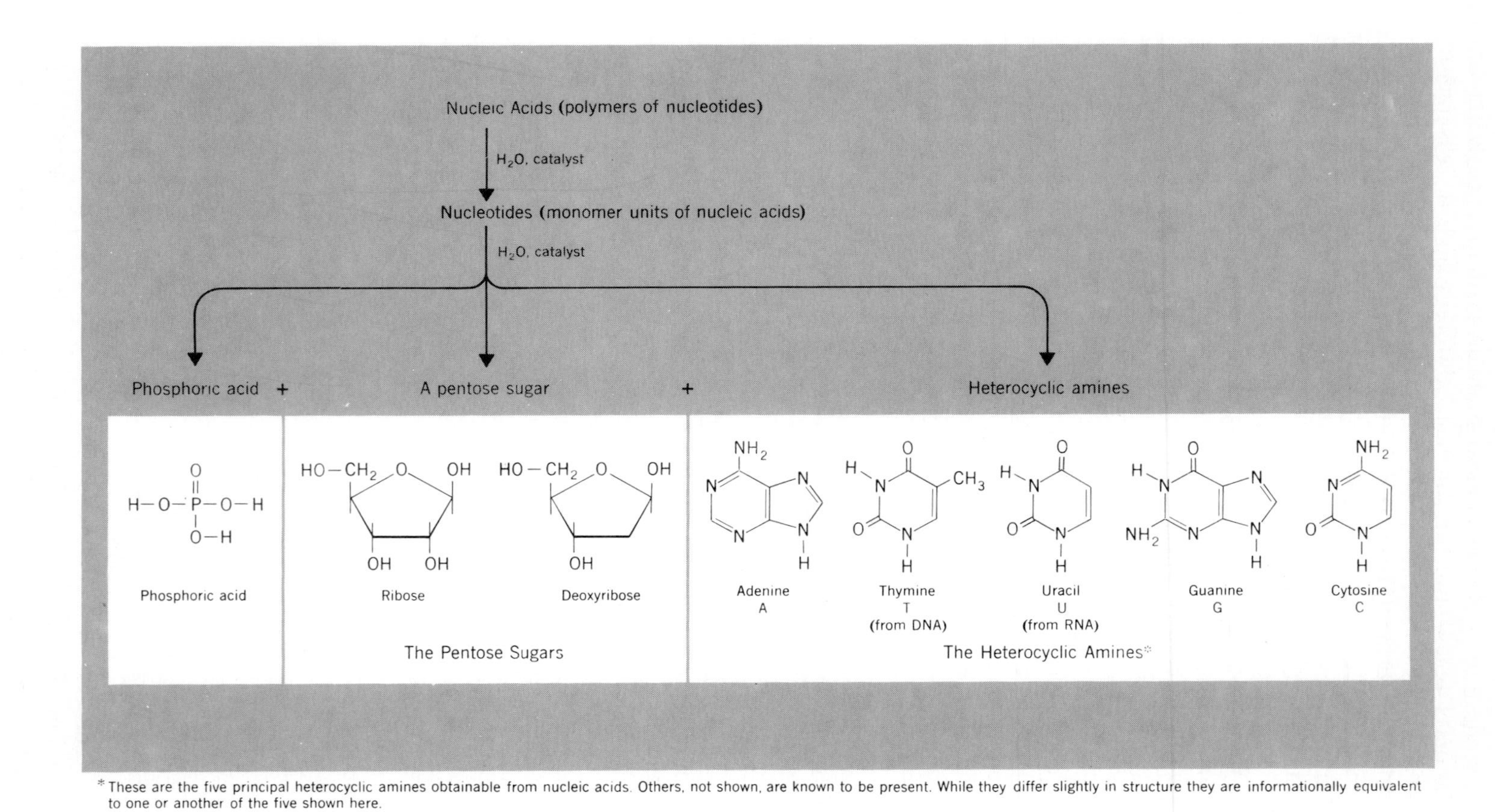

* These are the five principal heterocyclic amines obtainable from nucleic acids. Others, not shown, are known to be present. While they differ slightly in structure they are informationally equivalent to one or another of the five shown here.

Figure 14.3 Hydrolysis products of nucleic acids.

are either based on ribose, in which case they are called **ribonucleic acids,** abbreviated **RNA,** or on deoxyribose, in which case they are called **deoxyribonucleic acids, DNA.**

Figure 14.4 shows the assembly of a typical nucleotide, adenosine monophosphate (AMP), from phosphoric acid, a sugar, and an amine. The nucleotides of RNA are built from ribose; those of DNA, from deoxyribose.

How various monomer units—nucleotides—are assembled into nucleic acids is illustrated by Figure 14.5. Like the formation of nucleotides, the formation of nucleic acids in principle involves nothing more than the splitting out of water. The phosphoric acid unit on one nucleotide splits out water with an alcohol unit of the next nucleotide to form a phosphate ester link between the two nucleotides. The process is repeated until hundreds of nucleotide units are incorporated into the polymer.

The "backbone" of a nucleic acid consists of alternating phosphate and pentose units. Projecting from this backbone are various heterocyclic amines. Knowing this, we may now condense the structure of a nucleic acid, as is illustrated in Figure 14.6. The distinctiveness of any one nucleic acid (and, therefore, each of its sections, the genes) lies in the order in which the nucleic acid units are selected and assembled.

Nature was not capricious in her selection of the heterocyclic compounds that project from the backbone of a nucleic acid. Their functional groups and geometries are such that they fit together in pairs by means of hydrogen bonds. Analysis of DNA, for example, reveals that adenine and thymine are present in a ratio of 1:1. This ratio suggests that the two somehow may be "paired" in DNA. When models of these compounds are examined, they have functional groups situated in precisely the proper locations to make hydrogen bonding between them possible. They "fit" together quite nicely, as illustrated in Figure 14.7; we say that **base pairing** exists between them. A purine base always pairs with a pyrimidine base.

14.6 Crick-Watson Theory

One of the most important scientific developments of the twentieth century occurred when F. H. C. Crick and J. D. Watson proposed a structure for DNA that made it possible for the first time to develop a molecular basis of information or heredity. In the early 1950s these two toyed with molecular

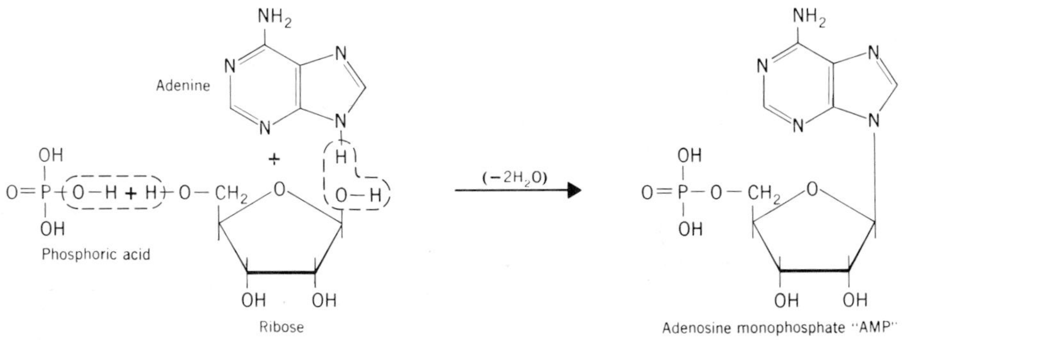

Adenine + Ribose + Phosphoric acid $\xrightarrow{(-2H_2O)}$ Adenosine monophosphate "AMP"

Figure 14.4 A typical nucleotide and the smaller units that are used to make it.

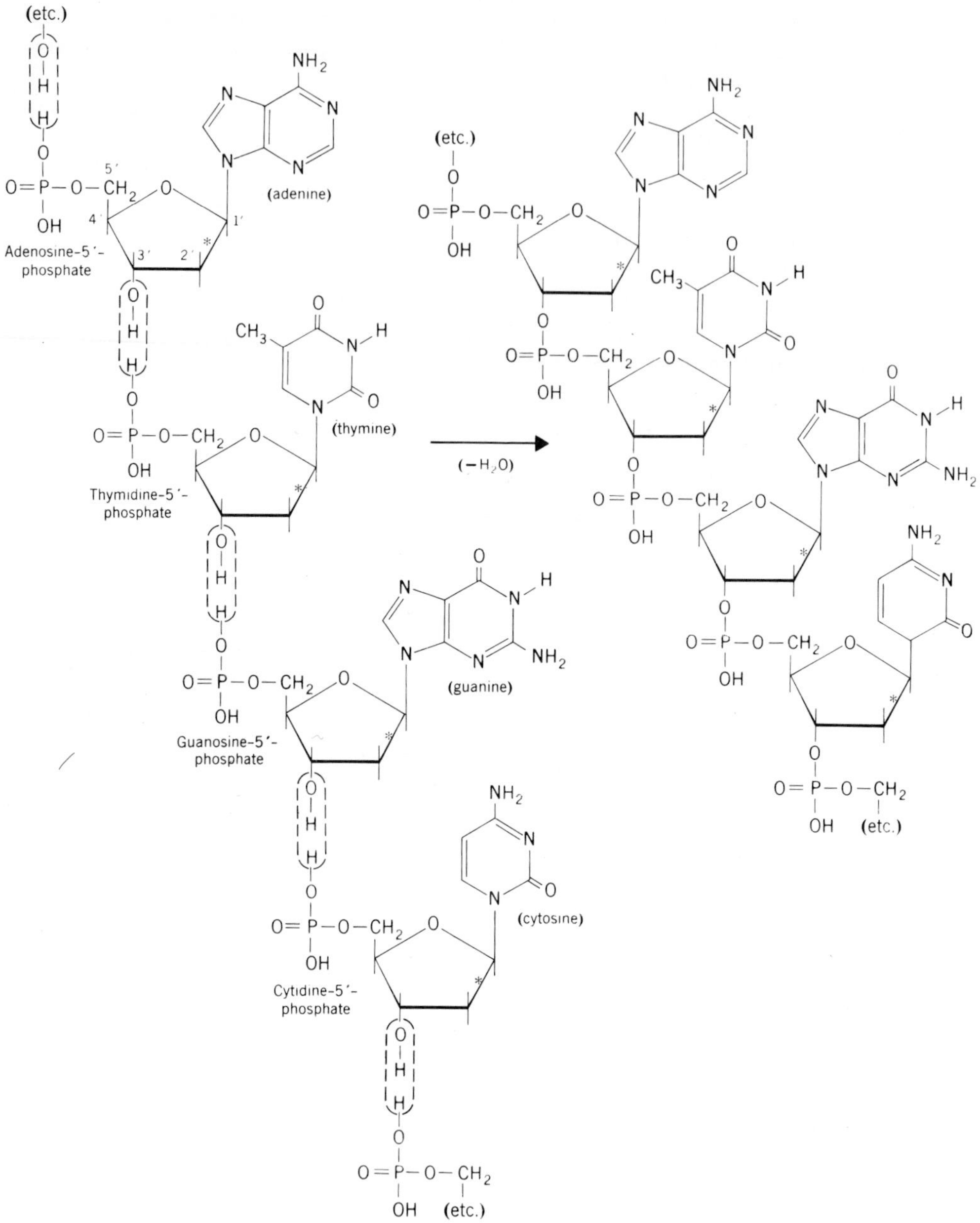

Figure 14.5 How a nucleic acid is related to its nucleotide monomers. Shown here is a segment of DNA. If sites marked by the asterisks carried —OH groups, the segment would be of RNA (provided uracil replaced thymine). Deoxyribonucleic acid strands of several million nucleotide units are known. Individual segments averaging 1500 pairs of nucleotides (of a double helix) generally make up one distinct gene.

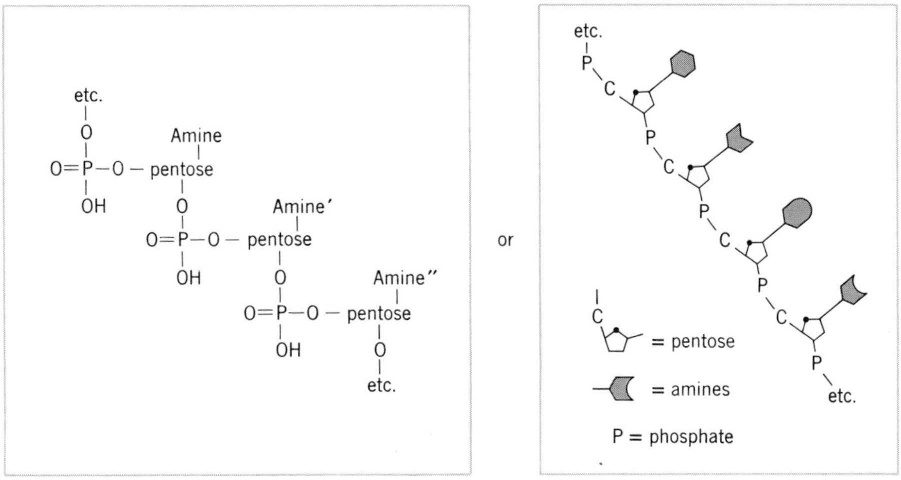

Figure 14.6 Condensed structures illustrating essential features in molecules of all nucleic acids. In DNA the pentose is deoxyribose; in RNA, ribose.

models of nucleotides, knowing that base-pairing occurred, and exploited X-ray data of DNA taken by R. Franklin and M. Wilkins to construct a model of DNA in which two coiled chains of DNA intertwined to form a double helix. In this **double-helix theory** of Crick and Watson, the two strands are complementary. A thymine unit of one chain fits or pairs, via hydrogen bonds, to an adenine unit opposite to it on the second chain. Guanine also pairs via hydrogen bonds to cytosine on the other chain.

The **DNA double helix,** represented in Figure 14.8, is reminiscent of a

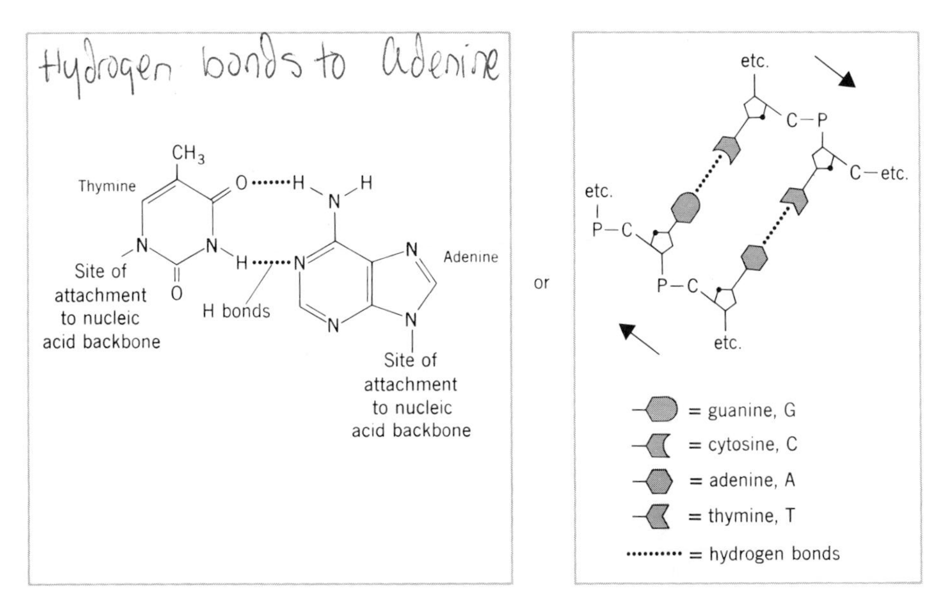

Figure 14.7 Pairing of thymine and adenine units by hydrogen bonds. Uracil can pair to adenine in the same way. Guanine and cytosine also form a pair.

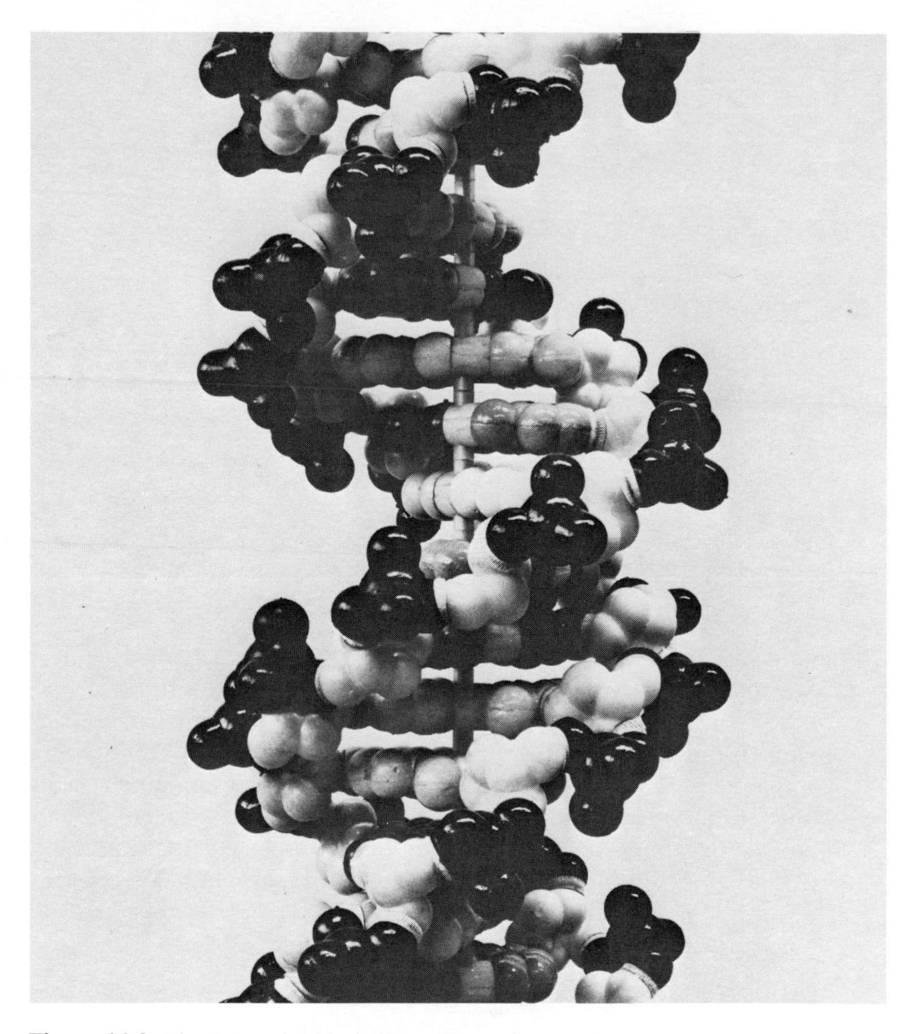

Figure 14.8 The DNA double helix, molecular model. The darkest parts are phosphate diester groups; the lightest are deoxyribose units; and the heterocyclic amines appear as the steps in the spiral. (Reprinted from PSYCHOLOGY TODAY Magazine, (May, 1967). Copyright © Ziff-Davis Publishing Co. Photo by John Oldenkamp.)

spiral staircase, in which the steps are the paired heterocyclic amines projecting toward each other from the backbones. These relations are a little clearer in the schematic drawing of the DNA double helix in Figure 14.9. Another feature, shown in Figures 14.7 and 14.9 and amplified in Figure 14.10, is that the backbones of the two strands run in opposite directions.

14.7 Replication of DNA

Replication occurs by a semiconservative process. This means that each of the two, new "daughter" double helices has one of the original strands intact and entwined with a newly made, complementary strand. The original strands do not fragment and disperse. Replication is extraordinarily accurate thanks both to the replicating enzymes and to the requirements imposed by the "fitting" of adenine to thymine or of guanine to cytosine.

A very general picture that accounts for these facts is given in Figure 14.11. This figure emphasizes only one aspect of replication, the pairing of nucleotides that insures that daughter helices will be identical with the parent. As seen in Figure 14.12, replication is discontinuous. Fragments of DNA are made along each of the two chains of the parent, both in a 5′ to 3′ direction (see Figure 14.10). Then the fragments are knitted together to make the complete daughter strands. The first gene to be made entirely artificially but yet able to function in a living cell (a bacterium) was announced in 1976 by a team led by Nobel laureate H. G. Khorana at the Massachusetts Institute of Technology.

14.8 The Genetic Code and DNA

An individual gene is a section of a DNA molecule that is "coded" for a specific genetic message. Since all DNA molecules have the same "backbones," differences between DNA molecules and their gene sections rest on the order in which the amines are strung along the backbones. It is this order of amines

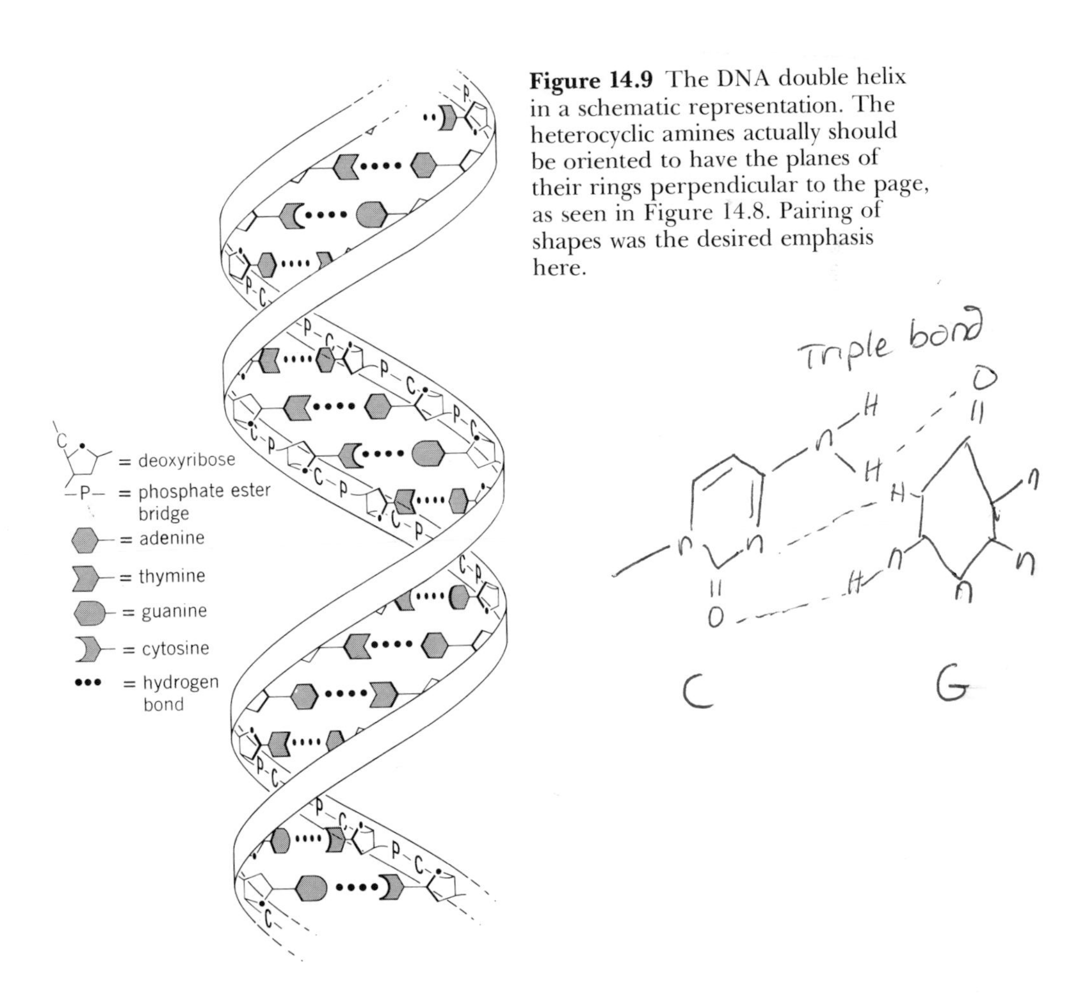

Figure 14.9 The DNA double helix in a schematic representation. The heterocyclic amines actually should be oriented to have the planes of their rings perpendicular to the page, as seen in Figure 14.8. Pairing of shapes was the desired emphasis here.

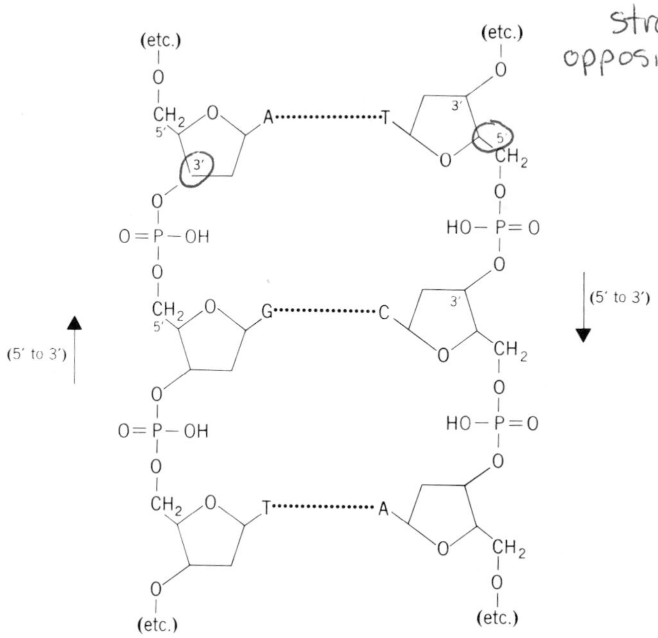

Figure 14.10 Strands of a DNA double helix run in opposite directions. The phosphate diester joins the 5′ carbon of one deoxyribose unit to the 3′ carbon of the next. (The primes on the numbers go with positions in the sugar ring; unprimed numbers would be in the heterocyclic rings. A = adenine, T = thymine, G = guanine, and C = cytosine.)

or bases that is the **genetic code** for the genetic message of a given gene. We now ask, "What exactly is this code?" Genes are chemicals, not computer tapes. How is the code related to biochemical processes in cells?

The answer to this question is simply stated. Each gene is coded to specify an amino acid sequence in a specific enzyme. A one-gene–one-enzyme relation exists. Hence, by directing the syntheses of the catalysts necessary for life, the DNA of genes controls the biochemical processes in cells. The general scheme relating DNA to enzymes is illustrated in Figure 14.13. To go into more detail, we need to learn about RNA.

14.9 Ribonucleic Acids (RNA)

Three types of RNA are known, and we must first distinguish among them. All three types are synthesized under the control of DNA. (A fourth type of RNA associated with virus particles is also known.)

14.10 Ribosomal RNA (*r*RNA)

Studding portions of the endoplasmic reticulum (Figure 14.1) are granules that might vary from 7 to 20 nm in diameter. Called **ribosomes,**

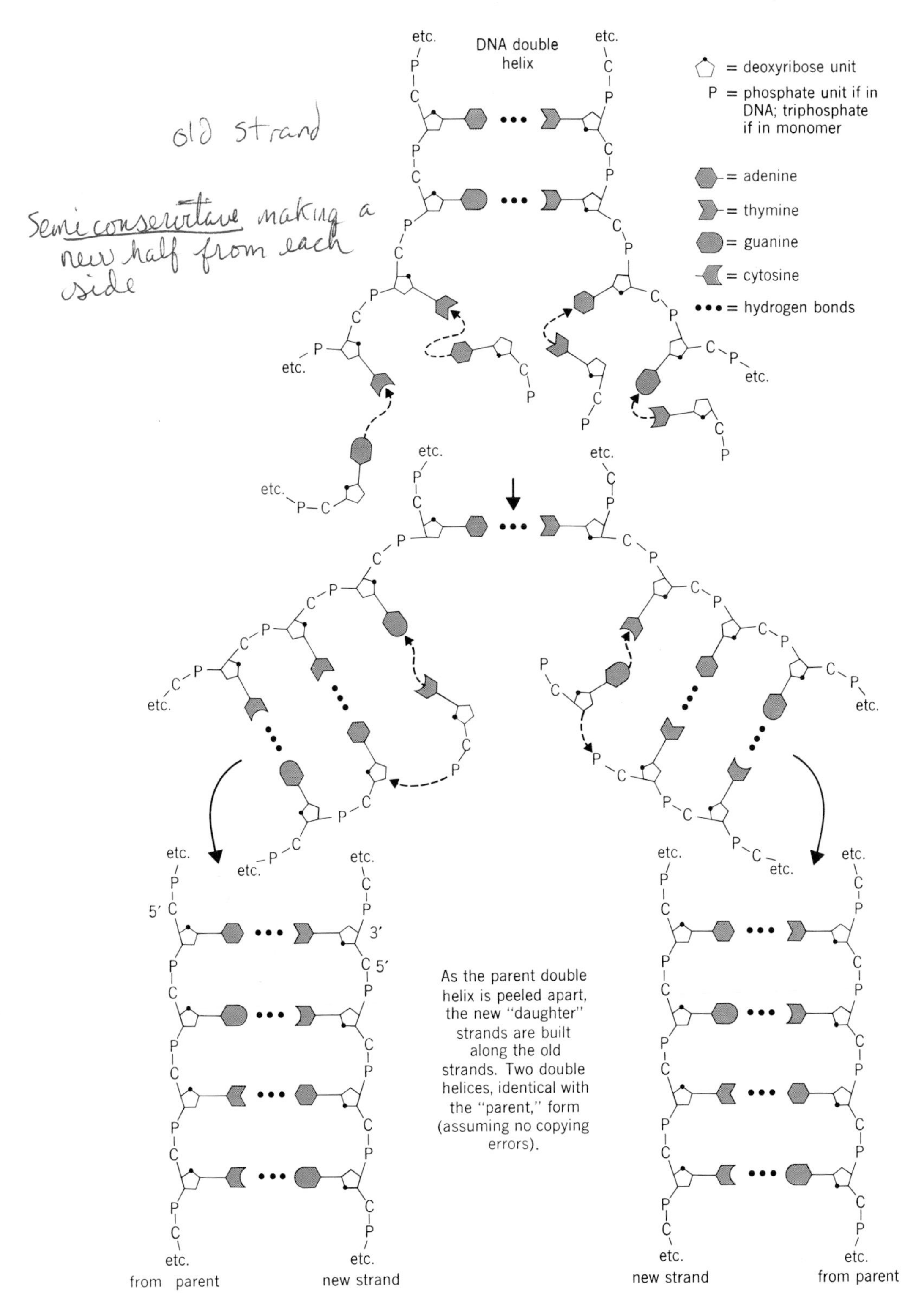

Figure 14.11 Replication of DNA. Not all can be shown in one figure. The discontinuous nature of replication is illustrated in the next figure.

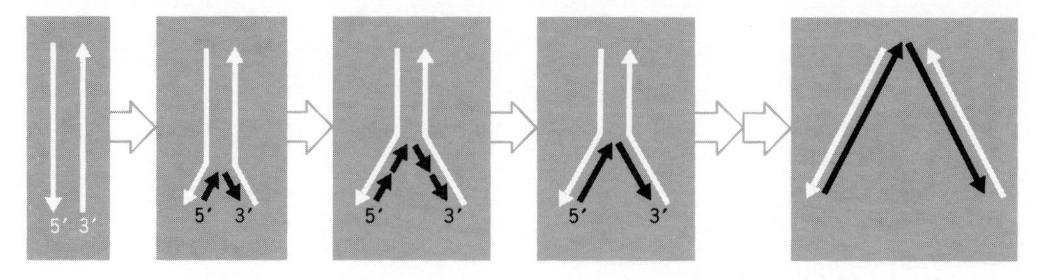

Figure 14.12 Replication proceeds by a discontinuous process. Fragments of daughter DNA strands are assembled along the parent strands. Assembly on both sides goes in a 5′ to 3′ direction. Fragments are then joined to each other to make the complete strands. As the new strands are made the parent double helix unwinds.

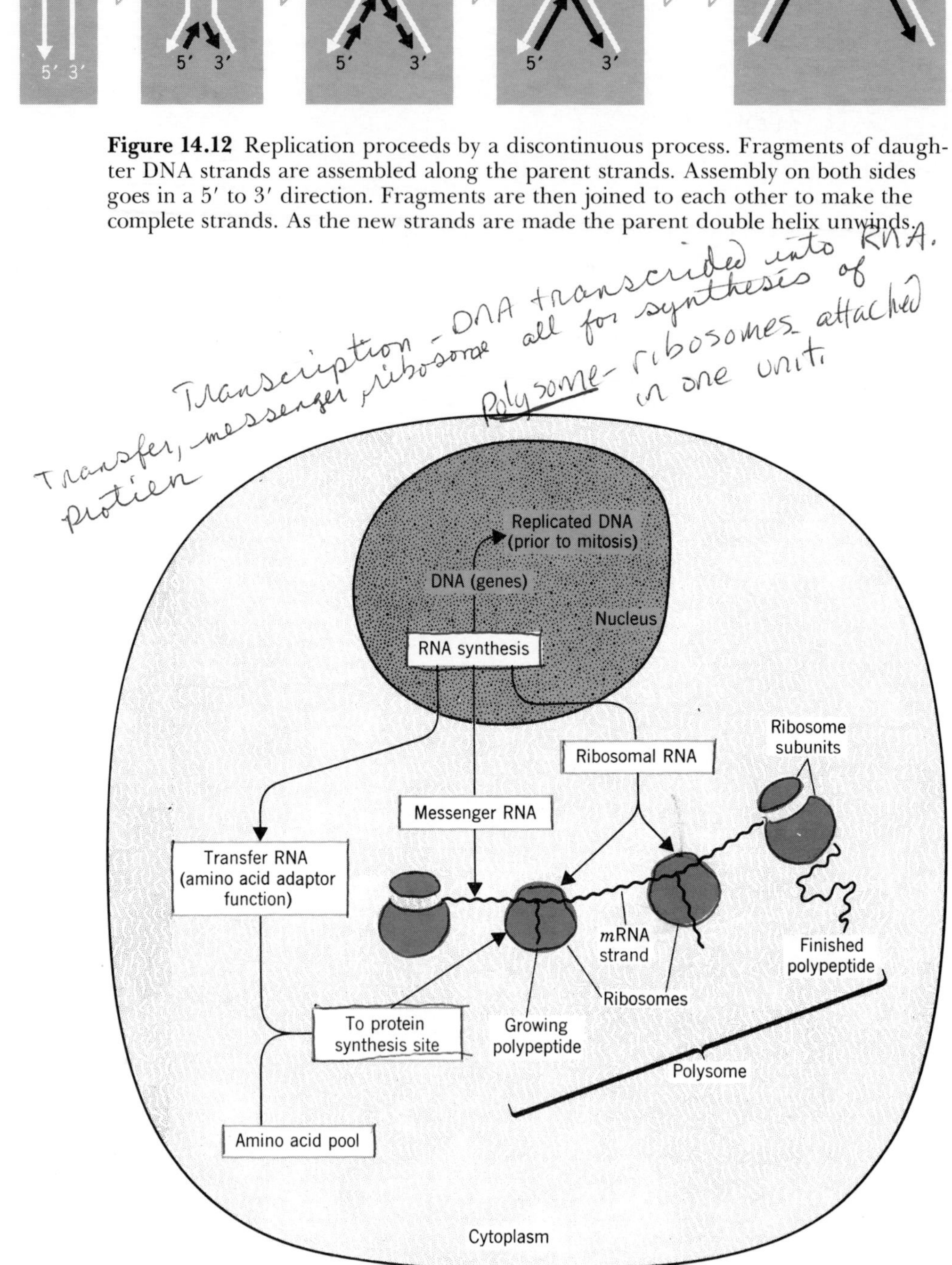

Figure 14.13 The relations of DNA to various RNAs and protein synthesis.

these granules are sometimes found free in the cytoplasm. They are made of RNA and protein, about 50% of each, and they contain most of the RNA in the cell. Subunits of ribosomes, as indicated in Figure 14.13, form a complex with messenger RNA, the second type of RNA we shall study. Ribosomes are the sites of protein synthesis, and yet the RNA in a ribosome does not itself appear to direct this work. Messenger RNA does this.

14.11 Messenger RNA (*m*RNA)

Messenger RNA molecules vary considerably in length, and they account for about 5 to 10% of the total RNA in a cell. They are not very stable and are remade as needed. As we shall see, *m*RNA molecules bear the genetic code because they are synthesized under the direct supervision of DNA. A very general picture of how this is brought about is given in Figure 14.14. Once made, *m*RNAs move out of the nucleus into cytoplasm, where they hook ribosomes to themselves at intervals along their chains. Such an assembly of many ribosomes along an *m*RNA chain is called a polysome (poly-ribosome). A molecule

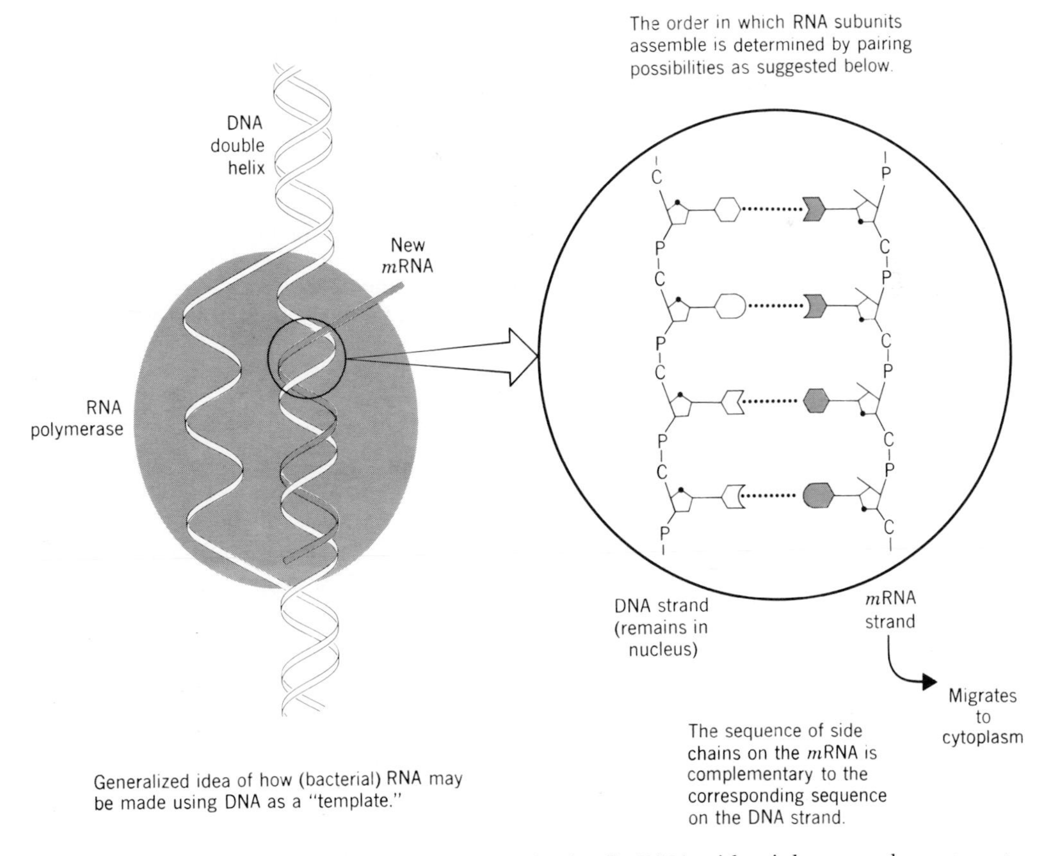

Figure 14.14 Transcription—the synthesis of *m*RNA with triplets complementary to those on DNA. (Because uracil and thymine are informationally equivalent, we may use the same geometric shape to represent both, uracil on RNA and thymine on DNA.)

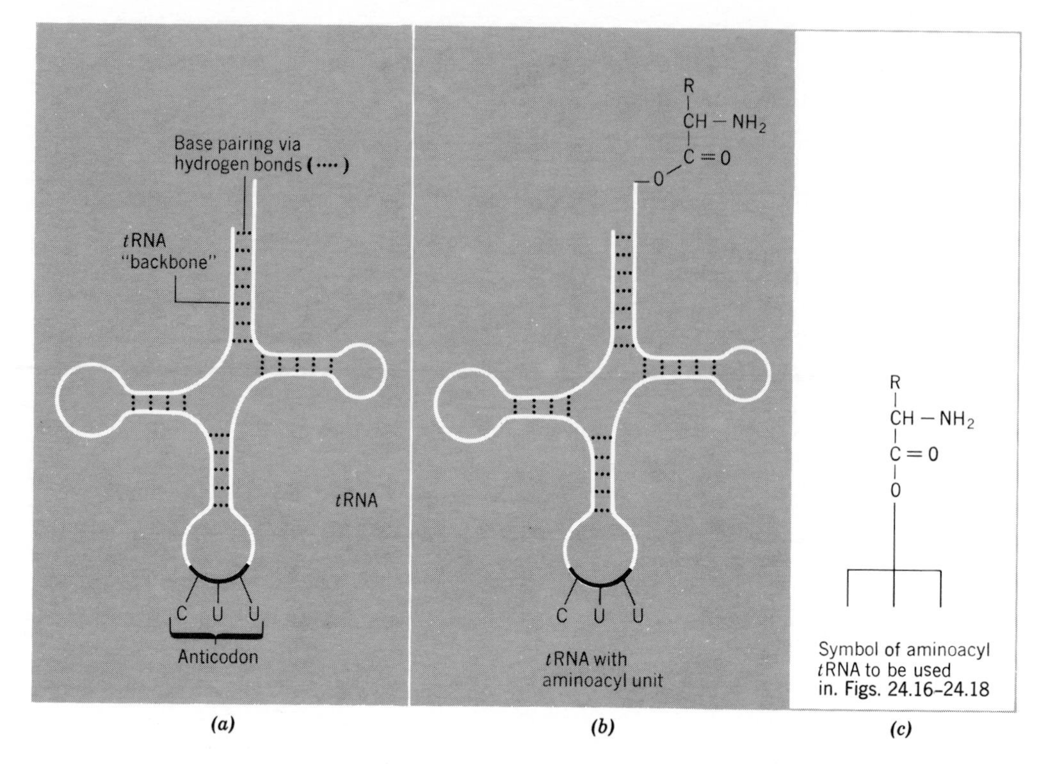

Figure 14.15 *t*RNA. The anticodon at the bottom is made of the triplet CUU. (*a*) The *t*RNA molecule without its unique amino acyl group. (*b*) The *t*RNA-amino acyl system. (*c*) A condensed, schematic symbol for the *t*RNA-amino acyl system for use in subsequent figures. The flat cloverleaf designs given in (*a*) and (*b*) are meant only to show that *t*RNA has three loops, that considerable base-pairing occurs, and that both an anticodon and an aminoacyl binding site are present. The actual three-dimensional shape is much more twisted than shown here.

of *m*RNA can dissociate from ribosomes, and ribosomes can associate with different *m*RNA molecules of different molecular weight. There is evidence that a ribosome moves along an *m*RNA chain while the synthesis of a polypeptide, directed by the *m*RNA, takes place. We shall return to this later. Protein synthesis requires, of course, that amino acids be brought to the *m*RNA site in the order in which they are to appear in the final polypeptide. For this task, the cell uses the third type of RNA, transfer RNA.

14.12 Transfer RNA (*t*RNA)

At least 60 different species of *t*RNA molecules exist. **Transfer RNA** molecules are the smallest of the RNAs. The first ones to be characterized had from 70 to 80 nucleotide units. Being small, *t*RNAs are more soluble and more mobile in the cytoplasmic fluid. Each *t*RNA has an amino acid binding site and an *m*RNA coding site called an anticodon, also indicated in Figure 14.15.

The function of *t*RNA is to attach to itself the particular amino acid for which it is coded and carry it to a protein-synthesis site on a polysome (specifically, an *m*RNA site in contact with a ribosome). It delivers its amino acid to the growing end of a polypeptide chain at just that moment called for by the genetic code.

14.13 Codons and Anticodons

The "alphabet" of the genetic code has 4 "letters," the 4 bases: A, T (or U), G, and C. (Uracil, U, can substitute for thymine, T.) The alphabet of protein structure, however, has 20 letters, the 20 amino acids. We obviously cannot translate on a one letter to one letter basis from the genetic to the protein systems. To solve this, the system uses the four genetic letters in groups of three. Each amino acid has associated with it at least one three-letter "word" made of the genetic alphabet. Each triad of bases on a messenger RNA strand (where U does substitute for T) is one coded "word," called a **codon,** that corresponds to a particular amino acid. A series of codons on *m*RNA, therefore, programs a series of amino acid residues in an enzyme. Known codon assignments are given in Table 14.1. Phenylalanine, for example, is coded either by

TABLE 14.1
Codon Assignments in the Genetic Code[a]

First	Second				Third
	U	C	A	G	
U	phenylalanine	serine	tyrosine	cysteine	U
	phenylalanine	serine	tyrosine	cysteine	C
	leucine	serine	ochre	CT	A
	leucine	serine	amber	tryptophan	G
C	leucine	proline	histidine	arginine	U
	leucine	proline	histidine	arginine	C
	leucine	proline	glutamine	arginine	A
	leucine	proline	glutamine	arginine	G
A	isoleucine	threonine	asparagine	serine	U
	isoleucine	threonine	asparagine	serine	C
	isoleucine	threonine	lysine	arginine	A
	methionine, or formylmethionine	threonine	lysine	arginine	G
G	valine	alanine	aspartic acid	glycine	U
	valine	alanine	aspartic acid	glycine	C
	valine	alanine	glutamic acid	glycine	A
	valine	alanine	glutamic acid	glycine	G

[a] These codon assignments have been obtained largely from research with *E. coli,* but there is mounting evidence that they apply universally to all organisms. Thus the codon UUC, one of two that specifies phenylalanine, would specify this amino acid in any organism. GGU is one of four codons for glycine. The codons designated *ochre, amber,* and *CT* specify, when encountered on an *m*RNA strand during polypeptide synthesis, that the chain should be terminated. The codon that specifies formylmethionine, at least in *E. coli,* is needed to start polypeptide synthesis. For speculations about the fact that the same amino acid may be specified by more than one codon, see this reference (the source of the data for this table): J. M. Lewin. *The Molecular Basis of Gene Expression,* page 81 and following. Wiley-Interscience, New York, 1970. Used by permission.

UUU or UUC. Alanine is coded by any one of four: GCU, GCC, GCA, or GCG. These codon assignments appear to be universal; that is, they apply to all forms of life.

A codon on an *m*RNA strand cannot arise unless a complementary triplet, called an **anticodon,** is on the DNA strand that guides the synthesis of the *m*RNA. The anticodon for a GAA codon on *m*RNA is CTT on DNA, since G

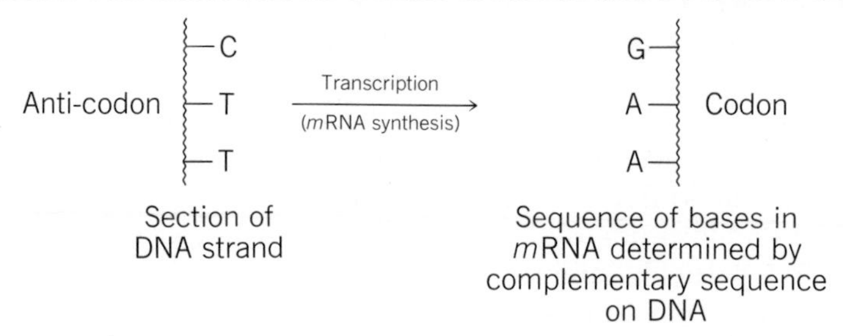

pairs with C and A pairs with T. The anticodon for a UUC codon would be AAG, since U (in *m*RNA) pairs with A (on DNA) and C pairs with G.

Exercise 14.1 Using Table 24.1, what amino acids are specified by each of the following codons on an *m*RNA molecule?
(a) CCU (b) AGA (c) GAA (d) AAG

Exercise 14.2 What amino acids are specified by the following anticodons on DNA?
(a) GGA (b) TCA (c) TTC (d) GAT

The series of triads on DNA are a series of anticodons. When these are transcribed into a series of codon triads on *m*RNA, the genetic message is moved from the gene to the messenger for that gene, *m*RNA. The messenger, with the aid of *r*RNA, *t*RNA, enzymes, and sources of chemical energy then translates that genetic message into a distinctive sequence of amino acids on a polypeptide that will be an enzyme or part of an enzyme. **Transcription** is DNA-directed synthesis of *m*RNA. **Translation** is *m*RNA-directed synthesis of a polypeptide. Thus we have in the most general terms an explanation for the one-gene–one-enzyme relation proposed several years before Crick and Watson by George Beadle and Edward Tatum (co-winners of the 1958 Nobel Prize in physiology and medicine). Crick and Watson showed how the structure of DNA could be correlated with its biological function. The discovery that a special messenger RNA was required for enzyme synthesis came from the work of two French scientists, F. Jacob and J. Monod (co-winners of the 1965 Nobel Prize in physiology and medicine). We next see how *m*RNA works together with other substances to make distinctive amino acid sequences.

14.14 Polypeptide Synthesis

The principal steps are the following.

1. Amino acids are activated—picked up by their individual *t*RNAs.
2. The elongation complex forms—the first *t*RNA with its amino acid, the

*m*RNA strand, and the subunits of the ribosome come together at one end of the *m*RNA.
3. Elongation proceeds—the polypeptide strand grows as one amino acid after another, carried into place on *t*RNAs, is incorporated.
4. Polypeptide synthesis is terminated—from start to finish takes about 10 sec per polypeptide strand.

Figure 14.15 shows a convenient symbol for *t*RNA joined to its particular amino acid. To put the two together, the chemical energy of ATP is needed as well as special enzymes. The first amino acid to go into the polypeptide synthesis is, at least in the bacterium *Escherichia coli* (*E. coli*), a methionyl unit that is modified by a formyl group to make it resemble an amide (or peptide). Instead of an amino group, $—NH_2$, it has been changed to an *N*-formylamino group, —NH—CH=O. At the end of the polypeptide synthesis, this group is removed.

Each ribosome consists of two subunits, one smaller and one larger. The smaller unit attaches at the end of the *m*RNA strand together with the first *t*RNA-amino acid unit, as illustrated in Figure 14.16. Then the larger subunit locks into place and the ribosome rolls down the *m*RNA strand toward the other end as polypeptide synthesis occurs at the ribosomal surface, a surface with enzymes needed for translation.

The sequence of events in elongation consists of a repetition of the following steps, illustrated in Figure 14.17. We shall symbolize a *t*RNA-amino acyl unit by *t*RNA-aa.

1. The first *t*RNA-aa moves from the acceptor site on the ribosome to the donor site as the ribosome shifts slightly to bring the second codon over the donor site (Figure 14.17*a*). This translocation needs an enzyme on the ribosome surface, and the diphtheria toxin works by inactivating that enzyme.
2. The second *t*RNA-aa, attracted by fitting possibilities, arrives at the second codon of the *m*RNA (Figure 14.17*b*).
3. The aminoacyl part of this newly arrived particle is broken out and given to the carbonyl carbon of the first aminoacyl unit as the *t*RNA of the latter is released (Figure 14.17*c*). The first peptide bond has now formed and we have a dipeptide dangling from the ribosome.
4. This dipeptide unit, with the second *t*RNA still attached, shifts over to make room for the next *t*RNA-aa as the ribosome moves down the *m*RNA to bring the third codon into the proper position over the acceptor site (Figure 14.17*d*).
5. We now repeat steps 1 to 3 until the end of the *m*RNA strand is reached or a nonsense codon—a chain terminating codon—is reached.

Figure 14.18 shows what happens at termination. Either the *N*-formyl group of the first amino acid is removed, leaving methionine at the N-terminal end of the polypeptide, or the whole methionyl unit is removed leaving a different N-terminus. The new polypeptides move off the ribosome into the tubules of the rough endoplasmic reticulum, migrate to the Golgi complex, are put into a "package" (a secretory granule), which goes to the cell membrane and discharges the polypeptides outside the cell.

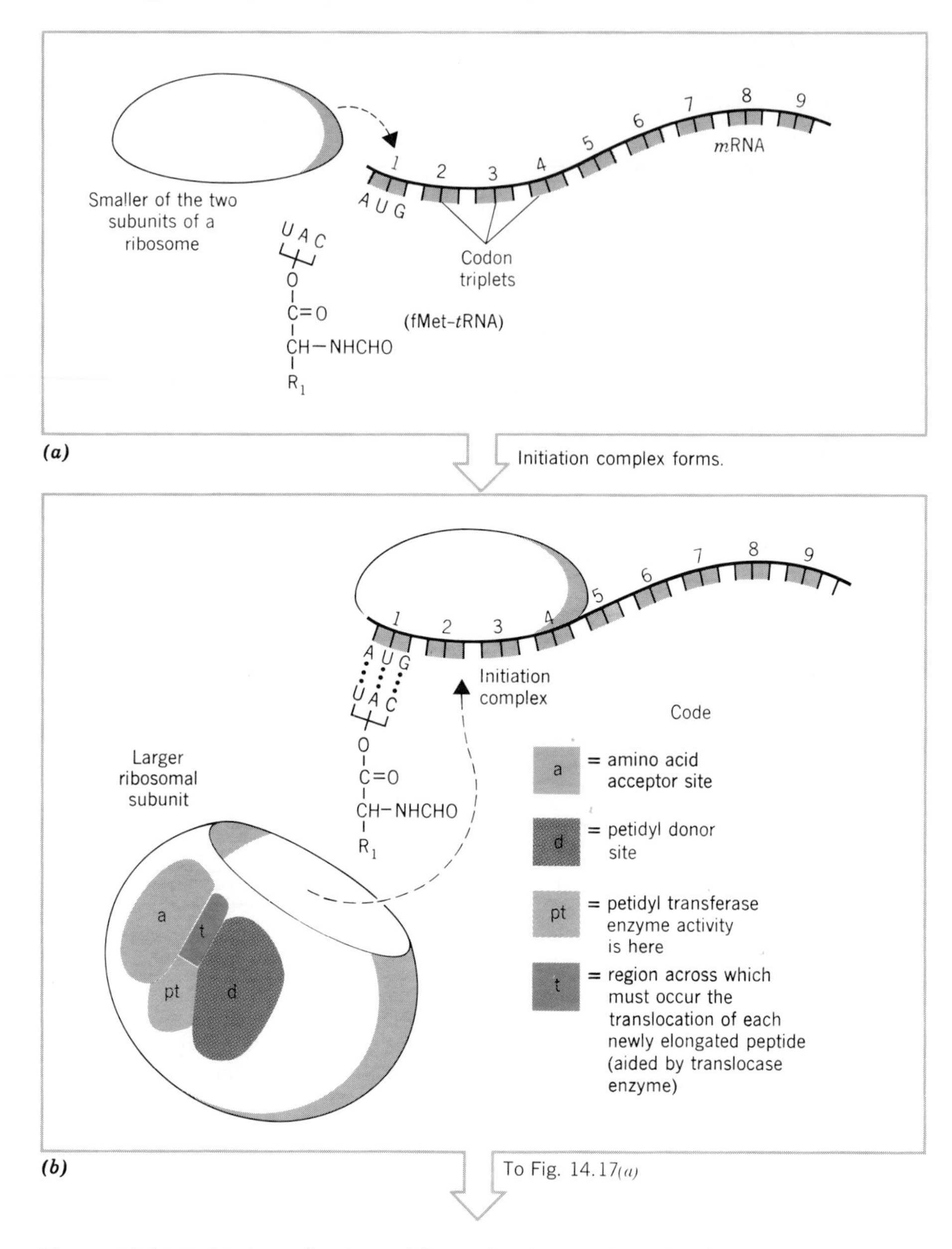

Figure 14.16 Initiation of polypeptide synthesis. (*a*) The subunits move to combine. (*b*) The larger ribosomal subunit completes the formation of the initiation complex.

14.15 Control of Polypeptide Synthesis

What switches genes off or on is an intensively investigated question. From studies with bacterial genes the evidence indicates that genes are repressed until particular events release just those genes that the cells in a given tissue should have activated. A **repressor** molecule, probably a polypeptide made by the direction of its own gene, binds to a small segment of the whole DNA strand. See Figure 14.19. That prevents transcription, the synthesis of *m*RNA.

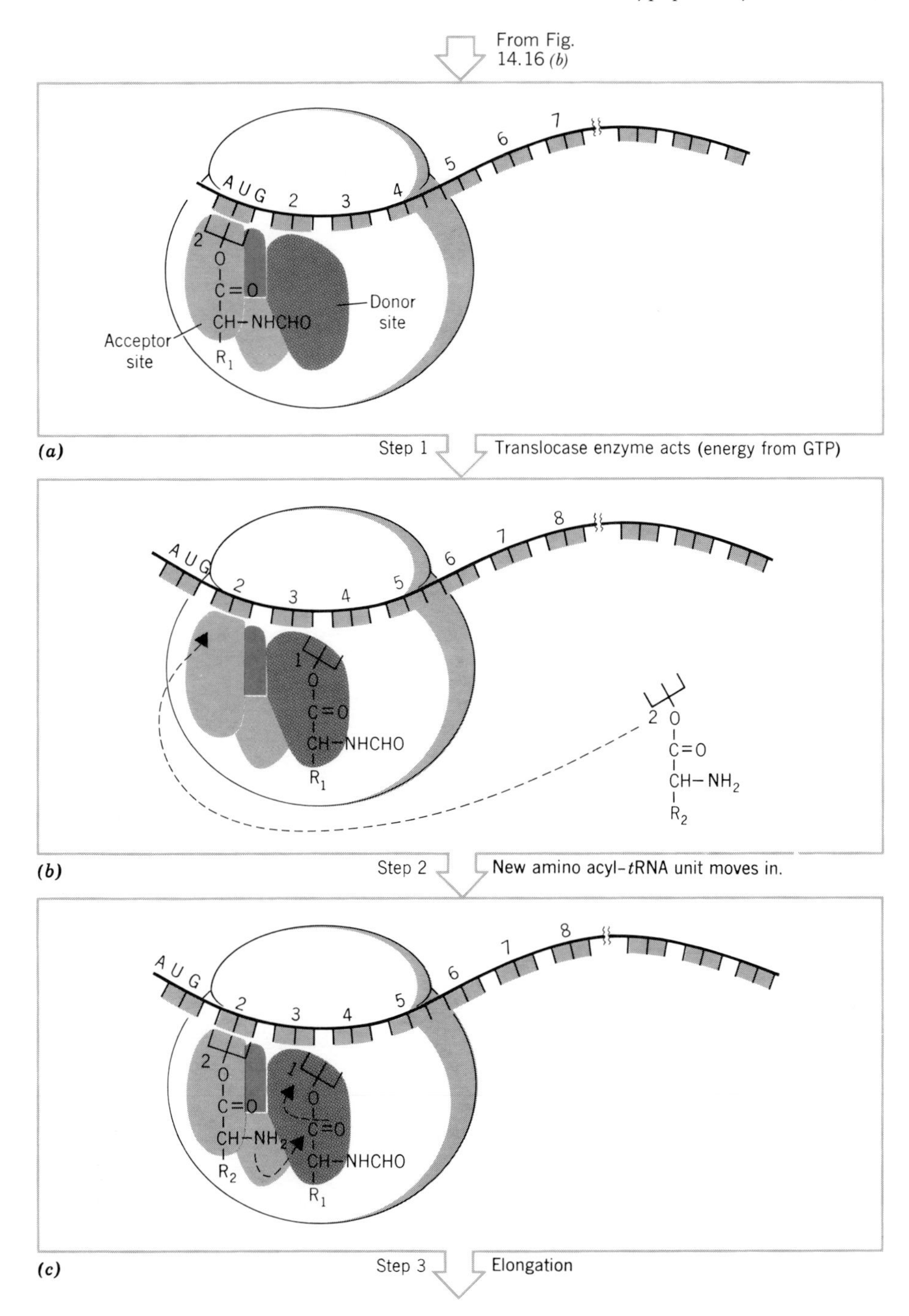

Figure 14.17 Elongation of the polypeptide chain. Individual steps are discussed in the text.

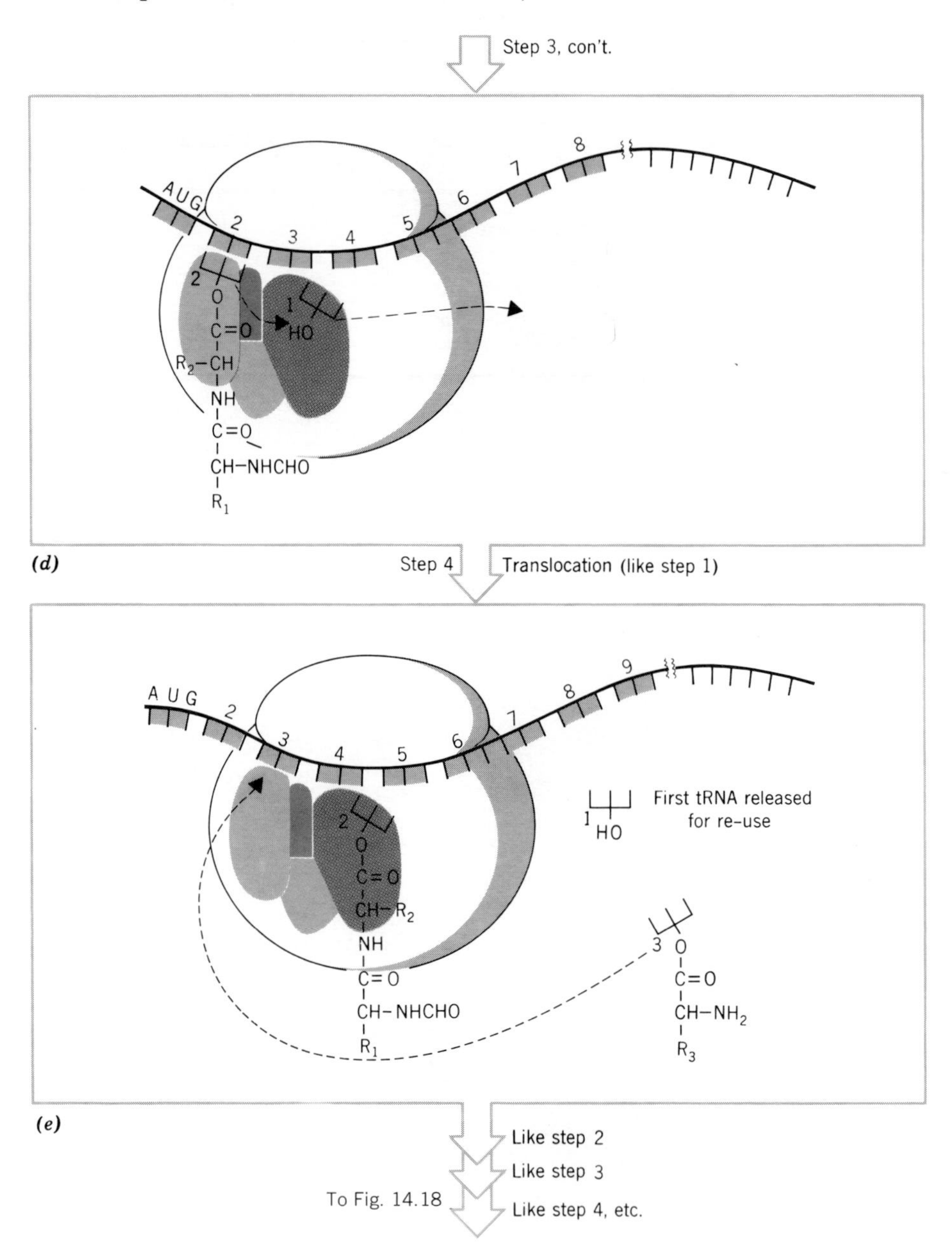

Figure 14.17 (continued)

When that *m*RNA is needed, an **inducer** molecule combines with repressor, changes its shape, and makes it drop off the DNA strand. This, in effect, activates the gene because now it is free to help make more *m*RNA.

The enzyme β-galactosidase, for example, is needed by *E. coli* to hydrolyze lactose to galactose and glucose. Lactose is the inducer that gets the repressor molecule off the DNA strand whose information is needed to make this enzyme. When the lactose disappears because of the work of the enzyme, the repressor is free to go back on the DNA and repress further synthesis of the

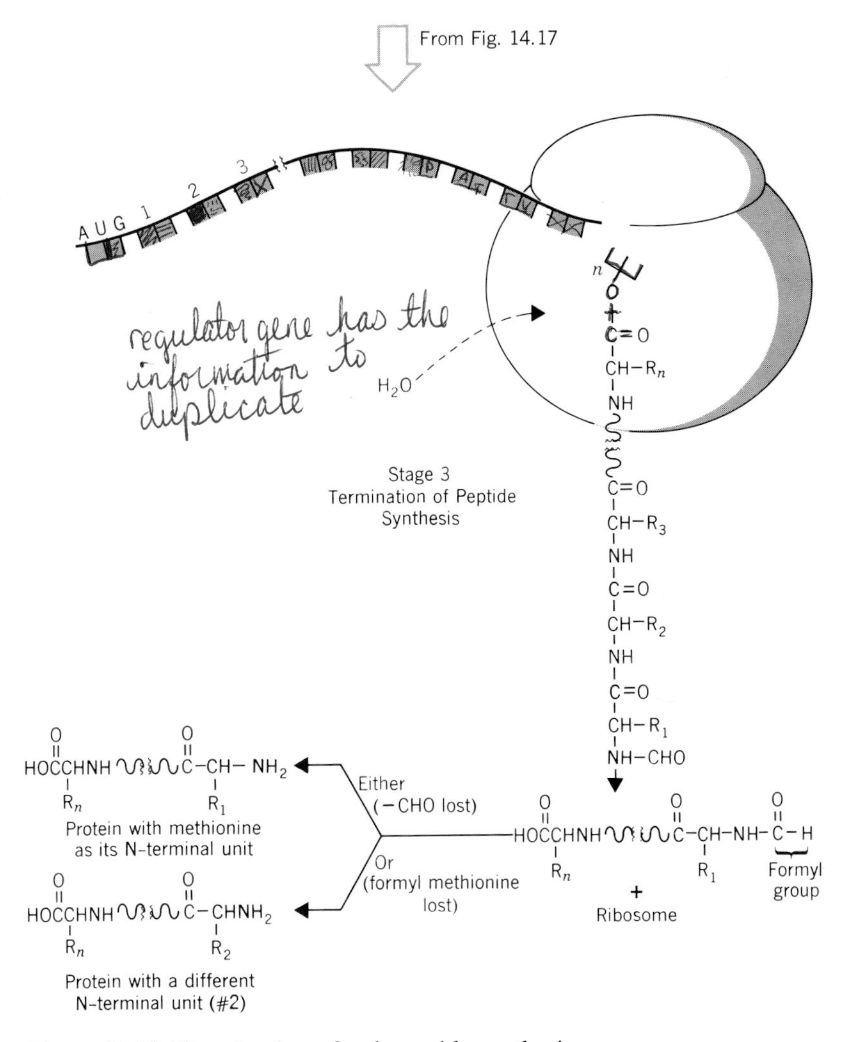

Figure 14.18 Termination of polypeptide synthesis.

enzyme (Figure 14.19). We have here a type of feedback inhibition that illustrates one way by which genes might be controlled and through genes, enzymes. F. Jacob and J. Monod of France shared the 1965 Nobel Prize in Physiology and Medicine (with A. Lwoff) for originating this theory.

14.16 Inhibition of Polypeptide Syntheses in Bacteria by Antibiotics

Several important antibiotics kill bacteria by inhibiting the syntheses of polypeptides that these organisms are trying to make in order to multiply. Streptomycin inhibits the initiation of polypeptide synthesis. Chloramphenicol inhibits the ability of transferring the newly arrived amino acyl unit to the elongating strand. The tetracyclines inhibit the binding of *t*RNA-aa units

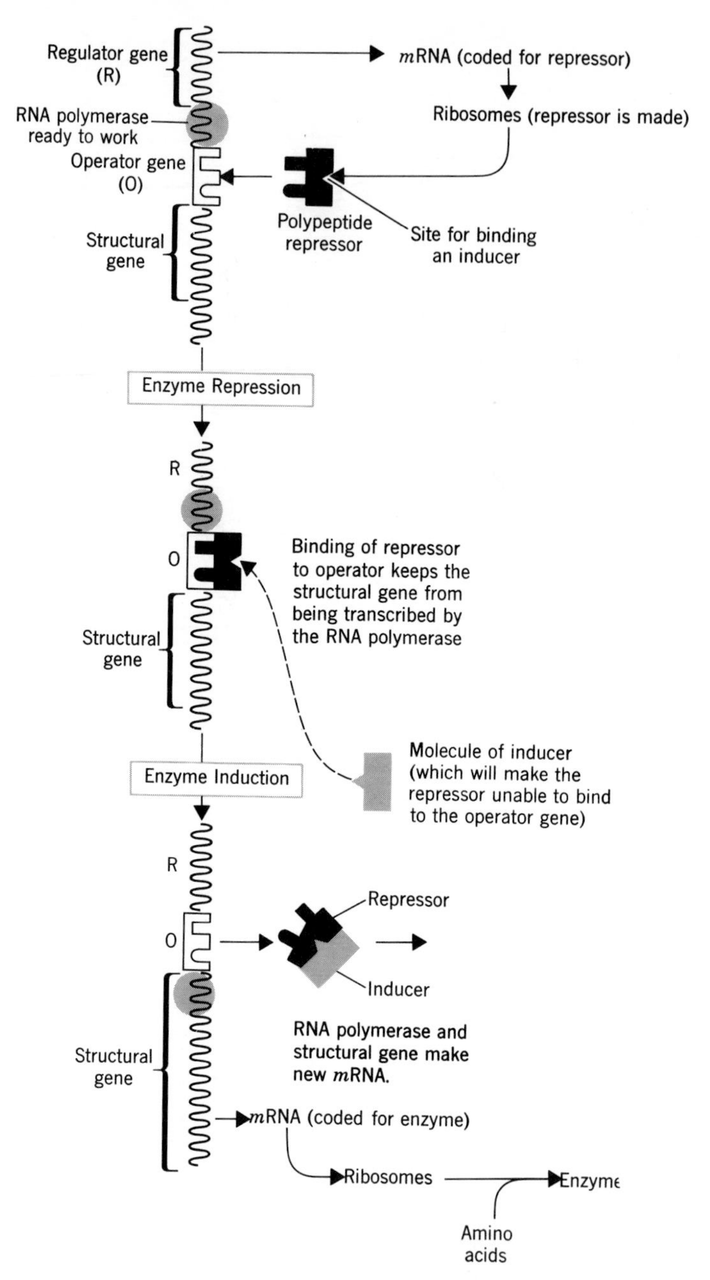

Figure 14.19 Enzyme repression and induction. The enzyme β-galactosidase in the bacterium *E. coli* is repressed by a small polypeptide (top). Lactose, the inducer, however, forces the repressor off (middle), and this lets the structural gene proceed to make the enzyme. Thus, when lactose arrives in the cell, the enzyme needed to metabolize it is made. When the lactose is used up, the gene is switched off because the repressor no longer is kept from binding to the operator gene.

when they arrive at the ribosome. Erythromycin, Puromycin, and cycloheximide also interfere with elongation.

14.17 Viruses

Viruses are agents of infection made of nucleic acid surrounded by an overcoat of protein and, in some viruses, other materials. Unlike a cell, a virus has either RNA or DNA, but not both. Viruses can neither synthesize proteins nor generate their own energy for metabolism. Viruses are at the borderline of living systems and generally are considered to be unique packages of dead chemicals except when they are within the kinds of cells that they infect. The very simple viruses have only three genes; the most complex have roughly 250 genes.

Viruses are unusually selective in the kinds of cells they infect. Once inside their host cells, virus particles take over both the metabolic and genetic machinery of the host to manufacture new virus particles. Only a few viruses remain dormant in their host cells.

Some evidence exists that some human cancers might be caused at least indirectly by viruses, but clinching direct evidence has not yet been discovered. Indirect evidence is plentiful. Some human cancers contain RNA units extremely similar to RNAs of viruses that cause tumors in mice. Examples are lymphomas, breast adenocarcinomas, and leukemias. Some human cancer cells, unlike normal cells, have enzymic activity by which DNA synthesis can be directed by RNA. (Normally, it is the other way around; DNA directs the synthesis of RNA.) Some viruses have the same enzymic activity, an enzyme called reverse transcriptase. Finally, some human cancers have DNA sequences not found in normal cells but found in some viruses.

14.18 Hereditary Diseases

About 2000 diseases in man are believed to be caused directly or indirectly by a genetic disorder. Cystic fibrosis is one of the most common. An estimated one person in 20 carries the gene for this disease. It hits about one of every 1000 born. Sickle-cell anemia, a condition almost entirely confined to the population of Negroes, afflicts roughly 300 000 people in the United States, and is widespread in certain parts of Africa (particularly Ghana) where, at least among those with mild forms, it seems to provide some protection against malaria. Sickle-cell anemia is the result of a defective gene.

In albinism, the pigments normally giving distinctive colors to hair, skin, and eyes do not get made because one gene is defective, the gene that directs the synthesis of an enzyme needed to make the pigments.

In phenylketonuria, or **PKU,** one defective gene results in the absence of an enzyme needed to handle the amino acid phenylalanine. As a result another metabolic event occurs to phenylalanine more frequently than normal, its conversion to phenylpyruvic acid. Phenylpyruvic acid builds up in the bloodstream, and since it is a ketone that has a phenyl group, the condition is called phenylketonemia. Eventually, the renal threshold for phenylpyruvic

$$\underset{\text{Phenylalanine}}{C_6H_5CH_2\overset{NH_2}{\overset{|}{C}}H-\overset{O}{\overset{\|}{C}}-OH} \qquad \underset{\text{Phenylpyruvic Acid}}{C_6H_5CH_2\overset{O}{\overset{\|}{C}}-\overset{O}{\overset{\|}{C}}-OH}$$

acid is reached and the kidneys put increasing quantities of this material into the urine, a condition called phenylketonuria, or PKU for short. If not discovered shortly after birth, a phenylketonuric infant will become permanently mentally retarded. How this happens in humans is not clear, but it may be caused by a reduced rate of protein synthesis in the developing brain. At least, it is known that when PKU is experimentally induced in young rats, the polysomes in their brains (but not in their livers) are broken back to their separate ribosome units and *m*RNA, an event that would certainly affect the synthesis of enzymes the brain needs. Congenital heart disease and microcephaly (small head) can also result.

A simple blood test taken four or five days after birth will detect PKU. The treatment of PKU has been through the control of the diet in such a way that the infant gets proteins especially low in phenylalanine. This is extremely difficult (one slice of bread has all of this amino acid a PKU infant can handle) and, therefore, the diet is potentially dangerous itself. The situation calls for the most expert medical attention. A special protein hydrolisate low in phenylalanine, marketed as Lofenalac®, is available for dietary use. If the infant can be brought through the most critical periods of rapid brain growth and development, the prospects for a normal later life are quite promising.

Cystic fibrosis, sickle-cell anemia, albinism, PKU—the list could be extended—are all dramatic examples of the dependence of physical and emotional health on events at the molecular level where information for living is translated and transcribed.

14.19 Amniocentesis and Enzyme-Gene Behavior

Because of our growing knowledge of the molecular basis of heredity, careful genetic sleuthing, and the ability to culture human cells, amniocentesis has emerged as a powerful tool for prenatal diagnosis. The amniotic fluid is fluid surrounding the fetus, which is in a cavity called the amniotic cavity. In **amniocentesis,** some of that fluid is withdrawn at about the 14th week of gestation by means of a hypodermic needle and syringe. Cells in the fluid include those of the fetus, and these are separated and grown as a tissue culture. In 2 to 6 weeks enough cells have developed to examine for gross chromosome defects or for particular enzymes, enzymes produced by defective genes known to be associated with certain hereditary diseases. If such are found, the possibility of a therapeutic abortion is discussed with the parents, who must make the final decision. Parents with genetic histories that might cause them not to take the risk of having any babies at all may, because of amniocentesis, determine if their developing fetus will be born with the defect. The procedure is controversial because of the abortion issue.

14.20 Radiations and Gene Damage

Some, perhaps all of the inherited, inborn errors of metabolism—hereditary diseases—arose in the past when specific genes became damaged either by chemicals or by radiations. Atomic radiations, particularly X rays and gamma rays, go right through soft tissue. They knock electrons and nuclei about as they go, creating unstable ions and free radicals (particles with unpaired electrons). New covalent bonds may form. Should that happen to a segment of a DNA molecule, the gene ever after may transcribe for a slightly different enzyme or it may not transcribe at all. Yet, depending on the nature and extent of the damage, the cell might be able to replicate that DNA and pass it, defective, on to succeeding cells. If that occurs among germ cells, then the next offspring will inherit the defect.

Since the primary site of radiation damage is in the cell's genetic apparatus, the first symptoms of exposure to radiation will occur in tissue whose cells most frequently divide. Cells in bone marrow are examples. Since they make white cells, an early sign of radiation damage is a fall in the white cell count. Cells in the intestinal tract also divide frequently, and even moderate exposure to X rays or gamma rays (as in cobalt ray therapy for cancer) produces intestinal disorders.

Chemicals that mimic radiations in their effect on human tissue are called **radiomimetic substances.** Those that cause mutations in experimental animals are called **mutagens.** Those that cause cancer are **carcinogens.** Some will cause birth defects and are named **teratogens.** Some pesticides and food additives have been found to be one or more of these types of compounds or are suspected of being so.

Brief Summary

Hereditary Information. The genetic apparatus of a cell is mostly in its nucleus and consists of chromatin, a complex of DNA and proteins. Strands of DNA, a polymer, carry segments that are individual genes. Chromatin (both its DNA and protein) replicates prior to cell division, and the duplicates segregate as the cell divides. Each new cell thereby inherits exact copies of the chromatin of the parent cell. If copying errors are made, the daughter cells are mutants. They may be reproductively dead—incapable of themselves dividing—or they may transmit the mutant character to succeeding cells. The expression of this character might be as a cancer, a tumor, or a birth defect. Atomic radiations, particularly X rays and gamma rays, are potent mutagens, but many chemicals mimic these rays—are radiomimetic.

DNA. Complete hydrolysis of this polymer gives phosphoric acid, deoxyribose, and a set of four heterocyclic amines or organic bases: adenine (A), thymine (T), guanine (G), and cytosine (C). The molecular backbone of the DNA polymer is a series of deoxyribose units joined by phosphate diester units. Joined to each deoxyribose is one of the four bases. The series of bases is the genetic information. Each group of three succeeding bases makes up one anticodon. The series of anticodons corresponds to a specific series of amino acids in a polypeptide. Each gene with its sequence of anticodons specifies one polypeptide (one enzyme, usually). The DNA strands exist in the cell nucleus as double helices with the backbones forming right-handed, intertwining spirals and the bases from one strand attracted to the bases of the other by hydrogen bonds. A always pairs with T by hydrogen bonds; G with C. The Crick-Watson theory uses this structure of DNA to explain how replication occurs.

Replication of DNA. Strands of DNA unwind from each other as the monomers for new DNA (the nucleotides) assemble along the exposed bases. Segments of new DNA are made, and these are linked to form the complete DNA. The process is semiconservative. Each new double helix has one of the parent DNA strands and one new, complementary strand.

RNA. RNA is similar to DNA except for these features. RNA is found mostly in the cytoplasm; DNA in the nucleus. In RNA ribose replaces deoxyribose; uracil (U) replaces thymine (T), but the two are informationally equivalent. (U can pair to A just as T can pair to A.) Three types of RNA are based on both size and function. *r*RNA is in ribosomes and polysomes. A ribosome contains both *r*RNA and protein having enzyme activity needed in polypeptide synthesis. *m*RNA is the carrier of the genetic message from DNA to the site where polypeptide is assembled. *t*RNAs are the smallest of the RNAs and are carriers of amino acids.

Polypeptide Synthesis. Genetic information is first transcribed when DNA directs the synthesis of *m*RNA. Each anticodon on DNA specifies a codon on *m*RNA. The *m*RNA moves to the cytoplasm to form an initiation complex when it joins subunits of a ribosome and the first *t*RNA-aa unit to become part of the polypeptide. The ribosome then rolls down the *m*RNA as *t*RNA-aa units come to the *m*RNA codons as the latter become properly aligned over enzymic sites on the ribosome. Elongation proceeds to the end of the *m*RNA (or to a terminating codon). At termination the polypeptide strand leaves and may be "groomed" to give it its final N-terminus. It then adopts whatever secondary and higher structures naturally form. The operation may be controlled by a feedback mechanism in which an inducer molecule removes a repressor of the gene letting the gene work. Several antibiotics inhibit bacterial polypeptide syntheses.

Viruses. These packages of DNA or RNA in a protein overcoat can take over metabolic and genetic apparatus in their host cells. Some are implicated in human cancer.

Selected References

Books

1. L. Stryer. *Biochemistry.* W. H. Freeman and Company, San Francisco, 1975. Part IV, "Information," consists of seven chapters on the storage, transmission, and expression of genetic information.
2. W. L. Nyhan, ed. *Heritable Disorders of Amino Acid Metabolism.* John Wiley & Sons, Inc., 1974. A technical reference with emphasis on screening, diagnosis, and treatment of a number of metabolic diseases.

Articles

1. D. D. Brown. "The Isolation of Genes." *Scientific American,* August 1973, page 21. A discussion of how it is now possible to find and purify that particular stretch of DNA that encodes a particular RNA.
2. O. L. Miller, Jr. "The Visualization of Genes in Action." *Scientific American,* March 1973, page 34. A beautifully illustrated discussion of electron microscope studies that support theories of gene action.
3. T. Maniatis and M. Ptashne. "A DNA Operator—Repressor System." *Scientific American,* January 1976, page 64. Concerning the control of genetic activity.
4. J. Cairns. "The Cancer Problem." *Scientific American,* November 1975, page 64. A discussion of the variety of possible causes of cancer.

5. P. Justice and G. F. Smith. "PKU." *American Journal of Nursing,* August 1975, page 1303. The nature of PKU, how to screen for it, and how to use the diet to control it.
6. C. J. Witkop, Jr. "Albinism." *Natural History,* October 1975, page 48. Striking color photos illustrate this article.
7. R. O. Brady. "Inherited Metabolic Diseases of the Nervous System." *Science,* 27 August 1976, page 733.
8. T. H. Maugh II. "The Artificial Gene: It's Synthesized and It Works in Cells." *Science,* 1 October 1976, page 44.

Questions and Exercises

1. How are all DNA molecules structurally alike?
2. How do different DNAs differ, structurally?
3. How are all RNA molecules structurally alike?
4. What are the principal differences, structurally, between DNA and RNA?
5. What is the difference between a nucleotide and a nucleic acid?
6. What is the difference between a naturally occurring DNA strand and an individual gene?
7. What are the products of the complete hydrolysis of RNA? (Name them.)
8. What are the products of the complete hydrolysis of DNA? (Name them.)
9. When DNA is hydrolyzed the proportions of A to T and of G to C are always very close to 1 : 1, regardless of species. Explain.
10. What is the principal noncovalent force in a DNA double helix?
11. How did Crick and Watson explain the relation of gene structure to gene function?
12. What was the contribution of Beadle and Tatum to the chemistry of heredity?
13. What was the main contribution of Jacob and Monod to genetically directed polypeptide synthesis?
14. What are the different functions of *r*RNA, *m*RNA, and *t*RNA?
15. If this sequence appeared on a DNA strand: A G T C G G A

    what sequence would appear in each case?

    (a) On the DNA strand opposite to it in a double helix

    (b) On the *m*RNA strand made under the direction of the given segment
16. If a tripeptide has the structure: Try·Try·Try (where Try is the three-letter symbol for tryptophan), what short nucleic acid segment would be coded for it in each case? (Use the kind of symbol employed in the previous question.)

    (a) In the *m*RNA strand (b) In the DNA strand

chapter 15

Extracellular Fluids of the Body

What a cascade of internal chemistry is released by the sight and smell of good food and even by the sound of its preparation! The entire digestive system mobilizes to do its thing. Thanks to dozens of enzymes these chocolate chip cookies will soon be the kinetic energy of a little boy. (Photo by D. Fricke/Photo Researchers)

15.1 The Internal Environment

The French physiologist Claude Bernard (1813–1878) was the first to point out that all higher animals have not only an external environment of wind and weather but also an internal environment of fluids and solutions. We handle huge variations of external conditions because we have nearly perfect control over the internal. Our purpose in this chapter is to study some of the better understood chemical features of the internal environment that participate in that control.

The **internal environment** consists of all fluids not actually inside cells. The fluids of the internal environment make up about 20% of body weight. Fluids in the interstices or spaces between cells—the **interstitial fluids**—make up three-quarters of the internal environment. The bloodstream constitutes nearly all the rest, but other fluids having small contributions (by volume) are the lymph, cerebrospinal fluid, synovial fluid, and the digestive juices. We shall concentrate on the chemical processes of digestion and of the bloodstream. Because the operation of the kidneys and the formation of urine, diureses, are actually a feature of blood chemistry we must investigate them, too.

Digestion

15.2 The Digestive Tract

The names and locations of the principal parts of this tube running through the body are given in Figure 15.1. At various places different solutions of enzymes, the digestive juices, hydrolyze different kinds of food molecules. Our strategy will be to study each digestive juice in turn in the order in which they act on food.

15.3 Saliva

Saliva flow is stimulated by the sight, smell, taste, and even the thought of food. The pH of saliva varies from 5.8 to 7.1. Besides water (99.5%) it includes a mucin, a glycoprotein that lubricates food, and **α-amylase** (ptyalin). α-Amylase catalyzes the partial hydrolysis of starch to dextrins and maltose. Proteins and lipids pass through the mouth essentially unchanged.

15.4 Gastric Juice

When food arrives in the stomach, the cells of the gastric glands lining the inner surface of the stomach are stimulated by hormones to release **gastric juice** (pH 0.9 to 2.0). One kind of gastric gland secretes **mucin,** a viscous glycoprotein that coats the stomach to protect it against digestive juices and acid. Mucin is continuously produced and is only slowly digested. If for any reason its protection of the stomach is hindered, part of the stomach may be digested causing an ulcer. Another gastric gland secretes hydrochloric acid at a concentration of roughly 0.1 molar—about one million times more concentrated than acid in the bloodstream. A third gastric gland secretes the zymogen, **pep-**

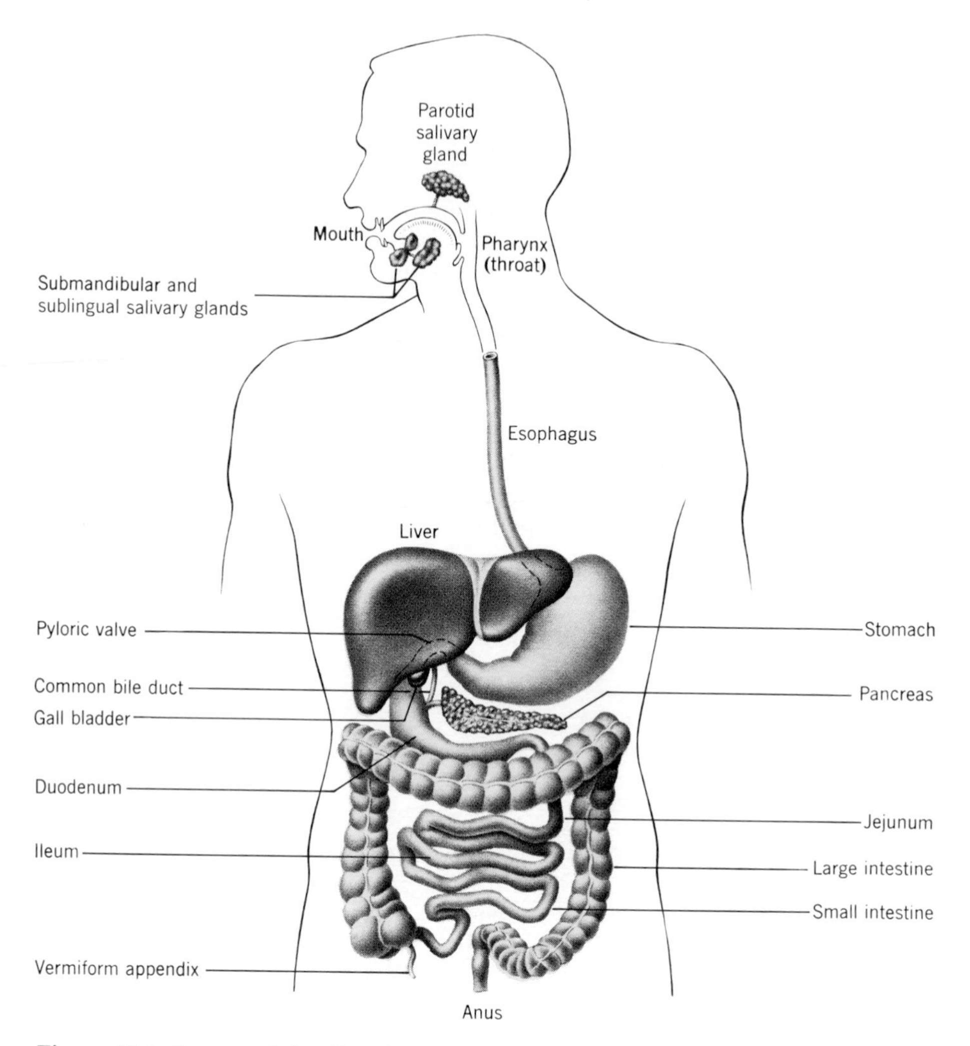

Figure 15.1 Organs of the digestive tract. (Adapted by permission from E. D. Wilson, K. H. Fisher, and M. E. Fuqua. *Principles of Nutrition,* 2nd ed., 1965. John Wiley & Sons, Inc., New York.)

sinogen. Pepsinogen is changed by the action of hydrochloric acid and traces of pepsin into the proteolytic enzyme, **pepsin,** whose optimum pH is 1 to 1.5. That enzyme plus another, gastricsin, catalyze the only important digestive work in the stomach, hydrolysis of some peptide bonds of proteins. Churning and digesting activities in the stomach produce a liquid mixture called **chyme.** Chyme is released in portions through the pyloric valve into the **duodenum,** the first 12 in. of the upper intestinal tract. The acidity of chyme is neutralized by mildly alkaline digestive juices in the upper intestinal tract.

15.5 Intestinal Juice

When chyme enters the duodenum, glands in its walls release **intestinal juice,** an enzyme-rich secretion important to the digestion of all foods. The enzymes sucrase, lactase, and maltase catalyze the hydrolysis of sucrose, lactose, and maltose. Lecithinase, phosphatase, and a lipase act on both complex and simple saponifiable lipids. Enzymes for hydrolyzing nucleic acids are present—nucleases and nucleotidases. Enzymes for hydrolyzing very small polypeptides are present. One of the most important enzymes, however, is **enterokinase,** a "master switch" that turns on digestive power among several important proteolytic enzymes secreted by the pancreas.

15.6 Pancreatic Juice

The entry of chyme into the duodenum also signals hormone action that releases a digestive juice made by the pancreas. **Pancreatic juice** (pH 7.1 to 8.2) contains a lipase, α-amylase; a maltase; and a nuclease. Besides these enzymes are several **zymogens:** trypsinogen, chymotrypsinogen, procarboxypeptidase, and proelastase. Enterokinase, from intestinal juice, activates **trypsinogen:**

$$\text{trypsinogen} \xrightarrow{\text{enterokinase}} \text{trypsin}$$

Besides digesting polypeptides in chyme, **trypsin** (optimum pH 7) activates the other **proteolytic enzymes,** changing their zymogens into enzymes—chymotrypsin, carboxypeptidase, and elastase, all most active at pH = 7 to 8. Trypsin can also activate its own zymogen, trypsinogen. Enterokinase, therefore, turns on all this proteolytic activity.

Trypsin and chymotrypsin act generally on peptide bonds that are near particular side chains in large polypeptides. Elastase is especially effective in helping to degrade elastin, a fibrous protein. Carboxypeptidase helps to break off amino acid residues at the C-terminus of a polypeptide. Another enzyme, aminopeptidase, is present to remove residues from the N-terminus.

Pancreatic lipase is activated by components in still another intestinal secretion, bile.

If the pancreatic lipase and zymogens are activated prematurely within the pancreas, this vital organ will be destroyed by digestion, an effect that occurs in acute pancreatitis.

15.7 Bile

Bile, assembled in the gall bladder from substances made by the liver, is secreted into the duodenum when chyme enters. It contains no enzyme, yet it is essential to the proper digestion of lipids. Bile contains salts called the **bile salts** that are surface-active agents and powerful detergents. (The structure of cholic acid, a steroid whose anion is one of the anions of the bile salts, is given in Table 11.3.)

Without bile salts, pancreatic lipase cannot work effectively. The detergent action of these salts break fat globules into microglobules. Although the total volume of fat, of course, is not affected, the total surface area is increased enormously. Water and enzymes, therefore, can get at the molecules of lipids much more readily and quickly. Bile salts are also important in helping the migration of vitamins A and D from the digestive tract into the blood stream.

The body releases bile not only as a secretion, but also as a means of excretion. Color pigments from the breakdown of heme called the bile pigments leave the internal environment via the bile and feces. Cholesterol is also excreted via the bile. Sometimes cholesterol and some bile salts precipitate in the gall bladder forming gallstones. If the gall bladder has to be removed by surgery, the patient's dietary lipid allowance must be carefully controlled because without bile salts these lipids cannot be digested effectively.

15.8 The Large Intestine

As the chyme moves through the duodenum and digestion continues, monosaccharides, fatty acids, glycerol (plus some monoacylglycerols), and amino acids leave the intestinal tract and enter general circulation. No digestive functions are performed in the large intestine. However, large numbers of microorganisms in residence there, as by-products of their own metabolism, produce vitamins K and B plus some amino acids. These are absorbed by the body, but their contribution to overall nutrition in humans is not large. Undigested matter, including nearly all cellulosic fiber from vegetables and whole grain cereals in the diet, plus the digestive secretions and water make up the feces.

Exercise 15.1 Name the enzymes and the digestive juices responsible for providing them (or their zymogens) that handle the digestion of each of the following.
(a) Large polypeptides (b) Triacylglycerols (c) Amylose (d) Sucrose

Blood Chemistry

15.9 Circulatory System

The circulatory system (Figure 15.2) is our main line of chemical communication between the external and internal environments. All the veins and arteries together are called the **vascular compartment.** The **cardiovascular compartment** is this system plus the heart. The pulmonary branches carry blood to the lungs for the oxygenation of **hemoglobin** and the removal of carbon dioxide. The systemic branches carry blood around the rest of the system. At the intestinal tract, blood picks up products of digestion. Most are immediately monitored by the liver, which among a number of duties checks for foreign molecules and tries to process them into forms that can be used or excreted. In the kidneys the bloodstream is purified of nitrogen wastes, particularly urea. The kidneys also are vital to adjusting the pH of blood and the concentrations of electrolytes. At various endocrine glands the blood picks up and circulates whatever hormones these glands release. Always present in blood are white cells that attack invading bacteria, red cells or **erythrocytes** that carry hemoglobin, and platelets needed in clotting. Finally, blood carries several zymogens that participate in the clotting mechanism.

The principal types of substance in whole blood are outlined in Figure 15.3. One kilogram of blood has about 80 g of proteins: albumins, globulins, and fibrinogen. **Albumins** help carry hydrophobic molecules such as fatty

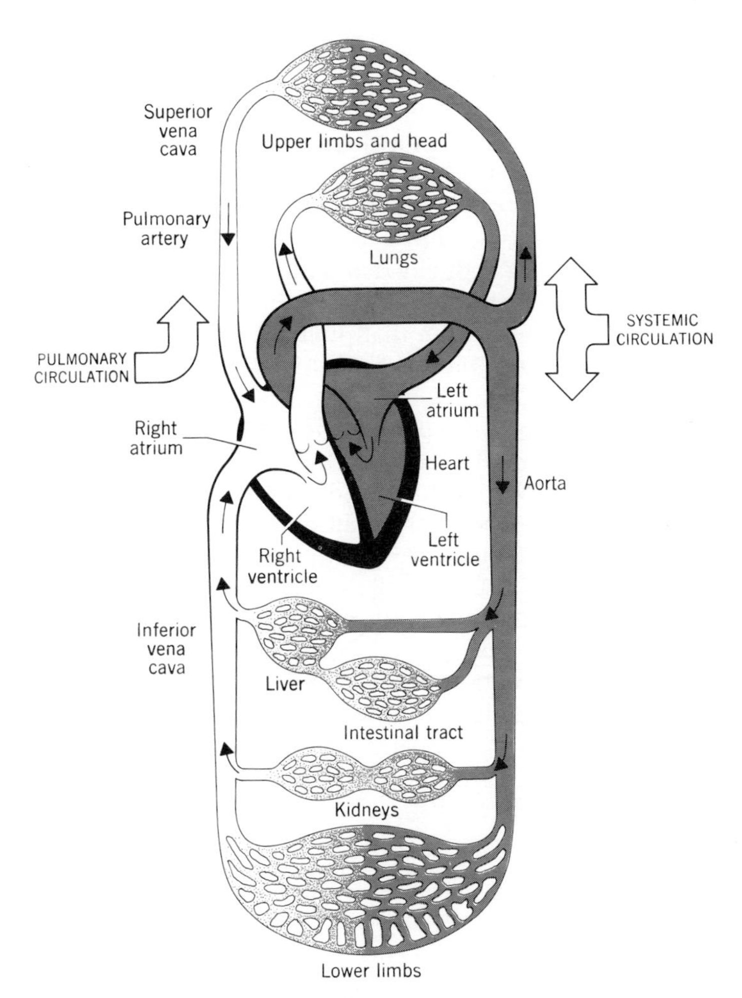

Figure 15.2 Human circulatory system. The more darkly shaded areas denote oxygenated blood. Venous blood (light areas) returns to the heart, which pumps it into the capillary beds around the alveoli of the lungs. Oxygenated blood then leaves for the rest of the system.

WHOLE BLOOD		
Formed Elements (45%)		**Blood Plasma** (55%)
Red cells (erythrocytes)	Oxygen carriers	**Blood Serum**; Fibrinogen
White cells (leukocytes)	Bacteria fighters	Blood Serum: Electrolytes, Water; Proteins
Platelets	Needed in blood clotting	Proteins: Albumins, Globulins

Figure 15.3 Major components of blood.

acids, and they are necessary to the maintenance of the net osmotic pressure of blood. Some globulins carry ions that otherwise could not be soluble in a fluid of pH slightly greater than 7 (e.g., Fe^{2+} and Cu^{2+}). The **γ-globulins** help protect against infectious disease. **Fibrinogen** is converted to an insoluble form, **fibrin,** when a blood clot forms.

The **electrolytes** in blood are given in Figure 15.4. They supply ions the body needs, help maintain osmotic pressure relations, and serve as buffers. The sodium ion dominates in both blood and interstitial fluid, and the potassium ion dominates inside cells. Sodium ions that get in cells are steadily pumped out by the same mechanism that pulls potassium ions back into the cell. The sodium pump occurs within the cell membrane as discussed in an earlier chapter.

The greatest difference between plasma and interstitial fluid is the protein content of blood. The proteins in blood are chiefly responsible for the blood's having a higher osmotic pressure relative to the interstitial fluid.[1] The total osmotic pressure of blood is derived from all dissolved solutes—electrolytes, organic substances, and proteins. However, the walls of the bloodstream are dialyzing membranes. Electrolytes as well as small molecules can move somewhat freely from blood to interstitial fluid and back again. The proteins colloidally dispersed in blood, however, stay in the bloodstream. Their presence gives blood an effective concentration greater than interstitial fluid, and they generate what is called the **colloidal osmotic pressure** of blood. Therefore, the natural flow of water is from interstitial fluid into blood if that flow is controlled solely by dialysis. Why this natural flow does not always and everywhere occur, we shall study later in this chapter.

15.10 Hemoglobin and Exchange of Respiratory Gases

Because so much of human well-being depends on the exchange of oxygen and carbon dioxide, and so many clinical situations arise when this exchange is hindered, we shall examine it in some detail. Our discussion will include a numbered list of facts and descriptions about oxygen-carbon dioxide exchange followed by a figure that summarizes the events.

1. *Each molecule of hemoglobin in an erythrocyte (red blood cell) can carry four molecules of oxygen.*

Adult human hemoglobin, Hb, has quaternary structure as a conjugated protein. It consists of four subunits, each a polypeptide with one heme molecule, for a total of four oxygen-binding sites, one on each heme. We define **oxygen affinity** as the percent to which all hemoglobin molecules in blood are saturated with oxygen. Two subunits in hemoglobin are identical, and we shall symbolize them by α. The other two are identical and have the symbol β. In deoxygenated hemoglobin, as it exists within normal red cells, a small particle, the **diphosphoglycerate ion** (DPG), nestles in a cavity between the four

[1] As a reminder and a useful memory aid, "high solute concentration means high osmotic pressure"; the solvent flows in osmosis (or dialysis) from a region where the solute is dilute to the region where it is concentrated, with the "goal" being to even out the concentrations everywhere.

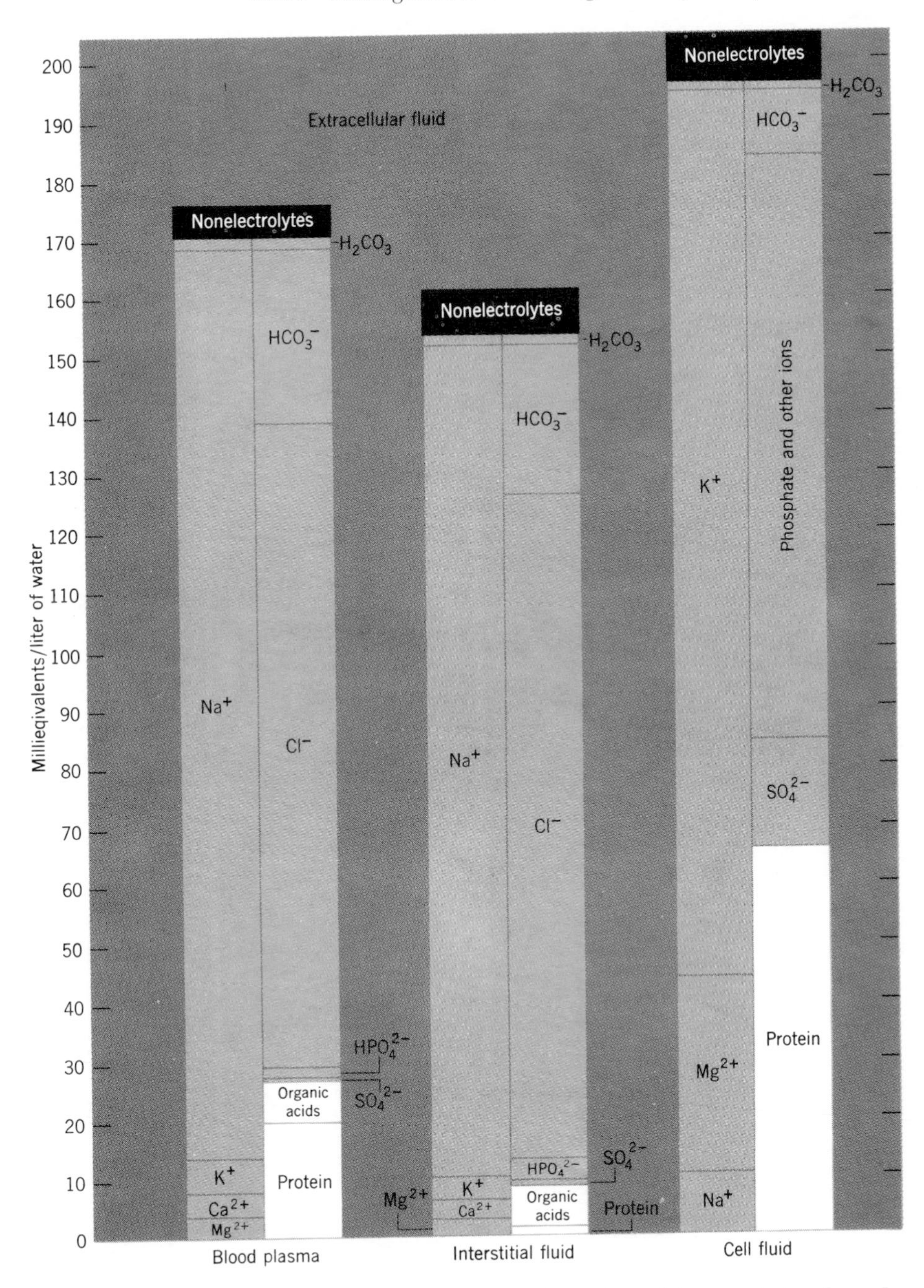

Figure 15.4 Electrolyte composition of body fluids. (Adapted by permission from J. L. Gamble, *Chemical Anatomy, Physiology and Pathology of Extracellular Fluids,* 6th ed., 1954. Harvard University Press, Cambridge, Mass.)

subunits. We may represent deoxygenated hemoglobin by **1**, where each

$^{2-}O_3POCH_2CHCO_2^-$
|
OPO_3^{2-}
Diphosphoglycerate
(DPG)

circle is a subunit without oxygen, and the centered diamond is DPG. Several salt bridges hold hemoglobin together.

2. *The binding of oxygen to hemoglobin is cooperative.*

The first molecule of oxygen to bind to a completely deoxygenated hemoglobin molecule (**1** → **2**) has an **allosteric effect** (see Section 13.15). When it binds, it changes the subunit's shape (represented as the change of a circle to a square), and it changes the affinity of a neighboring site for oxygen. The first molecule, in a sense, breaks a small molecular dam letting the remaining oxygen molecules flood in more easily. The evidence suggests that the first oxygen molecule either expels or starts to expel the DPG unit. (We shall assume that DPG is expelled fully at this stage.) Departure of DPG lets a neighboring subunit change its tertiary structure. It switches to a slightly different shape, one that exposes its heme unit to easy binding by oxygen. (We shall represent a subunit without oxygen but with a changed tertiary form by a square. If it holds oxygen, we shall include the formula for oxygen.) As the second subunit picks up an oxygen, it changes a bit more in shape causing the remaining subunits to change their tertiary structures (as if the initial break in the "dam" becomes wider), **2** to **3**. They now easily bind oxygen, **3** to **4**, and the hemoglobin is saturated, **4**.

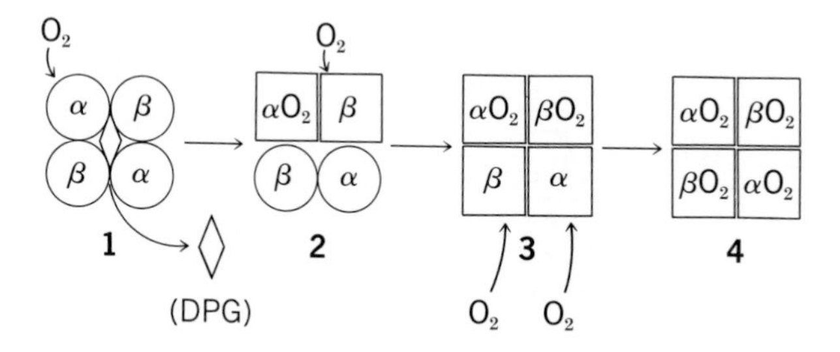

The **hemoglobin-oxygen dissociation curve** in Figure 15.5 has a "lazy S" or sigmoid shape, diagnostic of the allosteric effect just described (and also discussed in Section 13.15). This curve is the result of plotting the percent of hemoglobin saturation—the oxygen affinity—versus the partial pressure of oxygen in the gases in contact with blood. In region A of the figure a small change in partial pressure does not make the percent saturation increase nearly as much as a small change in region B. In region A we have the first molecules of oxygen moving in, "breaking the dam." Then, rather soon, it takes very little increase in pO_2 to get all the hemoglobin nearly saturated. By the same token, in region B it takes only a slight drop in partial pressure to get oxygen off hemoglobin. Thus at the partial pressure of oxygen in alveoli of

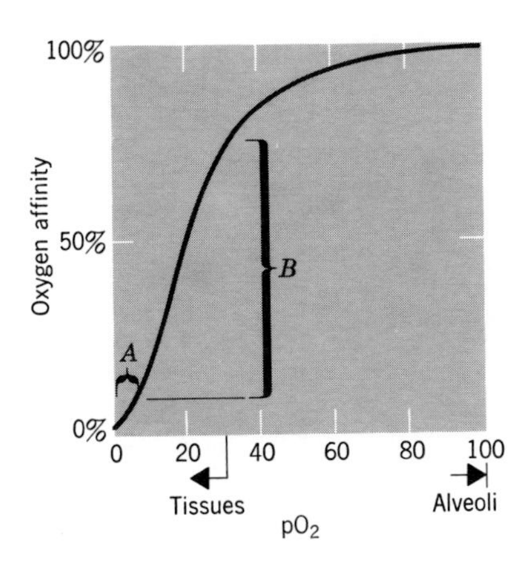

Figure 15.5 Hemoglobin-oxygen dissociation curve. (Regions *A* and *B* of the curve are discussed in Section 15.10.)

lungs, $pO_2 = 100$ mm Hg, hemoglobin is 100% saturated. But at actively metabolizing cells, where $pO_2 = 40$ mm Hg or less (and falling), oxygen easily comes off—just where it is needed.

Without this cooperative action hemoglobin could deliver only half as much oxygen, and our heart-lung-circulatory system would have to be considerably larger or work much harder.

3. *The hemoglobin-oxygen dissociation curve is dependent on pH.*

Hemoglobin is a very weak acid, but it becomes stronger as it accepts oxygen. The oxygenation of hemoglobin (at the lungs) generates hydrogen ions, in other words. Deoxygenation (at active tissue) requires hydrogen ions. To write this as a chemical equation, we shall simplify by writing hemoglobin as HHb, the first H representing a potential hydrogen ion. Moreover, we shall symbolize oxygenated hemoglobin simply as HbO_2^-, call it oxyhemoglobin, and ignore the fact that each hemoglobin unit actually carries four oxygens. We may therefore write:

$$HHb + O_2 \rightleftharpoons HbO_2^- + H^+ \qquad (15.1)$$

With any equilibrium Le Châtelier's principle can operate: If stress is put on the equilibrium, it will shift to relieve the stress. **Alveoli** are small, tissue-thin sacs of blood capillaries in lungs where blood and air exchange gases. At alveoli the stress is the relatively high partial pressure of oxygen, and the equilibrium shifts to the right to take up oxygen (trying to reduce its pressure but never really succeeding). This shift also produces oxyhemoglobin and hydrogen ion.

If the stress in the equilibrium of Equation 15.1 is a drop in pH (i.e., the concentration of H^+ rises), the equilibrium shifts to the left to use up H^+, which also releases oxygen. If the stress is a rise in pH (i.e., the concentration of H^+ declines), the equilibrium shifts to the right to make more H^+ (and also more oxyhemoglobin).

Figure 15.6 shows two hemoglobin-oxygen dissociation curves at values of pH within the physiological pH range. The average pH of blood is 7.35 to 7.45. Below pO_2 of 100 mm Hg, at every value of pO_2 the oxygen affinity is

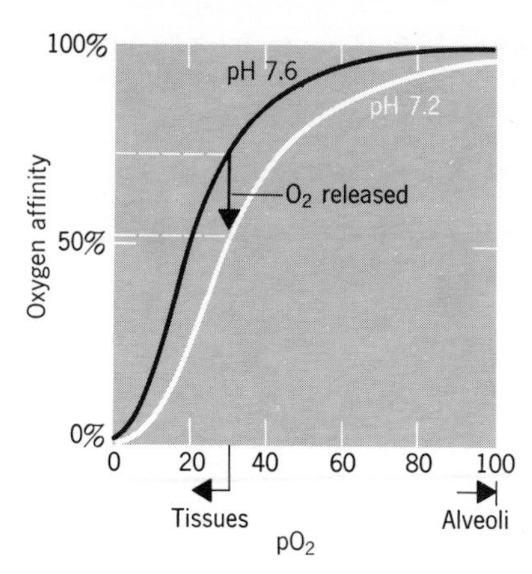

Figure 15.6 Hemoglobin-oxygen dissociation curves at two different values of pH of blood. In active tissues, where pO_2 is 40 mm Hg or less and hydrogen ions are generated to shift the curve down, oxygen is readily released. In alveoli, where hydrogen ions are used to help release carbon dioxide, the curve shifts up, raising the affinity of hemoglobin for oxygen.

less when the blood is more acidic than average (lower curve, at pH 7.2). The enormous importance of this effect is that at active tissues needing oxygen, acids are generated that help release oxygen. As acids help expel oxygen, the DPG (released earlier) moves back, helping to expel the last oxygen molecule. This role of DPG amounts to keeping a lid on how high oxygen affinity can go. If it is too high, oxygen cannot be released. Evidence indicates that without DPG the oxygen affinity of hemoglobin would be too high for life as we know it.

4. *The hemoglobin-oxygen dissociation curve is dependent on* pCO_2.

About 60% of carbon dioxide made in active tissue moves to the lungs chemically bound to hemoglobin. (The rest moves as bicarbonate ion.) Each subunit in hemoglobin has a free amino group that can react with carbon dioxide according to Equation 15.2.

$$R{-}NH_2 + CO_2 \rightleftharpoons R{-}NH{-}CO_2^- + H^+ \qquad (15.2)$$

We may write this as Equation 15.3, a further simplification that ignores the exact number of carbon dioxide molecules held by each hemoglobin.

$$HHb + CO_2 \rightleftharpoons Hb{-}CO_2^- + H^+ \qquad (15.3)$$

Carbamino-
hemoglobin

Under relatively high values of the partial pressure of carbon dioxide, pCO_2, oxygen affinity is less, as seen in Figure 15.7.

Carbon dioxide also combines with water to give carbonic acid, H_2CO_3, that ionizes slightly to release hydrogen ions:

$$CO_2 + H_2O \rightleftharpoons H_2CO_3 \rightleftharpoons HCO_3^- + H^+ \qquad (15.4)$$

Thus, not only do actively metabolizing tissues generate hydrogen ions

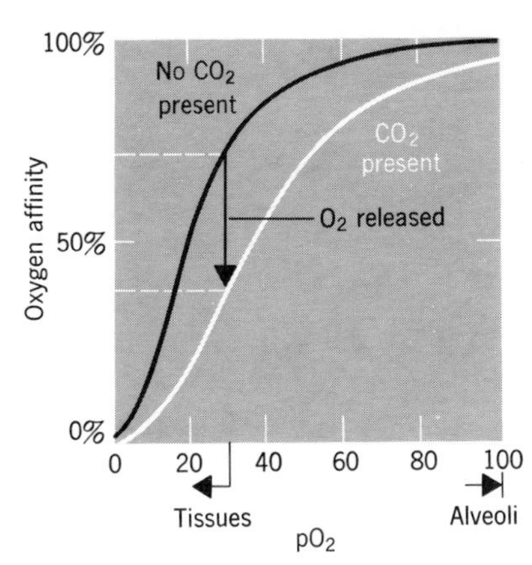

Figure 15.7 Hemoglobin-oxygen dissociation curves with and without the presence of carbon dioxide. At active tissue, where carbon dioxide is released, the oxygen affinity is less (lower curve) and oxygen is released. At alveoli, where carbon dioxide is leaving the bloodstream and less is present in red cells (upper curve), the oxygen affinity is greater, helping to pull oxygen in.

directly (Equation 15.4), they also make them indirectly (Equation 15.3). These ions help shift the equilibrium of Equation 15.1 to the left, and this movement of hydrogen ions has been traditionally called the **isohydric shift.**

Bicarbonate ions generally leave the red cell and return in the serum to the lungs. To balance the (−) charge lost from the red cell, a chloride ion moves into the cell. This exchange is called the **chloride shift.**

The combined action of H^+ and CO_2 to lower the oxygen affinity of hemoglobin is known as the **Bohr effect.**

Exercise 15.2 In what two ways does the oxygenation of hemoglobin in red cells in alveoli help release CO_2?

Exercise 15.3 In what way does waste CO_2 at active tissues help release oxygen from the red cell?

Exercise 15.4 In what way does extra H^+ at active tissue help release oxygen from the red cell?

5. *Myoglobin has a higher affinity for oxygen than hemoglobin.*

Myoglobin is also a hemoprotein, but unlike hemoglobin it has only one polypeptide-heme unit. Moreover, no cooperative allosteric effect helps it bind oxygen. As the myoglobin-oxygen dissociation curve of Figure 15.8 indicates, tissues (such as muscles) containing myoglobin, in a sense, help pull oxygen off oxyhemoglobin by tying up released oxygen and keeping the partial pressure of oxygen in such tissue low. Ignoring the facts that the oxygen must migrate from the red cell all the way into the tissue cell and the shuttling of protons, we could write the following equation (where My = myoglobin):

$$HbO_2^- + My \rightleftharpoons Hb + MyO_2^- \qquad (15.5)$$

The cycle of events in the exchange and transport of oxygen and carbon

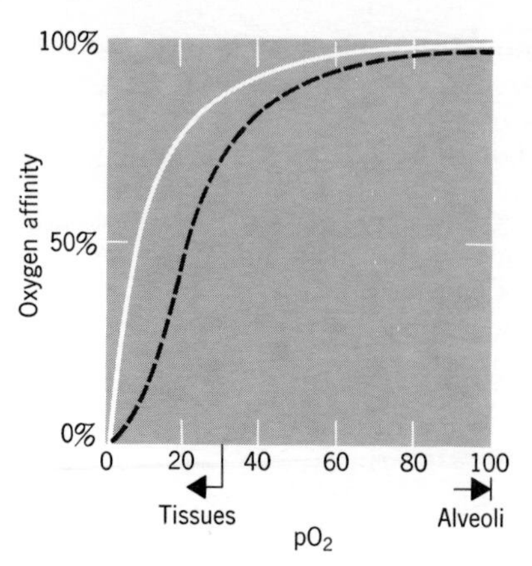

Figure 15.8 Myoglobin-oxygen dissociation curve (solid line). Even at very low partial pressures of oxygen the oxygen affinity of myoglobin is substantially higher than that of hemoglobin (dashed line). At tissues containing myoglobin, therefore, oxyhemoglobin cannot compete with myoglobin for oxygen. Oxygen is released into the tissue.

dioxide between lungs and actively metabolizing tissue is illustrated in Figure 15.9. On the left a red cell is passing through a capillary in an alveolus of the lungs. It carries free HHb, $HbCO_2^-$, and HCO_3^- and is in a region where partial pressure gradients favor a loss of CO_2 and a gain of O_2. On the right in Figure 15.9, this same cell has moved to a capillary in an active tissue, one needing oxygen and having wastes (H^+ and CO_2) to remove. Now the red cell carries HbO_2^- and has the potential for absorbing CO_2 and H^+, the latter needed to help release oxygen. Figure 15.9 summarizes most of the discussion in this section. It should be studied carefully, using the boxed numbers in the legend to follow the events.

Acid-Base Balance of Blood

15.11 Acidosis and Alkalosis

When measured at room temperature, the pH of blood is 7.35 to 7.45. **Acid-base balance** exists when the blood pH is kept in that range. **Acidosis** is a condition defined by a drop in pH toward the more acid side. **Alkalosis** is a rise in pH toward the more alkaline side. Either condition is serious and must receive medical attention if mechanisms of the body do not appear able to handle it. Either may be caused by a derangement in metabolism giving us metabolic acidosis or metabolic alkalosis. Either may be caused by some defect in the respiratory system giving us respiratory acidosis or respiratory alkalosis.

Respiration is controlled by nerve signals emanating in parts of the central nervous system called the respiratory centers. These centers continuously monitor the concentrations of carbon dioxide and oxygen in the blood as well as the blood pH. For example, if the partial pressure (i.e., the concentration) of carbon dioxide in blood rises and the partial pressure of oxygen declines (as in heavy exercise), the respiratory centers induce a faster and deeper rate of breathing. (We leave further details to other courses and books.) Clinically, however, pCO_2, pH, and [HCO_3^-] are easily and rapidly measured, and their values are used to diagnose a number of conditions.

15.12 Carbonate Buffer Revisited

The principal buffer system in blood uses dissolved carbon dioxide—in the form of H_2CO_3, carbonic acid—and dissolved bicarbonate ion.

To neutralize excess acid,

$$H^+ + HCO_3^- \longrightarrow H_2CO_3 \longrightarrow H_2O + CO_2 \quad (15.6)$$

To neutralize excess base,

$$HO^- + H_2CO_3 \longrightarrow H_2O + HCO_3^- \quad (15.7)$$

Since carbon dioxide can be excreted at the lungs, its formation and loss means an irreversible removal of a hydrogen ion. Bicarbonate ion can be removed by kidneys or by diarrhea, a relatively ineffective mechanism of control. The loss of HCO_3^- irreversibly removes a hydroxide ion, the one that produced HCO_3^- (Equation 15.7). Respiratory centers normally can sense when the buffer is generating excess carbon dioxide and can induce more rapid and deeper breathing, a response we call **hyperventilation** or **Kussmaul breathing.** Likewise, if they sense that the pH is rising, meaning a rise in [OH^-], the respiratory centers act to conserve what will neutralize OH^-, namely carbonic acid, which requires that carbon dioxide be conserved. Hence **hypoventilation** occurs—slow, shallow breathing.

As we shall see later, healthy kidneys respond to acidosis by trying to put hydrogen ion into the urine for excretion and by trying to retain bicarbonate ion and sodium ion. They also try to remove any anions of the acids responsible for metabolic acidosis, but in so doing the kidneys cannot help but remove sodium ions as well. If a negative ion leaves, a positive ion must also leave to balance the charge. To remove all these ions requires the removal of considerable water as well, which may lead to dehydration. If the kidneys are injured or diseased and cannot perform these services, wastes build up in the blood, a condition known generally as **uremia** or uremic poisoning ("ur-" of the urine; "-emic" in the blood).

With these facts as background let us now look at several clinical situations involving respiration and blood chemistry. We shall use the following symbols:

pH	the pH of whole blood
pCO_2	the partial pressure of carbon dioxide in blood, in mm Hg
[HCO_3^-]	the concentration of bicarbonate ion in blood measured in milliequivalents per liter (mEq/liter).

The normal values are: pH, 7.35 to 7.45; pCO_2, 35 to 40 mm Hg; [HCO_3^-], 25 to 30 mEq/liter. If a clinical situation is characterized by a drop in one of these values, we shall use an arrow pointing down; if a rise in one of these values, then the arrow will point up. Following the arrow, in parentheses, will

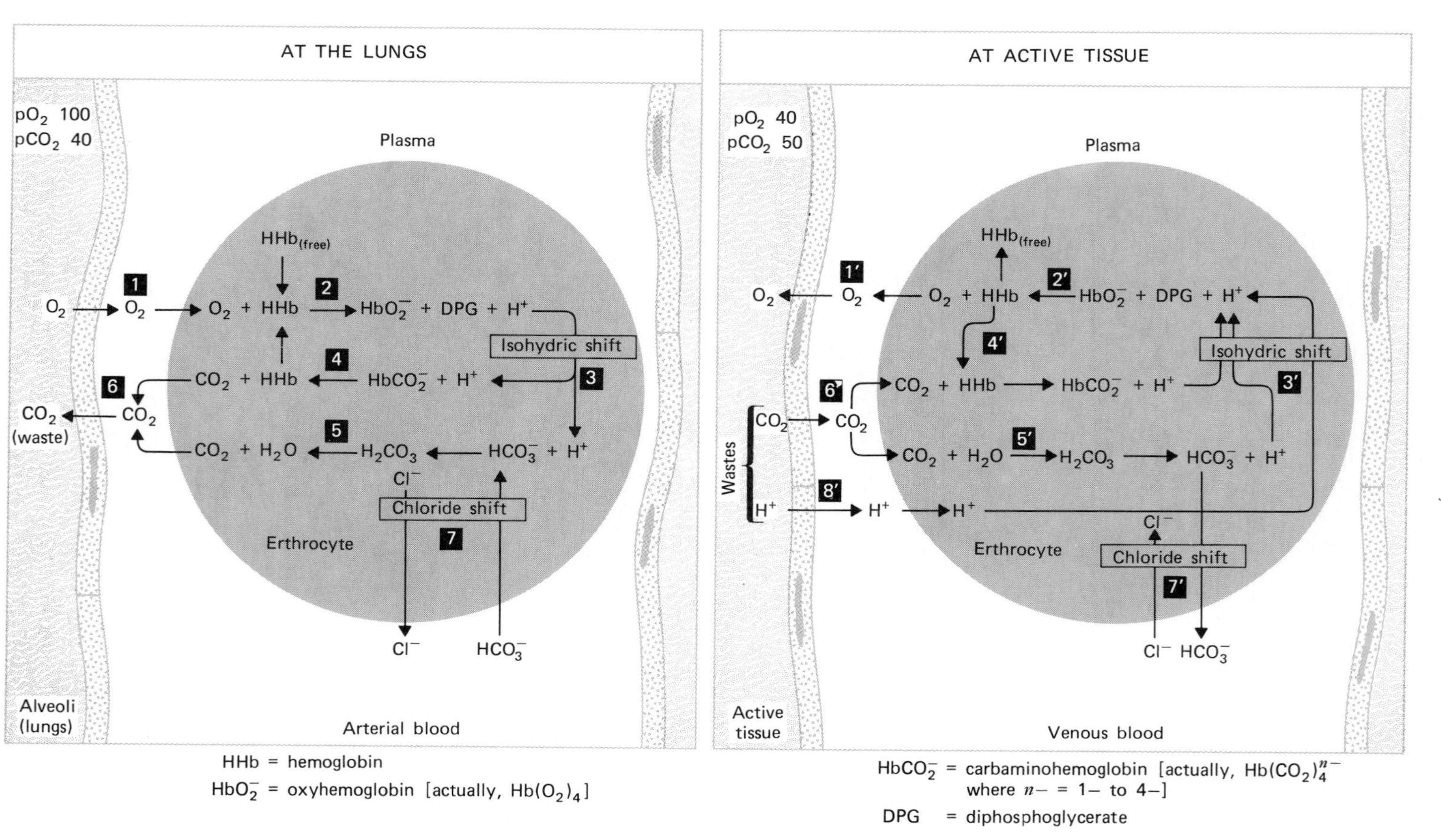

HHb = hemoglobin

HbO_2^- = oxyhemoglobin [actually, $Hb(O_2)_4$]

$HbCO_2^-$ = carbaminohemoglobin [actually, $Hb(CO_2)_4^{n-}$ where $n-$ = 1− to 4−]

DPG = diphosphoglycerate

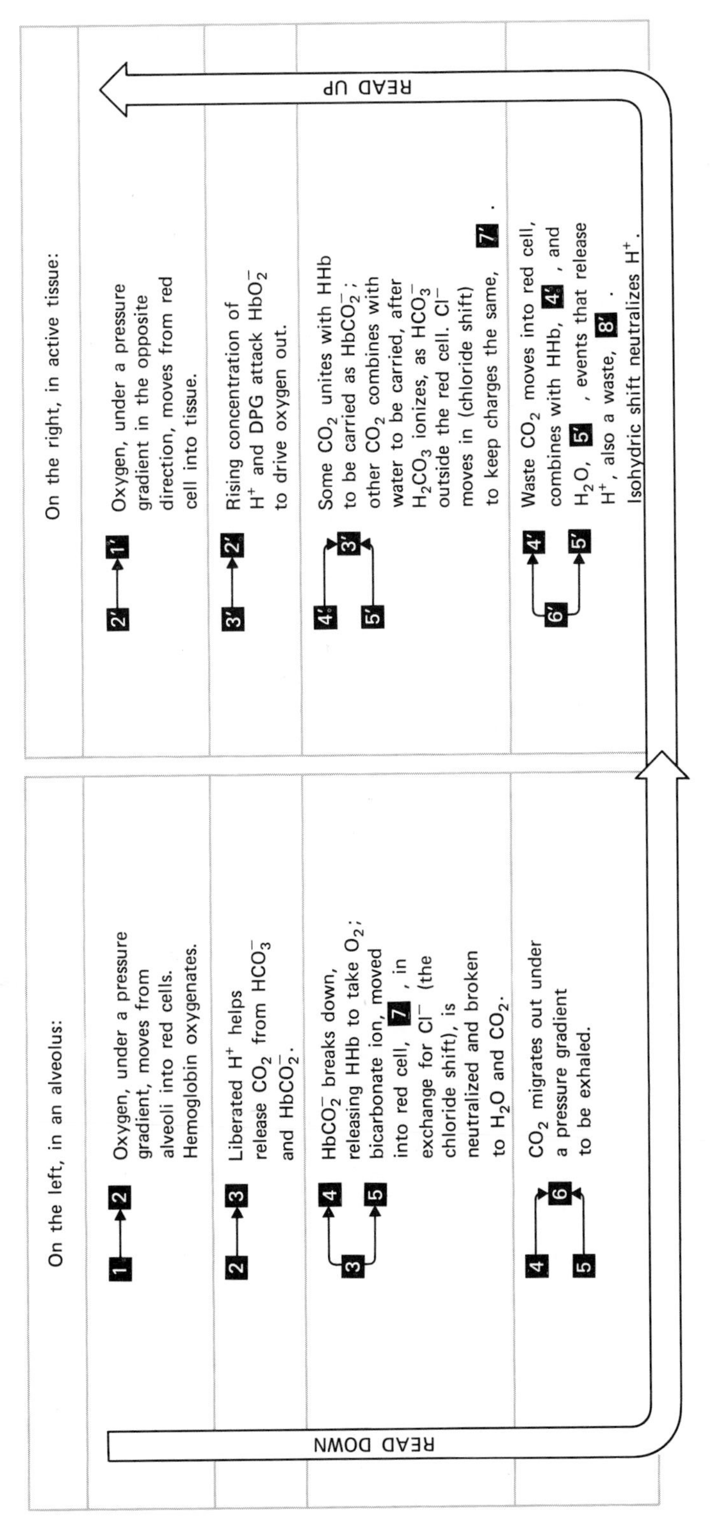

Figure 15.9 Oxygen—carbon dioxide transport in blood.

be a typical value in the clinical situation, which you may compare with the normal range just given.

15.13 Metabolic Acidosis

Lab results: pH ↓ (7.20); pCO_2 ↓ (30 mm Hg); $[HCO_3^-]$ ↓ (14 mEq/liter)

Typical patient: An adult comes to the clinic with a severe infection. Unknowingly, he has diabetes.

Range of causes of lab results: Diabetes mellitus, severe diarrhea (with loss of HCO_3^-), kidney failure and uremic poisoning, prolonged starvation, severe infection, salicylate (e.g., aspirin) overdose.

Symptoms: Hyperventilation (because respiratory centers are trying to get CO_2 out of lungs), increased urine formation (to remove acid), thirst (to replace water lost via urine), drowsiness, headache, restlessness, disorientation.

Treatment: If kidneys function—isotonic HCO_3^- intravenously to restore HCO_3^- level, thereby raising pCO_2 and neutralizing H^+. Restore water. In diabetes—insulin therapy. If kidneys do not function—hemodialysis.

15.14 Metabolic Alkalosis

Lab results: pH ↑ (7.56); pCO_2 ↑ (50 mm Hg); $[HCO_3^-]$ ↑ (36 mEq/liter)

Typical patient: Postsurgery patient with persistent vomiting.

Range of causes of lab results: Prolonged loss of contents of stomach (including acid)—vomiting or removal via nasogastric tract; overdose of bicarbonate (e.g., to ease pain of a stomach ulcer); loss of potassium ion (and with it, loss of chloride ion) in severe exercise or stress.

Symptoms: Hypoventilation (to retain CO_2 and H_2CO_3 to neutralize base), numbness, tingling, headache, possibly convulsions.

Treatment: Isotonic ammonium chloride (NH_4Cl), intravenously. ($NH_4^+ + OH^- \rightarrow H_2O + NH_3$, which reduces pH). Replace K^+ loss.

15.15 Respiratory Acidosis

Lab results: pH ↓ (7.10); pCO_2 ↑ (68 mm Hg); $[HCO_3^-]$ ↑ (40 mEq/liter)

Typical patient: Chain smoker with emphysema.

Range of causes of lab results: Any pulmonary obstruction—emphysema, pneumonia, asthma, or any cause of shallow breathing such as overdose of narcotics, barbiturates, or anesthetics; anterior poliomyelitis or cardiopulmonary disease.

Symptoms: Shallow breathing; patient is hypoventilating because it cannot be helped; is trying to blow off CO_2 but cannot; respiratory centers do not respond to pCO_2 and lungs cannot handle removal of CO_2.

Treatment: Underlying problem must be treated—possibly sodium carbonate intravenously; possibly hemodialysis.

In severe obstructive pulmonary emphysema, hemoglobin is not fully saturated with oxygen in the lungs. Oxygen delivery to active tissues, therefore, is reduced—a condition called **hypoxia** ("hypo-" reduced or lowered or under; "-oxia," oxygen). To compensate, the system manufactures more diphosphoglycerate (DPG) in red cells, which helps more efficiently to unload what oxygen does arrive, as we studied in Section 15.10.

15.16 Respiratory Alkalosis

Lab Results: pH ↑ (7.54); pCO_2 ↓ (32 mm Hg); $[HCO_3^-]$ ↓ (20 mEq/liter)

Typical patient: Someone nearing surgery and experiencing anxiety.

Range of causes of lab results: Prolonged crying, rapid breathing at high altitude, overbreathing, fever, hysterics, disease of central nervous system, improper management of a respirator.

Symptoms: Hyperventilation (because it cannot be helped). Convulsions may occur.

Treatment: Rebreathe one's own exhaled air (breathe into a sack); administer carbon dioxide; treat underlying causes.

Adjustment to high altitude and its lower partial oxygen pressure involves a number of changes. Given time the system will manufacture more red cells to deliver oxygen. Each red cell synthesizes a higher level of diphosphoglycerate (DPG) to make it easier for oxygen to be released by the red cells. If not given time, if the move from sea level to a high altitude occurs in the course of strenuous backpacking throughout one day, overbreathing may occur (to try to increase pO_2 in lungs). This blows out CO_2 excessively and may lead to alkalosis. In the meantime the individual may experience a set of symptoms called "mountain sickness" or "high altitude sickness"—rapid pulse, shortness of breath, headache, fatigue, mental depression, chest pain, cold extremities, and possibly sleeplessness, nausea, and vomiting. For some low-landers the only cure is prompt return to lower elevations. For most the problem will pass in 1 to 2 days, but reduced work output will be required.

The starvation of tissues for oxygen, called anoxemia or **anoxia,** occurs also in severe anemia (too little hemoglobin), and carbon monoxide poisoning. Carbon monoxide also binds heme, but 200 times more strongly than oxygen. When carbon monoxide binds to heme, oxygen cannot be picked up. The treatment is the administration of nearly pure oxygen.

Careless use of pure oxygen can be dangerous. Enough molecular oxygen will be carried simply in solution in the serum to supply active tissues. The release of oxygen from hemoglobin that is needed to help take waste acid and carbon dioxide out of the tissue does not occur and acidosis results.

15.17 Stored Whole Blood

When whole blood has been stored several days, the level of DPG in the red cells declines. The oxygen affinity of hemoglobin thereby increases. If stored blood with reduced DPG were transferred, its red cells would unduly retain oxygen as they passed through active tissue. The net effect could be like that of hypoxia. To prevent this, a special nutrient (e.g., inosine) is added to whole blood that must be stored. Inosine, unlike DPG, can enter red cells from the outside. (DPG is too electrically charged.) Also, DPG can be made from inosine in red cells, and the whole blood retains a normal oxygen affinity.

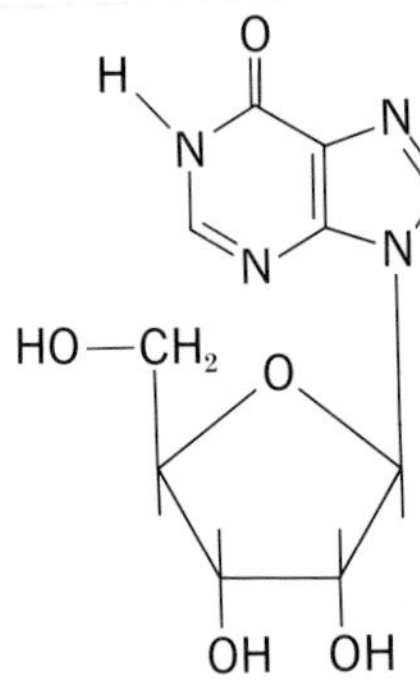

Inosine

Exercise 15.5 If respiratory centers are functioning normally, what kind of acid-base disfunction is probably present if an individual is doing each activity? What does each activity accomplish (or try to do)?
(a) Hyperventilation (b) Hypoventilation

Exercise 15.6 Why does shallow breathing in emphysema lead to acidosis? Why is the patient not hyperventilating to counteract acidosis?

Exercise 15.7 Why does hyperventilation in hysterics cause alkalosis?

15.18 Kidney Function and the Blood

Diuresis (Greek "to make water or urine") in the kidneys is an integral part of the system's control over concentrations of electrolytes in blood and the blood pH. Figure 15.10 shows the parts of a kidney that participate. Huge quantities of fluids, electrolytes, organic wastes, amino acids, and blood sugar leave the bloodstream by diffusion each day at the hundreds of thousands of glomeruli. All of the glucose and amino acids and most of the fluids and electrolytes return. Some return by passive transport—dialysis—others by active transport under the direction of hormones. Most (but not all) of the wastes stay in the urine. Urea is the principal nitrogen waste (30 g/day), but creatine (1 to 2 g/day), uric acid (0.7 g/day), and ammonia (0.5 g/day) are also excreted.

Vasopressin, an octapeptide, is a hormone that helps regulate the overall concentration of substances in blood. The hypophysis, which makes vasopressin, will release this hormone if the osmotic pressure of blood goes up by

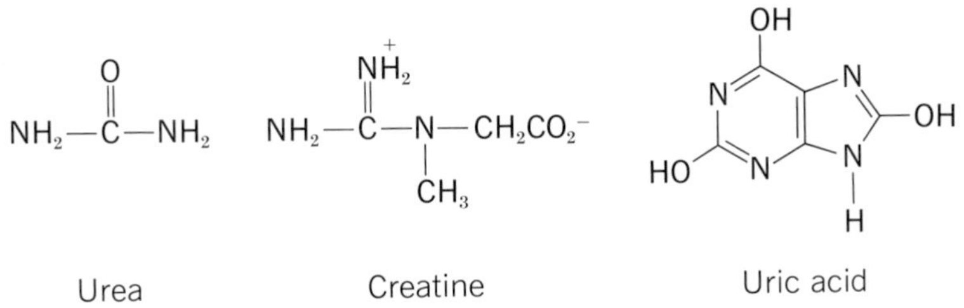

as little as 2%. At the kidneys, vasopressin promotes the reabsorption of water and therefore is often called the antidiuretic hormone (ADH). A higher osmotic pressure (called hypertonicity) means a higher concentration of solutes and colloids in blood. Vasopressin helps blood to retain water to keep the concentration from going higher. The thirst mechanism is also stimulated to bring in water to dilute the blood. Conversely, if the osmotic pressure of blood drops (hypotonicity) by as little as 2%, the hypophysis retains vasopressin. Water that has left the bloodstream at the glomeruli now will not return as much. A low osmotic pressure means a low concentration; the absence of vasopressin lets urine form, reduces the amount of water in blood, and raises the concentration of its dissolved matter. With the help of vasopressin a normal individual may vary the intake of water widely and yet preserve a stable overall concentration of substances in blood.

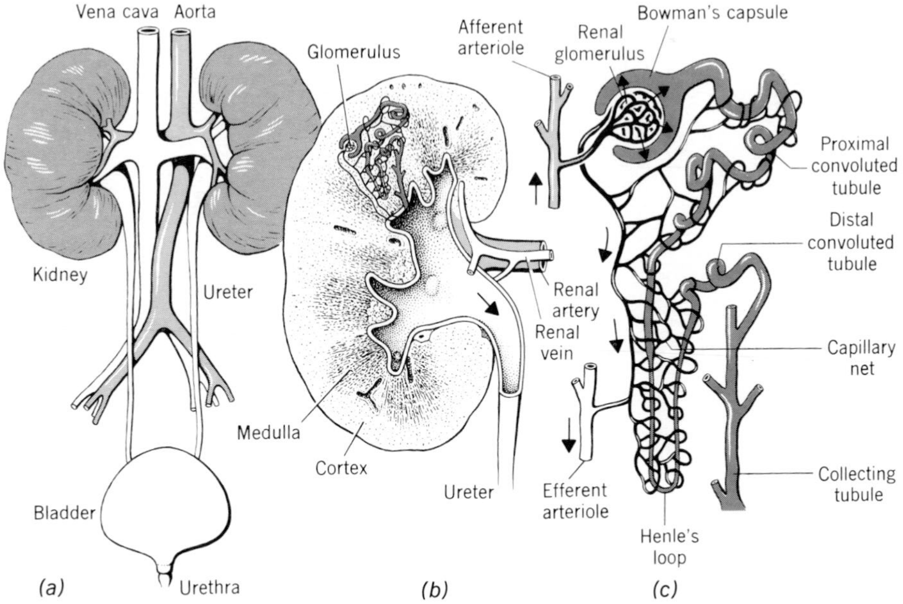

Figure 15.10 The kidneys. (*a*) Principal parts. (*b*) Location of nephrons. (*c*) Details of nephron and associated capillary bed. Each kidney has about 1 million nephrons. Fluids and solutes (but not colloids, lipids, or cells) leave the blood in the glomerulus and selectively return in the promixal and distal tubules. Final electrolyte adjustments are made primarily in the distal tubules. (Adapted from G. E. Nelson, G. G. Robinson, and R. A. Boolootian, *Fundamentals Concepts of Biology,* 2nd ed., 1970. John Wiley & Sons, Inc., New York. Used by permission.)

In diabetis insipidus, a rare disorder, the hypophysis cannot release vasopressin. Diuresis proceeds to produce upwards of 5 to 12 liters of urine per day compared to 0.6 to 2.5 liters per day.

Aldosterone helps stabilize the sodium ion level in blood. This steriod hormone, made in the adrenal cortex, is secreted if the sodium ion level drops. At the kidneys, aldosterone signals reactions that help retain sodium ions in blood. To do that requires some retention of water, also. If the sodium ion level rises, aldosterone secretion stops and sodium ions are let out of blood (together with water) and excreted in the urine.

The kidneys also help regulate the bicarbonate ion in blood. They can manufacture and retain this important part of the blood buffer system or they can let some of this ion leave in the urine. When the blood is battling acidosis, the kidneys produce an acidic urine, an effort that helps reduce the acidosis. A rough outline of how the kidneys do this is given in Figure 15.11. Freshly voided urine has a pH of about 6. It may go as low as 4 in severe acidosis, and at that pH the hydrogen ion concentration in urine is 1000 times that in normal blood. In alkalosis the pH of urine may rise to 8.2 or higher as bicarbonate ion is excreted.

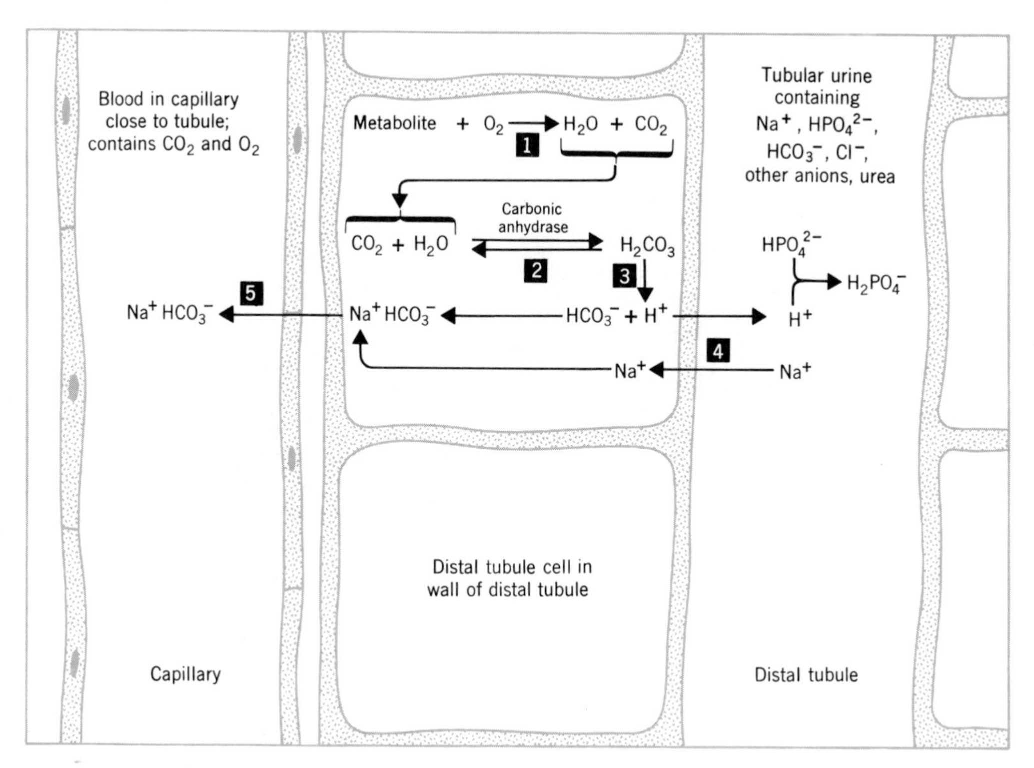

Figure 15.11 Acidification of urine. [1] Catabolism generates some CO_2. [2] CO_2 changes to carbonic acid. [3] Hydrogen ions leave the cell to go into the urine (in distal tubule) to replace the positive charge lost when sodium ion, [4], is pumped out of urine. (Aldosterone aids this exchange.) [5] Bicarbonate ion and sodium ion are returned to the bloodstream. Thus losses of bicarbonate ion elsewhere are made up, and the kidneys help control acidosis. (After R. F. Pitts, *American Journal of Medicine,* Vol. 9, page 356, 1950.)

The kidneys can also respond to blood pressure. If the blood pressure drops (as in hemorrhaging), the kidneys secrete a trace of **renin,** an enzyme, into the blood. Renin acts on one of the globular proteins in blood, angiotensinogen (a proenzyme) to convert it to another enzyme, **angiotensin** I. This, in turn, converts still another protein in blood to angiotensin II. Angiotensin II is the most potent vasoconstrictor known. By constricting the blood vessels, it causes an increase in blood pressure, which also insures that some semblance of proper filtration at the kidneys continues. If blood pressure rises (hypertension), normal kidneys retain renin.

Angiotensin II also triggers the release of aldosterone, which, as we learned above, will help the blood retain water.

Transport of Nutrients by Blood

15.19 Fluid Exchange at Tissue Cells

At the narrowest part of a capillary loop (Figure 15.12) arterial blood becomes venous blood as oxygen and nutrients diffuse into surrounding cells and wastes diffuse out. Fluids carry these substances, and fluid diffusion at capillary loops is estimated at 400 gal/min. What fluid leaves on one side of the loop must exactly return on the other. Lymph ducts, thin-walled, closed-end capillaries in soft tissue, help carry some fluids back to circulation. (They also have white cells, make antibodies, and generally aid the body's defense.)

Water and dissolved solutes (but not colloidal macromolecules such as proteins) are forced out on the arterial side by the higher arterial blood pressure. Chemical exchanges occur in the tissue. On the venous side, where the opposing blood pressure is less, water and dissolved solutes (now wastes) dialyze back into the bloodstream. Since blood has an extra colloidal osmotic pressure that interstitial fluid does not have (Section 15.9), the natural direction of dialysis is always from the interstitial compartment into the capillary. The extra pressure of arterial blood over venous blood makes the difference on the arterial side and reverses the natural direction of dialysis. These relations, first explained by E. H. Starling (1866–1927), a British physiologist, and now known as **Starling's hypothesis,** are summarized in Figure 15.12.

15.20 Traumatic Shock

In trauma such as sudden severe injury, major surgery, or extensive burns, capillary walls become more permeable to proteins. These leak into interstitial spaces and the colloidal osmotic pressure of blood drops. Interstitial fluid then cannot dialyze back into the bloodstream as effectively, and the blood volume quickly drops. Oxygen transport is impaired. The individual goes into traumatic shock. The restoration of blood volume is mandatory to control it.

15.21 Edema

If kidneys fail to keep serum proteins in the bloodstream, the colloidal osmotic pressure of the blood drops. As in shock (but developing much more

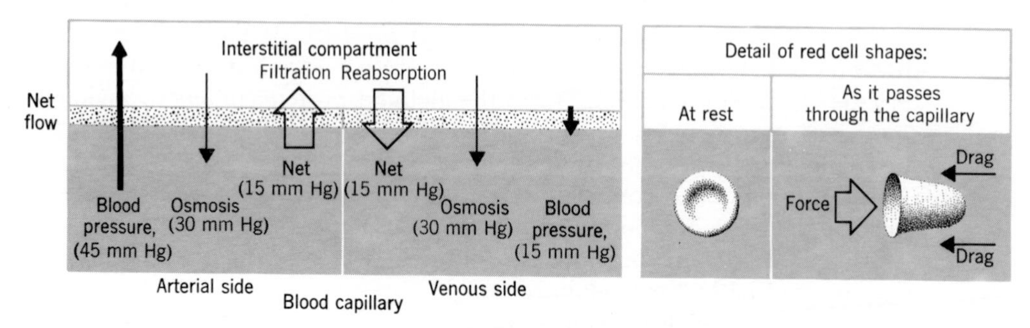

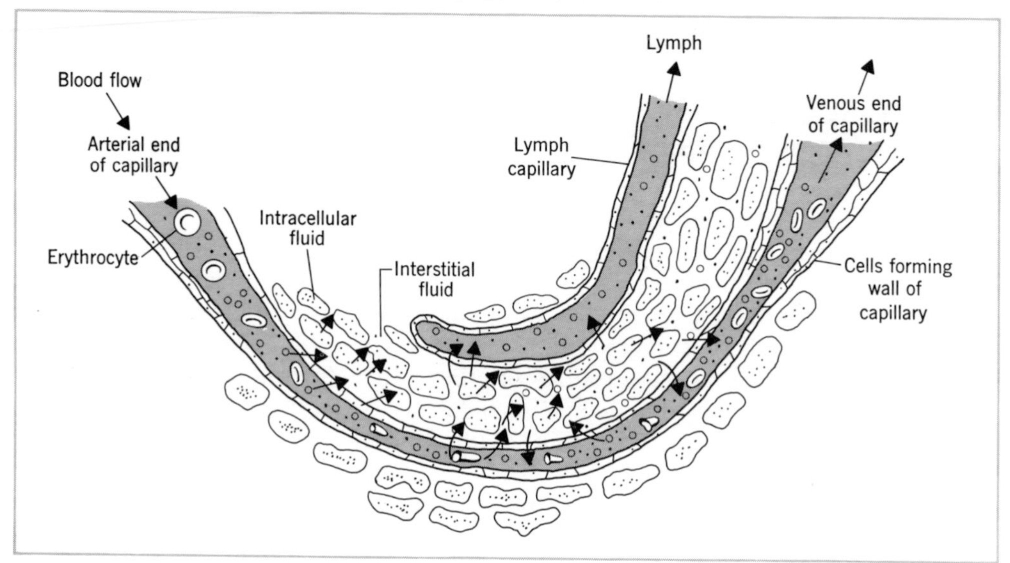

Figure 15.12 Exchange of nutrients and wastes at capillaries. As indicated at the top (with some exaggeration), blood pressure counteracts dialysis and osmosis on the arterial side and forces fluid to move from capillary into interstices. On the venous side blood pressure is lower, letting dialysis and osmosis bring fluids back out of the interstices. The distortion of a red cell as it goes through the capillary loop is believed to aid in the release of oxygen.

slowly), interstitial compartments retain fluid and the individual appears puffy and waterlogged. The condition is edema (Greek, "swelling"). Edema occurs at a stage in starvation when blood proteins have been consumed enough to lower the colloidal osmotic pressure of blood. Edema may arise if veins are obstructed as in varicose veins and certain cancers. The venous blood pressure rises. This back pressure hinders dialysis and the interstitial compartments retain fluids.

Mechanical injury to tissue may cause localized edema by injuring capillaries and reducing the return of fluids by the veins.

The importance of the lymph ducts in returning some fluid to circulation is seen in a particularly ugly tropical disease, elephantiasis. Small worms (filari) enter these ducts and reduce drainage through them. The swelling, particularly in the legs, is grotesque.

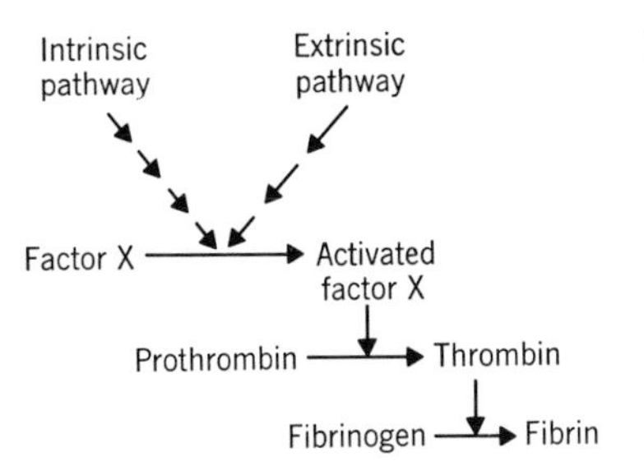

Figure 15.13 Outline of clotting mechanism.

15.22 Clotting of Blood

A number of substances that are a part of the clotting mechanism are present in the blood and others are released when tissue is cut. A whole series of zymogen activations are launched by the initial trauma. Fourteen factors including calcium ion have been identified as part of the scheme. Most are proteins made in the liver. The final step in clotting is the conversion of fibrinogen to fibrin, catalyzed by the enzyme **thrombin.** The whole function of the elaborate mechanism to control clotting is to insure that thrombin is never around except when and where a clot is needed. Thrombin exists in an inactive form, **prothrombin,** a proenzyme. The synthesis of prothrombin requires vitamin K. Prothrombin is activated by factor X, and the activation of factor X is the culmination of two cascades of steps that converge. See Figure 15.13. One series involves materials that are intrinsically present in blood and are set into reaction by contact with the alien surface causing the cut. Their transformations make up the **intrinsic pathway** in blood clotting. The other cascade involves materials in the injured tissue, and their reactions are called the **extrinsic pathway.** These two pathways converge for the final steps, as outlined in Figure 15.13. A detailed discussion of these pathways is beyond the scope of this book.

When blood is removed, it is not possible to prevent contact with an alien surface. The clotting of drawn blood is prevented simply by adding something to the blood to tie up calcium ions, one of the clotting factors. Sodium oxalate or sodium citrate may be used. In the first case, calcium precipitates as calcium oxalate. In the second, calcium ions complex with citrate ions and are effectively removed.

Vitamin K deficiency (as in some hemorrhagic diseases) reduces the formation of prothrombin in the liver. The same effect can be achieved by dicumarol, a chemical formed in decaying sweet clover, which is sometimes given to prevent thrombosis, the unwanted formation of a clot within a blood vessel. Heparin, an extremely potent anticoagulant, is found in cells located near the walls of the tiniest capillaries. It acts by slowing the conversion of prothrombin to thrombin.

Brief Summary

Digestion. The α-amylase in saliva begins the digestion of starch. Pepsin in gastric juice starts the digestion of proteins. In the duodenum trypsinogen (from the pancreas) is activated by enterokinase (from the intestinal juice) and becomes trypsin. Trypsin not only helps digest proteins it also activates trypsinogen and several other pancreatic zymogens to proteolytic enzymes—to chymotrypsin, carboxypeptidase, elastase and aminopeptidase. Intestinal juice supplies enzymes for the digestion of disaccharides, nucleic acids and lipids. The pancreas also supplies an important lipase. Bile salts are detergents essential to the digestion of lipids and the absorption of fat-soluble vitamins A and D. The end products of the digestion of proteins are amino acids; of carbohydrates: glucose, fructose and galactose; and of lipids: fatty acids and glycerol. Complex lipids are hydrolyzed to their components and nucleic acids yield phosphate, pentoses and sidechain amines.

Blood. Various proteins in blood give it a colloidal osmotic pressure needed in nutrient exchange at capillary loops. Albumins supply carriers for hydrophobic molecules and serum-insoluble metallic ions. The γ-globulins help in defense against infection. Fibrinogen is the precursor of a blood clot. Among the electrolytes in blood some are used in acid-base balance and they generally are involved in osmotic pressure relations. Blood transports oxygen and products of digestion to all tissue; organic wastes to the kidneys; excess carbon dioxide to lungs; and hormones to target cells. Lymph, another fluid, helps return some substances to blood from tissues. Sudden failure to retain protein in blood leads to shock. Slower loss of protein as in kidney disease leads to edema. A cut launches both an intrinsic and extrinsic cascade of zymogen conversions that lead to the change of fibrinogen to fibrin.

Respiration. When saturated fully with oxygen, each molecule of hemoglobin carries four of oxygen. The process of becoming saturated is cooperative and works by an allosteric effect. Release of O_2 at active tissues is also cooperative. Waste CO_2 and H^+ at active tissue plus DPG inside well-oxygenated red cells plus a favorable partial pressure gradient help drive O_2 out of red cells and into active tissue. If myoglobin is present this movement of O_2 is further aided. At alveoli CO_2 leaves under a partial pressure gradient. H^+ ions needed to release CO_2 from $HbCO_2^-$ and HCO_3^- are produced as O_2 binds to hemoglobin. Respiration is controlled principally by respiratory centers that monitor changes in blood levels of O_2, CO_2 and H^+.

Acid-Base Balance. The system uses the carbonate buffer to inhibit acidosis by irreversibly removing extra H^+ when it releases extra CO_2 at the lungs. The extra CO_2 is made by using the excess H^+ to neutralize HCO_3^-. HCO_3^- is replaced by the kidneys, which also help control acidosis by putting H^+ into urine. The carbonate buffer system controls alkalosis when H_2CO_3 neutralizes OH^- and changes to HCO_3^- (and H_2O). Metabolic acidosis (relatively common) with hyperventilation as well as metabolic alkalosis (rarer) with hypoventilation arise from disfunctions in metabolism but the respiratory centers are working normally. Respiratory acidosis and alkalosis occur when respiratory centers are not working. The difference in operation of respiratory centers accounts for differences in pCO_2 and $[HCO_3^-]$ between metabolic and respiratory disfunctions.

Diuresis. The kidneys with the help of hormones and changes in blood pressure, blood osmotic pressure, and concentrations of ions, monitor and control the concentrations of solutes in blood. Vasopressin tells the kidneys to keep water in the bloodstream. Aldosterone tells the kidneys to keep sodium ion (and, therefore, water as well) in the bloodstream. A drop in blood pressure tells the kidneys to release renin which will activate a vasoconstrictor and aldosterone—to raise blood pressure and retain water. In acidosis the kidneys put H^+ into urine and replace HCO_3^- lost from blood. In alkalosis the kidneys put some HCO_3^- into urine.

Selected References

Books

1. G. F. Filley. *Acid-Base and Blood Gas Regulation,* 1971. Lea & Febiger, Philadelphia.
2. R. M. Cherniack, L. Cherniack, and A. Naimark. *Respiration in Health and Disease,* 2nd ed., 1972. W. B. Saunders Company, Philadelphia.
3. J. P. West. *Respiratory Physiology—The Essentials,* 1974. The Williams & Wilkins Company, Baltimore.
4. L. Stryer. *Biochemistry,* 1975. W. H. Freeman and Company, San Francisco. Chapters 3 and 4 are discussions of oxygen transport.
5. J. L. Kee. *Fluids and Electrolytes with Clinical Applications,* 1971. John Wiley & Sons, Inc., New York. An excellent programmed learning study for nurses and a source consulted for Sections 25.11 through 25.16.

Articles

1. J. E. Sharer. "Reviewing Acid-Base Balance." *American Journal of Nursing,* June 1975, page 980. An outstanding discussion and review, one also used in preparing Sections 25.11 through 25.16.
2. A. R. Frisancho. "Functional Adaptation to High Altitude Hypoxia." *Science,* 31 January 1975, page 313.

Questions and Exercises

1. What are the end products in the digestion of each of the following substances?
   (a) Proteins (b) Carbohydrates (c) Triacylglycerols
2. What functional groups are hydrolyzed when each of the substances in Question 1 is digested? (Refer to other chapters, too.)
3. What are the principal zymogens, if any, in each?
   (a) Saliva (b) Gastric juice (c) Intestinal juice (d) Pancreatic juice (e) Bile
4. How is enterokinase a "master switch" in digestion?
5. What happens if pancreatic zymogens are prematurely activated?
6. What services are performed by bile salts?
7. Why is bile described as both a secretion and an excretion?
8. List some services each performs in blood (to the extent of our study).
   (a) Albumins (b) Globulins (c) Hemoglobin
   (d) H_2CO_3 (e) HCO_3^- (f) Electrolytes in general
   (g) Fibrinogen (h) Angiotensinogen
9. In the binding of oxygen to hemoglobin in what way is the allosteric effect important to life and health?
10. How does a fall in blood pH reduce oxygen affinity?

11. How does a rise in blood pCO_2 reduce oxygen affinity?

12. How do each of the following participate in the exchange of respiratory gases?
    (a) Bohr effect
    (b) Isohydric shift
    (c) Chloride shift
    (d) DPG

13. In what direction are each of these gradients? ("Direction" means the direction from high to low pressure. Answer the question in terms of specific cells or tissues or organs. For example, one answer might be, "from blood to lungs.")
    (a) pO_2 at lungs
    (b) pCO_2 at lungs
    (c) pO_2 at active cells
    (d) pCO_2 at active cells
    (e) Colloidal osmotic pressure of blood at a capillary loop
    (f) Na^+ ion concentration between interstitial compartment and intracellular fluid
    (g) H^+ concentration between the urine being made and the bloodstream in the kidney.

14. The oxygen affinity of hemoglobin in a fetus is higher than that of adult hemoglobin. How is this important to the fetus?

15. The oxygen affinity of adult hemoglobin is reduced by a factor of 26 in the absence of DPG. In what specific way is the effect of DPG on the oxygen affinity important to life?

16. Construct a table using arrows (↑) or (↓) and typical laboratory data that summarize the changes observed in respiratory and metabolic acidosis and alkalosis. Column headings should be

Condition	pH	pCO_2	$[HCO_3^-]$

17. Compare and contrast respiratory and metabolic acidosis with respect to changes in the values for the following quantities.
    (a) pH (b) pCO_2 (c) $[HCO_3^-]$

    In what way are the two types of acidosis the same? In what ways do they differ?

18. If the osmotic pressure of blood has increased, what has changed to cause that increase?

19. How does the system respond to an increase in the osmotic pressure of blood?

20. If the sodium ion level of blood falls, how does the body respond?

21. What is the response of the kidneys to a decrease in blood pressure?

22. Alcohol in blood suppresses the secretion of vasopressin. How does this affect diuresis?

23. Explain how malnutrition upsets the net osmotic pressure of blood.

24. What do the kidneys do to reduce developing acidosis?

25. In fluid exchange between plasma and active tissue
    (a) What provides a net filtration pressure?
    (b) What provides a net reabsorption pressure?

26. How does the sudden loss of albumins from plasma lead to shock?

27. How can kidney disease lead to general edema?

28. How does vitamin K assist the clotting mechanism (and sometimes, therefore, is administered to pregnant women just before delivery)?

29. How can the clotting of drawn blood be prevented?

30. How are fibrinogen and fibrin related?

chapter 16

Biochemical Energetics

On her way to her second gold medal at the Montreal Olympic Games Romania's Nadia Comaneci displays perfect form and balance. One of the many factors in her success, and in that of all athletes, is the ability of her body to mobilize chemical energy on sudden demand. Stored carbohydrates are part of that system, and we learn more of this next. (United Press International.)

16.1 Energy for Living

We cannot use steam energy, like a locomotive. We cannot use solar energy, like plants. We need chemical energy for living. We transfer chemical energy in food and oxygen to make high-energy molecules that are suited to our particular internal needs. As these molecules then react with other systems in our tissues, they provide energy for three main tasks: mechanical work, the work of active transport (and associated electrical activities in nerves), and the work of making macromolecules—our own proteins, polysaccharides, lipids, and nucleic acids. We also need some energy as heat.

We transfer chemical energy from food and air during the **catabolism**[1] of any product of digestion: hexoses, fatty acids, glycerol, amino acids, pentoses, amines—anything with carbon-hydrogen bonds. Our chief sources of chemical energy, however, are carbohydrates and fatty acids. On high-protein diets we also "burn" amino acids because we have no other use for them and cannot excrete them. On severe diets or during starvation, we also turn to our own tissue amino acids for energy.

We do not, of course, "burn" anything internally, but the word is commonly used because the end products of the combustion and the end products of the catabolism of glucose and fatty acids are identical—carbon dioxide and water. Therefore, we may measure the maximum energy available in theory for bodily needs by measuring the heat of combustion. Given the law of conservation of energy, the maximum energy from catabolism could be neither more nor less than that from combustion, since reactants and products are identical. As equations 16.1 and 16.2 show, 1 mole of fatty acid is almost four times as rich in energy as 1 mole of glucose.[2]

$$\underset{\text{Glucose}}{C_6H_{12}O_6} + 6O_2 \longrightarrow 6CO_2 + 6H_2O \qquad \Delta H_{comb} = -673 \text{ kcal/mole} \qquad (16.1)$$

$$\underset{\text{Palmitic acid}}{CH_3(CH_2)_{14}CO_2H} + 23O_2 \longrightarrow 16CO_2 + 16H_2O \qquad \Delta H_{comb} = -2400 \text{ kcal/mole} \qquad (16.2)$$

When we catabolize instead of burn these substances in open air, roughly half of the energy is still released as heat. The other half becomes part of the chemical energy of an important family of compounds, the high-energy phosphates.

Exercise 16.1 Using a straight mole calculation, what is the energy yield per *gram* of glucose; per gram of palmitic acid?

[1] Metabolism, as we have learned, is any chemical process and the sum total of all of them in a living system. The two categories of metabolism are anabolism, reactions that make big molecules from smaller; and catabolism, reactions that break molecules down (from the Greek: *ana,* up; *cata,* down; and *ballein,* to throw or cast).

[2] The symbol (Δ) means "change in"; "H" roughly means heat or energy content; "comb" means combustion. Thus ΔH = change in energy content in going from reactants to products, all at 1 atm pressure and a constant temperature. The minus sign means that the heat is liberated; a plus sign would have meant that heat is required to make the change go.

16.2 High-Energy Phosphates in Metabolism

The names of the principal phosphates involved in the mobilization of energy in the body are given in Table 16.1. Each has a **phosphate-group transfer potential,** a relative ability to transfer a phosphate group from itself to some acceptor, as in Equation 16.1. We shall usually write the phosphate group as $—OPO_3^{2-}$, but its state of ionization varies with the pH of the medium. At physiological pH the phosphate group exists mostly as the singly and doubly

$$\underset{\text{higher potential}}{R—O \sim PO_3^{2-}} + R'O—H \longrightarrow R—O—H + \underset{\text{lower potential}}{R'—O \sim PO_3^{2-}} \quad (16.3)$$

ionized forms, $—PO_3H^-$ and $—PO_3^{2-}$ (see Equation 16.3). We shall use a squiggle (~) for the phosphate-oxygen bond that breaks in the reaction. We reserve the squiggle mostly for the **high-energy phosphate** compounds: ATP and above in Table 16.1. In those compounds that bond is often called the **high-energy phosphate bond,** although the energy is really a property not just of that bond but of the whole molecule and the reaction in which the energy is released.

16.3 The Energy from Phosphates

The energies recorded in Table 16.1 are not heats of combustion. Instead they are energies obtained when these compounds react with water, not that we want their reactions with water internally. We do not, in fact. These compounds have to exist in water, and if they reacted rapidly with water all their chemical energy would be wasted as heat. However, in the reaction with water a phosphate group transfers from each to the same reactant (Equation 16.4). Hence data obtained thereby give us a means of comparing phosphate-group transfer potentials, the absolute values in kilocalories per mole of the free energy of hydrolysis,

$$R—O \sim PO_3^{2-} + H—OH \longrightarrow R—OH + HPO_4^{2-} \quad (16.4)$$

Another special feature of the data in Table 16.1 is that they are not total energies of hydrolysis. Rather they are that part of the total that could be used for useful work. Of the total energy of a reaction only a part can ever be tapped for useful work in any real process. Some is inevitably lost as waste heat. The maximum part not wasted and theoretically available for useful work is called the **free energy**—better, free energy change—and symbolized by $\Delta G°$ [the "G" is to honor Josiah Willard Gibbs (1839–1903) one of the greatest scientists ever produced by the United States; the degree sign (°) signifies special standardized conditions that need not concern us here.] "Free" means basically just that, free from nature for our use. A large negative value of free energy means a very high amount of free energy available from that compound. It means nothing about how fast that energy can be released; only that, rapidly or slowly produced, the amount of free energy is large. In the body each use of a high-energy compound requires a special enzyme. No en-

TABLE **16.1**
Some Organophosphates in Metabolism[a]

Phosphate	Structure	Free Energy of Hydrolysis[b]	Phosphate Group Transfer Potential[c]
Phosphoenolpyruvate	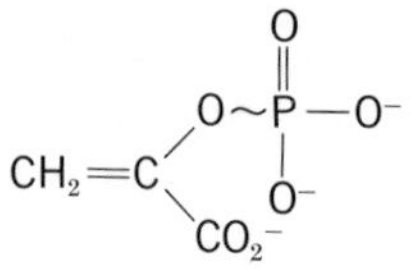 $CH_2{=}C(CO_2^-)-O{\sim}P(=O)(O^-)-O^-$	−14.8	14.8
1,3-Diphosphoglycerate	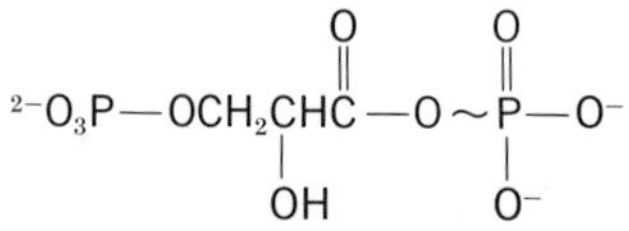 $^{2-}O_3P-OCH_2CH(OH)C(=O)-O{\sim}P(=O)(O^-)-O^-$	−11.8	11.8
Phosphocreatine	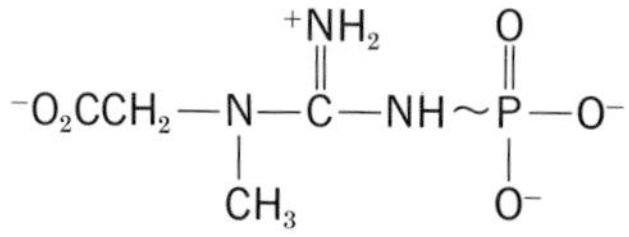 $^-O_2CCH_2-N(CH_3)-C(={}^+NH_2)-NH{\sim}P(=O)(O^-)-O^-$	−10.3	10.3
Acetylphosphate	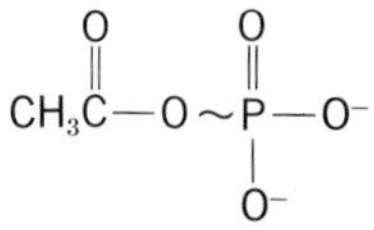 $CH_3C(=O)-O{\sim}P(=O)(O^-)-O^-$	−10.1	10.1
Adenosine triphosphate ATP	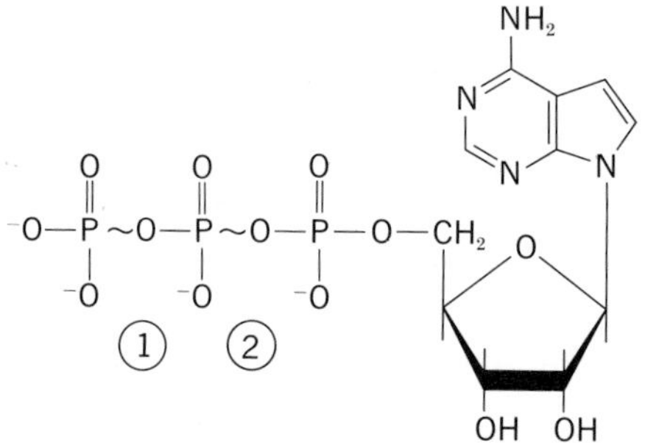 $^-O-P(=O)(O^-){\sim}O-P(=O)(O^-){\sim}O-P(=O)(O^-)-O-CH_2$ (ring O; OH OH; base N, N, N, NH_2); bonds ① and ②	−7.3 ① −7.3 ②	7.3 7.3
Glucose 1-phosphate	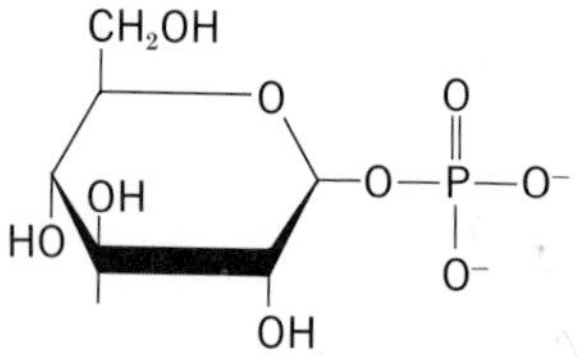 ring with CH_2OH, O, OH, HO, OH; $-O-P(=O)(O^-)-O^-$	−5.0	5.0
Fructose 6-phosphate	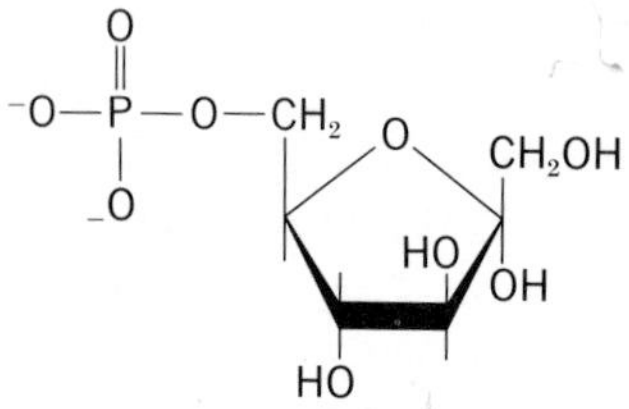 $^-O-P(=O)(O^-)-O-CH_2$; ring with O, CH_2OH, HO, OH, HO	−3.8	3.8

TABLE 16.1 (*Continued*)

Phosphate	Structure	Free Energy of Hydrolysis[b]	Phosphate Group Transfer Potential[c]
Glucose 6-phosphate	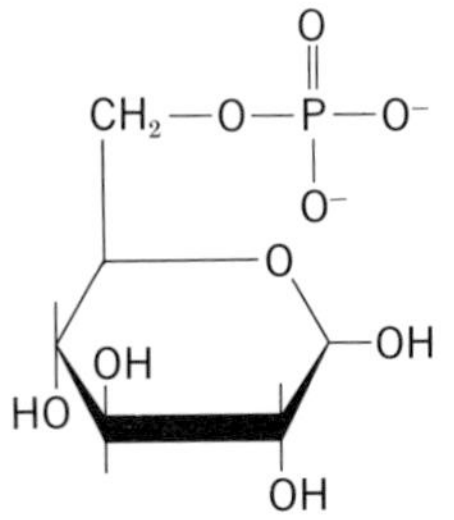 (glucose ring with CH_2—O—PO_3^{2-} at C-6; OH groups; HO)	−3.3	3.3
Glycerol 3-phosphate	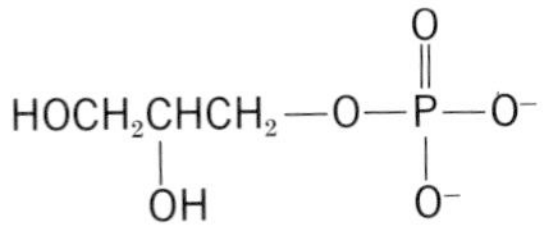 $HOCH_2CH(OH)CH_2$—O—P(=O)(O^-)—O^-	−2.2	2.2

[a] Full ionized forms of the structures are shown, but the actual state of ionization varies with the pH of the medium.
[b] In kilocalories per mole.
[c] Simply the free energy of hydrolysis (in kcal/mole) taken as the positive.

zyme exists for simple hydrolysis, except for digestive phosphatases, and the rates of hydrolysis are very low in cells, fortunately.

Exercise 16.2 State whether or not each of these reactions can occur, based on phosphate-group transfer potentials.
(a) 3-Phosphoglycerate + ATP ⟶ 1,3-Diphosphoglycerate + ADP
(b) Glucose + ATP ⟶ Glucose 1-phosphate + ADP

16.4 Adenosine Triphosphate (ATP)

One compound in Table 16.1 is especially important: **adenosine triphosphate** or **ATP.** From the lowest to the highest forms of life, ATP is universally used as the principal carrier of free energy for bodily functions. Virtually all of biochemical energetics comes down to the syntheses and uses of this substance. Almost any energy-demanding activity consumes ATP. Muscle contraction, for example, may be very simply written as Equation 16.5. ATP reacts with the proteins in relaxed muscle (or with something within the proteins). As a result, changes in tertiary protein structure occur making the protein fiber contract. Simultaneously, the ATP changes to ADP and P_i.

$$\text{"Relaxed" muscle} + \text{ATP} \longrightarrow \text{"contracted" muscle} + \text{ADP} + P_i \quad (16.5)$$

ADP is **adenosine diphosphate,** and $\mathbf{P_i}$ stands for the inorganic phosphate ion in whatever state of ionization allowed by the pH of the medium; ADP and P_i are the same products formed when ATP reacts with water.

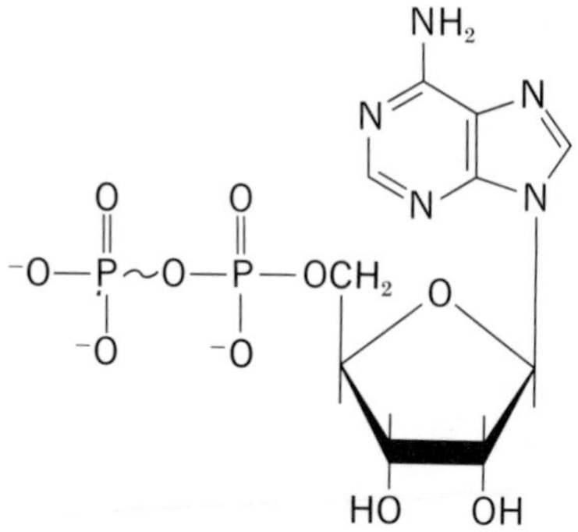

ADP
Adenosine
diphosphate

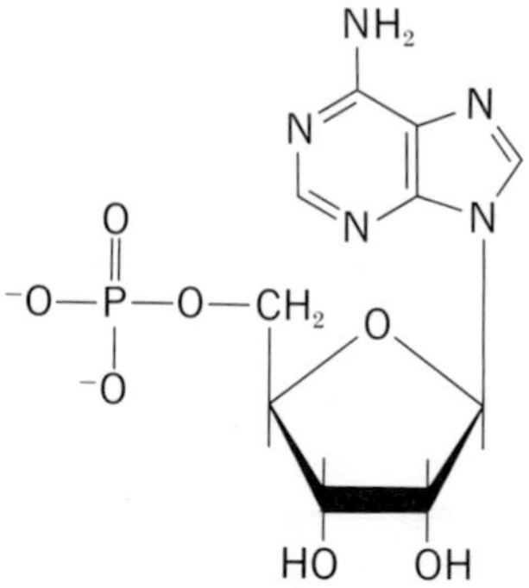

AMP
Adenosine
monophosphate

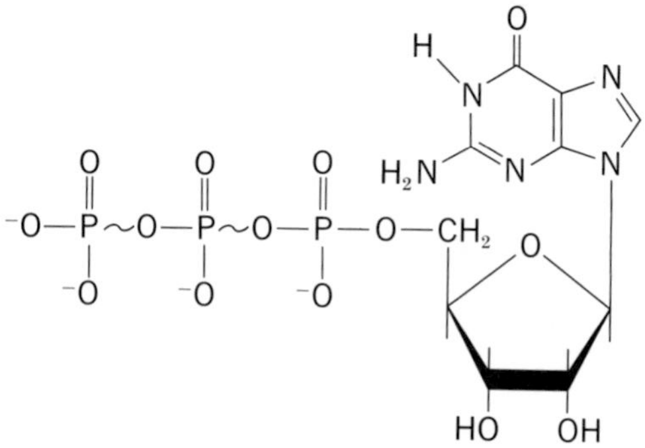

GTP
Guanosine
triphosphate

Occasionally, other nucleotide triphosphates are carriers of free energy. We shall encounter guanosine triphosphate (GTP) later in the chapter. But without small risk of serious error we shall ignore them, or mention them only as needed.

Once ATP is used, it must be remade or no more internal work can be done. The chief purpose of catabolism is to transfer chemical energy in such a way that ATP is remade from ADP and P_i. Any compound in Table 16.1 above ATP can, in principle, transfer its phosphate group to ADP to make ATP, provided that the enzyme is available. When ATP is made by transferring a phosphate group from another phosphate to ADP, the process is called **substrate phosphorylation** (where ADP is the substrate). The transfer has to be direct or through a common intermediate. The transferred phosphate group does not move in two steps, first to water (becoming phosphoric acid, or its salts) with release of heat, and second back to ADP with the aid of the heat. We do not operate by using heat to drive uphill reactions. We can use inorganic phosphate, of course, to change ADP back to ATP, but we do not accomplish that by the use of heat. Instead, even that change involves the direct transfer of chemical energy through common intermediates and is driven by a major pathway, the respiratory chain (see the next section).

Adenosine triphosphate has two high-energy phosphate bonds, and occasionally the second bond (see Table 16.1) is broken in some transfer of chemical energy. The products are then **AMP, adenosine monophosphate,** and pyrophosphate ion symbolized by **PP_i**. We can remake ATP from AMP, too.

The ATP in muscle tissue will sustain muscle activity only a fraction of a second. To provide immediate regeneration of ATP, muscle tissue makes and stores phosphocreatine (Table 16.1). This high-energy phosphate (higher than ATP) is used as an electrical condenser or storage battery of phosphate bond energy. Phosphocreatine reserves, built up during inactivity, are switched on to transferring their high-energy phosphates to ADP to regenerate ATP while the system cranks up the respiratory chain to take over that function.

16.5 Respiratory Chain

Phosphate-group transfer potential is one source of ATP. A phosphate of very high potential gives its phosphate group to ADP in substrate phosphorylation. Another way for making ATP is to tap the energy in compounds with high-electron transfer potential. These are compounds that bring us close to the use of the chemical energy in the oxygen we breathe, and ATP is made by a process called **oxidative phosphorylation.**

Respiration, physiologically understood, includes not just breathing but also the consumption of oxygen by cells. Oxygen is reduced to water, and Equation (16.6) is a bare-bones statement of what must happen in that reaction.

$$\underset{\text{Pair of electrons}}{(:)} + \underset{\text{Pair of protons}}{2H^+} + \underset{\text{Atom of oxygen}}{\cdot\ddot{\underset{..}{O}}\cdot} \longrightarrow \underset{\text{Molecule of water}}{H-\ddot{\underset{..}{O}}-H} + \text{energy} \tag{16.6}$$

The electrons and protons come from energy metabolites, intermediates in the catabolism of sugars and fats. Generally the electrons of C—H bonds are used. Often the transfer is via a hydride ion, $H:^-$, passed directly from donor to acceptor with a proton, H^+, trailing along later.

The chief receptors of hydride are the **respiratory enzymes** that participate in a series of reactions called the **respiratory chain.** The flow of electrons from initial donor into initial receptor is down an energy hill all the way to oxygen. The flow is irreversible. At various places special complexes of enzymes occur that can take some of the free energy in that flow and use it to put ADP and P_i back together as ATP. Before we study any details, let us step back for a broad look at the principal oxidative pathways in the body.

16.6 Major Oxidative Pathways. An Overview

We ask first, what initiates ATP synthesis? How does the body know when to oxidize energy metabolites to make ATP? The answer, on the surface, is quite simple: whenever the levels of ADP and P_i rise and the level of ATP drops. The synthesis of ATP is under feedback control (see Section 13.17);

ATP actually inhibits certain key enzymes needed to make it. Thus if the supply of ATP is high, no oxidation occurs. When ATP is used up, the enzymes are free to catalyze reactions that bring the supply of ATP back. We say that a condition of **homeostasis** is maintained. Homeostasis, a name coined by Walter Cannon (1871–1945), an American physiologist, means the behavior of an organism to stimuli that starts a series of metabolic events to restore the system as much as possible to its original state (just before the stimulus). We might think of it as a special application of Le Châtelier's principle, except that the latter speaks only of restoring equilibrium, not necessarily the original equilibrium.

When the level of ADP and P_i rise, the body is stimulated to increase its rate of breathing. The respiratory chain moves into action to reduce the oxygen to water and simultaneously manufacture ATP. To provide electrons for that reduction, another grand metabolic sequence, the **citric acid cycle,** begins to churn. This cycle also needs "fuel," and it uses acetyl groups, acetyl groups joined to a coenzyme, coenzyme A. Virtually all products of digestion can be catabolized to supply acetyl groups to the citric acid cycle.

Glucose breaks down in a series of changes called **glycolysis** ("glycose—or glucose—loosening or breaking"). Glycolysis generates two pyruvate ions from each glucose molecule (Equation 16.7). (We write "pyruvate" rather than "pyruvic acid" because at physiological pH, the acid is neutralized and it exists as its anion.)

$$\underset{\text{Glucose}}{C_6H_{12}O_6} \xrightarrow[\text{(several steps; not balanced)}]{} \underset{\text{Pyruvate}}{2\,CH_3\overset{\overset{O}{\|}}{C}CO_2^-} \rightleftharpoons \underset{\text{Lactate}}{2\,CH_3\overset{\overset{OH}{|}}{C}HCO_2^-} \qquad (16.7)$$

Each pyruvate is broken down to an acetyl group. From glucose to pyruvate no oxygen is required, and glycolysis is often called the **anaerobic sequence** of glucose catabolism ("an-," without; "-aerobic," oxygen, without oxygen). We shall study the details of this pathway in the next chapter.

Fatty acids break down in a series of reactions called the fatty acid cycle (better, the fatty acid spiral). Each "turn" of the cycle clips one acetyl group away from the long fatty acid chain. This sequence will be studied under the metabolism of lipids.

Most amino acids can be catabolized to acetyl units or to intermediates in the citric acid cycle itself, and we shall study these under the metabolism of proteins.

The major pathways of biochemical energetics are outlined in Figure 16.1. Our interest in this chapter is in the two oxidative pathways that can accept breakdown products from any food, the citric acid cycle and the respiratory chain. Their operation requires the presence of oxygen within the cell, and together they make up the **aerobic sequence.**

16.7 Respiratory Enzymes

The chief respiratory enzymes are those that include as coenzymes NAD^+ and FAD (both described in Section 13.4), coenzyme Q (Section 4.16) and one resembling FAD called flavin mononucleotide, FMN. The latter is simply

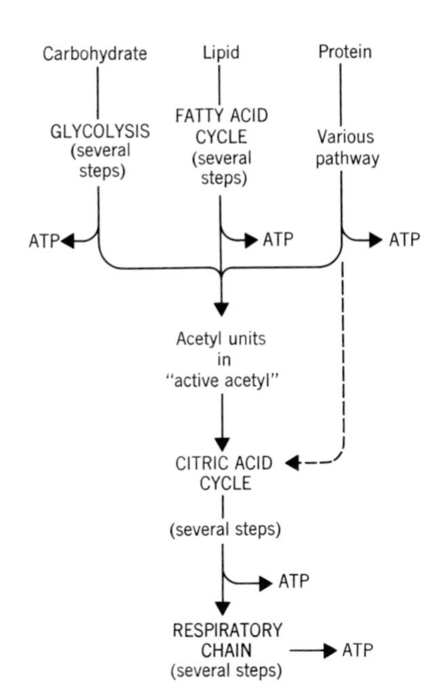

Figure 16.1 The major pathways in biochemical energetics.

the phosphate ester of riboflavin (a B-vitamin), whereas FAD is riboflavin linked by a phosphate to an AMP unit. The nicotinamide unit in NAD^+ accepts hydride ion as follows, where we represent the donor as an alcohol:

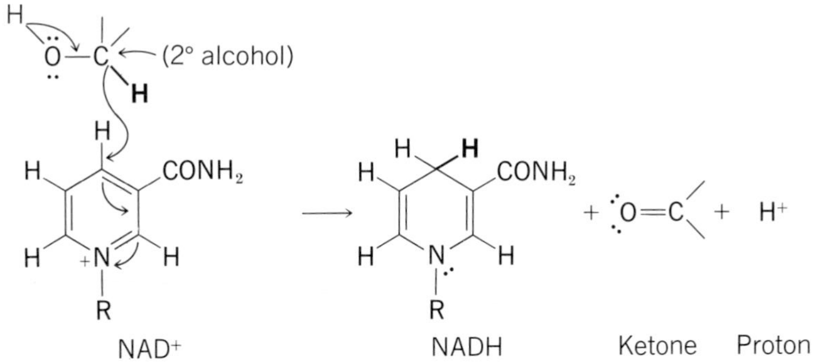

(2° alcohol)

NAD+ → NADH + Ketone + Proton

The riboflavin unit is the same in both FAD and FMN and a dienelike network in this unit accepts hydride as follows, where we show an alkanelike unit as the donor and just part of the riboflavin system:

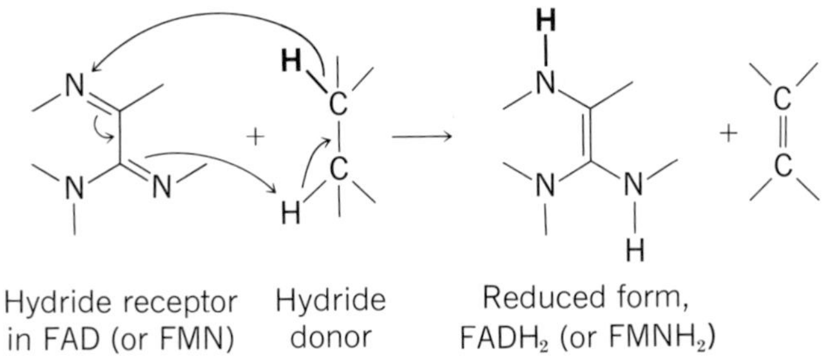

Hydride receptor in FAD (or FMN) + Hydride donor → Reduced form, $FADH_2$ (or $FMNH_2$) + alkene

Coenzyme Q undergoes a quinone-hydroquinone conversion as follows, where we represent the donor simply as MH_2 (M for metabolite) and give only the bare essentials of the coenzyme:

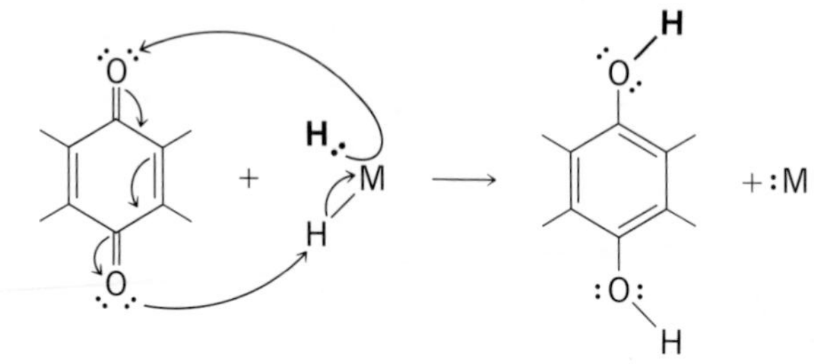

Coenzyme Q (oxidized form) + Metabolite ⟶ Coenzyme Q (reduced form) + :M

A second set of respiratory enzymes follows coenzyme Q farther down the energy hill of the respiratory chain. This is a series of heme-containing enzymes having iron ions; these enzymes are called cytochromes. The last one holds a copper ion (accounting for our use of copper as a trace element). These enzymes accept only electrons, not hydride ions. The cytochromes pass electrons from metal ion to metal ion until at the final enzyme they are delivered to oxygen. Protons are taken from the surrounding buffer system and water forms.

The respiratory enzymes occur and work together as a team in packages called respiratory assemblies that spangle the inner surfaces of mitochondria (Figure 16.2). On adjacent surfaces of these tiny "powerhouses of the cell" are teams of enzymes for the citric acid cycle and the fatty acid cycle. An individual cell in a tissue that is normally very active in using ATP may have thousands of mitochondria per cell. The flight muscle of the wasp, for example, has about 100 000 mitochondria per cell.

16.8 Electron Transport in the Respiratory Chain

When NAD^+ accepts $H:^-$, we may write the equation as follows.

$$MH_2 + NAD^+ \longrightarrow M: + NADH + H^+ \qquad (16.8)$$

The reduced form, NADH then passes the hydride ion to FMN:

$$NADH + FMN + H^+ \longrightarrow NAD^+ + FMNH_2 \qquad (16.9)$$

Reduced $FMNH_2$ then gives up hydrogen to coenzyme Q, CoQ:

$$FMNH_2 + CoQ \longrightarrow FMN + CoQH_2 \qquad (16.10)$$

The reduced coenzyme Q passes the electrons it received to cytochrome *b* and drops off the protons to the surrounding solution to be buffered.

Equations 16.8 to 16.10 and others like them are often written in a simpler and visually clearer way as follows.

$$\begin{array}{cccc} MH_2 & NAD^+ & FMNH_2 & CoQ \\ M: & NADH + H^+ & FMN & CoQH_2 \end{array}$$

Pairs of curved arrows connect reactants to the chief products they become. Where the arrows nearly touch, we signify that something passes or crosses over—hydride ion or electrons. At the first pair of curved arrows in the previous equation we say "as MH_2 changes to M: it passes $H:^-$ to NAD^+ and releases a proton."

The whole respiratory chain from the NAD^+-bearing enzyme to the final cytochrome is outlined in Figure 16.3. A fork occurs near the beginning of the

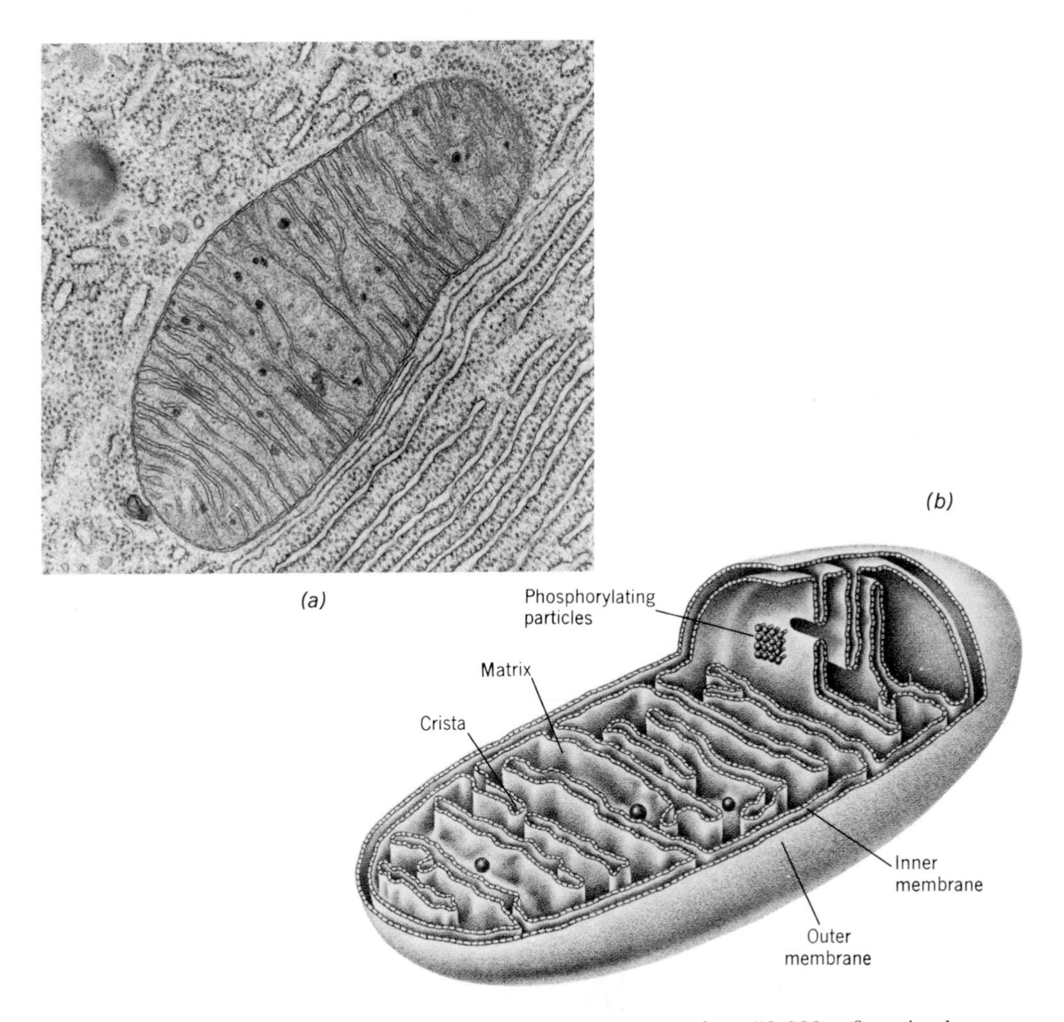

Figure 16.2 A mitochondrion. (a) Electron micrograph (×53 000) of a mitochondrion in a pancreas cell of a bat. (b) Perspective showing the interior of a mitochondrion. Each phosphorylating particle or respiratory assembly contains the respiratory enzymes. (Micrograph courtesy of Dr. Keith R. Porter. Drawing courtesy of A. Nason and R. L. DeHaan, *The Biological World,* 1973. John Wiley & Sons, Inc., New York. Used by permission.)

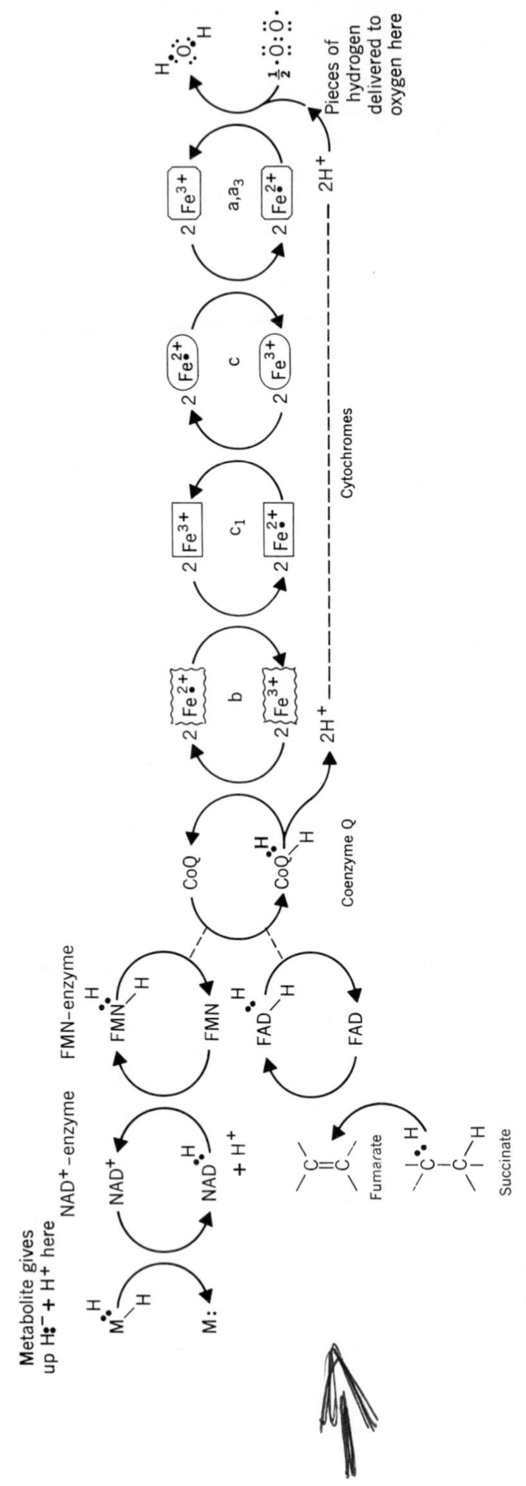

Figure 16.3 The respiratory chain.

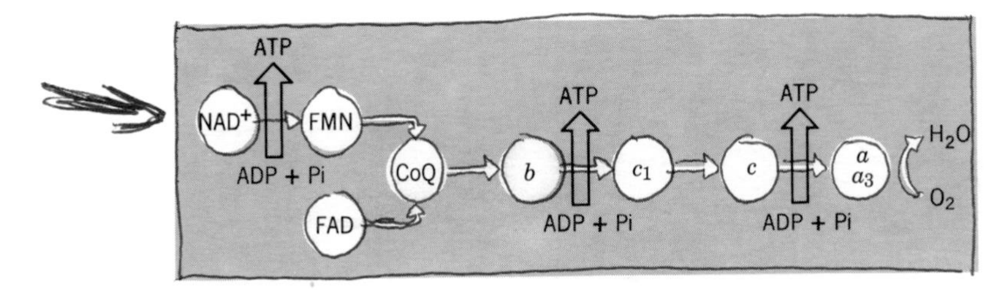

Figure 16.4 Oxidative phosphorylation takes place at three sites along the respiratory chain, shown here by the broad arrows.

pathway, where the FAD-bearing enzyme can accept hydrogen directly from a metabolite, MH_2, and feed it to coenzyme Q, bypassing NAD^+.

16.9 Oxidative Phosphorylation

The electron flow in the respiratory chain (oxidation) provides energy for the synthesis of ATP (phosphorylation) at three places. One site is between NAD^+ and FMN. Another is between cytochrome b and cytochrome c_1. The third is located near cytochrome c and the complex of cytochromes a and a_3. See Figure 16.4.

Each metabolite that delivers hydrogen to NAD^+ leads to three ATPs. A metabolite that gives hydrogen to FAD instead leads to two ATPs.

Exactly how enzymes and energy cooperate in oxidative phosphorylation is still not fully answered and remains a major problem of molecular biology. We shall not survey the theories currently in contention. Much more clearly understood is the citric acid cycle that fuels the respiratory chain.

16.10 Citric Acid Cycle

Acetyl units are torn apart by the citric acid cycle, Figure 16.5, with their carbons and oxygens emerging as carbon dioxide. Their hydrogens and the electrons that bond them go into the respiratory chain. For an acetyl group to enter the citric acid cycle it must be joined to coenzyme A, which may be represented as CoA—SH.

$H—S—CH_2CH_2NHC(=O)CH_2CH_2NHC(=O)CH(OH)C(CH_3)_2CH_2O—P(=O)(O^-)—O—P(=O)(O^-)—O—CH_2$ (ribose–adenine, 3′-phosphate $^-O—P(=O)—O^-$)

← Pantothenate unit →

Coenzyme A (CoA-SH)

$CH_3C(=O)—S—CoA$

Acetyl coenzyme A
"active acetyl"
(condensed symbol)

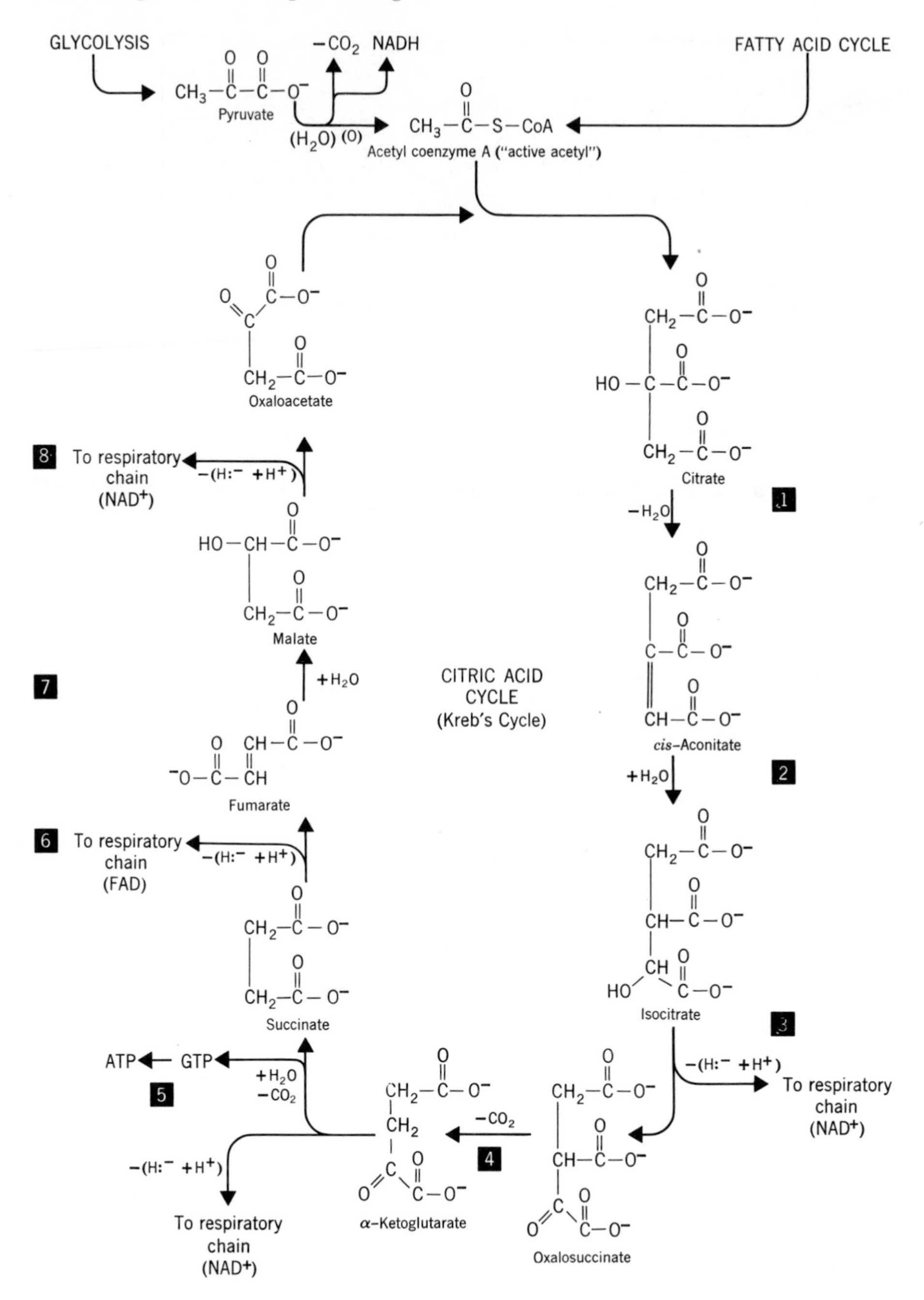

Figure 16.5 Citric acid cycle. The boxed numbers refer to the discussion in the text. Hans Krebs, a German-born British biochemist, is credited with fitting the last pieces of this puzzle together, many of the pieces discovered by him. To honor that work the cycle is often called the Kreb's cycle. He shared the 1953 Nobel Prize in medicine and physiology with Fritz Lipmann, who made major contributions to biochemical energetics, particularly the important roles of high energy phosphates.

This coenzyme requires pantothenic acid, one of the vitamins. The bond with acetyl group is a carbonyl-sulfur bond, and acetyl coenzyme A is a thio ester. ("Thio" signifies the formal replacement of O by S; e.g., thio alcohol, R—S—H.) This bond is much more easily broken than a carbonyl-oxygen bond, and acetyl coenzyme A, often called "active acetyl," has a high acetyl-group transfer potential. Coenzyme A is the universal carrier of acyl groups in general.

Exercise 16.3 The enzyme assembly for changing pyruvate to acetyl coenzyme A requires thiamine, one of the B-vitamins. In the deficiency disease for this vitamin, beriberi, what substance would one expect to accumulate at an abnormally high level in blood (and for which an analysis could be performed as part of the diagnosis)?

In the first step of the citric acid cycle, acetyl coenzyme A transfers its acetyl group to a keto dicarboxylate ion, oxaloacetate, and citric acid (rather, citrate ion) forms.[3] The reaction resembles an aldol condensation (Section 6.15) because one compound with an alpha hydrogen (acetyl coenzyme A) adds to a keto group (in oxaloacetate). Now begins a series of reactions by which citrate is degraded bit by bit until oxaloacetate is replaced. (The numbers below match those in Figure 16.5.)

1. Citrate is dehydrated to give the double bond in *cis*-aconitate.

2. *cis*-Aconitate is hydrated, water adding to the double bond to give isocitrate. These two steps accomplish no more than moving the —OH group.

3. The 2° alcohol in isocitrate is now dehydrogenated to oxalosuccinate; NAD^+ accepts the hydrogen. Three ATPs will be made.

4. Oxalosuccinate decarboxylates to give α-ketoglutarate. Look carefully in the structure of oxalosuccinate. The carboxyl group that splits out is beta to a keto group, and beta-keto carboxylic acids readily lose CO_2 (Section 7.21).

5. α-Ketoglutarate now undergoes a very complicated series of reactions involving water as a reactant, all catalyzed by one team of enzymes including coenzyme A. The result is decarboxylation and dehydrogenation. The hydrogen goes to NAD^+; three more ATPs will be made. The product, not shown in the Figure 16.5, is the coenzyme A derivative of succinic acid. Its conversion to succinate ion generates guanosine triphosphate, GTP, a high-energy phosphate mentioned in Section 16.4 as being very similar to ATP. Guanosine triphosphate gives a phosphate unit to ADP to make one ATP. Thus the citric acid cycle includes one substrate phosphorylation. The details of this remarkable series of reactions, α-ketoglutarate to succinate, giving one hydrogen to NAD^+ (for making three ATPs) besides making one ATP directly are beyond the scope of this book.

6. Succinate donates hydrogen to FAD not to NAD^+ because the electron transfer potential of succinate is too low. Two ATPs will be made; fumarate forms.

7. Fumarate adds water to its double bond and malate forms.

8. The 2° alcohol in malate gives up hydrogen to NAD^+. Three more ATPs will be made. The cycle has returned. We have remade one molecule of oxaloacetate, which now is ready to accept an acetyl group from active acetyl. If the ATP demand is still high, the cycle will go around again.

[3] At physiological pH the acids in the cycle exist largely as their anions.

Exercise 16.4 The enzyme for converting isocitrate to oxalosuccinate (see [3], Figure 16.5) is stimulated by one of these two—ATP or ADP—but not by the other. Which is the likely activator? Explain.

16.11 Sources of Acetyl Coenzyme A

Several substances supply acetyl coenzyme, an intermediate at one of the major metabolic crossroads in metabolism. As outlined in Figure 16.1, several amino acids may be broken down to acetyl units. Fatty acids degrade in two-carbon acetyl units. Glycolysis makes pyruvate ions, which can be changed to active acetyl.

The conversion of pyruvate ion to acetyl coenzyme A needs both decarboxylation and oxidation. A team of enzymes handles the task, and the several changes are very similar to the α-ketoglutarate to succinate conversion in the citric acid cycle.

The overall change is

$$\underset{\text{Pyruvate}}{CH_3-\overset{\overset{\large O}{\|}}{C}-\overset{\overset{\large O}{\|}}{C}-O^-} + \underset{\text{Coenzyme A}}{CoA-SH} + NAD^+ \longrightarrow \underset{\text{Acetyl coenzyme A}}{CH_3\overset{\overset{\large O}{\|}}{C}-S-CoA} + CO_2 + NADH \quad (16.11)$$

NAD^+ picks up $H:^-$, and therefore three ATPs will be made.

16.12 ATP Yields

The ATP yield from pyruvate ion through the citric acid cycle and the respiratory chain are summarized in Table 16.2.

As we shall see in Chapter 17, pyruvate ion is made from lactate ion, the true end-product of glycolysis. The change of lactate back to pyruvate is sim-

TABLE 16.2
ATP from Pyruvate in the Citric Acid Cycle and Respiratory Chain

Steps	Receiver of ($H:^-$ + H^+) in the Respiratory Chain	Molecules of ATP Formed
Pyruvate $\longrightarrow$ acetyl CoA	NAD^+	3
Isocitrate $\longrightarrow$ α-ketoglutarate	NAD^+	3
α-Ketoglutarate $\rightarrow$ succinyl-CoA	NAD^+	3
Succinyl-CoA $\xrightarrow{GDP}$ succinate	—	1
Succinate $\longrightarrow$ fumarate	FAD	2
Malate $\longrightarrow$ oxaloacetate	NAD^+	3
Total yield per pyruvate		15 ATP

ply the oxidation (dehydrogenation) of a 2° alcohol; NAD^+ takes the hydrogen, and three ATPs will be made. The total ATP yield by the aerobic sequence from one molecule of lactate is therefore 18 ATPs per lactate, 3 from lactate to pyruvate and 15 from pyruvate to carbon dioxide and water. Since each glucose gives 2 lactates, the aerobic part of glucose oxidation gives $2 \times 18 = 36$ATP. The anaerobic part of glucose catabolism gives 2 more ATP, for a grand total of 38ATP per glucose, anaerobic and aerobic sequences combined.

Brief Summary

High Energy Compounds. The family of biological organophosphates includes several whose free energies of hydrolyses are high, and they have high phosphate-group transfer potentials. Those with potentials equal to or higher than those of ATP are "high-energy phosphates." The rest are low-energy phosphates. Adenosine triphosphate is the most widely used transfer agent for chemical energy in biological energetics. It can be made by substrate phosphorylation from higher-energy phosphates or by oxidative phosphorylation from ADP and P_i.

Oxidative Phosphorylation. A series of electron transfer enzymes called the respiratory enzymes occur as a group in respiratory assemblies in mitochondria. The metabolites for the respiratory chain of reactions are intermediates in the citric acid cycle or other energy metabolites. NAD^+ and FAD accept $H:^- + H^+$ from some metabolite (MH_2). Their reduced forms, NADH and $FADH_2$, then pass hydride on; NADH to FMN, and $FADH_2$ to coenzyme Q. Reduced FMN (i.e., $FMNH_2$) feeds to coenzyme Q, also. Reduced coenzyme Q passes electrons to the cytochromes, which carry them to oxygen, which is reduced to water.

Citric Acid Cycle. Acetyl groups from acetyl coenzyme A are transferred to a four-carbon carrier, oxaloacetate, to make citrate. This six-carbon compound is then degraded bit by bit as $H:^-$ and H^+ are fed to the respiratory chain to power oxidative phosphorylation. (One substrate phosphorylation also occurs during the cycle.) Each acetyl unit helps make 12 molecules of ATP. Fatty acids and glucose are important suppliers of acetyl groups. By glycolysis glucose is broken to lactate. The oxidation of lactate to pyruvate sends some $H:^-$ to the respiratory chain. Oxidative decarboxylation of pyruvate to acetyl coenzyme A sends more $H:^-$ to the respiratory chain.

Selected References

Books

1. L. Stryer. *Biochemistry,* 1975. W. H. Freeman and Company, Publishers, San Francisco. Part II is about the generation and storage of metabolic energy.
2. H. N. Christensen. *Dissociation, Enzyme Kinetics and Bioenergetics,* 1975. W. B. Saunders Company, Philadelphia. A programmed learning introduction for students of biological and medical sciences.

3. A. L. Lehninger. *Bioenergetics,* 2nd ed. W. A. Benjamin, Inc., Menlo Park, Calif., 1971. The flux of energy from one storage form to another is discussed in this paperback.

Questions and Exercises

1. What products of digestion can be used for chemical energy to make ATP?
2. The complete catabolism of glucose gives what products?
3. What are the end-products of the complete catabolism of fatty acids?
4. Why are phosphoenolypyruvate and ATP called high-energy phosphates, but glycerol 3-phosphate is not?
5. Complete this structure of ATP:

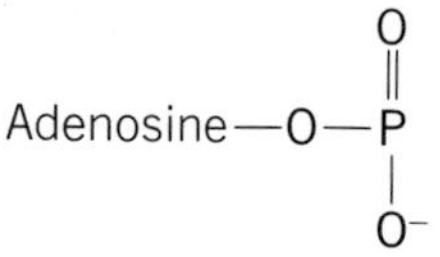

$$\text{Adenosine}-O-\overset{\displaystyle O \atop \displaystyle \|}{\underset{\displaystyle | \atop \displaystyle O^-}{P}}$$

6. In the manner of Exercise 5 write structures for ADP and AMP.
7. What coenzyme most frequently accepts hydride as the citric acid cycle operates?
8. What vitamin is needed to make the coenzyme of Question 7?
9. What coenzyme accepts hydrogen from a donor that is then left with a carbon-carbon double bond?
10. What vitamin is needed to make the coenzyme of Question 9?
11. Write the following in the form of an equation.

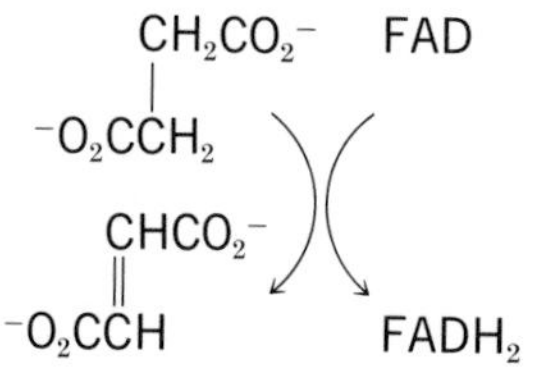

$$^-O_2CCH_2CH_2CO_2^- \;(\text{FAD} \rightarrow \text{FADH}_2)\; \longrightarrow \; ^-O_2CCH{=}CHCO_2^-$$

12. Write the following equation in the form used in Exercise 11.

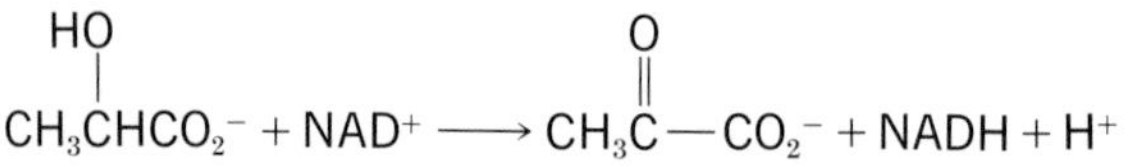

$$CH_3CH(OH)CO_2^- + NAD^+ \longrightarrow CH_3\overset{O}{\overset{\|}{C}}-CO_2^- + NADH + H^+$$

13. What is the difference between substrate and oxidative phosphorylation?
14. What throws the respiratory chain into action?
15. What makes the citric acid cycle start up?
16. How does glycolysis participate in making ATP (in general terms)?
17. How does the fatty acid cycle help make ATP (in general terms)?
18. Make a flow chart to show the general connections between products of digestion and the respiratory chain.

19. What is the general name for the Fe^{3+}/Fe^{2+}-containing enzymes in the respiratory chain?
20. How does ATP help control its own synthesis?
21. Write the condensed symbol we use for active acetyl.
22. What is the name of the universal carrier of acyl groups in metabolism?
23. What are the three principal uses for chemical energy in the body?

chapter 17

Metabolism of Carbohydrates

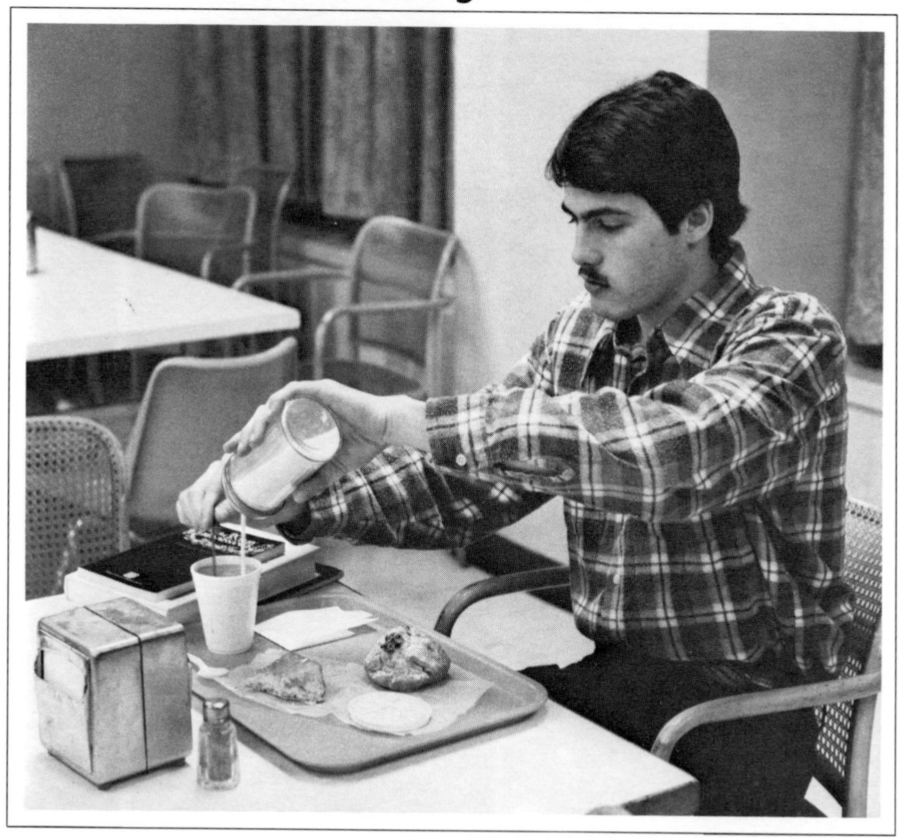

Look out pancreas, here it comes! Sugar at breakfast, at every coffee break, with most junk food. Is there a connection between the American diet and the high incidence of diabetes? We shall start an examination of that question in this chapter. (Photo by K. T. Bendo)

Catabolism of Glucose

17.1 Glycolysis

Glycolysis is a series of reactions that change glucose to lactate while a small but important amount of ATP is made. The series can run without oxygen and therefore is called the anaerobic sequence of glucose catabolism. The overall result is given by Equation 17.1

$$\underset{\text{Glucose}}{C_6H_{12}O_6} + 2ADP + 2P_i \longrightarrow \underset{\text{Lactate}}{2CH_3\overset{\overset{\large OH}{|}}{C}HCO_2^-} + 2H^+ + 2ATP \qquad (17.1)$$

The intermediate steps in glycolysis are given in Figure 17.1. From glucose to fructose 1,6-diphosphate some input of ATP occurs, but that investment of chemical energy pushes intermediates up an energy hill for a long, downhill slide that will return the investment with "interest," more ATP. A slightly smaller initial input of ATP is used if the starting point for glycolysis is glycogen instead of glucose, as the top part of Figure 17.1 shows.

The steps from glucose to lactic acid are discussed next, where the numbers refer to the boxed numbers in Figure 17.1.

1. Glucose is phosphorylated by ATP. This step is essentially irreversible, and it traps glucose inside the cell.

2. Glucose 6-phosphate undergoes a remarkable isomerization to fructose 6-phosphate, catalyzed by an isomerase. Without implying that the following describes the exact mechanism of the work of this particular enzyme, we yet may better understand the isomerization in terms of relatively simple and reasonable rearrangements of bonds:

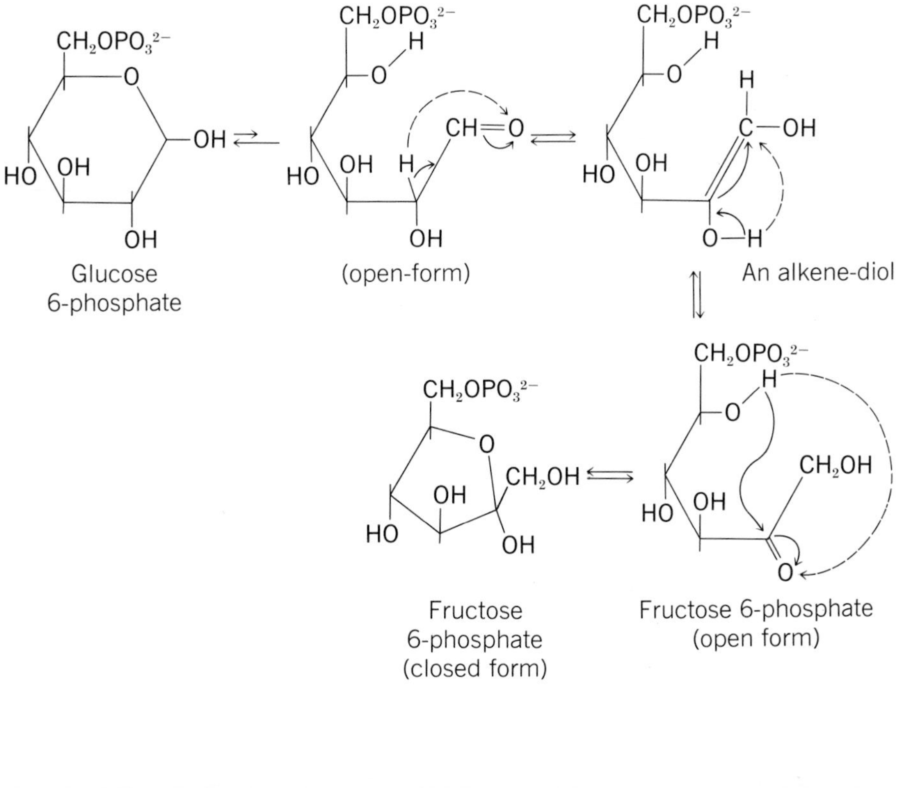

Glucose 6-phosphate ⇌ (open-form) ⇌ An alkene-diol ⇌ Fructose 6-phosphate (open form) ⇌ Fructose 6-phosphate (closed form)

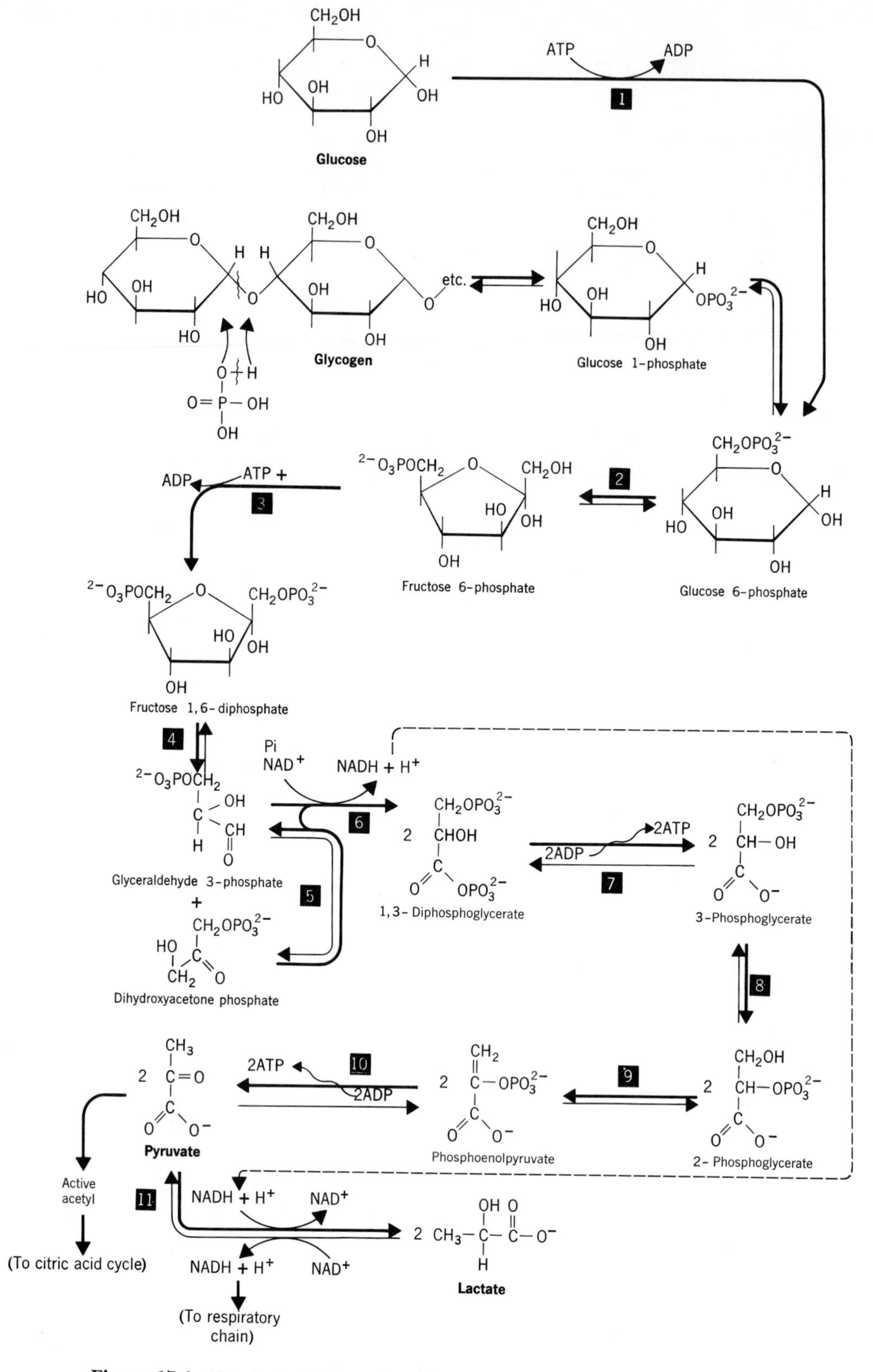

Figure 17.1 Glycolysis. The numbered steps are discussed in the text. Steps 1, 3, and 10 are essentially irreversible.

3. ATP transfers a phosphate unit to fructose 6-phosphate to make fructose 1,6-diphosphate. This step is essentially irreversible.

4. A hexose diphosphate breaks apart into two triose monophosphates. The reaction is a reverse of the aldol condensation (Section 6.15) and is catalyzed by aldolase. We may visualize the overall reaction in terms of a few simple and reasonable shifts of electrons and protons:

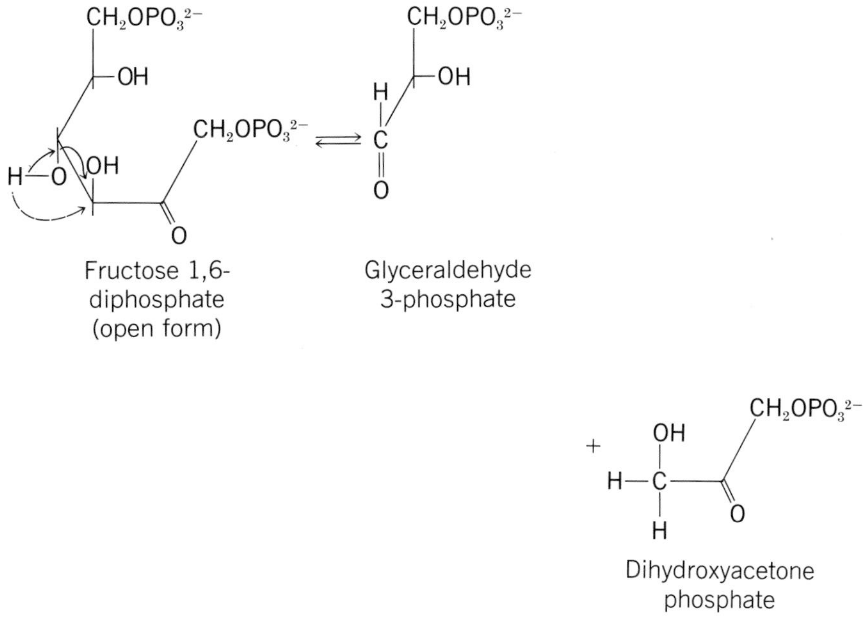

Fructose 1,6-diphosphate (open form) $\rightleftharpoons$ Glyceraldehyde 3-phosphate + Dihydroxyacetone phosphate

5. Dihydroxyacetone phosphate isomerizes to glyceraldehyde 3-phosphate. We may imagine an intermediate alkene-diol similar to one we wrote as part of the explanation for step 2.

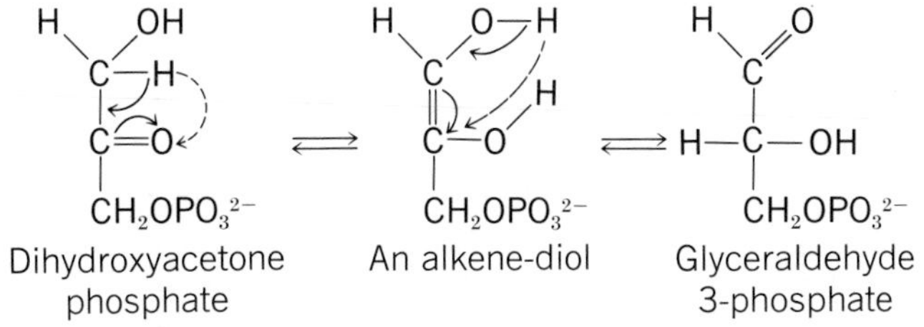

Dihydroxyacetone phosphate $\rightleftharpoons$ An alkene-diol $\rightleftharpoons$ Glyceraldehyde 3-phosphate

This isomerization insures that all the chemical energy in glucose can be obtained, because the main path continues with glyceraldehyde 3-phosphate.

6. An oxidative phosphorylation occurs and NAD^+ is reduced. While the aldehyde group in the starting material is oxidized to the carboxyl state, an inorganic phosphate unit enters to make a mixed anhydride between a carboxylic acid and phosphoric acid. 1,3-Diphosphoglycerate is one of the high-energy phosphates (Table 16.1). The enzyme system temporarily forms a thio ester, and we may visualize how this oxidative phosphorylation occurs roughly as follows:

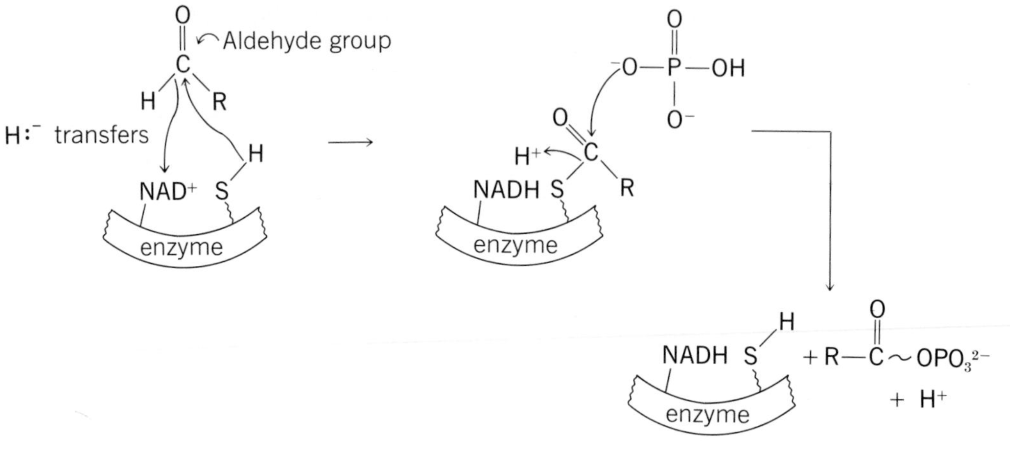

Aldehyde group

$H:^-$ transfers

NAD^+ S

enzyme

$^-O-P(=O)(O^-)-OH$

H^+

NADH S

enzyme

NADH S

enzyme

$+ R-C(=O)\sim OPO_3^{2-}$

$+ H^+$

7. 1,3-Diphosphoglycerate has a higher phosphate-group transfer potential than ATP, and it gives a phosphate to ADP. We get back the original investment of one ATP plus one more ATP besides. (Remember that each glucose gives *two* triose phosphates.)

8. The phosphate group in 3-phosphoglycerate shifts to the 2-position.

9. The dehydration of 2-phosphoglycerate to phosphoenolpyruvate has been compared to cocking a huge bioenergetic gun. This simple reaction converts a low-energy phosphate into the highest-energy phosphate in all metabolism, the phosphate ester of the enol form of pyruvate: phosphoenolpyruvate.[1]

10. More ATP is made as a phosphate group transfers from phosphoenolpyruvate to ADP. The loss of the phosphate group leaves, only temporarily, the enol form of pyruvate. This promptly and irreversibly rearranges to the keto form. The instability of enols (see footnote 1) is the driving force for the step.

11. If the mitochondrion is running aerobically when pyruvate is made, pyruvate ion changes to acetyl coenzyme A, which enters the citric acid cycle. To restore the reduced enzyme with NADH at step 6, NADH can pass off hydrogen to the respiratory chain directly, assuming that oxygen is available.

If the mitochondrion is not running aerobically, glycolysis must shut down unless the NADH-enzyme at 6 can be changed back to the oxidized form, NAD^+. That change in the absence of oxygen and the operation of the respiratory chain occurs by the reduction of pyruvate to lactate. Thus, under anaerobic conditions, lactate is the end-product of glycolysis.

[1] When a carbon of an alkene group holds an alcohol group, the compound is called an enol ("ene" + "-ol"). Enols are generally unstable and spontaneously rearrange to carbonyl forms.

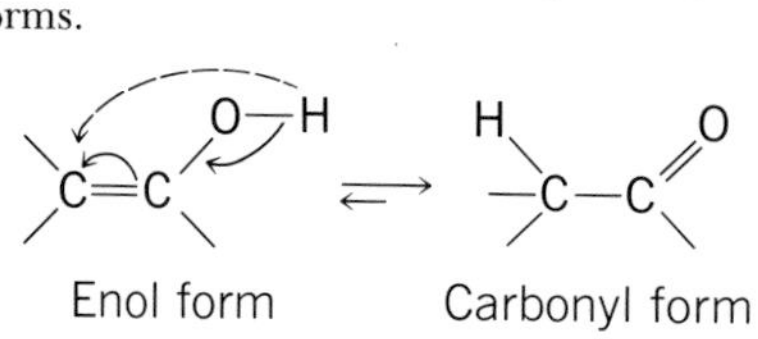

Enol form ⇌ Carbonyl form

17.2 Importance of Anaerobic Sequence

The supply of ATP in a resting muscle can sustain cellular work only about half a second. Supplies of phosphocreatine (Section 16.4) do not last much longer. Hence mechanisms for regenerating ATP must work swiftly and smoothly, and we must have a way for making ATP if oxygen delivery falls short of the needs for the aerobic sequence. Glycolysis provides that way. When a tissue receives oxygen slower then it needs it for the aerobic sequence, we say that the tissue has an **oxygen debt.** As a tissue accumulates an oxygen debt from running anaerobically for some time, lactate accumulates, and the tissue has to get back to aerobic operation. The tissue, indeed the whole organism, must pause to "get its breath" or "repay" the oxygen debt. Excess lactate is sent to the liver for other uses (see Section 17.7).

17.3 Other Carbohydrates in Glycolysis

Figure 17.2 discloses the centrality of glucose 6-phosphate to the catabolism of all carbohydrates. Glycogen stored in muscles, liver, and kidneys can release glucose and give the 6-phosphate. Galactose is changed eventually to this glucose derivative. Fructose may enter glycolysis directly at the fructose 6-phosphate stage and it may be changed to glucose 6-phosphate in a reverse of step 2. Thus all dietary hexoses catabolize by virtually the same pathway, that of glucose.

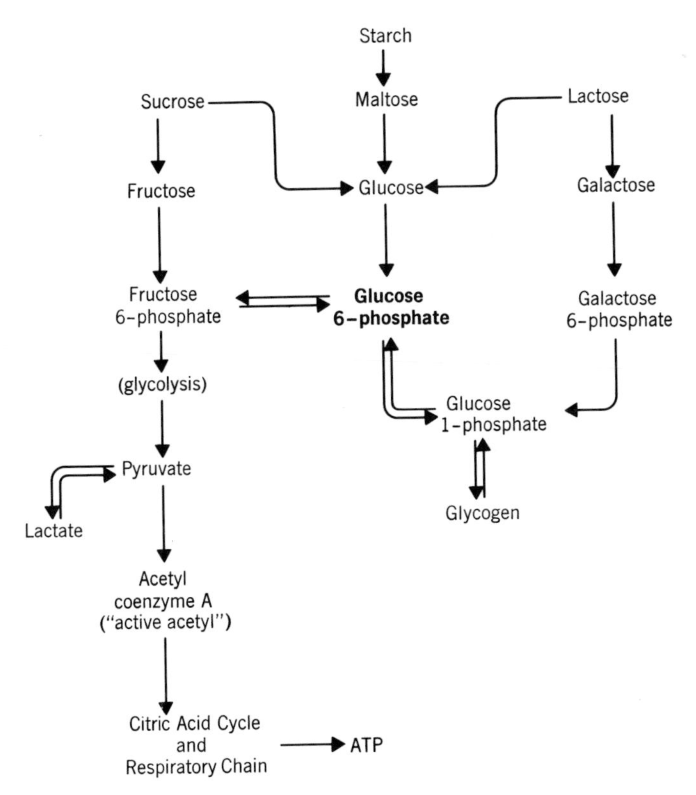

Figure 17.2 Convergence of pathways in the metabolism of dietary carbohydrates.

17.4 Pentose Phosphate Pathway of Glucose Catabolism

The biosynthesis of some macromolecules requires a reducing agent. Fatty acids, for example, are almost entirely alkanelike, and alkanes are the most reduced types of organic compounds. The reducing agent for fatty acids is the reduced form of $NADP^+$, another nicotinamide-containing coenzyme (Section 13.4).

Coenzymes NADPH and NADH are used in quite different ways in the body. NADH sends $H:^-$ down the respiratory chain to drive oxidative phosphorylation and make ATP. The hydride ion in NADH eventually reduces oxygen. The hydride ion in the close relative, NADPH, reduces organic substances instead.

The body's principal route to NADPH is the **pentose phosphate pathway** of glucose catabolism. This complicated series of reactions, which we shall not study in any detail, is very active in adipose tissue, where fatty acid synthesis occurs. Adipose tissue is fatty tissue surrounding internal organs and is the immediate culprit in the bulging waistline. Skeletal muscles have very little activity of the pentose phosphate pathway.

The oxidative reactions in the pentose phosphate pathway convert a hexose phosphate to a pentose phosphate; hence the name of the series.

$$\text{glucose 6-phosphate} + 2NADP^+ + H_2O \longrightarrow \text{ribose 5-phosphate} + 2NADPH + 2H^+ + CO_2 \quad (17.2)$$

Ribose 5-phosphate may be used to make the pentose systems in nucleic acids. If not needed for that, this compound may undergo a series of isomerizations and group-transfer reactions, the nonoxidative series of the pentose phosphate pathway. These reactions have the net effect of converting three pentose units into two hexose units (glucose) and one triose unit (glyceraldehyde). The hexoses may be catabolized via glycolysis or recycled to the oxidative series of the pentose phosphate pathway. Glyceraldehyde, as we shall soon see, can be converted to glucose and recycled as well. Thus glycolysis, pentose phosphate reactions, and the resynthesis of glucose all interconnect; specific bodily needs determine what is run.

Overall, with pentose recycle, the balanced equation for the complete oxidation of one glucose molecule via the pentose phosphate pathway is

$$6 \text{ glucose 6-phosphate} + 12NADP^+ \xrightarrow{\text{pentose phosphate pathway}} 5 \text{ glucose 6-phosphate} + 6CO_2 + 12\,NADPH + 12H^+ + P_i \quad (17.3)$$

Other names for the pentose phosphate pathway that may be used in other references are "hexose monophosphate shunt" and "phosphogluconate pathway" (after gluconic acid 6-phosphate, an intermediate).

17.5 Fermentation

The reactions of glycolysis occur in certain yeasts as well as in us, up to a point. At the pyruvate stage, some yeasts will decarboxylate pyruvate to acetaldehyde, Equation 17.4.

$$CH_3{-}\overset{\overset{\large O}{\|}}{C}{-}C(=O){-}O{-}H \xrightarrow{\text{certain yeasts}} CH_3{-}\overset{\overset{\large O}{\|}}{C}{-}H + CO_2 \qquad (17.4)$$

Pyruvic acid — Acetaldehyde

They use the reduction of the aldehyde group to regenerate the NAD^+-bearing enzyme required for step 6 (Figure 17.1):

$$CH_3{-}\overset{\overset{\large O}{\|}}{C}{-}H + NADH + H^+ \xrightarrow{\text{certain yeasts}} CH_3CH_2OH + NAD^+ \qquad (17.5)$$

Acetaldehyde — Ethyl alcohol

In this manner selected yeasts carry out alcohol **fermentation,** and the carbon dioxide released by Equation 17.4 causes the frothing and foaming that accompanies fermentation.

Anabolism of Glucose

17.6 Gluconeogenesis

Gluconeogenesis ("glucose-new-genesis") is the synthesis of glucose from starting materials that are not carbohydrates, that is, from sources other than starch, glycogen, disaccharides, or other monosaccharides. The overall scheme is outlined in Figure 17.3. Nearly all amino acids can be partly degraded to give a raw material for gluconeogenesis. Normally, the brain uses glucose for its energy needs, glucose taken directly from the bloodstream. The body's ability to make its own glucose from noncarbohydrate sources is an important protection for the brain's resources of energy.

When lactate, not amino acids, provides the starting material for gluconeogenesis, the process is more properly called simply glucogenesis, but we shall use the term gluconeogenesis. In anaerobic operations, skeletal muscles generate pyruvate faster than it can be consumed by its conversion to acetyl coenzyme A. Oxygen is not being delivered fast enough; an oxygen debt is building. The excess pyruvate is reduced to lactate to generate the NAD^+-bearing enzyme at step 6 in glycolysis (Figure 17.1). Gluconeogenesis is a means of disposing of lactate.

Lactate leaves skeletal muscle cells readily and is carried in the bloodstream to the liver, the principal site of gluconeogenesis. (Some of this activity also occurs in the kidneys.) Some of the lactate is converted to glucose, which is then put back into the bloodstream for distribution back to the muscles as well as to the brain.

Gluconeogenesis is not exactly the reverse of glycolysis. Three steps in glycolysis cannot be exactly reversed, steps 1, 3, and 10, Figure 17.1. These irreversible steps must be bypassed, and the liver (or kidneys) have special enzymes for these bypasses.

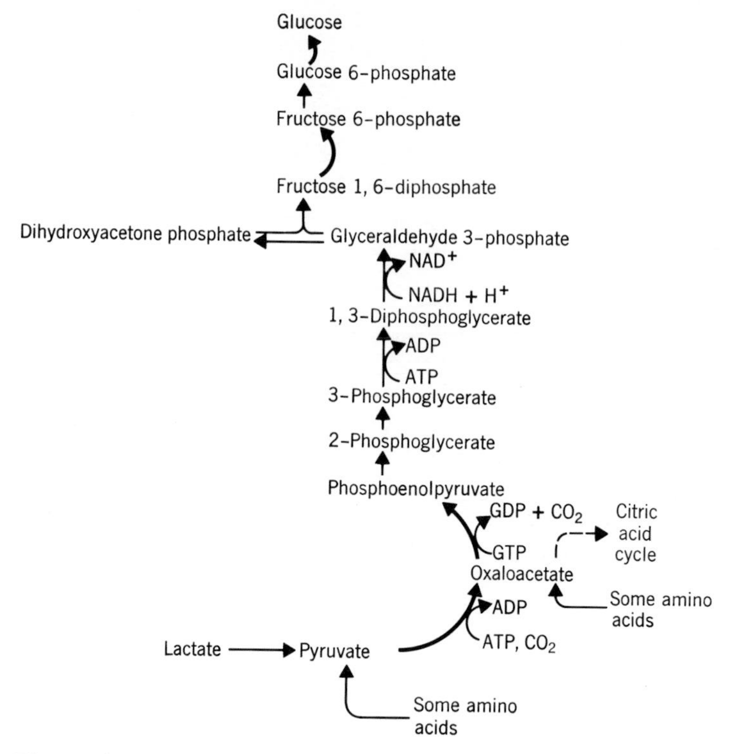

Figure 17.3 Gluconeogenesis. Straight arrows signify steps that are simply the reverse of corresponding steps in glycolysis (Figure 17.1). Heavy curved arrows signify steps that are unique to gluconeogenesis and are not the reverse of steps in glycolysis. Carbon skeletons from nearly all amino acids can be used in one way or another to make glucose.

In the bypass to reverse step 10 of glycolysis—make phosphoenolpyruvate from pyruvate—the first reaction uses carbon dioxide as a raw material to change pyruvate to oxaloacetate (Equation 17.6). The enzyme for the bypass, pyruvate carboxylase, requires the vitamin biotin. Adenosine triphosphate is required, which should come as no surprise.

$$\underset{\text{Pyruvate}}{CH_3-\overset{\overset{\displaystyle O}{\|}}{C}-CO_2^-} + CO_2 + ATP + H_2O \longrightarrow$$

$$\underset{\text{Oxaloacetate}}{{}^-O_2C-CH_2-\overset{\overset{\displaystyle O}{\|}}{C}-CO_2^-} + ADP + P_i + 2H^+ \qquad (17.6)$$

If glycolysis is downhill, gluconeogenesis has to be uphill in energy terms. We get ATP from glycolysis; hence we must use ATP (or GTP) to go back.

Oxaloacetate is the salt of a beta-keto acid, and these readily decarboxylate (see Section 7.21). As oxaloacetate kicks out CO_2, it also kicks out a bond to

GTP (like ATP) to take off a phosphate group, leaving GDP, and making the highest-energy phosphate of all, phosphoenolpyruvate.

$$^{-}:\ddot{O}-\overset{O}{\overset{\|}{C}}-\overset{\alpha}{CH_2}-\overset{:O:}{\overset{\|}{C}}_{\beta}-CO_2^- + {}^{-}O-\overset{O}{\overset{\|}{\underset{O^-}{\underset{|}{P}}}}\sim O-\overset{O}{\overset{\|}{\underset{O^-}{\underset{|}{P}}}}\sim O-\overset{O}{\overset{\|}{\underset{O^-}{\underset{|}{P}}}}-O-\text{Guanosine} \longrightarrow$$

Oxaloacetate (a β-keto acid, anion form)

GTP

$$CO_2 + CH_2=\overset{O\sim PO_3^{2-}}{\overset{|}{C}}-CO_2^- + GDP$$

Phosphoenolpyruvate

Since each glucose to be made requires two pyruvates, and each pyruvate uses two high-energy phosphates in gluconeogenesis, this bypass cost the equivalent of four ATP (2 ATP + 2 GTP) per glucose molecule made. At the reverse of step 7 (Figure 17.1) two more ATP will be used per molecule of glucose made. A total of six ATPs, therefore, are needed to make each glucose, which compares with two ATPs produced by glycolysis.

The bypasses to reverse steps 3 and 1 of glycolysis require only specific enzymes for hydrolyzing phosphate ester groups, and the enzyme team for gluconeogenesis has these enzymes. The other steps in gluconeogenesis are basically the reverse of corresponding steps in glycolysis.

Exercise 17.1 Amino groups of amino acids can be replaced by keto groups. Which amino acids would give these keto acids that participate in carbohydrate metabolism?
(a) Pyruvic acid (b) Oxaloacetic acid

Exercise 17.2 Referring to data in the previous chapter, how much ATP is made from the chemical energy in one lactate? If all this ATP were used to drive gluconeogenesis, how many molecules of glucose could be made? Assume that ATP may substitute for GTP.

17.7 Cori Cycle

The energy demand of gluconeogenesis is met by letting some lactate be used to make acetyl coenzyme A for the citric cycle and the respiratory chain. Of every six or seven molecules of lactate arriving in the liver, roughly one is catabolized to make ATP for the chemical energy needed to take the remaining five or six back to glucose. Lactate is like good rummage sent to the liver for salvage at a relatively low cost when oxygen intake returns to normal and the oxygen debt is repaid. Lactate has chemical energy that is recycled, some to make glucose, some to make ATP. The main features of this cycle, called the **Cori cycle** (after Carl and Gerty Cori, Nobel Prize, 1947, with B. Houssay), are shown in Figure 17.4.

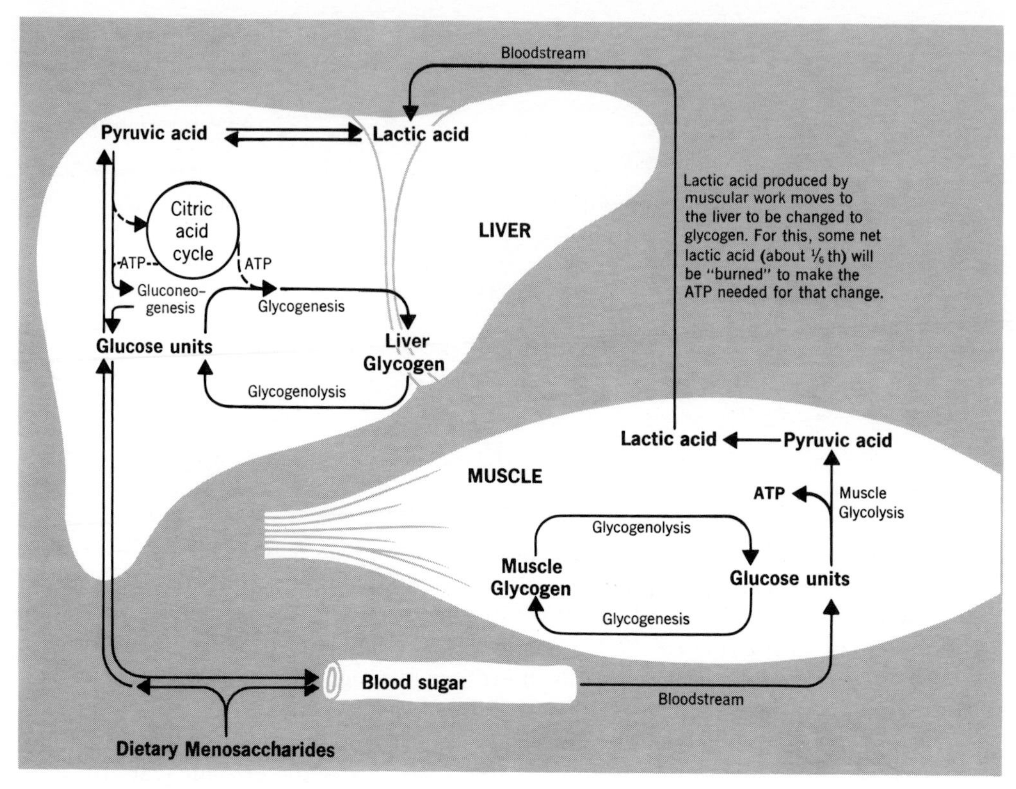

Figure 17.4 The Cori cycle.

17.8 Glycogenesis and Glycogenolysis

If needs for glucose in the bloodstream are met, glucose will be stored as glycogen in the liver and muscles, a reserve for later needs. The synthesis of glycogen is called **glycogenesis** ("glycogen-creation"). When glycogen reserves are used to release glucose, that reaction is called **glycogenolysis** ("glycogen-hydrolysis"). These events are included in Figure 17.4.

Glycogenolysis, the tapping of glycogen reserves for more glucose, may be initiated by a number of conditions—a drop in the level of glucose in the blood, caused by physical exercise or fasting; or release of hormones such as epinephrine, thyroxin, or glucagon. When glycogen reserves play out, the body may make new glucose and it may draw more and more on reserves of fatty acids for energy.

A number of inherited diseases involve the storage of glycogen. In Von Gierke's disease the liver enzyme, glucose 6-phosphorylase, is absent. It catalyzes the hydrolysis of glucose 6-phosphate, which cannot get out of the cell, to inorganic phosphate and glucose, which easily diffuses from the cell. Unless this hydrolysis occurs, glucose units cannot leave the liver. Thus in Von Gierke's disease glucose accumulates in the liver and remains as glycogen in such quantities that the liver becomes very large. At the same time the blood sugar level falls (Section 17.9), glycolysis in the liver accelerates, and the liver releases more and more pyruvate and lactate.

In Cori's disease the liver lacks the enzyme needed to debranch glycogen, an enzyme that catalyzes the hydrolysis of α-1, 6-glycosidic links. The clinical symptoms resemble those of Von Gierke's disease but are less severe.

In McArdle's disease the individual lacks the enzyme needed specifically in muscles to carry out glycogenolysis of muscle glycogen. The patient is not capable of much physical activity but otherwise is well developed and normal.

In Andersen's disease the enzyme for putting together the branches in glycogen is missing in the liver and the spleen. Liver failure from cirrhosis usually causes death by age 2.

17.9 Blood Sugar Level

The concentration of glucose in whole blood remains fairly constant at a value known as the **blood sugar level.**[2] It is expressed in milligrams of glucose per 100 ml of blood. After 8 to 12 hr of fasting, the blood sugar level of a normal adult is usually in the range of 60 to 100 mg/100 ml. This range is called the **normal fasting level.**

Hypoglycemia is a condition wherein the concentration of glucose in the blood is below the normal fasting level. Concentrations of blood glucose above the fasting level constitute a condition of **hyperglycemia.** If the blood sugar level is too high, the kidneys begin to transfer some of the excess glucose to the urine. The blood sugar level above which this occurs is called the **renal threshold** for glucose. It is normally at about 140 to 160 mg of glucose per 100 ml of blood, sometimes higher. Whenever detectable quantities of glucose appear in the urine, a condition of **glucosuria** is said to exist.[3]

The only nutrient that the brain normally uses for energy is glucose, taken directly from the bloodstream. The brain does not store glycogen as do muscles. Severe hypoglycemia, therefore, starves brain cells. Severe disorders of the central nervous system, such as convulsions and shock, can occur. Even mild hypoglycemia may be accompanied by dizziness and fainting spells.

Prolonged hyperglycemia at a glucosuric level is not itself immediately serious, but it indicates that the body is not able to withdraw glucose from the blood in a normal fashion.

The proper control of the blood sugar level by the body is a measure of its **glucose tolerance;** that is, its ability to utilize glucose in a normal way. The blood sugar level is controlled by a number of factors.

1. Cells normally withdraw glucose from circulation at a faster rate the higher the blood sugar level. A mass action effect operates. As tissue cells become "filled" with glucose (or glycogen)—as their needs for glucose are

[2] "Blood sugar" is not actually just glucose; it includes any monosaccharides that can reduce Benedict's reagent. However, most of that mixture is glucose, and we shall use the terms "blood sugar" and "blood glucose" as synonyms.

[3] The terms in this paragraph can be easily mastered if meanings for the following word parts are noted: hyper-, above, excessive (hypertension); hypo-, below, subnormal (hypoglycemia); glyco-, referring to a "glycose" (generic name for a monosaccharide); gluco-, referring to glucose; -emia, referring to blood; and -uria, referring to urine.

met—glucose will be converted to fat. Figure 17.5 illustrates various responses to a rising blood sugar level.
2. A number of hormones participate in regulating the blood sugar level.

Epinephrine is secreted by the adrenal medulla when the individual faces sudden danger or stress and must mobilize energy reserves promptly to meet the emergency. This hormone particularly affects muscle cells where it stimulates glycogenolysis, getting glucose ready for making ATP. The hormone also affects glycogenolysis in the liver, but glucagon is more active in that organ.

Glucagon is secreted by the alpha cells of the pancreas when the blood sugar level drops. By stimulating glycogenolysis in the liver, it helps put more glucose into circulation.

Growth hormone has a number of effects. One is the stimulation of the release of glucagon, which leads to a rise in blood sugar level.

Thyroxine affects several enzyme systems, including the activities of some involved in the oxidation of glucose. To replace the glucose, some glycogenolysis occurs.

A few steroid hormones made in the cortex of the adrenal gland aid in gluconeogenesis.

Insulin acts in several ways but principally by helping get glucose out of the bloodstream. We shall study more about insulin later in this chapter and in the next.

17.10 Cyclic AMP

Epinephrine (a relatively simple molecule) and glucagon (a polypeptide) cannot enter cells. To get their messages inside, therefore, the target cells use

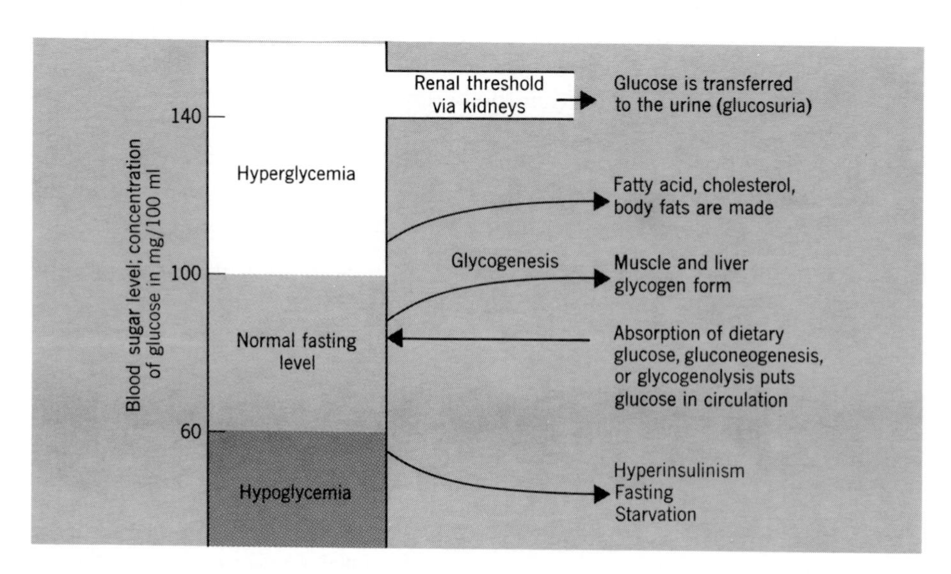

Figure 17.5 Factors affecting the blood sugar level.

an intermediary, a secondary messenger, one of the prostaglandins (Section 13.28). According to current theory, when the primary messenger, the hormone, arrives at the target cell, it stimulates a prostaglandin within the cell membrane to activate an enzyme also in the cell membrane but on the inside. This enzyme, adenyl cyclase, catalyzes the change of some of the cell's ATP to **cyclic AMP.**

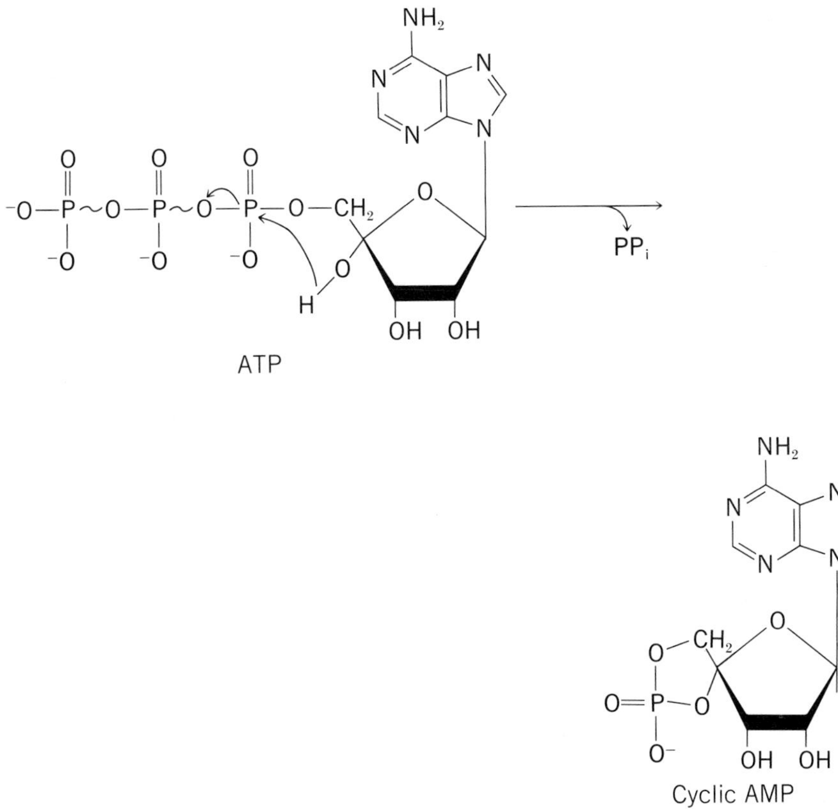

ATP $\xrightarrow{\quad} PP_i$

Cyclic AMP

Cyclic AMP then activates an enzyme, whatever enzyme corresponds to the message delivered by the hormone. These events are illustrated in Figure 17.6.

Besides glucagon and epinephrine, other hormones also work by stimulating the formation of cyclic AMP, including vasopressin, ACTH, and insulin. Some work by a "cascade" effect. The arrival of one molecule of epinephrine, for example, is believed to trigger the release of 30 000 molecules of glucose. The initial message of epinephrine is magnified by the generation of an enzyme activator: cyclic AMP. That enzyme activates another activator and that, in turn, another until the enzyme for glycogenolysis is activated. Each step magnifies the initial impact of epinephrine. The effect of epinephrine in some people is to raise their blood sugar levels above the renal threshold and cause glucosuria.

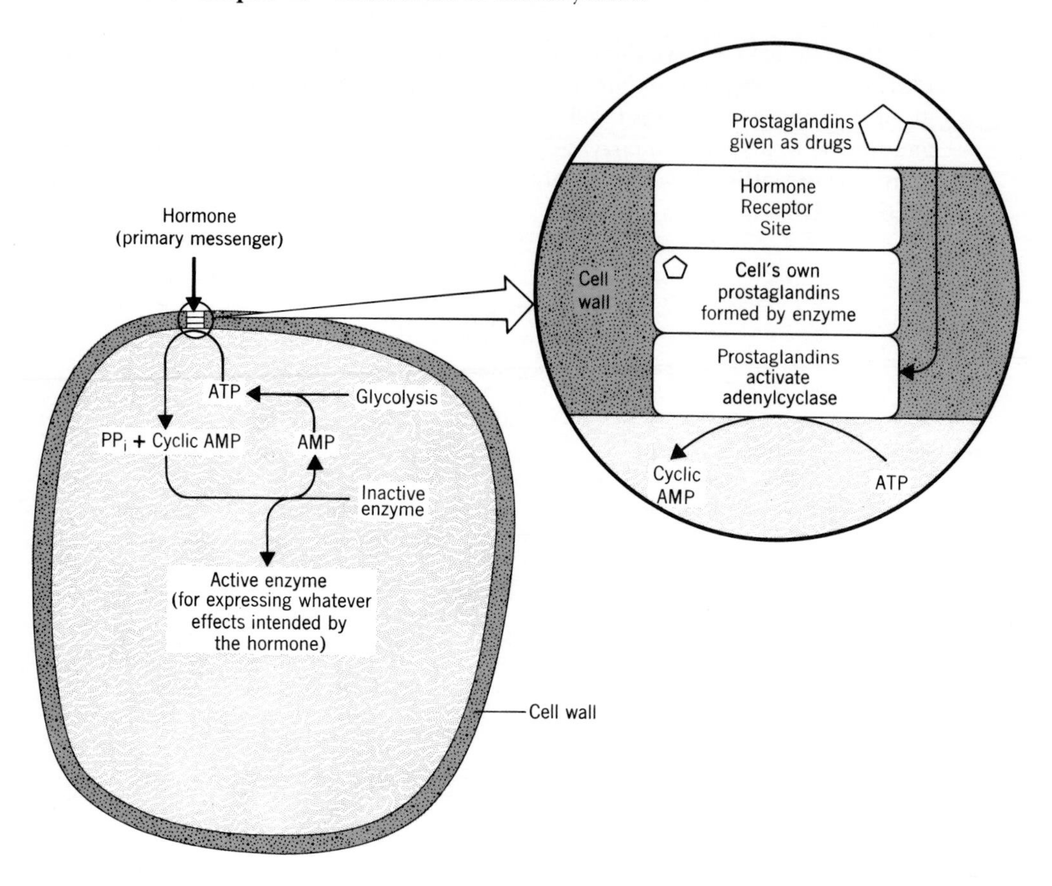

Figure 17.6 Translating the hormone's message into cellular activity through message transmitters—prostaglandins and cyclic AMP.

17.11 Diabetes Mellitus

The name for this disorder comes from the Greek, *diabetes,* to pass, and *mellitus,* honey-sweet, meaning to pass urine containing sugar. We shall call it diabetes for short.

According to the U.S. National Commission on Diabetes about 10 million American citizens are directly affected by diabetes. Its prevalence in the United States increased by more than 50% between 1965 and 1973, and it appears to be increasing at a rate of 6% per year. The average U.S. citizen born today has a higher than one in five chance of getting diabetes, assuming no means of prevention is found. Diabetes and its many complications rank third behind cardiovascular disease and cancer as the cause of death in the United States.

In spite of being known for nearly 4000 years and of being treated by insulin since 1921 so much remains unknown about diabetes that it is still difficult to define. The definitions of most diseases include statements of specific causes, but the cause of diabetes is still not known. The decision to begin treatment for diabetes, therefore, rests on the interpretation of certain clinical observations. Clinically defined, **diabetes** is a disorder associated with a blood sugar level that is too much above normal for the existing metabolic situation of the individual and which is accompanied by a relatively specific vascular disease, microangiopathy. **Microangiopathy** is a change in the thickness, composition, and metabolism of the basement membrane of blood capillaries. (The basement membrane is the protein support structure encasing the single layer of cells of a capillary.) Microangiopathy is believed to lead to kidney problems, other vascular disorders, and eye problems of many diabetics, and it is most easily observed by ophthalmologists during an eye examination. Diabetic microangiopathy in the retina of the eye, called diabetic retinopathy, is the second most common cause of blindness in the United States and is the leading cause of new cases of blindness among adults. In adults retinopathy can be detected before abnormalities in sugar metabolism surface.

17.12 Diabetes and Insulin

In Section 17.9 we learned that insulin is a hormone whose effect is to help get glucose out of circulation and into cells of certain target tissues. The normal signal for the release of this hormone from the beta cells of the pancreas is a rise in the blood sugar level following the digestion and absorption of carbohydrates in the diet. The more glucose pouring into circulation, the more insulin is released. As the blood sugar level drops back down, insulin secretion shuts down.

Not all tissues need insulin to obtain glucose from circulation. The brain, the kidneys, and the intestinal tract do not, for example. The tissue most in need of insulin is adipose tissue.

17.13 Glucose Tolerance Test

The most common means for detecting the response of the pancreas to glucose is the **glucose tolerance test.** The patient receives a large dose of glucose, 100 g or more, and his blood sugar level is monitored over a period of a few hours. An individual with good glucose tolerance—one who handles glucose and presumably releases insulin normally—will have blood sugar levels as shown by the lower curve in Figure 17.7. The upper curve is typical for a patient with moderate to severe diabetes. Specialists emphasize that glucose tolerance data must be interpreted conservatively. Not everyone showing a high blood sugar level has diabetes, and a misdiagnosis can cause grave harm. The release of epinephrine, for example, in a stressful situation may promote enough glycogenolysis to sent a blood sugar level soaring, but that is not diabetes. Some specialists insist that consistent hyperglycemia during fasting must be observed before the diagnosis of diabetes mellitus can be made.

17.14 Types of Diabetes

The more severe form of diabetes generally occurs among those who contract the disease when young, prior to the end of their teens. Called juvenile onset diabetes, it is characterized by a severe or total lack of effective insulin, and administration of insulin is mandatory. Occasionally, an older person will develop this form.

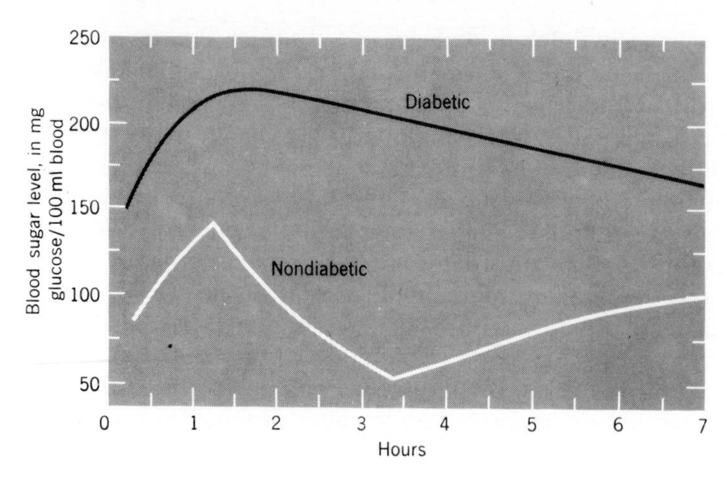

Figure 17.7 Glucose tolerance curves.

The less severe type of diabetes surfaces generally after age 30 or 40 and is called adult onset (or maturity onset) diabetes. About 20% of those with this form require insulin therapy. The rest manage by controlling their diets, by exercising, and sometimes by using a drug that is designed to stimulate the pancreas to release insulin.

17.15 Hyperinsulinism

The clinical definition of diabetes obscures the possibility that quite different mechanisms might be working in the juvenile and the adult-onset forms. A number of significant developments occurred in diabetes research in the early 1970s that are beginning to put some light onto the question of multiple causes of diabetes, clinically defined.

Individuals with juvenile-onset diabetes, about 10% of all diabetics in America, cannot make enough insulin. The problem is a defective pancreas. Why it is defective, of course, is another question, and the answer may be genetic causes. It has long been known that diabetes tends to run in families. Some with the juvenile-onset type may have suffered damage to the pancreas as the result of an infection by mumps virus or measles virus. Either may lodge in the pancreas and after a few years make it unable to manufacture or release insulin.

People with adult-onset diabetes generally make and release normal amounts of insulin—sometimes more than normal amounts—and the problem is insulin resistance by fat and muscle cells. Insulin resistance is the lack of sensitivity of the target cells to the action of insulin. That action is transmitted through the cell wall by an insulin-receptor protein in the outer part of the cell wall. The receptor protein was first isolated from rat liver membranes by scientists at Johns Hopkins University in 1971. Individuals with adult-onset diabetes are believed to have fewer receptor protein molecules per cell than normal. In obese test animals, the higher the concentration of insulin in circulation the fewer the insulin-receptor molecules in liver, fat, muscle, and blood cells. Exceptions to this correlation occur and they are not understood. However, more and more evidence is pointing to the possibility that the high insulin levels *cause* the reduced number of receptors. Insulin does have some ability to catalyze the breakdown of protein, and that may be the reason for the correlation between high insulin levels and reduced receptor molecules.

In the adult-onset diabetic, glucagon activity is usually normal. Since glucagon acts in opposition to insulin—it helps glucose get out of cells, whereas insulin helps glucose get in—the effect of reduced insulin activity in the face of normal glucagon activity is to deplete the target cells of glucose. They must turn to gluconeogenesis and make glucose internally.

About 70% of all adult-onset diabetics are overweight, and obesity may possibly be the cause of their diabetes. The incidence of diabetes is high in overfed, industrialized populations and very low among the undernourished. Overeating, particularly sweets, keeps the pancreas working and the insulin-level up. Perhaps this higher than normal exposure of receptor proteins in target cells to circulating insulin increases the rate of breakdown of the receptors. Then insulin-resistance has developed along with the symptoms of diabetes. The practice, at least in American homes, of consuming relatively large quantities of sugar-coated breakfast cereals and sugar-rich soft drinks, desserts, snacks, and candies not only promotes tooth decay but may also contribute to the incidence of diabetes. Pregnancy and prolonged emotional stress also contribute to the onset of diabetes in some individuals.

The traditional methods of controlling diabetes have been very careful regulation of the diet, exercise, and the administration of insulin. Current research at control is aimed at reducing or eliminating the need for insulin injections. One approach being studied is the implantation of healthy pancreas tissue from a donor. The beta cells may lodge anywhere. Provided they "take" and endure, they may make insulin anew for the patient. Another approach is to have a device implanted that will continuously monitor the blood sugar level and release insulin as needed. Considerable research on both approaches remains to be done.

If not controlled, diabetes eventually leads to metabolic acidosis, and if that is not treated the patient will go into coma and die. The development of acidosis in diabetes involves the metabolism of lipids; we shall delay further discussion of this aspect of the disease until the next chapter.

17.16 Possible Causes of Diabetes

Hyperinsulinism is the appearance of too much insulin in the blood. It may arise from injecting too much insulin or from some overstimulation of the pancreas. It causes excessive removal of glucose from circulation, and brain cells accustomed to using blood sugar for energy suddenly are undernourished. The result is **insulin shock.** The remedy is the immediate ingestion of sugar: syrup, honey, candy bar, or sweetened fruit juice, for example.

Moderate hypoglycemia with less severe disorder to the brain may occur in those who have sugar-rich breakfasts and coffee breaks with little protein or lipid. As the lower curve in Figure 17.7 shows, a strong stimulation to the pancreas can bring the blood sugar level quite low after about 3 hr, low enough to give the brain a problem. The individual may even faint, but usually is groggy, irritable, bleary-eyed, and may fall asleep in class. A sugar-rich breakfast practically guarantees in many people a midmorning "sag," to which most respond by taking a sugar-rich coffee break. That perks them up until lunch, which too often is also poorly balanced nutritionally. A mid-afternoon "sag" then develops that sends them back to yet another sugar-rich coffee break. The dreary sequence continues until evening dinner, very likely the only meal of the day coming close to balance. The prevention is simple: a good, balanced breakfast with moderation in the use of sugar at all times. A breakfast with protein and lipid besides carbohydrate means a slower entry of glucose into circulation and no overstimulation of the pancreas.

Brief Summary

Glycolysis. The anaerobic conversion of one molecule of glucose to two of lactate also generates two molecules of ATP. Therefore, glycolysis permits tissues to operate even when their oxygen supply falls short of normal needs for a period and they run an oxygen debt. Galactose and fructose, other dietary hexoses, enter glycolysis at particular points. Certain yeasts run glycolysis, too, but at the pyruvate stage go on to give carbon dioxide and ethyl alcohol, not lactate.

Pentose Phosphate Pathway. Glucose may be catabolized through pentose intermediates to make NADPH, instead of ATP. The NADPH is used when some biosynthesis of a macromolecule requires a reducing agent.

Gluconeogenesis. Lactate or pieces of amino acids can be converted to glucose by reactions that in all but three steps are the reverse of glycolysis. Gluconeogenesis consumes ATP but, by salvaging chemical energy in lactate (or amino acids), it reserves the potential for making much more ATP when the tissue goes back to an aerobic basis.

Blood Sugar Level. The concentration of glucose in blood, normally 60 to 100 mg/100 ml, is raised by a carbohydrate-rich diet, sometimes to hyperglycemic levels (generally above 100 mg/ml, but that varies with the individual). Occasionally, the renal threshold is exceeded resulting in glucosuria. A rising blood sugar level stimulates the pancreas to release insulin, a polypeptide hormone that helps get glucose molecules out of circulation and into tissue (particularly adipose tissue). The brain gets its chemical energy principally from glucose taken from circulation (not from glycogen), and if the blood sugar level drops to a hypoglycemic level the brain's ability to function may be impaired. Several hormones respond directly or indirectly to a fall in the blood sugar level. Glucagon is one, released by the pancreas. Growth hormone is another. Gluconeogenesis, which will raise the blood sugar level, is stimulated by steroid hormones made in the adrenal cortex. In times of sudden stress epinephrine stimulates glycogenolysis. Epinephrine and glucagon (as well as some other hormones) act by stimulating prostaglandin in the target cell wall to activate adenyl cyclase. This enzyme catalyzes the conversion of ATP to cyclic AMP, which in turn activates an enzyme that finally expresses within the cell the "message" delivered to the cell by the hormone.

Diabetes Mellitus. Clinically, an insufficiency of active insulin, diabetes shows itself in poor glucose tolerance, a blood sugar level at glucosuric heights, and problems in the vascular compartment. In treating diabetes with insulin, an overdose must be avoided to prevent insulin shock. (The story of diabetes is continued in Chapter 18.)

Selected References

Books

1. A. L. Lehninger. *Bioenergetics,* 2nd ed. W. A. Benjamin, Inc., Menlo Park, Calif., 1971.
2. L. Stryer. *Biochemistry.* W. H. Freeman and Company, Publishers, San Francisco, 1975.
3. O. B. Crofford, Committee Chairman. *Report of the National Commission on Diabetes to the Congress of the United States,* Vol. 1, 10 December 1975. Early chapters provide statistical data on diabetes as well as surveys of possible causes and future directions for research.

Articles

1. I. Pastan. "Cyclic AMP." *Scientific American,* August 1972, page 97. How cyclic AMP may set in motion increases in levels of hormonal activities.
2. E. W. Sutherland. "Studies on the Mechanism of Hormone Action." *Science,* 4 August 1972, page 401. The Nobel Prize lecture of Sutherland, the discoverer of cyclic AMP. Three dozen known effects of cyclic AMP are tabulated.
3. T. H. Maugh, II. "Hormone Receptors: New Clues to the Cause of Diabetes." *Science,* 16 July 1976, page 220.

Questions and Exercises

1. In what way is glycolysis important?
2. Why does glycolysis under anaerobic conditions end in lactate, not pyruvate?
3. What substances are missing in the following equation? Also, balance the equation.

$$\underset{\text{Glucose}}{C_6H_{12}O_6} + 2ADP + \underline{\qquad\qquad} \longrightarrow \underset{\text{Lactate}}{2C_3H_5O_3^-} + \underline{\qquad\qquad} + 2ATP$$

4. To hold the pH steady as glycolysis occurs, what specific substance produced by glycolysis must be neutralized?
5. The pentose phosphate pathway uses $NADP^+$, not NAD^+. What forms from the $NADP^+$ and how does the body use that product (in general terms)?
6. What are two other names for the pentose phosphate pathway?
7. Why is the central nervous system rather quickly affected by hypoglycemia?
8. If the blood sugar level drops, what does glucagon do?
9. If glycogen reserves are depleted and the blood sugar level is low, what does the liver do (and to some extent the kidneys also)?

10. In a period of relatively brief fasting will the system normally go hypoglycemic? Explain.
11. During severe exercise an oxygen debt accumulates. Explain what that means and how it happens.
12. In a period of prolonged fasting or starvation what does the system do to try to maintain its blood sugar level?
13. How is a rising blood sugar level kept under control?
14. Amino acids are not excreted. They are not stored in the way glucose units can be stored as glycogen. What probably happens to the excess amino acids on a high protein diet in an individual not exercising very much?
15. What apparently stimulates the pancreas to release insulin? Glucagon?
16. What is the principal effect of insulin in circulation?
17. How do juvenile and adult-onset diabetes often differ?
18. List some factors that apparently contribute to the onset of diabetes.
19. Graphically compare the results of glucose tolerance tests on two individuals, one with a healthy glucose tolerance and the other with diabetes.
20. What can happen if too much insulin is injected (or secreted)?
21. What service does epinephrine perform and what triggers it?
22. What part does cyclic AMP have in the work of certain hormones?
23. Why does severe hyperinsulinism lead quickly to "insulin shock," whereas the absence of insulin, at least for a period, is not so serious?
24. Is glucosuria always caused by diabetes? If not, how else might it arise?

chapter 18

Metabolism of Lipids

During long, sustained periods of high energy demand, as in this New York marathon, the body turns to stored fat. How that works is studied in this chapter. (Photo by Gale Constable/Duomo.)

Storage of Lipids

18.1 Adipose Tissue

Lipids are deposited in nearly all parts of the body, but a specialized connective tissue called **adipose tissue** is the principal "depot." It occurs as subcutaneous tissue in the abdominal area and around certain organs (e.g., kidneys). "Depot fat" located in these areas cushions organs against sharp jolts and bumps and insulates them against sudden variations in temperature. In addition to giving passive insulation, active metabolic processes in the cells of this tissue generate a small amount of heat that helps offset losses of heat when the outside temperature drops.

Wherever adipose tissue is found, it is quite active metabolically. Its cells have large numbers of mitochondria, and nerves and capillaries extend in and among them. One of the problems associated with obesity is the extra load placed on the heart both to move the extra weight around and to pump blood throughout a much larger and more widely extended capillary network.

18.2 Lipid Reserves

After a few hours without food, the body's glycogen reserves are gone. The lipids in actual circulation plus the blood sugar could sustain metabolic activity only a few minutes. The fat in storage in the liver is good for about an hour. Other larger reserves are obviously needed, and the body normally has two. One is the unabsorbed food still in the digestive tract. The other is adipose tissue. The average adequately nourished adult human male has enough lipid in reserve to sustain life for 30 to 40 days, assuming that he has enough water. (In an abnormal situation such as starvation, amino acids released from tissue proteins constitute a third reserve of chemical energy.)

The lipids on deposit in adipose tissue come largely from the carbohydrates and the lipids of the diet. About 30% of dietary carbohydrate is converted to fat. The conversion of glucose into fatty acid costs some energy (ATP), and about a quarter of the glucose supply is sacrificed to furnish it. If 30% of dietary carbohydrate is changed to lipid, and if another 25% is used to power this change, a significant portion of all oxygen consumption and heat production in the body goes into just this operation.

Distinct advantages exist for the individual in storing chemical energy as lipid rather than as carbohydrate. Lipid has a much higher **energy density.** It stores more calories per gram of tissue. To store glucose as wet glycogen, it takes about 0.6 g to yield 1 kcal. But when glucose is converted into fat, only about 0.11 g is needed to yield 1 kcal.

When our energy requirements are "balanced" by our food intake, the amount of lipid present in the body remains fairly constant. This does not mean that depot fat just "rests" in inert storage. A constant coming and going of the molecules of depot fat occurs. Newly arrived molecules are deposited in storage as those that have been in storage are removed.

Exercise 18.1 To survive for 40 days on the energy of triacylglycerols alone, consuming an average of 2500 kcal/day, how much triacylglycerol would have to be in storage (in kilograms)?

Exercise 18.2 If we could not store triacylglycerols and had to store chemical energy as wet glycogen, how much wet glycogen would we have to have in reserve to supply the energy of Exercise 18.1 (in kilograms)?

18.3 Catabolism of Lipids

To mobilize the energy reserves in adipose tissue, several steps must take place; they are outlined in Figure 18.1.

1. Triacylglycerols in adipose tissue (or other "depots") must be hydrolyzed to produce free fatty acids and glycerol. "Free fatty acids" mean largely nonesterified acids. They may be bound to plasma proteins, however, as **lipoprotein complexes.** The lipase that acts to hydrolyze triacylglycerols in adipose tissue is activated by another enzyme that, in earlier turn, is activated by cyclic AMP (Section 17.10). Cyclic AMP forms in this situation when any one of a few hormones, including epinephrine and glucagon (Section 17.9) arrive at adipose cells. Insulin, on the other hand, inhibits the formation of cyclic AMP in adipose tissue. Hence insulin suppresses the hydrolysis of triacylglycerols, presumably because insulin makes glucose available for energy, and fatty acids are therefore not needed. The glycerol produced by the hydrolysis of depot fat is changed to dihydroxyacetone phosphate, which enters the glycolysis pathway (Section 17.1).

2. The free fatty acid molecules are carried to the liver, the principal site of fatty acid catabolism.

3. Fatty acids are broken down two carbons at a time to provide $FADH_2$,

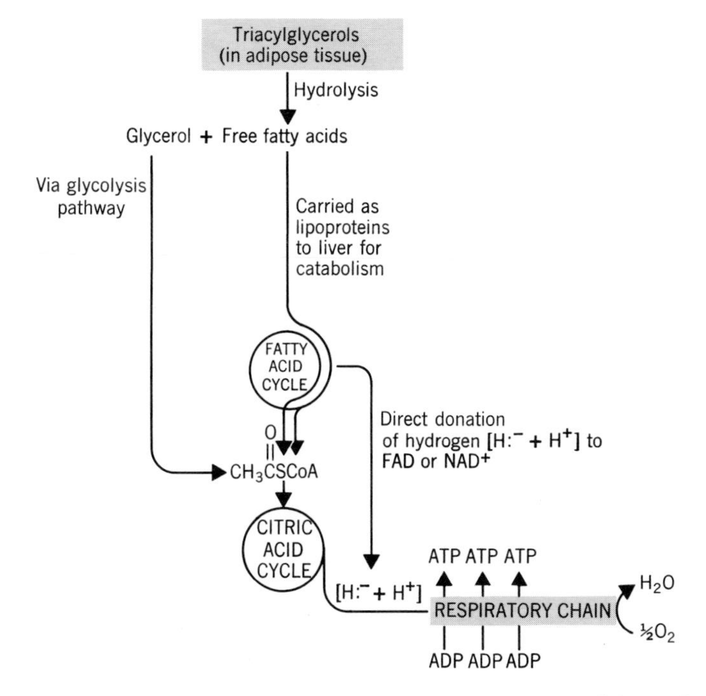

Figure 18.1 Mobilization of energy reserves in triacylglycerols of adipose tissue.

NADH, and acetyl coenzyme A. All these provide chemical energy for making ATP by the respiratory chain.

Although the main site of fatty acid oxidation is the liver, fatty acids and smaller fragments produced from fatty acids can be used for energy in the heart and in skeletal muscles. In fact, most of the energy needs of resting muscle are met by intermediates from fatty acid catabolism, not from glucose. The degradation of fatty acids occurs by a repeating series of steps known as the **fatty acid cycle** (Figure 18.2).

18.4 Fatty Acid Cycle. Beta Oxidation

Fatty acids are oxidized inside mitochondria, but they cannot get inside (especially when long-chain) until first activated by linking to coenzyme A. Activation consumes one molecule of ATP, but the return of this energy investment is huge, as we shall see. Catalyzed by acyl CoA synthetase, the activation occurs in two steps. In the first (Equation 18.1), the carboxylate ion of the fatty acid displaces pyrophosphate from ATP, one of the instances in which ATP breaks down to give PP_i instead of simply P_i. (The formation of cyclic AMP was another.)

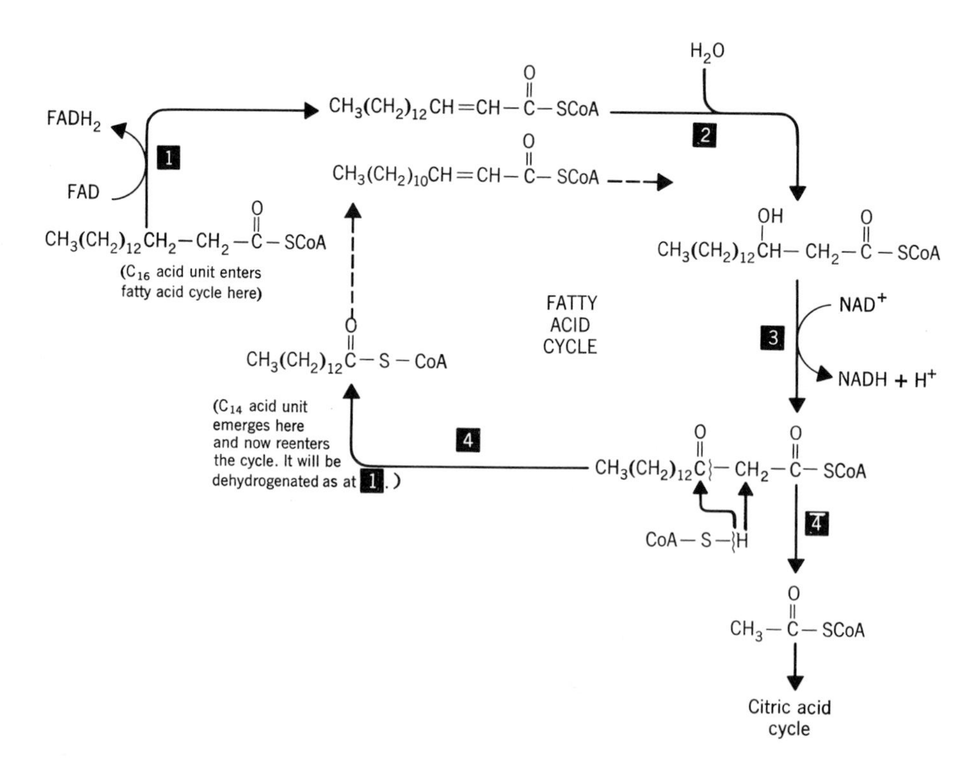

Figure 18.2 Beta oxidation by the fatty acid cycle. (Boxed numbers correspond to numbered discussion in Section 18.4.)

$$R{-}CO_2^- + ATP \rightleftharpoons R{-}\overset{\overset{\displaystyle O}{\|}}{C}{\sim}AMP + PP_i \xrightarrow[H_2O]{} 2P_i \qquad (18.1)$$

Carboxylate ion (from fatty acid) — Acyl adenylate

In the second step (Equation 18.2), coenzyme A displaces AMP from the acyl adenylate.

$$R{-}\overset{\overset{\displaystyle O}{\|}}{C}{\sim}AMP + H{-}S{-}CoA \rightleftharpoons R{-}\overset{\overset{\displaystyle O}{\|}}{C}{-}S{-}CoA + AMP \qquad (18.2)$$

Acyl CoA

Acyl CoA, once formed, transfers the acyl group to a substance in the mitochondrial membrane, which passes it inside the cell, where it is rejoined to a different molecule of coenzyme A and oxidized.

The oxidation of a fatty acyl unit occurs by a sequence of four steps that degrade the unit by two carbons at a time, producing one molecule of $FADH_2$, one of NADH, and one of acetyl coenzyme A in each sequence. The sequence reoccurs and the long chain of the fatty acyl unit is broken down until no more two-carbon units can be taken. $FADH_2$ and NADH directly fuel the respiratory chain and the acetyl groups enter the citric acid cycle. The four steps are identical in kind, but each reoccurrence involves molecules of two fewer carbons. Although the whole series is not exactly a cycle, we still call it the fatty acid cycle.

The steps in the fatty acid cycle are as follows, where the numbers refer to parts of Figure 18.2.

1. The first step is dehydrogenation.

$$CH_3(CH_2)_{12}\overset{\beta}{C}H_2{-}\overset{\alpha}{C}H_2{-}\overset{\overset{\displaystyle O}{\|}}{C}{-}SCoA + FAD \xrightarrow{\boxed{1}}$$

Palmitoyl coenzyme A

$$CH_3(CH_2)_{12}CH{=}CH{-}\overset{\overset{\displaystyle O}{\|}}{C}{-}SCoA + FADH_2 \longrightarrow [H{:}^- + H^+] \qquad (18.3)$$

An α,β-unsaturated acyl derivative of coenzyme A

($FADH_2$ → FAD; $[H{:}^- + H^+]$ → Respiratory chain: $\frac{1}{2}O_2$ → H_2O, 2ATP)

Several flavoprotein enzymes, each with a riboflavin unit, act as the catalysts, each handling an acyl coenzyme A unit of a different chain length. This step leads to two ATP units via the respiratory chain.

2. The second step is hydration of the double bond.

$$CH_3(CH_2)_{12}CH{=}CH{-}\overset{\overset{O}{\|}}{C}{-}SCoA + H_2O \xrightarrow{\boxed{2}} CH_3(CH_2)_{12}\overset{\overset{OH}{|}}{C}H{-}CH_2{-}\overset{\overset{O}{\|}}{C}{-}SCoA \tag{18.4}$$

β-Hydroxyacyl derivative of coenzyme A

3. The third step is another dehydrogenation; a 2° alcohol is oxidized to a keto group. The NAD^+-bearing enzymes handle this and the hydrogen removed is sent into the respiratory chain to drive the formation of three ATPs as shown in Equation 18.5:

$$CH_3(CH_2)_{12}\overset{\overset{OH}{|}}{C}H{-}CH_2{-}\overset{\overset{O}{\|}}{C}{-}SCoA + NAD^+ \xrightarrow{\boxed{3}}$$

$$CH_3(CH_2)_{12}\overset{\overset{O}{\|}}{C}{-}CH_2{-}\overset{\overset{O}{\|}}{C}{-}S{-}CoA + NADH + H^+ \tag{18.5}$$

β-Keto acyl coenzyme A

$NADH + H^+ \rightarrow NAD^+$ and $[H{:}^- + H^+]$; $[H{:}^- + H^+]$ → Respiratory chain; $\frac{1}{2}O_2 \rightarrow H_2O$; → 3ATP

The first three steps oxidize the beta position of the original acyl unit; therefore, the fatty acid cycle is often called **beta oxidation.**[1] The result of these steps is to weaken the bond between the alpha and the beta carbons.

4. The fourth step breaks off an acetyl coenzyme A unit.

$$CH_3(CH_2)_{12}\overset{\overset{O}{\|}}{C}{-}CH_2{-}\overset{\overset{O}{\|}}{C}{-}SCoA \xrightarrow[\text{CoA—S—H}]{\boxed{4}} CH_3(CH_2)_{12}\overset{\overset{O}{\|}}{C}{-}SCoA + CH_3{-}\overset{\overset{O}{\|}}{C}{-}SCoA \tag{18.6}$$

Myristoyl coenzyme A; Acetyl coenzyme A → Citric acid cycle → Respiratory chain → 12 ATP

The remaining acyl unit, the original now shortened by two carbons, undergoes the same sequence of dehydrogenation, hydration, dehydrogenation and cleavage. After seven repetitions of this sequence one molecule of palmitoyl coenzyme A will be broken into eight molecules of acetyl coenzyme A.

Table 18.1 summarizes the yield of ATP from the oxidation of one palmitoyl unit. Since the recovery of ATP from AMP requires input of two high-

[1] Because Franz Knoop pioneered the research that led to our present understanding of beta oxidation, the fatty acid cycle is also called "Knoop oxidation."

TABLE 18.1
Yield of ATP from Palmitoyl CoA Catabolized by the Fatty Acid Cycle

Seven Turns of the Cycle Produce:	ATP from Each Energy-Rich Intermediate	Total ATP Produced
$7FADH_2$	2	14
7NADH	3	21
$8CH_3\overset{O}{\overset{\|}{C}}SCoA$	12	96
		131 ATP
Deduct two high-energy phosphate bonds for activating acyl unit		−2
Net ATP yield per palmitoyl unit		129 ATP

energy phosphate bonds, the initial investment of one ATP to activate the acyl unit actually is equivalent to investing two ATP in the overall balance. The net yield is 129 ATP per molecule of palmitic acid.

Exercise 18.3 If the free energy available by the direct hydrolysis of ATP is 7.3 kcal/mole, and that available by the direct combustion of palmitic acid is 2400 kcal/mole, with what efficiency does the fatty acid cycle convert the free energy of palmitic acid to ATP?

18.5 Catabolism of Unsaturated Fatty Acids

Unsaturated acids are catabolized by reactions very similar to those of the saturated acids. At one stage an extra enzyme, an isomerase, is needed to catalyze the change in location of a double bond. We shall not study further details. Some unsaturated acids are essential to good health. See Section 20.7.

18.6 Anabolism of Lipids

We learned in an earlier chapter that acetyl coenzyme A stands at a major metabolic crossroads. It may originate in the catabolism of amino acids, glucose (and other sugars), and fatty acids. If the system needs ATP, acetyl coenzyme A is degraded by the citric acid cycle to provide fuel for the respiratory chain. However, when ATP demand is low (little exercise) and food intake is relatively high (overeating), the system has a problem. It cannot (normally) excrete glucose, fatty acids, or amino acids. Excess fatty acids in the diet are directly stored as fat; some glucose may be used to restore glycogen levels; if needed, tissues are rebuilt and repaired from the amino acids. That still usually leaves excess glucose and amino acids. To handle these, the system eventually degrades them to acetyl coenzyme A and then makes new fatty acid molecules that are put into storage in adipose tissue. An outline of these relationships is given in Figure 18.3.

18.7 Lipigenesis Cycle

Whereas fatty acids are catabolized inside mitochondria, they are made outside these organelles in the fluid of the cytoplasm (the cytosol). As a general rule nature segregates sequences of catabolism from those of anabolism. Even though some steps in anabolism appear to be the reverse of corresponding steps in catabolism, other steps are decidedly different and use

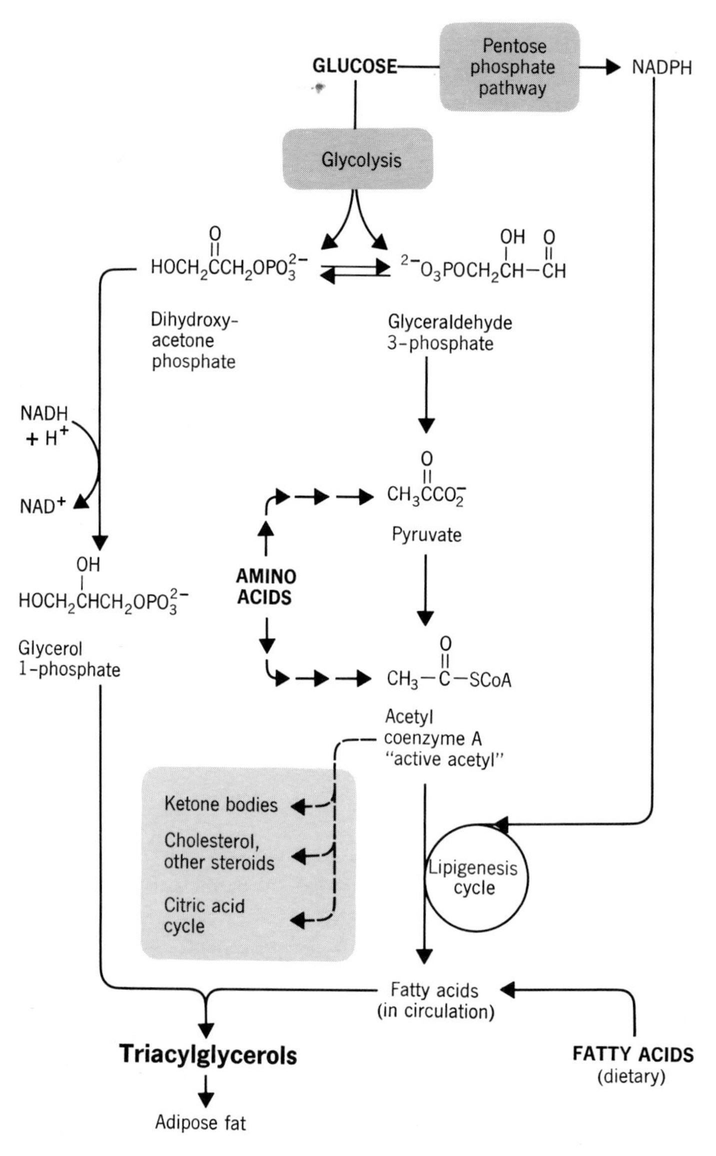

Figure 18.3 Principal sources of triacylglycerols in adipose tissue.

different enzyme systems. One such uniquely different step in fatty acid anabolism is the temporary use of carbon dioxide. Acetyl coenzyme A is carboxylated according to Equation 18.7, in which one high-energy phosphate is invested.

$$CH_3\overset{O}{\overset{\|}{C}}-SCoA + ATP + CO_2 \longrightarrow {}^-O-\overset{O}{\overset{\|}{C}}-CH_2-\overset{O}{\overset{\|}{C}}-SCoA + ADP + P_i + H^+ \quad (18.7)$$

Malonyl CoA

The product is malonyl CoA. Because its $-CH_2-$ group is flanked by two carbonyl groups, it is activated especially well for the next step, a reaction that will pin an acetyl group to the $-CH_2-$ group as carbon dioxide splits out. The next step strongly resembles the Claisen condensation (Section 7.22), but before it can happen the acyl units must be shifted from coenzyme A to a different enzyme called acyl carrier protein, ACP. Fatty acid anabolism is thus fully segregated from catabolism. The next step continues with acetyl ACP and malonyl ACP, and the condensation of these two, Equation 18.8, makes something like a β-keto ester. The driving force for the reaction is the loss of carbon dioxide.

$$CH_3-\overset{O}{\overset{\|}{C}}-ACP + {}^-O-\overset{O}{\overset{\|}{C}}-CH_2-\overset{O}{\overset{\|}{C}}-ACP \xrightarrow[\text{condensation}]{CO_2} CH_3-\overset{O}{\overset{\|}{C}}-CH_2-\overset{O}{\overset{\|}{C}}-ACP + ACP \quad (18.8)$$

Acetyl-ACP, Malonyl-ACP, Acetoacetyl ACP

The product of Equation 18.8 is a β-ketoacyl derivative. Its keto group is next hydrogenated, not by NADH but rather by NADPH made in part by the pentose phosphate pathway (Section 17.4). This hydrogenation, Equation 18.9, otherwise looks just like the reverse of Equation 18.5.

$$CH_3-\overset{O}{\overset{\|}{C}}-CH_2-\overset{O}{\overset{\|}{C}}-ACP \xrightarrow[\text{hydrogenation}]{NADPH + H^+ \rightarrow NADP^+} CH_3-\overset{OH}{\overset{|}{C}H}-CH_2-\overset{O}{\overset{\|}{C}}-ACP \quad (18.9)$$

β-Hydroxybutyryl ACP

Dehydration, Equation 18.10, next inserts a double bond.

$$CH_3-\overset{OH}{\overset{|}{C}H}-CH_2-\overset{O}{\overset{\|}{C}}-ACP \xrightarrow[\text{dehydration}]{} CH_3-CH=CH-\overset{O}{\overset{\|}{C}}-ACP + H_2O \quad (18.10)$$

Crotonyl ACP

The double bond is reduced by NADPH to give a butyryl derivative, Equation 18.11.

$$CH_3-CH=CH-\overset{\displaystyle O}{\overset{\|}{C}}-ACP \xrightarrow[\text{hydrogenation}]{NADPH + H^+ \quad NADP^+} CH_3-CH_2-CH_2-\overset{\displaystyle O}{\overset{\|}{C}}-ACP \quad (18.11)$$

Butyryl ACP

The overall effect is the elongation of one acetyl unit by two carbons. The sequence reoccurs as a cycle, called the **lipigenesis cycle.** The butyryl unit next condenses with a malonyl unit as carbon dioxide splits out to give a six-carbon keto acyl unit (as in Equation 18.8). This is hydrogenated, dehydrated, and hydrogenated, as in Equations 18.9 through 18.11, to give a hexanoyl system. The sequence once again occurs to make an octanoyl unit, and so on until the C_{16} palmitoyl unit is made. Any further elongation or any insertions of double bonds to make the unsaturated fatty acids is accomplished by other enzyme systems.

Up to the palmitoyl stage the synthesis of fatty acids is catalyzed by a complex of enzymes called fatty acid synthetase. The growing acyl unit swings on the ACP unit from one active site to another in this multienzyme complex.

As indicated in Figure 18.3, acetyl units that are not needed to make ATP not only are used to make fatty acids, but also are routinely used to synthesize the steroid hormones and cholesterol. Dietary cholesterol is not necessarily the culprit in rising levels of cholesterol in the blood. We make our own, and if we do that too much, then we are advised to reduce the dietary intake to help bring the blood cholesterol under control. Cholesterol biosynthesis involves nearly three dozen steps, which are beyond the scope of this text.

A third use of acetyl units not needed for ATP is the synthesis of a group of compounds, the ketone bodies, and with these we return to complete our study of the molecular basis of diabetes mellitus.

18.8 Formation of Ketone Bodies

In a well-nourished, healthy individual on a day-to-day caloric balance between energy intake and outgo, the catabolisms of carbohydrates and fatty acids are in balance. Acetyl coenzyme A made by the fatty acid cycle goes into the citric acid cycle as rapidly as it is made. If, however, carbohydrate metabolism goes awry, acetyl coenzyme A made from fatty acids tends to spill over into the other pathways indicated in Figure 18.3. One leads to metabolic acidosis. Let us see how.

Acetyl coenzyme A enters the citric acid cycle only as the carrier, oxaloacetate, is available (see Figure 16.5). The supply of oxaloacetate declines in two disorders involving carbohydrates. Both force the system to make new glucose—to gluconeogenesis—which requires oxaloacetate (Section 17.6). In one disorder, starvation, the blood sugar level drops and the body struggles to make its own glucose from carbon skeletons of amino acids. In the other disorder, diabetes mellitus, the blood sugar level is high, but glucose does not get into certain cells without insulin. To compensate for the lack of insulin, the body tries to make even more glucose. The reason is possibly to get a glucose concentration gradient high enough between the bloodstream and tissue so that glucose cannot help but diffuse even without insulin. Gluconeogenic activity is therefore high in diabetes as well as in starvation. At the same time, with glucose less available for energy, fatty acids are more rapidly catabolized in an effort to generate the needed ATP. The steps in the fatty acid cycle given by Equations 18.3 and 18.5 provide fuel for the respiratory chain. They also lead to acetyl coenzyme A, Equation 18.6, and that leads to trouble in diabetes.

When gluconeogenesis drains the supply of oxaloacetate, more and more acetyl coenzyme A will be processed in the liver to make a group of three compounds collectively known as the **ketone bodies:** acetoacetic acid and β-hydroxybutyric acid (both as anions) and acetone. One, acetoacetate, is the source of the other two. Acetoacetate is made in three steps from acetyl coenzyme A. The overall equation, without going into the step-by-step details, is given by Equation 18.12.

$$2CH_3\overset{O}{\overset{\|}{C}}{-}SCoA + H_2O \xrightarrow[\text{(three steps)}]{} \underset{\text{Acetoacetate}}{CH_3\overset{O}{\overset{\|}{C}}CH_2\overset{O}{\overset{\|}{C}}O^-} + 2CoASH + H^+ \tag{18.12}$$

Since acetoacetate is the salt of a β-keto acid (Section 7.21), it can spontaneously decarboxylate, giving acetone (Equation 18.13).

$$\underset{\text{Acetoacetate}}{CH_3{-}\overset{O}{\overset{\|}{C}}{-}CH_2{-}\overset{O}{\overset{\|}{C}}{-}O^-} + H_2O \longrightarrow \underset{\text{Acetone}}{CH_3{-}\overset{O}{\overset{\|}{C}}{-}CH_3} + HCO_3^- \tag{18.13}$$

Acetoacetate can also be hydrogenated by NADH to give β-hydroxybutyrate (Equation 18.14).

$$\underset{\text{Acetoacetate}}{CH_3{-}\overset{O}{\overset{\|}{C}}{-}CH_2{-}\overset{O}{\overset{\|}{C}}{-}O^-} \xrightarrow{NADH + H^+ \quad NAD^+} \underset{\beta\text{-Hydroxybutyrate}}{CH_3{-}\overset{OH}{\overset{|}{C}H}{-}CH_2{-}\overset{O}{\overset{\|}{C}}{-}O^-} \tag{18.14}$$

Acetoacetate and β-hydroxybutyrate leave the liver and enter general circulation. Skeletal muscles use acetoacetate to obtain much of their ATP for "resting" functions. Some tissues, for example, heart muscle, use acetoacetate and β-hydroxybutyrate for energy in preference to glucose. In starvation even the brain, given time, can adapt from using glucose to using these two ketone bodies for energy.

18.9 Ketosis

If disorders in carbohydrate metabolism go uncorrected for a prolonged period of time, the level of the ketone bodies in the bloodstream gradually rises. Normally, the levels of acetoacetate and β-hydroxybutyrate in blood, in units of micromoles/100 ml, are 2 and 4, respectively. In prolonged, undetected and untreated diabetes these values can increase as much as 200-fold. The condition of excess levels of ketone bodies in blood is called **ketonemia.**

The kidneys work to remove ketone bodies and put them into the urine as ketonemia advances. The condition of higher than normal concentrations of ketone bodies in urine is called **ketonuria.**

At high ketone body levels in blood the lungs pick up enough acetone to be detected by its odor on the breath, a condition we may call **"acetone breath."**

The combined occurrence of ketonemia, ketonuria, and acetone breath is called **ketosis.** Unchecked ketosis will lead to a general breakdown and death, as we shall now study.

18.10 Ketoacidosis

Acetoacetate and β-hydroxybutyrate are negative ions. To remove them at the kidneys, positively charged ions must also be removed to preserve electrical neutrality. Thus ketosis means particularly the loss of sodium ion, because it is the predominant metallic ion in blood (see Figure 15.4). The loss of sodium ion from the blood also reduces the availability of its counter ion in the chief blood buffer, the bicarbonate ion. The loss of bicarbonate ion means the loss of base and an increase in hydrogen ion in the blood—acidosis. The increased occurrence of the reactions of gluconeogenesis and the fatty acid cycle also produce protons that eventually get the better of the blood buffers. Thus untreated ketosis, becoming gradually worse, leads inevitably to acidosis—**ketoacidosis** to indicate the particular origin.

18.11 Diabetic Coma

Both the hyperglycemia and the ketosis in untreated diabetes cause physiological disturbances. As the kidneys struggle to remove both glucose and ketone bodies, they not only must remove positive ions, but they also must remove more water than is normal. All the ions and the glucose must be in the dissolved state as they are removed; the more solute removed, the more solvent (water) is needed. Nitrogen wastes (urea) form more rapidly as amino acids are metabolized to help make new glucose. The elimination of these wastes also requires water. Untreated diabetics commonly have huge thirsts and void large volumes of urine each day. If the thirst is not sufficient to bring in enough water, then the extracellular fluids including blood suffer a general loss in water and become more concentrated and more viscous (thick). The delivery of blood throughout the system thus becomes more difficult. Since the brain must have blood, some blood flow is diverted away from the kidneys, which only makes their task harder and ketosis continues to advance. Sometimes the individual involuntarily uses vomiting to get rid of some of its body acid in an effort to reduce acidosis, and vomiting means further loss of water. If dehydration begins to deplete fluids from intracellular spaces, the result can be coma. The series of events in diabetes are outlined in Figure 18.4.

The exact cause of diabetic coma has been attributed in the past variously to dehydration, circulatory failure, some poisoning action of the ketone bodies in brain cells, acidosis, and impaired handling of oxygen of brain cells.

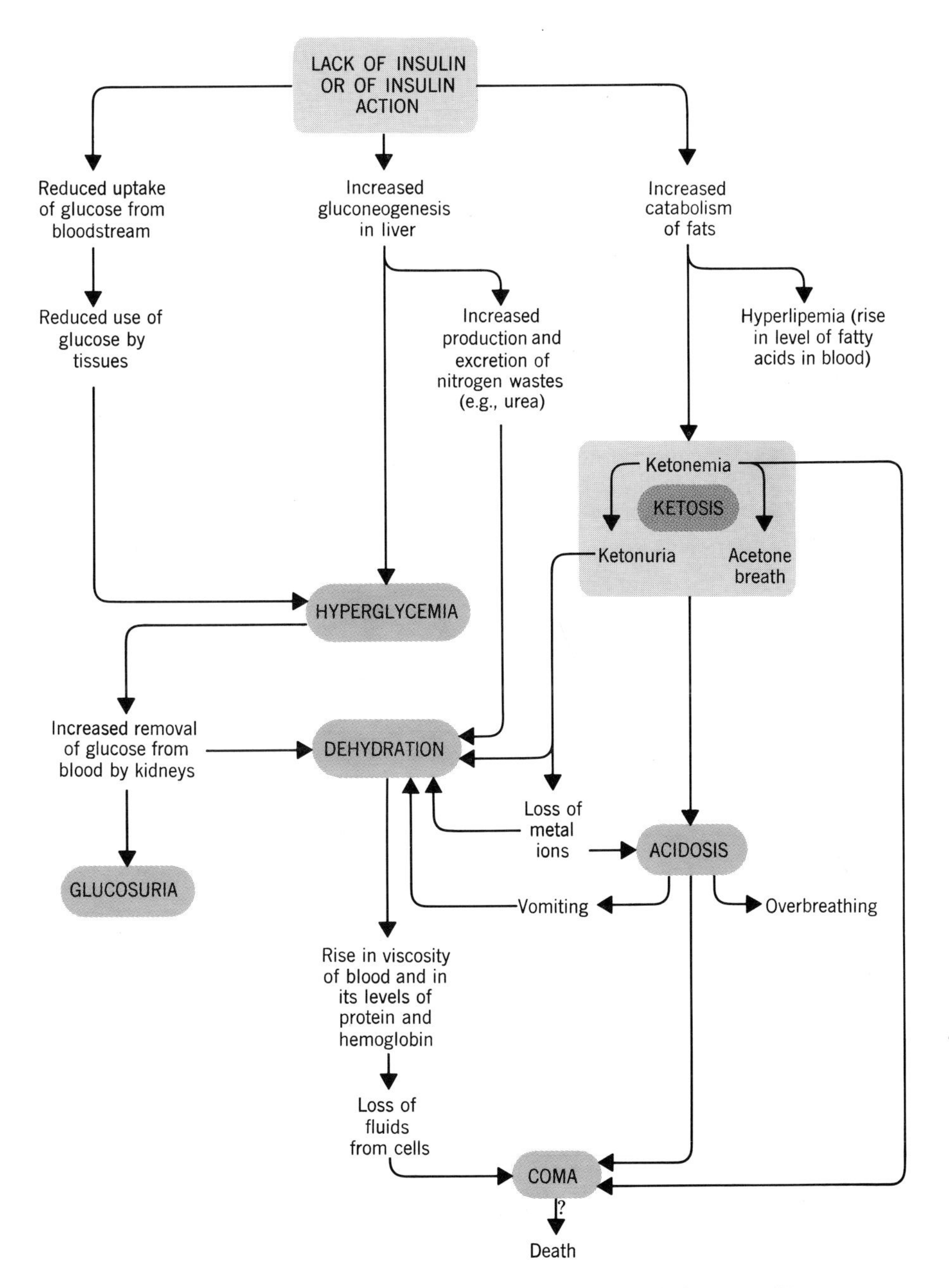

Figure 18.4 Principal sequence of events in uncontrolled diabetes mellitus.

Apparently, no single mechanism can be blamed, and the exact cause of the loss of consciousness is not known. Oxygen delivery in blood to the brain is essentially unimpaired in diabetes, but oxygen consumption by the brain is reduced almost by half in advanced stages of the untreated disease. Severe dehydration occurs in cholera without producing coma. Deliberate injection of isotonic acetoacetate in animals fails to cause coma, but hypertonic solutions will. No one cause of diabetic coma, in other words, has been proved. A combination of factors apparently causes diabetic coma, particularly the poorer use of both oxygen and glucose by the brain, cellular dehydration, and possibly the prolonged presence of acetoacetate.

The precoma as well as the coma stages of untreated diabetes are genuine medical emergencies requiring prompt action. Treatment involves the start of insulin therapy; efforts to correct ketoacidosis by intravenous feeding of fluids containing sodium ion, chloride ion, and lactate ion (to supply base); and the emptying of the stomach. When acidosis is particularly severe, isotonic sodium bicarbonate may be administered to resupply buffer.

Normally, the level of bicarbonate ion in blood is 22 to 30 millimoles/liter. (For bicarbonate ion, 22 millimoles are equivalent to 1.85 g sodium bicarbonate.) This drops to 16 to 20 millimoles in mild acidosis, 10 to 16 in moderate acidosis, and below 10 millimoles bicarbonate ion/liter in severe acidosis.

Acidosis is serious for many reasons, one being that it disrupts the mechanism for transporting oxygen (Section 15.11). In moderate to severe acidosis, the difficulties in taking in oxygen at the lungs give the individual severe "air hunger." Breathing becomes painful and difficult, a condition called dyspnea.

18.12 Lactic Acid Acidosis and Athletic Accomplishment

Anaerobic glycolysis is an important backup source of ATP when a cell does not receive oxygen at a rate to sustain aerobic catabolism of glucose. The end-product of anaerobic glycolysis is lactate. Hydrogen ions are also produced; they lower the bicarbonate level in the blood and cause acidosis, specifically **lactic acid acidosis.** Acidosis from any source is acidosis; it is the nature of the treatment that varies. The lactate level in blood may rise from its resting level of 1 to 2 millimoles/liter to 10 to 12 millimoles/liter after hard work. This will cut the bicarbonate level roughly in half, to a level of 12 to 14 millimoles/liter, meaning that the moderate form of acidosis results. All athletes have experienced the violent dyspnea, the difficult, painful, gulping air hunger that accompanies maximum effort in a contest. When the lactate level is about 10 millimoles/liter, further work is rendered virtually impossible regardless of "will power." This is obviously a limiting factor in the further improvement of athletic records.

Most young people acquire the physical stamina and endurance usually attributed to middle age by the time they are 25 to 30 years old. Not being athletes, they undervalue fitness. Being physically fit and being an athlete are not the same, however. All sports require talents and skills, and some sports require great muscle power. But to anyone, athlete or otherwise, the heart is the most important muscle in the body. Good lung capacity is also essential for fitness. If you improve your heart and your lungs—your entire respiratory system—you will have developed the endurance that will permit you to operate aerobically as you engage in an active life and save your anaerobic reserve for those intense, short-interval bursts you sometimes need. To improve your heart and lungs (and keep them in shape), you must put demands on them (working up slowly and carefully if you are now out of shape). Running, swimming, and biking, or playing games that involve these activities as well as a good, balanced diet are your best insurance for physical stamina all your life.

Brief Summary

Lipid Storage. Fatty acids not needed for energy or to make other substances (e.g., steroid hormones and cholesterol) are not excreted but are stored instead in adipose tissue. Carbohydrates and amino acids not needed elsewhere are converted to fat and stored also. Storing chemical energy as fat is the most energy-dense, lowest-weight means possible.

Catabolism of Lipids. When needed for energy or biosynthesis, adipose fat is hydrolyzed. Its glycerol is sent into the glycolytic pathway and its fatty acids are carried on proteins in blood to the liver where they are activated and systematically broken apart, two carbons at a time, by the fatty acid cycle (beta oxidation). Each turn of the cycle has four steps: dehydrogenation, hydration, dehydrogenation, and cleavage of acetyl coenzyme A. Each turn produces one $FADH_2$ and one NADH for the synthesis of ATP by the respiratory chain. Each turn also produces one acetyl coenzyme A, which normally goes into the citric acid cycle for making ATP. (The last turn makes two acetyl coenzyme A.) The net ATP production is 129 ATPs per palmitoyl unit.

Anabolism of Lipids. A cycle of reactions—the lipigenesis cycle—builds fatty acids from acetyl units using carbon dioxide temporarily to help activate acetyl units. Of course, ATP is also consumed. Nonsaponifiable lipids, steroids, are also made from acetyl units.

Ketone Bodies. Acetoacetate, β-hydroxybutyrate, and acetone build up in the blood—ketonemia—in starvation or diabetes. These two disorders take so much oxaloacetate for gluconeogenesis that too little is left to handle incoming acetyl units (from fatty acid catabolism) by taking them into the citric acid cycle. The excess acetyl units increasingly condense to give acetoacetate from which the other two ketone bodies are made. Acetoacetate and β-hydroxybutyrate are normal sources of energy in some tissue.

Ketoacidosis and Diabetes. The increased generation of ketone bodies leads to depletion of the carbonate buffer and acidosis. Complicating this is the loss of fluid at the kidneys as excess glucose (in diabetes), ketone bodies, and nitrogen wastes are washed out of the bloodstream together with positive metallic ions (principally Na^+) to keep overall electrical neutrality. The loss of metal ions further depletes the carbonate buffer of blood. As water leaves, the blood becomes more viscous and concentrated, hindering the delivery of blood. Cells may begin to lose fluid. The impairment of blood flow, loss of cellular fluid, continuous contact with acetoacetate by brain cells, interference with oxygen consumption caused in part by acidosis—all these factors probably combine to cause diabetic coma.

Selected References

Books

1. A. E. Renold. "Intermediary Metabolism in Diabetes Mellitus." Chapter 7 in *The Pathology of Diabetes Mellitus,* S. Warren, P. M. Le Compte, and M. A. Legg, eds. Lea & Febiger, Philadelphia, 1966.
2. W. G. Oakley, D. A. Pyke, and K. W. Taylor. *Clinical Diabetes and Its Biochemical Basis.* Blackwell Scientific Publications, Oxford, 1968.
3. J. F. Fazekas and R. W. Alman. *Coma. Biochemistry, Physiology and Therapeutic Principles.* Charles C Thomas, Publisher, Springfield, Ill., 1962.

4. B. S. Leibel and G. A. Wrenshall, eds. *On the Nature and Treatment of Diabetes.* Excerpta Medica Foundation, Amsterdam, The Netherlands, 1965.
5. L. Stryer. *Biochemistry.* W. H. Freeman and Company, Publishers, San Francisco, 1975.

Articles

1. M. M. Grant and W. M. Kubo. "Assessing a Patient's Hydration Status." *American Journal of Nursing,* August 1975, page 1306. Factors affecting hydration and symptoms associated with changes.
2. H. W. Ryder, H. J. Carr, and P. Herget. "Future Performance in Footracing." *Scientific American,* June 1976, page 109. The physiological limits have not been reached; the barriers are largely psychological.

Questions and Exercises

1. What are the functions of adipose tissue?
2. Compare and contrast the availability of chemical energy stored as glycogen and as fat.
3. In what three ways is glucose used to make triacylglycerols?
4. Discuss the mobilization of chemical energy from fat. How is the citric acid cycle necessary to it?
5. Starting with the coenzyme A derivative of butyric acid, write equations for all the reactions that take place as it is degraded through the fatty acid cycle to acetyl coenzyme A units.
6. What are the four principal ways by which acetyl coenzyme A is used?
7. Under what circumstances will amino acids be used to make glucose?
8. Why does the liver make "new" glucose in diabetes?
9. How does starvation cause ketosis?
10. Describe how the lack of insulin may lead to acidosis and, thence, to death. Pay particular attention to the connection between insulin and ketosis, and between ketosis and acidosis.
11. What causes lactic acid acidosis?
12. What happens during a rest period to correct lactic acid acidosis?

chapter 19

Metabolism of Proteins

Sleeked out and hunkered down, this cat is after meat. Whether animals and humans get their proteins from meat or from vegetables, we all need amino acids. What we do with them is studied in this chapter. (Photo by Walter Chandoha.)

19.1 Absorption and Uses of Amino Acids

The end-products of protein digestion, amino acids, are rapidly transported across the walls of the small intestine. Some very small, simple peptides are also absorbed. Individual amino acids may be used in one of the following ways.

1. To synthesize new tissue protein, to repair old, or to replace proteins of body fluids that have broken down.
2. To make nonprotein compounds that contain nitrogen, such as nucleic acids, heme, and creatine.
3. To provide chemical energy and be catabolized. Intermediates in amino acids catabolism may enter the citric acid cycle or they may be used to make glucose (gluconeogenesis) or fatty acids, which may be stored in adipose tissue. The principal end-products of the complete catabolism of amino acids are carbon dioxide, water, and urea.

19.2 Nitrogen Pool

All amino acids wherever they are found in the body plus other simple nitrogenous substances make up the **nitrogen pool.** Amino acids may enter this pool both from the diet and from proteins of the body tissues or those in body fluids; they constantly undergo degradation and resynthesis. Most tissue proteins are in a dynamic state. They experience a constant turnover of the amino acids of which they are made. The turnover is fairly rapid among proteins of the liver and blood plasma; it is very slow among muscle proteins. The interrelations that involve the nitrogen pool are summarized by the diagram of Figure 19.1.

19.3 Nitrogen Balance

A healthy, adequately nourished person who excretes as much nitrogen per day as he takes in by the diet is on a **nitrogen balance.** During the growing

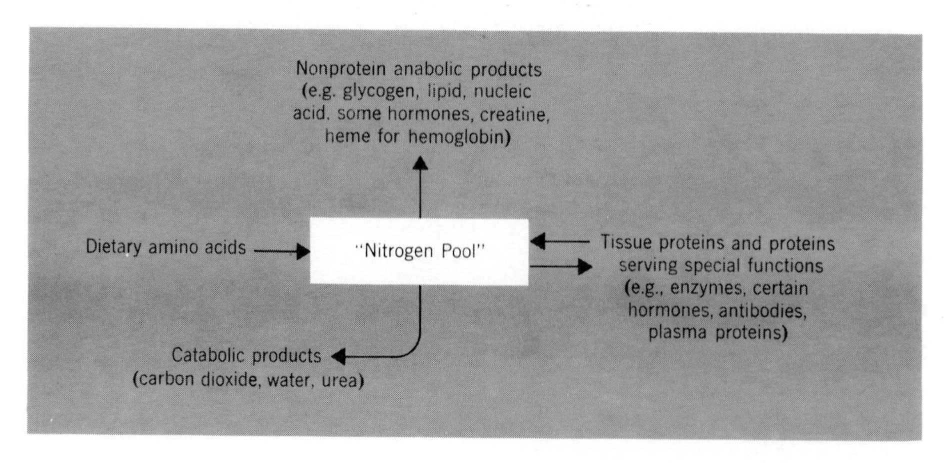

Figure 19.1 The nitrogen pool.

years the individual should on the average be on a positive nitrogen balance—taking in more nitrogen than excreting. In wasting diseases and those involving heavy gluconeogenesis the individual will be on a negative nitrogen balance as tissue proteins deliver up their amino acids for emergency uses and their nitrogen must be excreted in various forms.

19.4 Catabolism of Amino Acids

Amino acids not needed for biosyntheses are catabolized. Their amino groups emerge as parts of molecules of urea, which is excreted in the urine. Their carbon skeletons are processed and they eventually emerge, depending on the original amino acid, in acetyl CoA, acetoacetyl CoA, pyruvate, or an intermediate in the citric acid cycle—α-ketoglutarate, succinyl CoA, fumarate, or oxaloacetate.

The amino acids that lead to acetyl CoA or acetoacetyl CoA are called **ketogenic amino acids** (Table 19.1) because ketone bodies can be made from them. Amino acids leading to pyruvate or intermediates in the citric acid cycle are called **glucogenic amino acids** (Table 19.1) because their carbon skeletons can become parts of glucose molecules made by gluconeogenesis. Whenever full operations of the citric acid cycle and the respiratory chain are not needed to make ATP, one intermediate, oxaloacetate, together with pyruvate can be used to make glucose. (See Figure 17.3.) As Table 19.1 shows, some amino acids are both ketogenic and glucogenic. Only one, leucine, is exclusively ketogenic.

We cannot examine the catabolism of every amino acid. Lysine and tryptophan each take over a dozen steps to become acetoacetyl CoA. Certain kinds of reactions, however, commonly occur; we shall touch briefly on four—transamination, oxidative deamination, direct deamination, and decarboxylation. The liver is the major site of all amino acid catabolism.

TABLE 19.1
Amino Acids as Raw Materials for Glycogen and Fatty Acids

Glucogenic Amino Acids	
Alanine	Histidine
Arginine	Methionine
Aspartic acid	Proline
Asparagine	Serine
Cysteine	Threonine
Glutamic acid	Tryptophan
Glutamine	Valine
Glycine	
Ketogenic Amino Acids[a]	
Isoleucine	Phenylalanine
Leucine	Tyrosine
Lysine	

[a] All but leucine are also glucogenic.

19.5 Transamination

One of the amino acids, glutamic acid, is at a crossroads in processing the amino groups of other amino acids. Glutamic acid can be made in the body from an intermediate in the citric acid cycle, α-ketoglutarate. A number of other amino acids can transfer their amino groups to this intermediate in a complex reaction called **transamination,** Equation 19.1. Enzymes for this (19.1)

$$\underset{\alpha\text{-Amino acid}}{R{-}\underset{\displaystyle NH_3^+}{\underset{|}{CH}}{-}CO_2^-} + \underset{\alpha\text{-Keto glutarate}}{^-O_2CCH_2CH_2\overset{\displaystyle O}{\overset{\|}{C}}{-}CO_2^-} \longrightarrow$$

$$\underset{\text{New keto acid}}{R{-}\overset{\displaystyle O}{\overset{\|}{C}}{-}CO_2^-} + \underset{\text{Glutamic acid}}{^-O_2CCH_2CH_2\underset{\displaystyle NH_3^+}{\underset{|}{C}}HCO_2^-} \quad (19.1)$$

reaction are called transaminases and they carry one of the B-vitamins, pyridoxal, as a cofactor.

Exercise 19.1 Write the structures of the keto acids that would form from the transamination of each. (Refer to Table 12.1 as needed for structures of amino acids.)
(a) Phenylalanine (b) Leucine (c) Valine

19.6 Oxidative Deamination

The amino group of glutamic acid can be removed by a combined oxidation and hydrolysis in a reaction called **oxidative deamination,** Equation 19.2.

$$\underset{\text{Glutamate}}{^-O_2CCH_2CH_2\underset{\displaystyle NH_3^+}{\underset{|}{C}}HCO_2^-} + NAD^+ + H_2O \longrightarrow$$

$$\underset{\alpha\text{-Keto glutarate}}{^-O_2CCH_2CH_2\overset{\displaystyle O}{\overset{\|}{C}}CO_2^-} + \underset{\text{Ammonium ion}}{NH_4^+} + NADH + H^+ \quad (19.2)$$

The first step is dehydrogenation to give an imine—a compound with a carbon-nitrogen double bond, Equation 19.3.

Next, water adds to the double bond to give an unstable intermediate having both —OH and —NH_2 joined to the same carbon. Ammonia splits out from this to give the keto group in α-ketoglutarate, Equation 19.4. Ammonia

$$^{-}O_2CCH_2CH_2\underset{}{\overset{NH_2}{\overset{|}{C}}}H{-}CO_2H + NAD^+ \longrightarrow {}^{-}O_2CCH_2CH_2\overset{NH}{\overset{\|}{C}}{-}CO_2H + NADH + H^+ \qquad (19.3)$$

Glutamic acid (shown in nondipolar ionic form) — An imine

$$\underset{\text{Imine}}{H{-}\ddot{N}{=}C<} + \underset{\text{Water}}{H{-}\ddot{O}{-}H} \longrightarrow H_2\ddot{N}{-}\underset{|}{\overset{|}{C}}{-}\ddot{O}H \longrightarrow \underset{\text{Ketone}}{{-}\overset{}{C}({=}O){-}} + \underset{\text{Ammonia}}{NH_3} \qquad (19.4)$$

is changed into urea (Section 19.10) unless it is needed for other purposes. The overall funneling of several amino acids through the α-ketoglutarate–glutamic acid switch may be represented in the following flow scheme.

α-Amino acid → α-Keto acid; α-Ketoglutarate → Glutamate; Glutamate → α-Ketoglutarate; $NAD^+ + H_2O$ → $NADH + H^+ + NH_4^+ \longrightarrow$ Urea

Playing a small role in the oxidative deamination of α-amino acids, particularly lysine, is an amino acid oxidase having FMN as the hydrogen acceptor.

Amino groups not alpha to a carboxyl may be oxidized by still other enzymes called the monoamine oxidases and diamine oxidases, which use oxygen and generate ammonia and hydrogen peroxide. Some hormones (e.g., epinephrine) and drugs are processed in the liver by **amine oxidases.** The hydrogen peroxide is decomposed by the action of a special enzyme, catalase, to water and oxygen (Equation 19.5). Some drugs, such as amphetamines, inhibit the monoamine oxidases.

$$2H_2O_2 \xrightarrow{\text{catalase}} 2H_2O + O_2 \qquad (19.5)$$

19.7 Direct Deamination

Serine and threonine may both be changed to corresponding keto acids by first dehydrating to an unsaturated amino acid (Equation 19.6), which then rearranges as shown to an imine. The imine changes to the keto acid in a manner similar to that given in Equation 19.4. The overall result is a **direct deamination** (top of next page).

Exercise 19.2 Threonine dehydrase catalyses the conversion of threonine to α-ketobutyric acid. Assuming that the intermediate steps are analogous to those of serine (Equation 19.6), write the steps in this change of threonine.

$$HO{-}CH_2{-}\underset{NH_3^+}{\underset{|}{CH}}{-}CO_2^- \xrightarrow{-H_2O} CH_2{=}\underset{NH_2^+}{\underset{|}{C}}{-}CO_2^- \longrightarrow CH_3{-}\underset{NH}{\underset{||}{C}}{-}CO_2^- + H^+ \quad (19.6)$$

Serine — An imine

Hydrolysis as in Equation 19.4

$$CH_3{-}\overset{O}{\overset{||}{C}}{-}CO_2^-$$

Pyruvate

19.8 Decarboxylation

Some of the amines the body needs are made by decarboxylating amino acids. The synthesis of the hormone epinephrine, for example, begins with tyrosine. In the second step, **DOPA** (dihydroxyphenylalanine) is decarboxylated to dopamine; in two more steps epinephrine emerges.

Tyrosine → DOPA (Dihydroxyphenylalanine) $\xrightarrow{-CO_2}$ Dopamine $\xrightarrow{(O)}$ Norepinephrine $\xrightarrow{\text{Methylation}}$ Epinephrine

Tyrosine $\xrightarrow{-CO_2}$ Tyramine $\xrightarrow{(O)}$ Dopamine

An interesting sidelight involving DOPA is the fact that the level of its corresponding amine, dopamine, in the brain is reduced in Parkinson's disease. One successful treatment has been the oral administration of DOPA, which the system then decarboxylates and thereby helps bring back the level of dopamine in the brain.

19.9 Catabolism of Some Individual Amino Acids

We have just seen how some amino acids catabolize. The following are a few additional examples of how various amino acids enter ketogenic or glucogenic pathways. Transamination means the transfer of the amino group to the keto group of an alpha keto acid, usually to α-ketoglutaric acid (Section 19.5) or, when urea is being made, to oxaloacetic acid (Section 19.10).

Alanine

Transamination gives pyruvate that may go into the citric acid cycle, into gluconeogenesis, or be used for the biosynthesis of fatty acids.

$$CH_3CH(NH_3^+)CO_2^- \xrightarrow{\text{transamination}} CH_3C(=O)-CO_2^- \longrightarrow \text{Acetyl CoA} \longrightarrow \text{Citric acid cycle}$$

Alanine → Pyruvate → Gluconeogenesis

Aspartic acid

Transamination gives oxaloacetate, an intermediate in both gluconeogenesis and the citric acid cycle.

$$^-O_2CCH_2CH(NH_3^+)CO_2^- \xrightarrow{\text{transamination}} {^-O_2CCH_2C(=O)-CO_2^-} \longrightarrow \text{Citric acid cycle}$$

Aspartic acid → Oxaloacetate → Gluconeogenesis

Cysteine

The sulfhydryl group is oxidized to the sulfinic acid group, HO_2S- (one we have not studied, but bearing some resemblance to a carboxylic acid group, HO_2C-). The amino group is transferred, which leaves a keto group beta to the sulfinic acid unit. Sulfur dioxide splits out to make pyruvate. (Had the keto group been beta to a carboxyl group, carbon dioxide would have split out; the events are very similar.) The sulfur dioxide is then oxidized to sulfate ion, which is excreted by the kidneys.

$$HSCH_2CH(NH_3^+)CO_2^- \xrightarrow{(O)} HOS(=O)CH_2CH(NH_3^+)CO_2^- \xrightarrow{\text{transamination}} H-O-S(=O)-CH_2C(=O)CO_2^-$$

Cysteine → Cysteine sulfinic acid → β-Sulfinylpyruvate

$$\beta\text{-Sulfinylpyruvate} \longrightarrow [SO_2] \xrightarrow{(O)} SO_4^{2-}$$

$$\beta\text{-Sulfinylpyruvate} \longrightarrow CH_3C(=O)CO_2^- \text{ (Pyruvate)} \longrightarrow \text{Citric acid cycle, Gluconeogenesis}$$

Figure 19.2 puts these syntheses into the context of other metabolic pathways.

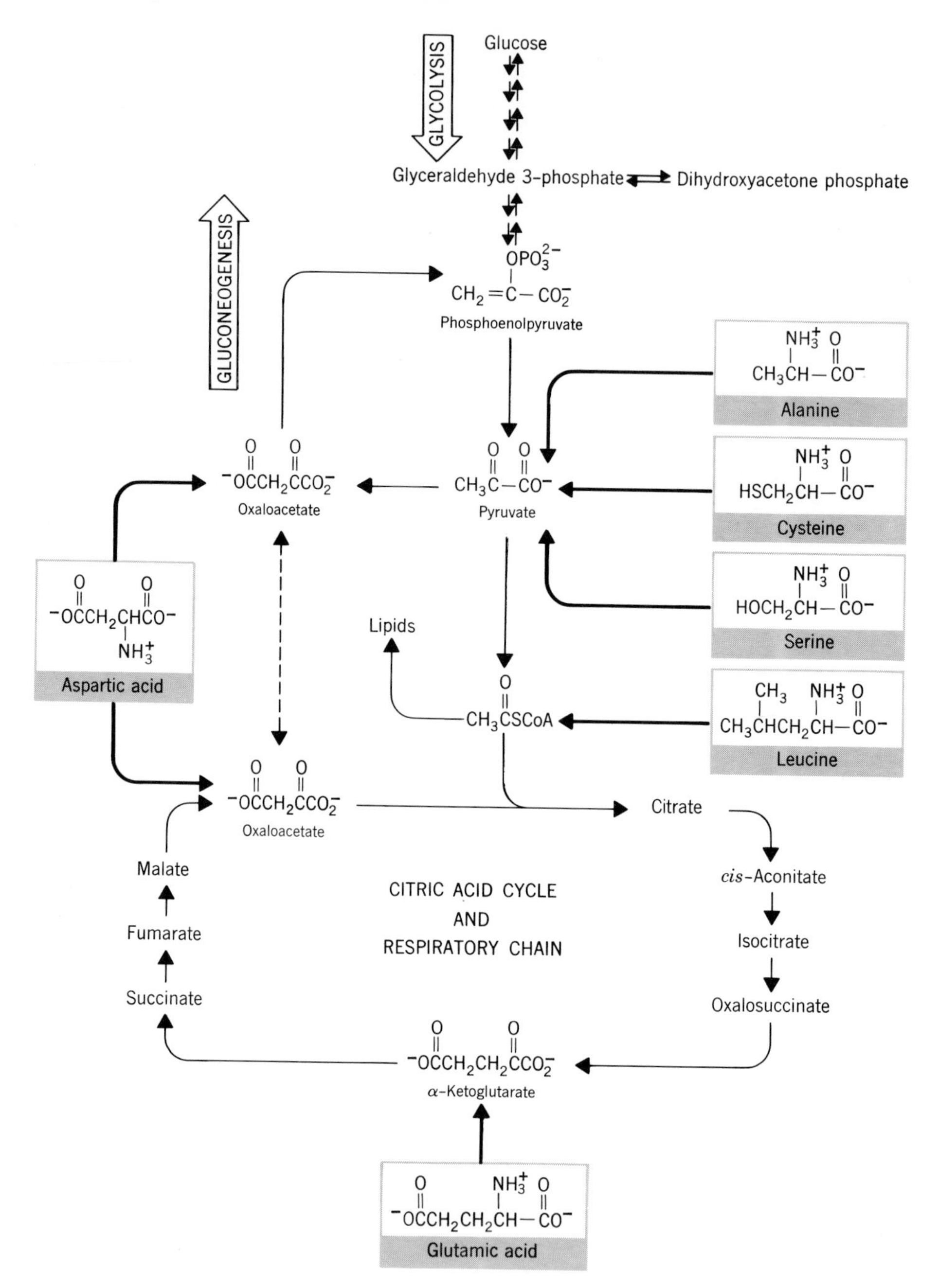

Figure 19.2 Catabolism of some amino acids—illustrative examples. Leucine is the only exclusively ketogenic amino acid. In humans there can be no net production of glucose via gluconeogenesis from acetyl coenzyme A. No extra oxaloacetate can be made from this compound.

19.10 Nitrogen Wastes—Urea

Urea, the principal nitrogen waste produced by the body, takes its nitrogen chiefly from amino acids and partly from the pyrimidines of nucleic acids (thymine and cystosine—T and C). Uric acid (or its anion, the urate ion) is the nitrogen waste from the purines of nucleic acids, adenine and guanine (A and G).

One nitrogen in urea comes from ammonia (or the ammonium ion) produced by oxidative deaminations. The other nitrogen comes from an amino acid, aspartic acid, that switches amino groups by transaminations involving oxaloacetate of the citric acid cycle. Urea emerges from a cycle of chemical reactions called the **urea cycle** or, sometimes, the Krebs' ornithine cycle, after the same scientist who later put the finishing touches on the citric acid cycle. Figure 19.3 gives most of the details. The principal steps are numbered to correspond to the following discussion.

1. Ammonia is activated in the form of carbamoyl phosphate, a high-energy mixed anhydride.
2. Carbamoyl group transfers to the carrier unit, ornithine, taking phosphate bond energy and rejecting P_i. Citrulline forms.
3. Citrulline condenses with the alpha amino group of aspartic acid to give argininosuccinate.
4. A fumarate forms from the original aspartic acid molecule as the amino group stays with the arginine that emerges. Fumarate is an intermediate in the citric acid cycle, and by the series of steps shown it is reconverted to aspartic acid at the expense of an amino group taken from some amino acid. Thus amino groups are fed through the aspartic acid—oxaloacetate switch—to urea.
5. Arginine is hydrolyzed. Urea forms and ornithine is regenerated to start another turn of the cycle.

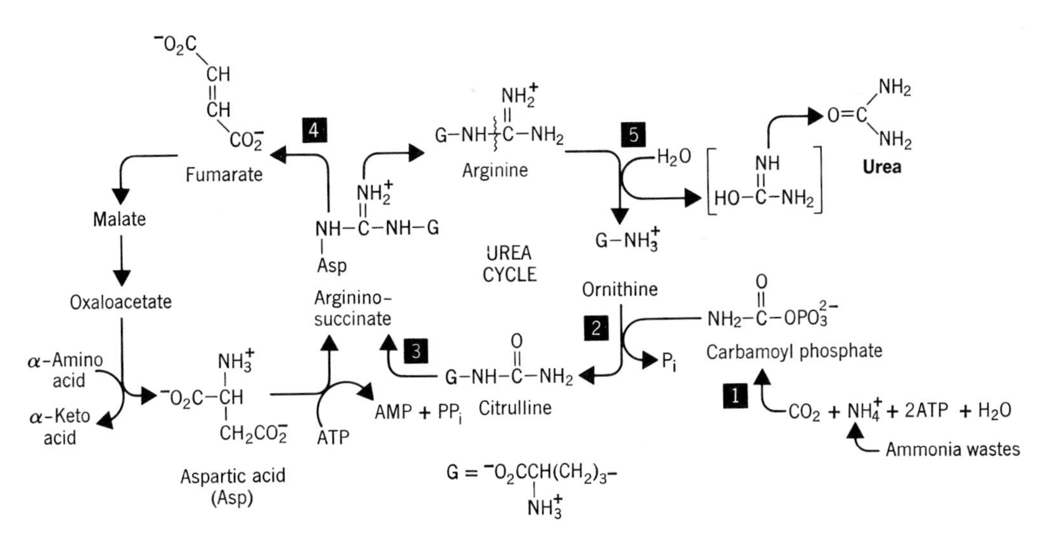

Figure 19.3 Urea Cycle. Boxed numbers refer to the discussion in Section 19.10.

The overall result of the urea cycle is given by Equation 19.7, which, as Figure 19.3 makes clear, is extremely oversimplified.

$$2NH_3 + H_2CO_3 \longrightarrow NH_2{-}\overset{\displaystyle O}{\overset{\|}{C}}{-}NH_2 + 2H_2O \tag{19.7}$$

Some inherited genetic defects produce enzymes for this cycle that have reduced activity. In such individuals the level of ammonium ion in blood rises, a condition called **hyperammonemia.** This ion is toxic, and hyperammonemia means mental retardation for some. Infants with this problem improve on low-protein diets.

19.11 Nitrogen Wastes—Uric Acid

The purine bases of the nucleic acids are catabolized to **uric acid.** We shall illustrate with adenosine monophosphate only. The numbered steps below correspond to steps in Figure 19.4.

1. A transamination removes the amino group of AMP.
2. Ribose phosphate is removed (and will enter the pentose phosphate pathway). The product is hypoxanthine.
3. An oxidation produces xanthine. (This intermediate also occurs in the catabolism of guanosine phosphate, another purine derivative in nucleic acid chemistry.)

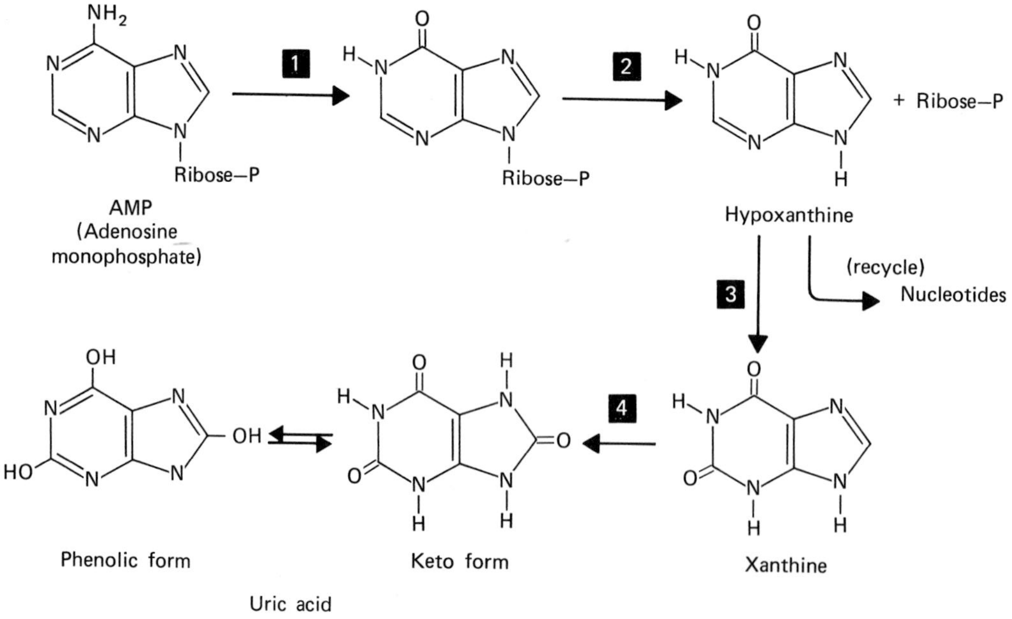

AMP (Adenosine monophosphate)

Hypoxanthine + Ribose–P

(recycle) Nucleotides

Xanthine

Keto form

Phenolic form

Uric acid

Figure 19.4 Catabolism of purine bases, illustrated by adenine in AMP. Boxed numbers refer to the discussion in Section 19.11.

4. Another oxidation produces the keto form of uric acid, which exists partly as the phenolic form. Uric acid has an acid dissociation constant of about 4×10^{-6} and therefore exists as a sodium salt, sodium urate, in the body.

In the disease known as **gout** the rate of formation of sodium urate is increased and crystals of this salt precipitate in joints causing painful inflammation and leading to arthritis. Kidney stones may form. Why sodium urate production increases is not known, but genetic factors may be responsible. Normally, some of the hypoxanthine made in step 2, above, is recycled back to nucleotide bases needed to make nucleic acids or high-energy phosphates. Some individuals with gout are known to have a partial deficiency of the enzyme system required for that recycling of hypoxanthine. Hence most if not all of their hypoxanthine ends up as more sodium urate than is normal.

In the Lesch-Nyhan syndrome, the enzyme for recycling hypoxanthine is totally lacking. The result is both bizarre and traumatic. Infants with the Lesch-Nyhan syndrome develop compulsive, self-destructive behavior at the age of two or three. Unless their hands are wrapped in cloth, they will bite themselves to the point of mutilation; they act with dangerous aggression toward others; and some become spastic and mentally retarded. Kidney stones develop early and gout comes later. A single enzyme missing—enormous tragedy.

19.12 Anabolism of Amino Acids

Humans cannot convert atmospheric nitrogen into any form usable for protein synthesis. We rely on plants, which rely on soil microorganisms. Scores of millions of microbes occur in each gram of most topsoil and some can "fix" nitrogen. **Nitrogen fixation** is the conversion of nitrogen by soil microorganisms into ammonia or ammonia derivatives. Some soil organisms can also change ammonia into nitrite and nitrate ions. Still others can use nitrite and nitrate instead of oxygen for metabolism and, as they do, generate elemental nitrogen. A nitrogen cycle, therefore, exists in nature and both plants and animals are parts of it. Plants can use ammonia or its derivatives, nitrite and nitrate ions, to make amino acids and proteins. Plant-eating animals get ready-made amino acids from plants. Other animals obtain needed amino acids from the plant-eaters. When plants and animals die and decay, soil microorganisms convert amino acids into inorganic forms, including nitrogen itself.

19.13 Essential Amino Acids

All 20 amino acids need not be provided directly in the diet because we have the ability to make roughly half of the 20 from the other half using transaminations and intermediates from other metabolic pathways. The 9 amino acids that must be in the diet are given in Table 19.2. They are called the **essential amino acids** not because they are any more important than the others, but rather because it is essential they be in the diet itself.

TABLE 19.2
Essential and Nonessential Amino Acids for Humans

Essential	Nonessential
Histidine[a]	Alanine
Isoleucine	Asparagine
Leucine	Aspartic acid
Lysine	Cysteine
Methionine	Glutamine
Phenylalanine	Glutamic acid
Threonine	Glycine
Tryptophan	Proline
Valine	Serine
	Tyrosine

[a] Believed to be essential only for infants.

19.14 Anabolism of Nonessential Amino Acids

Figure 19.5 indicates the principal routes to the **nonessential amino acids,** those we can make on our own. Glutamine and asparagine, not indicated in Figure 19.5, are made from glutamic acid and aspartic acid, respectively. Cysteine, also not shown, is made via complex reactions involving serine and methionine, one of the essential amino acids and the source of sulfur.

Figure 19.6 gives a broad overview of the major metabolic pathways we have studied.

19.15 Catabolism of Heme

Erythrocytes have a life span of only about 120 days. Eventually, they split open. The hemoglobin that spills out (Figure 12.10) is then degraded. Its breakdown products are eliminated through the bile in the feces and, to a slight extent, in the urine. The characteristic colors of bile, feces, and urine are caused by partially degraded heme molecules, the tetrapyrrole pigments or **bile pigments.**

The degradation of heme begins before the globin portion breaks away. One of the carbon "bridges" is removed, and we then have a linear chain of small rings rather than a large ring:

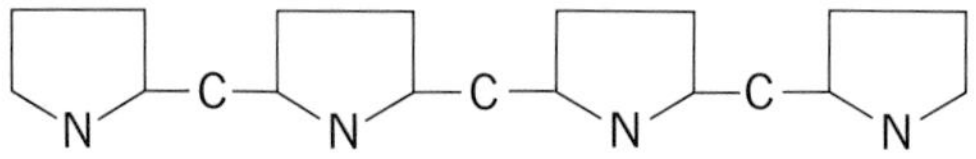

Carbon skeleton of bile pigments
(Double bonds are present in varying numbers and locations depending on the state of oxidation.)

The slightly broken hemoglobin molecule, now called verdohemoglobin, then

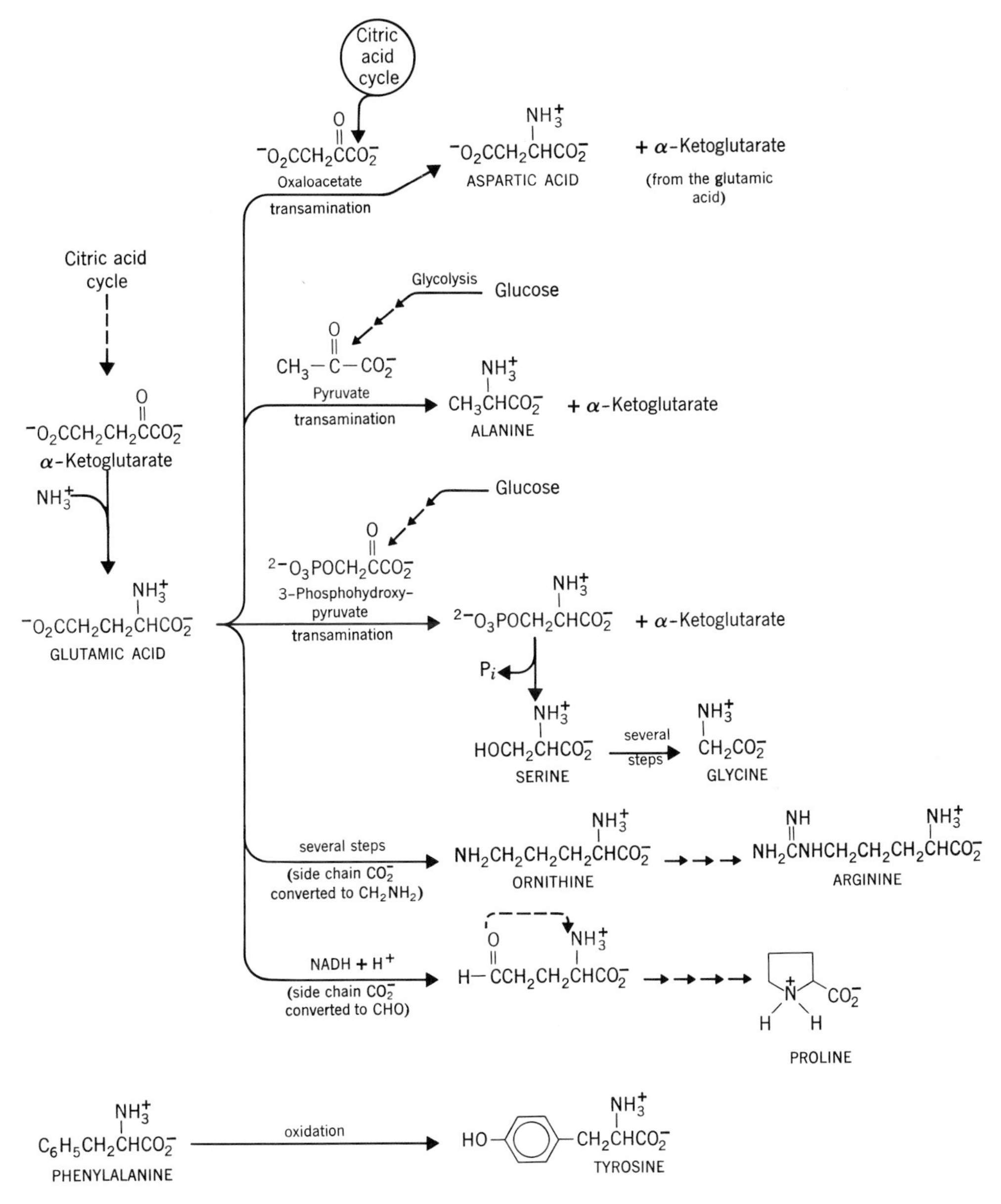

Figure 19.5 Biosynthesis of some nonessential amino acids.

splits into globin, ferrous ion, and the greenish pigment, **biliverdin** (Latin *bilis,* bile + *verdus,* green). Globin enters the nitrogen pool. Iron is conserved by the body in the form of a storage protein, ferritin, and is reused. In human beings biliverdin is converted by enzymes in the liver to a reddish-orange pigment called **bilirubin** (latin *bilis,* bile + *rubin,* red). Bilirubin is not only made by the liver but is also removed from circulation by the liver, which transfers it to bile. In this fluid it finally enters the intestinal tract.

The pathway from hemoglobin to bilirubin after the rupture of erythro-

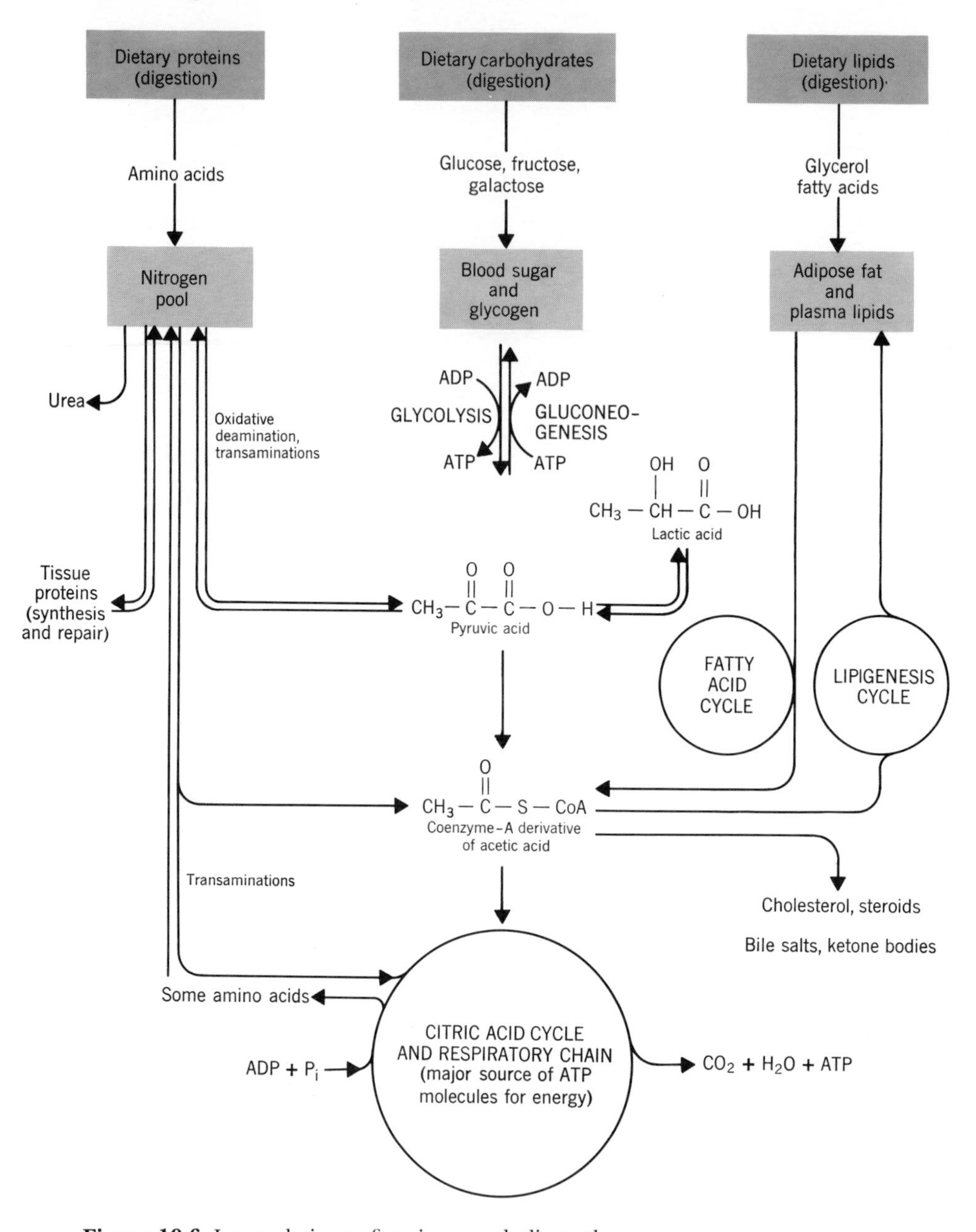

Figure 19.6 Interrelations of major metabolic pathways.

cytes (hemolysis) and the elimination of bilirubin is shown in the schematic diagram of Figure 19.7.

Bilirubin is the principal bile pigment in human beings. Routine flow of bile brings it to the intestinal tract, where the enzymes released by bacteria convert it to a colorless substance, mesobilirubinogen. This is further acted on to form **"bilinogen,"** which usually goes by other names signifying differences in destination rather than structure. Bilinogen that leaves the body in the feces is called stercobilinogen (Latin, *stercus,* dung). Urobilinogen is the

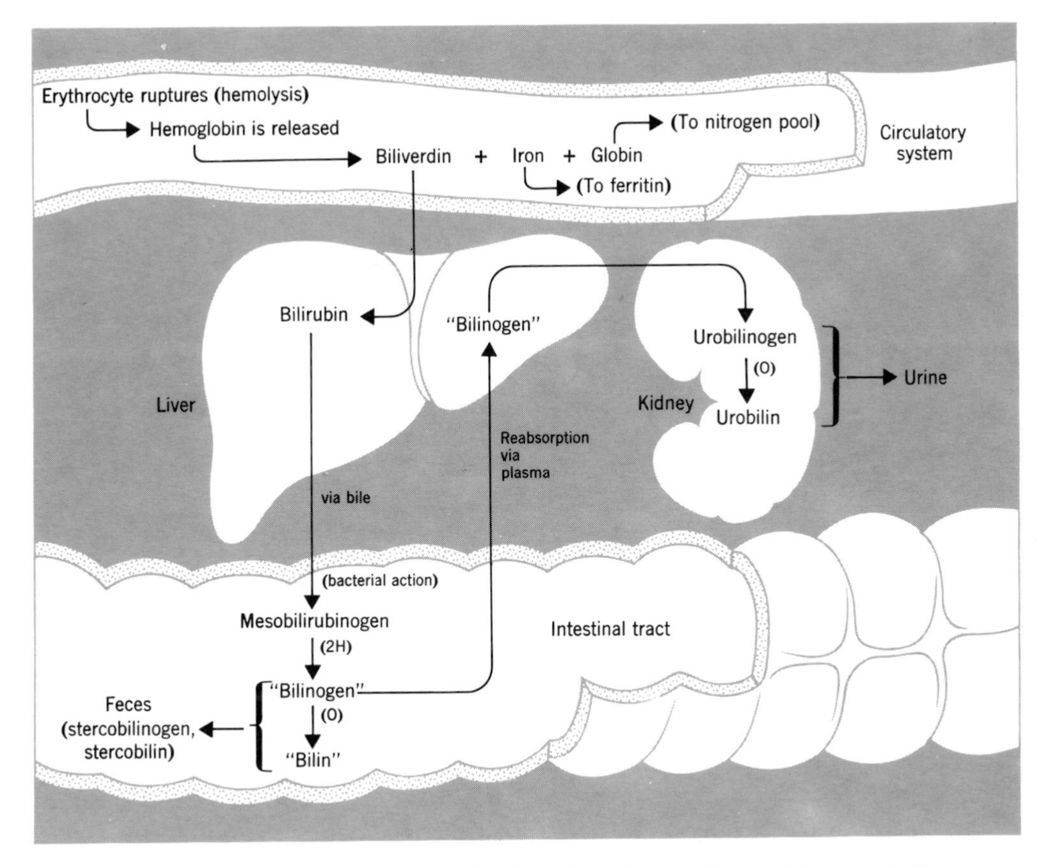

Figure 19.7 Formation and elimination of products of hemoglobin catabolism.

same compound, but it exists with the urine. Likewise, **bilin,** a reoxidized bilinogen, is properly called stercobilin or urobilin, depending on its destination. Because some bilinogen is reabsorbed from the intestinal tract through the bloodstream, small amounts of it appear in the urine. The characteristic brown color of feces and urine is caused by the presence largely of stercobilin or urobilin.

19.16 Jaundice

Jaundice (French *jaune,* yellow) is a condition that is symptomatic of malfunction somewhere along the pathway of heme metabolism. If bile pigments accumulate in the plasma in concentrations high enough to impart a yellowish coloration to the skin, the condition of jaundice is said to exist. Jaundice may result from one of three kinds of malfunctions.

Hemolytic jaundice results when hemolysis takes place at an abnormally fast rate. Bile pigments, particularly bilirubin, form faster than the liver can clear them. Hepatic diseases such as infectious hepatitis and cirrhosis sometimes prevent the liver from removing bilirubin from circulation. The stools are usually clay-colored, since the pyrrole pigments do not reach the intestinal tract.

Obstructions of bile ducts may prevent the release of bile into the intestinal tract, and the tetrapyrrole pigments in bile cannot be eliminated. Under these circumstances they tend to reenter general circulation. The kidneys remove large amounts of bilirubin, but the stools are usually clay-colored. As the liver works harder and harder to handle its task of removing excess bilirubin, it may weaken and become permanently damaged.

Brief Summary

Amino Acid Distribution. The nitrogen pool receives amino acids from the diet, from the breakdown of proteins in body fluids or tissues, and from any syntheses of nonessential amino acids that occur. Amino acids are used to build and repair tissue, replace proteins of body fluids, make nonprotein nitrogen compounds, provide chemical energy if needed, and supply molecular parts for gluconeogenesis or lipigenesis. In a healthy, well-nourished adult, nitrogen intake equals outgo and a nitrogen balance exists. The balance tips in favor of intake during years of growth and in favor of outgo during wasting diseases.

Amino Acid Catabolism. By reactions of transamination, oxidative deamination, direct deamination, and decarboxylation, the α-amino acids shuffle amino groups between themselves and intermediates of the citric acid cycle, or the synthesis of urea and nonprotein nitrogen compounds. Deaminated amino acids eventually become acetyl coenzyme A, acetoacetyl CoA, pyruvate, or an intermediate in the citric acid cycle. The glucogenic amino acids may be used to make glucose, and the ketogenic amino acids may be used to make ketone bodies or fatty acids. Of the heterocyclic amines from nucleic acids, the pyrimidines (cytosine and thymine, C and T) are broken to products that enter pathways of amino acids. Those of the purines (adenine and guanine, A and G) are changed to uric acid and excreted as sodium urate in the urine.

Amino Acid Anabolism. By nitrogen fixation molecular nitrogen (N_2) is changed to other inorganic nitrogen compounds that plants can use to make amino acids. Animals rely on these for their own use and can make 9 of the 20 amino acids they need from others. The 9 essential amino acids (8 for adults) must be in the diet.

Heme Catabolism. Heme is catabolized to tetrapyrrole pigments while its iron is reused. The pigments—first, biliverdin (green), then bilirubin (red), then mesobilirubinogen (colorless), and finally bilinogen and bilin (brown)—become stercobilin or stercobilinogen, urobilin or urobilinogen, depending on the route of elimination. If the liver cannot clear bilirubin and biliverdin from the blood, the patient will develop a jaundiced appearance.

Selected References

Books

1. A. White, P. Handler, and E. L. Smith. *Principles of Biochemistry,* 5th ed. McGraw-Hill Book Company, New York, 1973.
2. A. L. Lehninger. *Biochemistry*. Worth Publishers, Inc., New York, 1970.

3. L. Stryer. *Biochemistry.* W. H. Freeman and Company, Publishers, San Francisco, 1975.
4. W. L. Nyhan, ed. *Heritable Disorders of Amino Acid Metabolism.* John Wiley & Sons, Inc., New York, 1974.

Article

1. C. C. Delwiche. "The Nitrogen Cycle." *Scientific American,* September 1970, page 137. Included in this article are a number of informative figures, tables, and charts.

Questions and Exercises

1. What are the three general final outcomes of the α-amino acids in the body?
2. Why is not the nitrogen pool in one location in the way that triacylglycerols are generally localized in adipose tissue?
3. Why are growing infants in a positive nitrogen balance?
4. What are the principal end-products of the complete catabolism of the following?
   (a) α-Amino acids (b) The purine bases
5. Write an equation illustrating each of the following.
   (a) The transamination of alanine with α-ketoglutarate the acceptor
   (b) The oxidative deamination of glutamic acid
   (c) The direct deamination of serine
   (d) The decarboxylation of tyrosine
6. What is one service of the amino oxidases?
7. How does DOPA work in Parkinson's disease treatment?
8. In the biosynthesis of urea what are the sources of each of the following groups?
   (a) The two —NH_2 groups (b) The carbonyl group
9. In general terms what metabolic pathway is believed not to function well when each of these conditions develops?
   (a) Gout (b) Hyperammonemia
10. How may amino acids be used in starvation or in diabetes mellitus?
11. How is the nitrogen balance changed in starvation or in diabetes?
12. We make hemoglobin using glycine to make the heme unit. Yet glycine is not even listed as an essential amino acid. Explain.
13. In what particularly important way do we depend on plants for health?
14. Arrange the names of these substances in the order in which one produces another by placing the number of the substance first in line on the left, the number of the last-produced substance last in line on the right, and the numbers of the others in correct order between.

biliverdin	heme	hemoglobin	bilirubin	mesobilirubinogen
1	2	3	4	5
		bilin	bilinogen	
		6	7	

15. What are some conditions suggested by a jaundiced appearance?

chapter 20

Nutrition

Cassava, the starchy root of a semishrubby plant common to tropical regions, is the chief staple in many diets. This Nigerian woman is grinding the root to make a flour. Unhappily, cassava is a very poor source of good protein. Why, and what sources are better, are questions that are answered in this chapter. (FAO photo by C. Bavagnoli.)

20.1 What Is Nutrition?

Nutrition may be defined in both technical and personal terms. Technically, nutrition is the science of discovering the substances called **nutrients** needed for health; determining the amounts necessary to sustain health, to promote necessary growth, to repair damage, to replace loss and to provide energy for living; and identifying the natural products—food, air, water—that will supply these needs. In personal terms, nutrition is the sum of the actions you take to eat the right foods in the proper amounts at the best moments as you try to maintain a state of physical, mental, and social well-being and avoid dietary-related diseases and infirmities.

Nutrition is enormously vital to good health and it can serve as an important defense against potential problems lurking in both the genetic and environmental background. Moreover, good nutrition prevents a number of specific diseases and cures several. When James Lind discovered in 1852 that oranges, lemons, and limes could cure scurvy and K. Takaki in the 1880s found that diet could cure beri-beri, the science of nutrition was on its way to full stature.

20.2 Recommended Dietary Allowances

For over 30 years the Food and Nutrition Board of the National Research Council of the National Academy of Sciences has published **recommended dietary allowances.** In the judgment of this Board, based on a continuing study and evaluation of the latest findings, these allowances are the levels of intake of essential nutrients that are "adequate to meet the known nutritional needs of practically all healthy persons." The recommended dietary allowances of the Food and Nutrition Board as revised in 1974 are given in Table 20.1.

A number of points and qualifications about the recommended dietary allowances must be made and emphasized.

1. *The recommended dietary allowances or RDA are not the same as the United States Recommended Daily Allowances (USRDA).*

The latter are set by the U.S. Food and Drug Administration, based on the RDA, as standards for nutritional information on food labels.

2. *The RDA are not the same as minimum daily requirements for any one individual.*

Just as individuals differ widely in height, weight, and appearance, they also differ in biochemical needs. The recommended daily allowances, therefore, are set to exceed average requirements by enough to insure that "practically all people" will thrive. Most will receive more than they need; a few will actually not receive enough. Only in this kind of statistical framework are the RDAs meaningful.

3. *The RDA do not define therapeutic nutritional needs.*

People with chronic diseases, prolonged infections, or disorders of metabolism; people with daily intakes of certain medications; and prematurely born infants require special diets. The RDA do cover people according to age, sex, and size and they indicate special needs of pregnant and lactating women, but

TABLE **20.1**

Recommended Daily Dietary Allowances[a] of the Food and Nutrition Board, National Academy of Sciences, National Research Council, Revised 1974.

								Fat-Soluble Vitamins			
	Age	Weight		Height		Energy	Protein	Vitamin A Activity		Vitamin D	Vitamin E Activity[e]
	(years)	(kg)	(lb)	(cm)	(in.)	(kcal)[b]	(g)	(RE)[c]	(IU)	(IU)	(IU)
Infants	0.0–0.5	6	14	60	24	kg × 117	kg × 2.2	420[d]	1400	400	4
	0.5–1.0	9	20	71	28	kg × 108	kg × 2.0	400	2000	400	5
Children	1–3	13	28	86	34	1300	23	400	2000	400	7
	4–6	20	44	110	44	1800	30	500	2500	400	9
	7–10	30	66	135	54	2400	36	700	3300	400	10
Males	11–14	44	97	158	63	2800	44	1000	5000	400	12
	15–18	61	134	172	69	3000	54	1000	5000	400	15
	19–22	67	147	172	69	3000	54	1000	5000	400	15
	23–50	70	154	172	69	2700	56	1000	5000		15
	51+	70	154	172	69	2400	56	1000	5000		15
Females	11–14	44	97	155	62	2400	44	800	4000	400	12
	15–18	54	119	162	65	2100	48	800	4000	400	12
	19–22	58	128	162	65	2100	46	800	4000	400	12
	23–50	58	128	162	65	2000	46	800	4000		12
	51+	58	128	162	65	1800	46	800	4000		12
Pregnant						+300	+30	1000	5000	400	15
Lactating						+500	+20	1200	6000	400	15

[a] The allowances are intended to provide for individual variations among most normal persons as they live in the United States under usual environmental stresses. Diets should be based on a variety of common foods in order to provide other nutrients for which human requirements have been less well defined.
[b] Kilojoules (kJ) = 4.2 × kcal.
[c] Retinol equivalents.
[d] Assumed to be all as retinol in milk during the first six months of life. All subsequent intakes are assumed to be half as retinol and half as β-carotene when calculated from international units. As retinol equivalents, three-fourths are as retinol and one-fourth as β-carotene.

they do not go into any other special needs. Therapeutic needs for water and salt rise during strenuous physical activity and prolonged exposure to high temperatures. In some areas of the United States and the world intestinal parasites are common, and these organisms rob the affected people of some of their food intake each day. They, too, may need special diets.

4. *Recommended dietary allowances are obtained from a number of combinations and patterns of food and diet.*

No single food contains all nutrients. People take dangerous risks with their health when they go on a fad diet limited to one particular food—gelatin, or yogurt, or brown rice are some examples. The ancient wisdom of having a varied diet that includes meat, fruit, vegetables, grains, nuts, pulses (e.g., beans), and dairy products may seem to be based only on cultural and esthetic factors. These aspects no doubt are important, but a varied diet not

TABLE **20.1—*continued***

Water-Soluble Vitamins							Minerals					
Ascorbic Acid (mg)	Folacin[f] (μg)	Niacin[g] (mg)	Riboflavin (mg)	Thiamin (mg)	Vitamin B_6 (mg)	Vitamin B_{12} (μg)	Calcium (mg)	Phosphorus (mg)	Iodine (μg)	Iron (mg)	Magnesium (mg)	Zinc (mg)
35	50	5	0.4	0.3	0.3	0.3	360	240	35	10	60	3
35	50	8	0.6	0.5	0.4	0.3	540	400	45	15	70	5
40	100	9	0.8	0.7	0.6	1.0	800	800	60	15	150	10
40	200	12	1.1	0.9	0.9	1.5	800	800	80	10	200	10
40	300	16	1.2	1.2	1.2	2.0	800	800	110	10	250	10
45	400	18	1.5	1.4	1.6	3.0	1200	1200	130	18	350	15
45	400	20	1.8	1.5	2.0	3.0	1200	1200	150	18	400	15
45	400	20	1.8	1.5	2.0	3.0	800	800	140	10	350	15
45	400	18	1.6	1.4	2.0	3.0	800	800	130	10	350	15
45	400	16	1.5	1.2	2.0	3.0	800	800	110	10	350	15
45	400	16	1.3	1.2	1.6	3.0	1200	1200	115	18	300	15
45	400	14	1.4	1.1	2.0	3.0	1200	1200	115	18	300	15
45	400	14	1.4	1.1	2.0	3.0	800	800	100	18	300	15
45	400	13	1.2	1.0	2.0	3.0	800	800	100	18	300	15
45	400	12	1.1	1.0	2.0	3.0	800	800	80	10	300	15
60	800	+2	+0.3	+0.3	2.5	4.0	1200	1200	125	18+[h]	450	20
80	600	+4	+0.5	+0.3	2.5	4.0	1200	1200	150	18	450	25

[e] Total vitamin E activity, estimated to be 80% as α-tocopherol and 20% other tocopherols.
[f] The folacin allowances refer to dietary sources as determined by *Lactobacillus casei* assay. Pure forms of folacin may be effective in doses less than one-fourth of the recommended dietary allowance.
[g] Although allowances are expressed as niacin, it is recognized that on the average 1 mg of niacin is derived from each 60 mg of dietary tryptophan.
[h] This increased requirement cannot be met by ordinary diets; therefore, the use of supplemental iron is recommended.

only helps assure your receiving the major nutrients, it also makes most probable your taking in trace substances that may not yet even be discovered as important to health. The amounts of nutrients healthy people should have, on the average, are just part of the design of a diet. The nutrients must come in foods that are both acceptable and palatable and from a variety as well.

20.3 Human Nutritional Requirements

Biochemist E. F. Deatherage has compiled a list of the nutrients known to be required by every healthy adult person. This list, Table 20.2, includes what he has found to be a rough consensus of recommendations by both United States and United Nations scientists of the individual nutrients needed for good health. Some of the nutrients in the table are not included in official tabulations either because they are so common that a deficiency is impossible or because it has not been possible to determine precisely what the minimum requirements are. Trace elements are on the list according to their known

TABLE 20.2
Adult Human Nutritional Requirements

Nutrient	Daily Requirement: Grams (unless noted), Man[a]	Grams (unless noted), Woman[b]	Moles, Man[a]	Moles, Woman[b]
Oxygen	800	593	25.0	18.5
	560 liters[c]	414 liters[c]		
As air 21% O_2	2930 liters[d]	2173 liters[d]		
Water[e]	2800	2000	155	111
Energy	2700 kcal	2000 kcal		
If all from glucose	675	500	3.75	2.78
If all from fat	300	222	0.338	0.25
Protein	56	46	0.47[f]	0.38[f]
Essential amino acids				
Tryptophan	0.5		0.0025	
Phenylalanine[g]	2.2		0.0133	
Lysine	1.6		0.0110	
Threonine	1.0		0.0085	
Methionine[h]	2.2		0.0150	
Leucine	2.2		0.0168	
Isoleucine	1.4		0.0106	
Valine	1.6		0.0137	
Sodium chloride[i]	3.0		0.051	
Potassium	1.0		0.026	
Phosphorus	0.8		0.026	
Calcium	0.8		0.020	
Magnesium	0.35		0.015	
Essential fatty acids[j]	3.0		0.010	

[*Source:* F. E. Deatherage. *Food for Life.* Plenum Publishing Corporation, New York, 1975. Used by permission of author and publisher.]

[a] Man: weight 70 kg (154 lb), height 175 cm (5 ft 8 in.), age 22–50 years.
[b] Woman: weight 58 kg (128 lb), height 163 cm (5 ft 5 in.), age 22–50 years.
[c] At 0 °C (32 °F) and pressure of 1 atm or 760 mm of mercury; 19.8 and 14.6 ft^3, respectively.
[d] At room conditions of 20 °C (68 °F) and 740 mm of mercury, 103.5 and 76.8 ft^3, respectively.
[e] Includes water consumed directly and as food component and water produced from oxidation of food.
[f] Average molecular weight of amino acids is 120.
[g] 0.22 g if tyrosine is available.
[h] 0.35 g if cystine is available.
[i] 2–3 g is minimum but salt requirement as well as water requirement may go up in warm climates and in heavy work due to perspiration losses.
[j] As linoleic, linolenic, and/or arachidonic acid.

presence as cofactors in enzymes. The nutrients are arranged (except for water) in their order of daily requirements in moles.

Exercise 20.1 Why is histidine missing from Table 20.2 but included in Table 19.2?

Exercise 20.2 Offer a reason for the huge drop in molar requirements that occurs after the essential fatty acids in Table 20.2.

TABLE 20.2—*continued*

Nutrient	Daily requirement: Grams (unless noted), Man[a]	Woman[b]	Moles, Man[a]	Woman[b]
Iron[k]	0.018		3.2×10^{-4}	
Ascorbic acid, vitamin C	0.045		2.5×10^{-4}	
Zinc	0.015		2.2×10^{-4}	
Niacin, nicotinic acid amide, vitamin B_3	0.018		1.5×10^{-4}	
Fluorine	0.002		1.0×10^{-4}	
Vitamin E[l]	0.015		3.5×10^{-5}	
Copper	0.0015		2.4×10^{-5}	
Pantothenic acid,[n] vitamin B_5	0.005		2.3×10^{-5}	
Manganese[n]	0.001		1.8×10^{-5}	
Vitamin K[l]	0.005		1.1×10^{-5}	
Pyridoxine, vitamin B_6	0.002		1.0×10^{-5}	
Retinol, vitamin A[l]	0.0015		5.2×10^{-6}	
Riboflavin, vitamin B_2	0.0016		4.2×10^{-6}	
Thiamin, vitamin B_1	0.0014		4.2×10^{-6}	
Iodine	0.00012		9.4×10^{-7}	
Folic acid	0.0004		9.0×10^{-7}	
Biotin[n]	0.00015		6.0×10^{-7}	
Vitamin D[m]	0.00001		2.6×10^{-8}	
Chromium	0.000001 (1×10^{-6})		2.0×10^{-8}	
Cobalamin, vitamin B_{12}	0.000005 (5×10^{-6})		3.7×10^{-9}	
Cobalt, as vitamin B_{12}	0.0000002 (2×10^{-7})		3.7×10^{-9}	
Molybdenum[o]				
Vanadium[o]				
Selenium[o]				
Tin[o]				

[k] Recommended for menstruating women; for men about 0.01 g is recommended.
[l] A number of closely related chemical compounds may serve this vitamin function. 1 IU (International Unit) of A = 0.3 μg retinol or 0.6 μg β-carotene. 1 IU of E = 1.0 mg DL-α-tocopherol acetate or 0.81 mg D-α-tocopherol.
[m] Vitamin D may not be required by the adult but is by growing children; it is made from normal skin constituents by ultraviolet light or it may be fed in the diet. 1 IU of D = 0.025 μg D_3, activated 7-dehydrocholesterol.
[n] Although definitely required, it is difficult to establish the amount because this nutrient's daily requirement has not been established. Sufficient amounts are in many foods.
[o] It is known that this trace element is an integral part of vital enzyme systems, but the daily requirement has not been established. Sufficient amounts are in many foods.

Individual Nutrient Requirements

20.4 Water

Water is needed to provide a medium for chemical reactions in cells and tissues and as a fluid for transporting the constituents of blood and lymph. We take in water in both food and drink, and oxidation produces a small amount more. We have homeostatic mechanisms for maintaining the needed quantity of water and the concentrations of body fluids. The thirst mechanism is chiefly responsible for "demanding" water. Loss of water in the air we exhale, in both sensible and insensible perspiration, in the feces and in the urine all contribute to outgo. Figure 20.1 summarizes the relative amounts handled by various means.

Not included in Figure 20.1 are losses by sensible perspiration: sweat. Activity in the broiling sun of an arid desert may cause the loss of as much as 10 liters/day as the body battles to eliminate heat by the evaporation of water. Electrolytes are also lost. No amount of training can condition anyone to go without water, and if that need occurs after hot, sweaty toil you probably also need salts. If that need is not met you may suffer heat exhaustion, heat stroke, or heat cramps, or a combination.

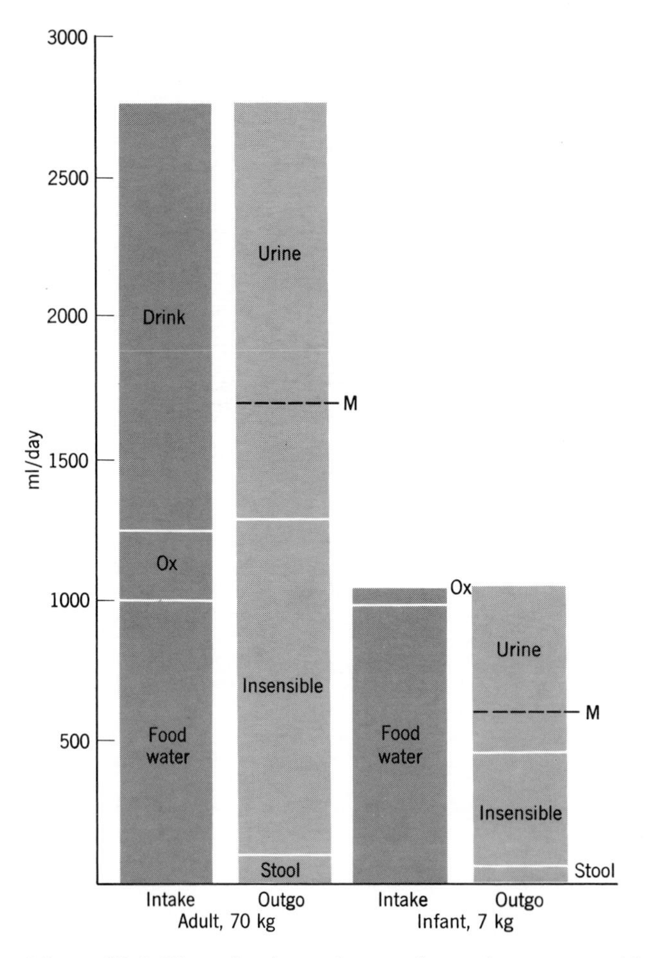

Figure 20.1 Water intake and outgo by major routes, without sweating. Dashed lines at M are the minimal volumes of urine at maximal concentrations of solutes. "Ox" means water formed by the oxidation of foods. (Source: *Recommended Dietary Allowances,* 8th ed., National Research Council, National Academy of Sciences, 1974.)

20.5 Energy

We customarily think only of foods as sources of energy rather than oxygen, probably because it is far more likely that we might know times of too little food than times of low oxygen supply. Both, however, supply us with chemical energy. An estimate of daily oxygen needs is given in Table 20.2. Table 20.1 gives the recommended daily food energy intake at various ages. Examples of how that intake is used by adults in light work occupations are given in Table 20.3.

Energy in food should not come exclusively either from carbohydrates or lipids. A zero-carbohydrate diet requires accelerated gluconeogenesis (Section 17.6), and the development of ketosis may occur. A zero-fat diet lacks the essential fatty acids, and such a diet makes the absorption of the fat-soluble vitamins more difficult. A diet of at least 15 to 25 g of food fat (e.g., 2–4 pats of margarine or butter) and 50 to 100 grams of digestible carbohydrate/day avoid the consequences described.

20.6 Proteins

The **essential amino acids** specified in Table 20.2 total only 12.7 g/day out of the 46 to 56 g/day of total protein judged to be required. The difference supplies our total nitrogen need. The total protein requirement given is more than the "bare minimum." It takes into account both an allowance of 30% extra to cover a wide range of individual variations in needs for amino acids and an assumption that proteins eaten in the diet are actually only 75% digested and absorbed into the bloodstream. The figures, in other words, provide some margin of error.

Not all proteins are easily and completely digested. They vary in digestibility, and that variation is expressed by a **coefficient of digestibility,** defined by Equation 20.1.

$$\text{coefficient of digestibility} = \frac{\text{nitrogen in food eaten} - \text{nitrogen in feces}}{\text{nitrogen in food}} \qquad (20.1)$$

Animal proteins are superior to those of plants and fruits in digestibility, as defined by this equation, having a coefficient of digestibility of 0.97 (i.e., 97% digestible). Proteins in fruits and fruit juices have a value of 0.85. Of the grains, whole grains rate relatively low: 0.79 for whole wheat flour and 0.67 for whole rye flour. Milling improves the digestibility coefficient: 0.83 for all purpose bread flour and 0.89 for cake flour. Milling, however, also reduces the quantities of vitamins and minerals in the flour as well as the fiber content. Milling does the same to rice. Brown rice has a digestibility coefficient of 0.75, but 0.84 is the value for polished white rice (with, however, less vitamins and minerals). The proteins in legumes and nuts have a digestibility coefficient of 0.78 and those of vegetables range from 0.65 to 0.74.

Even when an upward adjustment is made according to the protein source and digestibility, we still find that proteins have varying biological or nutritional values. In technical terms the **biological value** of a protein is the percent of its absorbed nitrogen that the body retains when the total protein intake is below the requirement level. (Absorbed nitrogen is not the same as ingested nitrogen; it is usually less.) The measurement of biological value is

TABLE 20.3
Examples of Daily Energy Expenditures of Mature Women and Men in Light Occupations

		Man (70 kg)		Woman (58 kg)	
		Rate		Rate	
Activity Category	Time (hr)	kcal/ min	Total (kcal) [kJ]	kcal/ min	Total (kcal) [kJ]
Sleeping, reclining	8	1.0–1.2	540 [2270]	0.9–1.1	440 [1850]
Very light[a]	12	up to 2.5	1300 [5460]	up to 2.0	900 [3750]
Light[b]	3	2.5–4.9	600 [2520]	2.0–3.9	450 [1890]
Moderate[c]	1	5.0–7.4	300 [1260]	4.0–5.9	240 [1010]
Heavy[d]	0	7.5–12.0			
Total	24		2740 [11 500]		2030 [8530]

Source: J. V. G. A. Durnin and R. Passmore, 1967. In *Recommended Dietary Allowances,* 8th ed., National Academy of Sciences, 1974.
[a] Seated and standing activities; driving cars and trucks, secretarial work; laboratory work; sewing, ironing; playing musical instruments.
[b] Walking (on the level, 2.5–3 miles per hour); tailoring and pressing; carpentry, electrical trades, restaurant work; washing clothes; light recreation—golf, table tennis, volleyball, sailing.
[c] Walking at 3.5–4 mph; garden work; scrubbing floors; shopping with heavy load; moderate sports—skiing, tennis, dancing, bicycling.
[d] Uphill walking with a load; pick and shovel work; heavy sports—swimming, climbing, football, basketball; lumbering.

done under carefully defined conditions that we shall not study. We shall view biological value as a measure of the efficiency with which the nitrogen of the actually absorbed amino acids of a protein are used by the body.

The largest single factor in a protein's biological value is its amino acid composition, and what most limits that value is the extent to which that protein supplies the essential amino acids in sufficient quantities for human use. When your tissues are manufacturing proteins, all the necessary amino acids must be present at the same time. If an essential amino acid such as lysine, for example, happens to be missing, when a lysine-containing protein is being made, protein synthesis has to stop. Thus a food whose protein is low in lysine would not have a high biological value.

Proteins can be rated according to their ability to supply essential amino acids all at the right time and in the right proportions. Nutritionists believe that human milk protein is the best of all with respect to digestibility and biological value, although whole egg protein is very close and is often taken as the reference for experimental work. According to various studies most 70-kg men could in principle be in nitrogen balance by ingesting 35 g/day of the proteins of human milk, which provides a standard of comparison.

Table 20.4 summarizes some information about most of the proteins that are prominent in various diets of the world's peoples. The essential amino acid most poorly supplied by the protein is called the **limiting amino acid,** and these amino acids are named in the second column of Table 20.4. The third column gives the number of grams of each food that a 70-kg man would have to digest and absorb per day to get the same amount of its limiting amino

TABLE 20.4
Comparison of Various Food Proteins with Human Milk Proteins

Food	Limiting Amino Acid	Food's Protein Equivalent to 35 g Human Milk Protein (g)	Digestibility Coefficient of Food's Protein	Amount of Food's Protein[a] (g)	Percent of Protein in the Food	Amount of Food Needed[b]	Calories of Food Received
Wheat	Lysine	80.4	79%	102	13.3%	767 g (1.7 lb)	2560 kcal
Corn	Tryptophan and lysine	72.4	60	120	7.8	1540 g (3.4 lb)	5660
Rice	Lysine	51.7	75	68.9	7.5	919 g (2.0 lb)	3310
Beans	Valine	50.5	78	64.8	24.0	270 g (0.59 lb)	913
Soybeans	Methionine and cysteine	43.8	78	56.2	34.0	165 g (0.36 lb)	665
Potatoes	Leucine	71.6	74	96.7	2.1	4600 g (10.1 lb)	3500
Cassava	Methionine and cysteine	82.4	60	137	1.1	12 500 g (27.5 lb)	16 400
Eggs	Leucine	36.6	97	37.8	12.8	295 g (0.65 lb)	477
Meat	Tryptophan	43.1	97	44.4	21.5	206 g (0.45 lb)	295
Cows' milk	Methionine and cysteine	43.8	97	45.2	3.2	1410 g (3.1 lb)	903

Source: Data from F. E. Deatherage. *Food for Life*. 1975. Plenum Press, New York.
[a] The grams of protein obtained from each food source needed to be nutritionally equivalent (with respect to essential amino acids) to 35 g of human milk protein, allowing for the poorer digestibility of that food's protein (its digestibility coefficient).
[b] The grams of each food equivalent in nutritional value (with respect to essential amino acids) to 35 g human milk protein allowing for the digestibility coefficient and the percent protein in the food.

acid as is found in 35 g of human milk protein. Such an intake, of course, would also supply all other needed amino acids. Thus 80.4 g of wheat protein—not wheat, but wheat protein—would have to be digested and absorbed to obtain the lysine available in 35 g of human milk protein. Taking into account a digestibility coefficient of 0.79 for wheat protein, this 80.4 g becomes 102 g of wheat protein, column 5, that must be eaten. Since wheat is, on the average, only 13.3% protein (column 6), the 70-kg man would have to eat 767 g of wheat (1.7 lb)—column 7—if his daily amino acid needs are to be met by wheat alone. If he does that, however, he unavoidably receives 2560 kilocalories of food energy, which is nearly as much total energy per day he should have, and his diet is monotonous besides.

The data in Table 20.4 reveal a truth of enormous importance to those concerned about supplying both the protein and total calorie needs of the world's burgeoning population. Neither children nor adults can eat enough corn, rice, potatoes, or cassava per day to meet both their protein and energy needs. Proteins of eggs and meat, however, are particularly good. We say that they are **adequate proteins**—they include all the essential amino acids in suitable proportions to make it possible to satisfy amino acid and total nitrogen requirements without excessive intakes of total calories. Soybeans are the best of the nonanimal sources of proteins. Cassava, a root, is a very inadequate protein, and corn (or maize) is also poor. Unhappily, huge numbers of the world's peoples, especially in Africa, Central and South America, and India and other countries of the Middle and Far East, rely heavily on these two foods. In these and other areas kwashiorkor and marasmus are common dietary-deficiency diseases and important public health problems.

Kwashiorkor, from the Ghan language of west Africa for "the sickness the older child gets when the next baby is born," is a protein-deficiency disease. Marasmus, from the Greek *marasmos* for "wasting," is chronic total undernutrition, with deficiencies in both protein and total calories. As long as an infant is breast-fed, he or she receives adequate protein and calories. If, when the next baby is born, the older child is placed on a diet rich in cassava, manioc (another root), or maize (corn), the child can no longer ingest in 1 day enough to supply all the essential amino acids.

Obtaining protein adequate in both quality and amount in the early years of life is absolutely vital both to physical and mental health. Not only are muscles, bones, and organs growing, but also brain development is occurring. An infant grows to just 20% of adult size in the first 3 years, but his or her brain grows to 80% of adult size. About 50% of the dry weight of the brain is protein.

Victims of marasmus are grossly underweight and may expect to have stunted growth. Victims of kwashiorkor also show retarded growth, mental apathy, depigmentation of skin and hair, and general edema (which often masks the true muscular wasting).

The data in Table 20.4 provide scientific support, if any were needed, for long-standing practices in all major cultures of including a wide variety of foods in the daily diet. Meat and eggs can insure adequate protein while leaving room for other foods pleasing to the palate. Milk, soybeans, and other beans also leave room for variety. As vegetarians well know, variety is essential as well as tasty. By including both rice (low in lysine) and beans (low in valine) in equal proportions, about 43 g of the combination is equal in protein value to 35 g of human milk protein. Thus less of the combined rice-bean diet is needed than rice alone or beans alone, and more room is left for the variety our taste buds crave. The rice and beans should, of course, be eaten fairly close together to insure that all essential amino acids are simultaneously available during the syntheses of proteins. Eating the rice of a rice-bean diet early in the morning and the beans in the evening works against efficient protein syntheses.

Exercise 20.3 The limiting amino acid in peanuts is lysine; 62 g of peanut protein is equivalent to 35 g of human milk protein. The coefficient of digestibility of peanut protein is 0.78. In order to match human milk protein in nutritional value:

(a) How much peanut protein must be eaten?
(b) How many grams of peanuts must be eaten if peanuts are 26.2% protein?
(c) How many kcals are also ingested with that many grams of peanuts if peanuts have 282 kcal/100 g?
(d) In respect to protein and energy, how do peanuts compare with rice? With soybeans?
(e) Could a child eat enough peanuts per day to satisfy protein needs and still have room for other foods?

20.7 Essential Fatty Acids

Linoleic acid and arachidonic acid (see Table 11.2) are the fatty acids essential for health and growth of human infants. Linoleic acid, evidently also essential to adults, is widely present in edible vegetable oils. Arachidonic acid occurs in animal fats in trace amounts, but it can be made in adults from linoleic acid (or linolenic acid, another polyunsaturated fatty acid). Deficiency conditions observed in humans include dermatitis and problems with transporting lipids in the bloodstream.

20.8 Vitamins

The term **vitamin** is applied to a compound if it meets these standards:

1. The compound is organic and cannot be synthesized at all (or at least made in sufficient amounts) by the metabolic reactions within the "host" and must be provided by the diet.
2. Its absence results in a specific **vitamin deficiency disease.**
3. Its presence is required for normal growth and health.
4. The compound is present in foods in small concentrations and is not a carbohydrate, a lipid, or a protein.

The two major classes of vitamins are those that are fat-soluble and those that are water-soluble. The former are mostly hydrocarbonlike and the latter are polar or ionic. Extra amounts of the former can accumulate in fatty tissue; those of the latter generally are excreted in the urine.

20.9 Fat-Soluble Vitamins

These are vitamins A, D, E, and K and they are found in the fatty fractions of living systems. Requirements for the adult are in Table 20.2. Excesses are generally dangerous partly because excesses build up in fatty tissue and, unlike the water-soluble vitamins, are not rapidly excreted. All of the fat-soluble vitamins are susceptible to destruction by prolonged exposure to the oxygen in air or to oxidizing agents such as those present in fats and oils that are turning rancid because of oxidation. The double bonds or phenolic systems in these vitamins are vulnerable to oxidation.

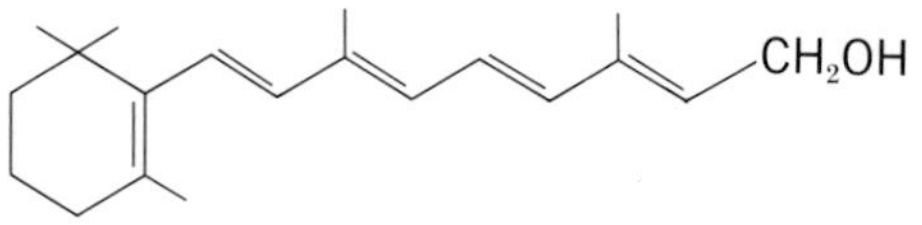

Retinol

Vitamin A is retinol, a polyunsaturated alcohol. The liver can make retinol out of certain plant pigments, particularly β-carotene, the yellow colored matter in carrots and other naturally yellow foods.

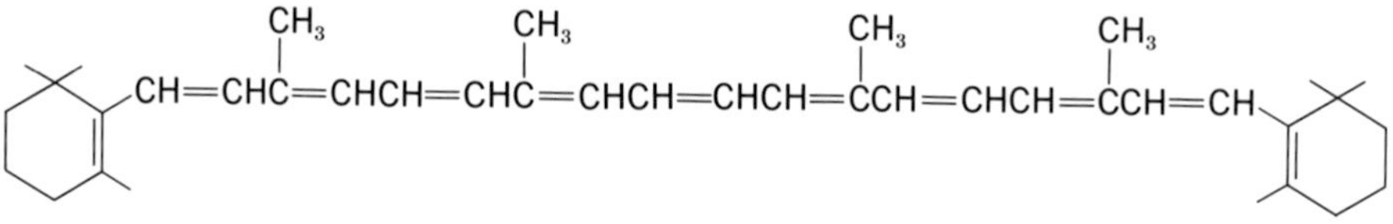

β-Carotene

We need retinol for healthy mucous membranes and eyes. Prolonged deficiency can lead to blindness, and an early sign of deficiency is impaired vision in dim light ("night blindness"). Excessive doses of retinol—2000 retinol equivalents or 6700 International Units (Table 20.2) per day above what normally is in the diet—are toxic.

Vitamin D exists in two forms, the naturally occurring cholecalciferol (D_3) and ergocalciferol (D_2), the synthetic form made from ergosterol, a plant steroid (see Table 11.3). Both forms of vitamin D are equally useful in humans, and each can be changed into hydroxy and dihydroxy derivatives, the forms active internally. Eggs, butter, liver, fatty fish, and fortified milk are good sources.

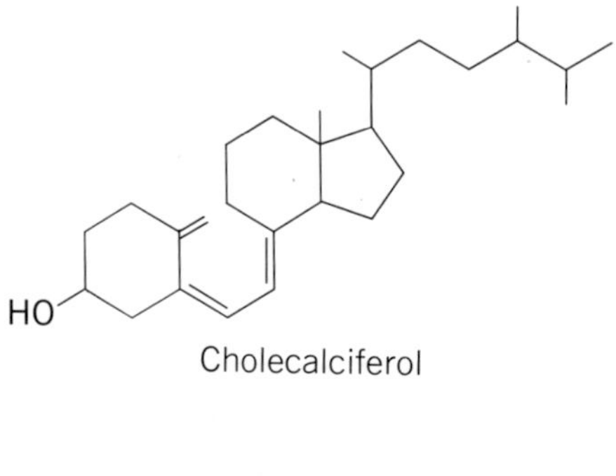

Cholecalciferol

Vitamin D is required during the years when bones and teeth are developing to promote the use of calcium and phosphate and assist in the formation of bone minerals. The chief natural sources of vitamin D are the steroids that we manufacture internally. These are converted by the action of direct sunlight on the skin into vitamin D. The deficiency disease, a bone disorder, is rickets. Youngsters working from dawn to dusk in dingy factories or mines during the early years of the Industrial Revolution were particularly prone to rickets simply because they saw little or no sun. No advantage is gained by large doses of vitamin D and if sufficiently excessive, the vitamin is dangerous. Excesses promote a rise in the calcium ion level of the blood, and that damages the kidneys and causes calcification of soft tissue.

Vitamin E is actually a mixture of various tocopherols, and they are especially present in vegetable oils. The Food and Nutrition Board finds no evidence that any general vitamin E inadequacy exists in the diet of U.S. citizens. Those whose diets are relatively high in polyunsaturated fatty acids have serum-vitamin E levels that are also high, higher than among those with diets having low levels of these fatty acids. Possibly we need more vitamin E the higher our intake of polyunsaturated lipids. The tocopherols are known to be fat preservatives. They inhibit direct attack by oxygen on unsaturated lipids and delay the onset of rancidity. Exactly what the tocopherols do for human adults is the subject of controversy. Edema and anemia have been reported in infants on formulas low in vitamin E. Low levels of vitamin E in the blood, when tested in laboratory vessels, results in a higher susceptibility of hemolysis of red cells, which may explain the anemia. According to some studies, the lack of vitamin E in diets of growing rabbits or rats leads to muscular dystrophy and paralysis and to increased susceptibility to sudden heart attacks, especially in males under stress. Proponents of vitamin E therapy (e.g., E. V. and W. E. Shute and Linus Pauling) view vitamin E as necessary to prevent heart diseases especially in an age when dietary levels of polyunsaturated fats are relatively high. Vitamin E is generally considered to be relatively nontoxic to humans, but because it is a fat-soluble vitamin the Food and Nutrition Board, National Academy of Sciences, recommends caution.

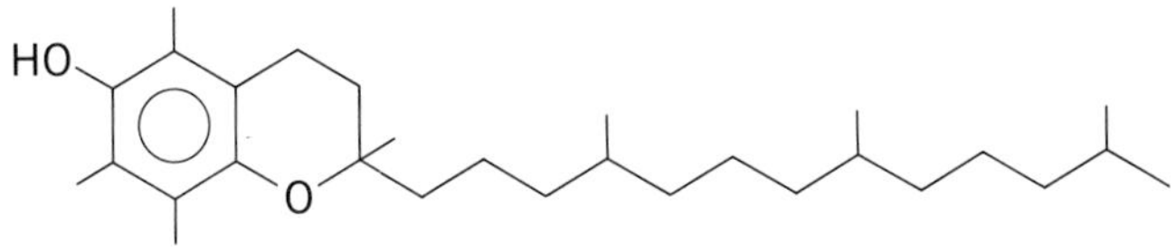

HO

O

α-Tocopherol

Vitamin K is an antihemorrhagic vitamin (see Section 15.22) present in green leafy vegetables. Only rarely does deficiency occur. Vitamin K activity is given by a small group of compounds, all quinones (e.g., Section 4.16). It is sometimes given to women just before childbirth and to infants just after birth to provide an extra measure of protection in the face of possible hemorrhaging.

20.10 Water-Soluble Vitamins

The water-soluble vitamins are identified by the Food and Nutrition Board as vitamin C, choline, thiamin, riboflavin, niacin, folacin, vitamin B_6, vitamin B_{12}, pantothenic acid, and biotin.

Vitamin C, or ascorbic acid, prevents scurvy, a possibly fatal disease in which collagen is not well made (see Section 12.15). According to some scientists (e.g., Linus Pauling), vitamin C does much more besides. A variety of studies have found that vitamin C is involved in the metabolism of amino acids, in the synthesis of some adrenal hormones, and in the healing of wounds. Since it has antioxidant properties, Pauling and others believe that vitamin C together with vitamin E is important in inhibiting the development of heart disease. To reduce the frequency and severity of colds, they also strongly urge relatively large doses, hundreds of milligrams to a few grams per day, compared with 35 to 45 mg/day recommended by the Food and Nutrition Board. The vitamin appears to be nontoxic at these higher levels, but some recent evidence points to the possibility that the prolonged presence of very high vitamin C levels may cause damage to genes. The vitamin is available in citrus fruits and juices, potatoes, leafy vegetables, and tomatoes. Extended cooking, heating over steam tables, or the prolonged exposure to air or iron or copper ions destroys it. Even when kept in capped bottles in a refrigerator, the vitamin C in commercial tablets (which contain a filler) slowly deteriorates.

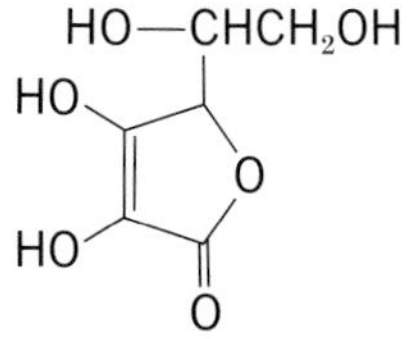

HO—$CHCH_2OH$ (ring: HO, HO, O, =O)

Ascorbic acid

$HOCH_2CH_2N^+(CH_3)_3$

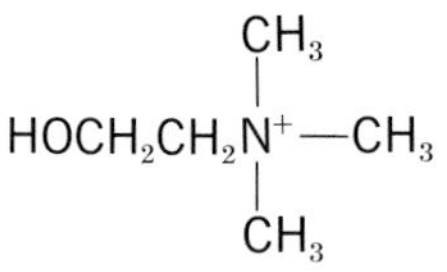

Choline

Choline is needed to make complex lipids (see Section 11.10) and acetylcholine serves in transmitting nerve signals (Section 13.20). Although the body can make choline, it is still classified as a vitamin by the Food and Nutrition Board because 10 species of animals, including dogs, pigs, and monkeys, have dietary requirements for it. It occurs widely in meats, egg yolk, cereals, and legumes. Although we make it, we may make it too slowly to meet all our needs and therefore we may require some supplement. (Deatherage—Table 20.2—does not list it.) No choline-deficiency disease has been demonstrated in humans, but in animals the absence of dietary choline leads to a fatty liver and hemorrhagic kidney disease.

Thiamin is part of an essential coenzyme for the catabolism of carbohydrates (see Section 23.4). Beri-beri is the thiamin-deficiency disease. Good sources are lean meats, legumes, and whole (or enriched) grains. Thiamin is not stored, and excesses are excreted in the urine. Our needs are related to our caloric intakes, and the Food and Nutrition Board recommends 0.5 mg/1000 kcal for children and adults. Thiamin is stable when dry but destroyed by alkaline conditions or by prolonged cooking.

Riboflavin is a greenish-yellow, fluorescent pigment necessary to the flavin coenzymes (see Sections 13.4 and 16.7) used in biological oxidations. Little if any is stored, and excesses in the diet are excreted. It is stable to heat and acid medium and is moderately stable to oxidation. Irradiation by light, the action of alkaline substances, or prolonged cooking destroy riboflavin. Deficiencies of riboflavin lead to the inflammation and breakdown of tissue around the mouth, nose, and the tongue, a scaliness of the skin and burning, itching eyes. Wound healing is impaired. The best source of riboflavin is milk, but certain meats (liver, kidney and heart) supply it. Cereals, unless enriched, are poor sources.

Niacin—meaning both nicotinic acid and nicotinamide (Section 13.4)—is essential to NAD^+ and $NADP^+$ (Section 16.7). Meat is a major source. Its deficiency disease is pellagra, characterized by skin problems and deterioration of the nervous system. Pellagra is a particular problem in areas where corn or maize makes up a major part of the diet because corn or maize is low in tryptophan. Niacin is one vitamin that we can to some extent make ourselves—from tryptophan, an essential amino acid. Each 60 mg of absorbed tryptophan in the diet produces about 1 mg of niacin, nearly the daily requirement for most people. If the diet is low in tryptophan, niacin must be provided in other foods (e.g., enriched grains). Niacin tends to be destroyed by prolonged cooking.

Folacin is the name used by the Food and Nutrition Board for folic acid (see Figure 13.7), pteroylmonoglutamic acid and related compounds that act biologically alike. It is needed for those enzymes that transfer one-carbon formyl groups (—CHO) in the syntheses of nucleic acids and heme. A deficiency disease is megaloblastic anemia. Folacin deficiency evidently is not a cause of pernicious anemia as was once thought. Fresh, leafy green vegetables, asparagus, and both liver and kidney are good sources. Folacin is relatively unstable to heat, air, and UV light, and its activity is often lost in both cooking and storage. Several drugs including alcohol promote folic acid deficiency.

Vitamin B_6 is actually a mixture—pyridoxine, pyridoxal, and pyridoxamine—and is needed by transaminases and other enzymes of amino acid metabolism. All three in the mixture are changed in the body to the coenzyme

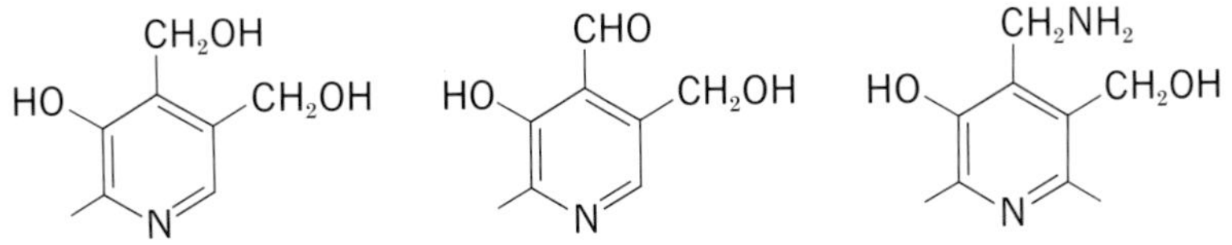

CH_2OH / HO / CH_2OH / N — Pyridoxine

CHO / HO / CH_2OH / N — Pyridoxal

CH_2NH_2 / HO / CH_2OH / N — Pyridoxamine

pyridoxal phosphate (see Section 13.4). The vitamin is present in meat, wheat, yeast, and corn. It is relatively stable to heat, light, and alkali. If deficient, the individual may develop hypochromic microcytic anemia and disturbances of the central nervous system. Roger Williams, a biochemist and nutritionist, believes that B_6 deficiency may also lead to atherosclerosis and heart disease. Some Danish researchers have accumulated evidence that this vitamin can be used to treat hangovers.

Vitamin B_{12}, or cobalamin, is a specific controlling factor for pernicious anemia. The deficiency disease, however, is very rare, and obtaining a diet that does not supply enough of this vitamin is difficult. Animal foods—particularly liver, kidney, lean meats, milk products, and eggs—are virtually the only sources, and the true vegetarians furnish most of the patients with B_{12} deficiency.

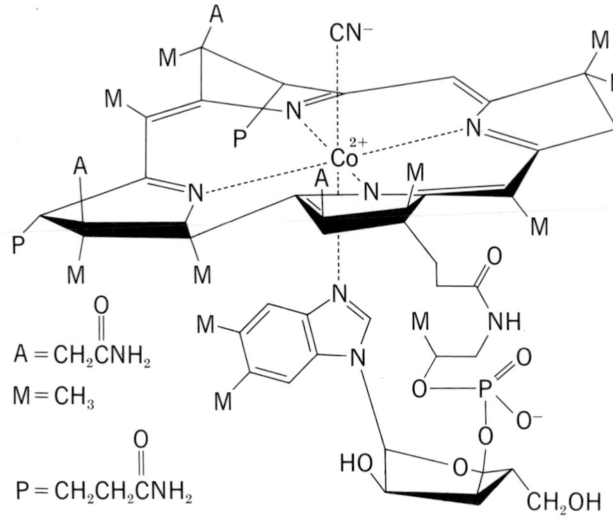

$A = CH_2CNH_2$ (C=O)

$M = CH_3$

$P = CH_2CH_2CNH_2$ (C=O)

Cyanocobalamin

Pantothenic acid is used to make the coenzyme A molecule (see Section 16.10). Signs of a deficiency disease have not been observed clinically in people, but the deliberate administration of panthothenic acid antagonists causes symptoms of cellular damage in vital organs. This vitamin is widely present in foods, particularly liver, kidney, egg yolk, and skim milk.

Biotin is required for those pathways where carbon dioxide is temporarily used, as in lipigenesis (Section 18.7). Signs of biotin deficiency are hard to find, but when the deficiency is deliberately induced, the individual experiences anorexia, nausea, pallor, dermatitis, and depression. When biotin is given again, the symptoms leave. Intestinal microorganisms probably make biotin. Egg yolk, liver, tomatoes, and yeast are good sources of biotin.

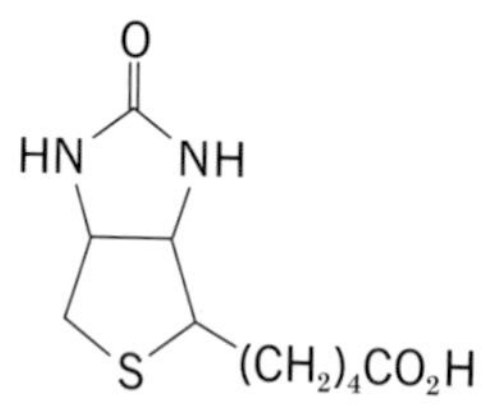

Biotin

20.11 Minerals

The minerals that we need at levels of 100 mg per day or more are calcium (Ca^{2+}), phosphorus (phosphate ion, P_i), magnesium (Mg^{2+}), sodium (Na^+), potassium (K^+), and chlorine (Cl^-). These are the principal **electrolytes** in the body. The other minerals are called the trace elements, and we need them in amounts of just a few milligrams per day.

Calcium is needed for bones, and about 1.2 kg of a 70-kg man is calcium with 99% of that in the skeleton. Only about 10 g calcium is found outside bone tissue, but it has several vital functions. Calcium ion is one key factor in blood clotting; it is essential to the mechanism whereby muscles contract and relax; and it aids in controlling the excitability of peripheral nerves. To insure that calcium is everywhere present to serve these purposes, the body has a homeostatic mechanism under hormonal control for regulating the calcium ion level in blood. Additional calcium must be in the diet during pregnancy and lactation. Milk is our best single source.

Magnesium is essential to several enzyme systems and inside the cell is, next to potassium, the most abundant cation. Some scientists believe that the magnesium ion is a master coordinator of metabolic events in some cells. As its availability is decreased, several cell functions decline all at the same time; as it is increased, these functions all accelerate. Magnesium deficiency in humans causes hyperexcitability, other behavioral disturbances, and may lead to convulsions. Fortunately, the ion is widely available in food and deficiencies are very rare.

Sodium, the chief metallic ion of extracellular fluids, is essential to osmotic pressure relations and total fluid volume. **Potassium,** the principal metallic ion inside cells, is involved with the functions of enzymes. The concentration gradients of sodium and potassium ions across cell membranes are essential to the conduction of nerve impulses. The normal potassium level in serum is 3.5 to 5.3 milliequivalents per liter. Either too much or too little causes death by cardiac arrest, but this ion at the right level is essential to good heart muscle action. During physical activity, potassium ions leave muscle cells (being replaced by sodium ions), and this leads to fatigue. When aldosterone instructs the kidneys to retain sodium ion (see Section 15.18), they generally excrete more potassium ion. Potassium deficiencies (which are more common than excesses) arise in a variety of conditions: stress, injury or surgery, burns, starvation, dehydration, vomiting, and diarrhea. These losses most commonly cause muscular weakness, a general malaise, dizziness, arrhythmia, and distention of the abdomen.

The serum level of sodium ion is normally 135 to 146 mEq/liter. This level drops as a result of burns, surgery, gastric suction, diarrhea (in adults), moderate vomiting, sweating, or drinking copious amounts of water. Signs are muscular weakness, abdominal cramps, irritability, headache, nausea, and vomiting. In severe vomiting more water is lost than sodium and the serum sodium level goes up. Diarrhea in babies also generally causes the loss of more water than sodium. Other causes of increased serum sodium ion level are dysfunction of the adrenal gland (producing too much aldosterone) or dysfunctions of the kidneys or heart. At too high serum sodium levels the skin is flushed and the body temperature is high, the tongue is rough and dry, and the heartbeat is fast (tachycardia).

The **chloride** ion is the principal negative ion in extracellular fluids. Its plasma level is normally 104 mEq/liter, whereas inside cells only traces are present: 1 to 4 mEq/liter. Chloride ion is essential for maintaining both fluid and electrolyte balance and is needed to make the hydrochloric acid of gastric juice. It normally is ingested together with sodium ion.

20.12 Trace Elements

The Food and Nutrition Board recognizes that 17 **trace elements** have been shown to have various biological functions in animals. Although all of them may some day be found to be essential to humans, the Board lists only 10 as thus far known to be required: fluorine, chromium, manganese, iron, cobalt, copper, zinc, selenium, molybdenum, and iodine. All are toxic in excessive amounts.

These elements occur quite widely in a variety of foods and drink, but many tend to be removed by food refining and processing.

Fluorine (as fluoride ion, F^-) is essential to the growth and development of sound teeth, and the Food and Nutrition Board recommends that public water supplies be fluoridated wherever natural fluoride levels are too low.

Chromium (Cr^{3+}) is required for normal glucose metabolism, and it may serve as a cofactor for the action of insulin. Chromium that occurs naturally in foods is significantly more easily absorbed than chromium given in simple salts. Most animal proteins and whole grains supply chromium.

Manganese (Mn^{2+}) is required for normal nerve function, for the development of sound bones, and for reproduction. It occurs in enzyme systems. Nuts, whole grains, fruits, and vegetables supply this element, but a recommended daily allowance cannot yet be set.

Iron (Fe^{2+}) is needed in heme and several enzymes. The intestines act to regulate how much dietary iron is actually absorbed in order to maintain a proper level in circulation. A high serum iron level apparently renders the individual more susceptible to infections.

Cobalt (Co^{3+}) is part of the vitamin B_{12} molecule and apparently has no other used in humans.

Copper (Cu^{2+}) occurs in a number of proteins and enzymes. If deficient in copper the individual synthesizes low-strength collagen and elastin and will suffer anemia, skeletal defects, and degeneration of the myelin sheaths of nerve cells. Ruptures and aneurysms of the aorta become more likely. The structure of hair is affected, and reproduction tends to fail. Fortunately, copper occurs widely in foods, particularly in nuts, raisins, liver, kidney, certain shellfish, and legumes; an intake of 2 mg/day assures a copper balance for nearly all people. Unhappily, however, the copper content of the foods of a typical U.S. diet has declined during the last 3 to 4 decades, and some scientists believe that the large increase in the incidence of heart disease in the United States during that same period has been caused partly by copper-low diets, or diets low in a proper balance of copper and zinc. (In high doses, copper salts are dangerous poisons.)

Zinc (Zn^{2+}) is present in many enzymes and in nucleic acids and bone. Without sufficient zinc in the diet loss of appetite, poor wound healing, and failure to grow soon appear. Where the deficiency is severe and chronic, as in the Middle East, dwarfism and poor gonadal development are observed. Some evidence exists that too little copper in relation to zinc contributes to heart disease. If the zinc to copper ratio is 14:1, for example, experimental animals suffer increased serum cholesterol levels that are believed to be responsible for hardening of the arteries. Thus too much zinc, in relation to copper, may be unhealthy. High lipid and sucrose diets favor a high zinc to copper ratio; high fiber foods (cereals, nuts, and legumes) lead to a better, lower ratio. The ratio in human milk is 6:1; in cow's milk, 38:1. Exercising tends to move some zinc out of the body in the sweat leaving a lower zinc to copper ratio. This may be one factor in the lower incidence of heart disease in those who exercise. The ideal ratio for humans is not known, and the connection between this ratio and heart disease is still in the speculative stage with research underway for more data.

Selenium is known to be essential in many animals including humans. How much we need is not known; too much is very toxic. A strong statistical relation exists between high levels of selenium in livestock crops and low incidences of human deaths by heart disease. Those who live in the United States where selenium levels are high—the great plains between the Mississippi River and the eastern Rocky Mountains—have one-third the chance of dying from heart attack and strokes as those who live where levels are very low—northeastern quarter of the United States, Florida, and the Pacific Northwest. Rats, lambs, and piglets on selenium-poor diets develop damage to heart tissue and abnormal electrocardiograms.

Molybdenum is in xanthine oxidase, an enzyme involved in nucleic acid metabolism. Deficiencies in humans are unknown meaning that almost any reasonable diet furnishes enough.

Iodine (I^-) is essential in hormones made by the thyroid gland, and iodine-deficient diets lead to thyroid enlargement known as goiter. About 1 μg/kg of body weight must be ingested each day to prevent goiter. Seafoods are excellent sources of iodine, but iodized salt with 75 to 80 micrograms iodine/g of salt is the surest way of getting the iodine that is needed.

Nickel, silicon, tin, and vanadium are possibly essential trace elements for humans because deficiency diseases for these elements have been induced in experimental animals.

20.13 Vitamins in Therapy

The classic vitamin-deficiency diseases—scurvy, pellagra, beri-beri and others—all respond well, of course, to supplemental vitamins. Of more recent interest is the emergence of a small group of scientists and physicians who believe that a number of other ailments, ranging from mental illness to the common cold, can be prevented or cured by the use of relatively massive doses of vitamins. Linus Pauling is the most popularly known figure in this movement, which has been identified as **megavitamin therapy** or orthomolecular therapy. Pauling's espousal of vitamin C was mentioned in Section 20.10.

According to Pauling, citing the research of others, mental illness can happen if the concentration of any of the following vitamins in the brain is too low: thiamin, niacin, pyridoxine (B_6), vitamin B_{12}, biotin, folic acid, or vitamin C. If these deficiencies are remedied by megavitamin therapy, according to Pauling, the treatment may be the best single approach for many patients. Pauling does not claim that it is the only approach.

The National Institute of Mental Health discourages the treatment of mental illness by orthomolecular therapy. The American Psychiatric Association has reported findings that the Pauling approach has uniformly negative value, and has expressed concern over long-range toxic effects of megavitamin doses. No doubt, more research on both sides of the controversy will be performed.

20.14 Food Additives

Controversy swirls around the food additives. A **food additive** is defined by the National Research Council-National Academy of Sciences as "a substance or mixture of substances, other than a basic food stuff, which is present in food as a result of any aspect of production, processing, storing, or packaging. The term does not include chance contaminants" (e.g., pesticide residues). The types of food additives are the following:

1. *Color additives*—food dyes added for aesthetic purposes. Some that were used for decades (e.g., Red Dyes Nos. 1 and 4) were banned in 1976, because they were found to be carcinogenic in test animals.
2. *Enzymes*—those used in making cheese, bread, crackers, for example, or meat tenderizers.
3. *Vitamins and amino acids*—Vitamin supplements are familiar. Amino acids are sometimes added to increase the availability of a limiting amino acid (e.g., lysine in corn products).
4. *Antimicrobial agents*—Some of these we have studied: sodium benzoate (Section 7.25), propionates (Section 7.26), sorbates (Section 7.27), and parabens (Section 7.28). Sodium nitrite and sodium nitrate are added to a variety of sausage and luncheon meats to inhibit bacteria (and give a redder color), but considerable evidence exists that these anions can cause cancer. They represent one area of great controversy, and they may be on their way out in the United States.
5. *Antioxidants*—BHA and BHT (Section 14.19) are examples we have studied.
6. *Acidulents*—These are organic acids, such as citric, malic, lactic, sorbic, succinic, and fumaric acids, added to enhance flavors, control acidities, viscosities and hardness, and aid in the curing of meat. The ions of all those listed are intermediates in metabolism.
7. *Sequestrants*—These sequester or tie up metal ions that otherwise might promote the deterioration of food. Trace metals, for example, often accelerate the onset of rancidity.
8. *Gums*—These are thickening agents: gum arabic, gum tragacanth as well as pectin. They also serve as glazes, bulking agents, crystal preventers in ice cream and syrup, clarifying agents, and suspending agents.
9. *Starch and modified starch*—Powdered sugar has some starch to prevent caking.
10. *Surface-active agents*—These are wetting agents, emulsifiers, lubricants, and thickeners. Monoacyl glycerols or sorbitols are often used.
11. *Polyhydric alcohols*—Glycols, glycerol, sorbitol, and mannitol are often added to fondants, caramels, and marshmallows to retard crystallization and help them stay soft.
12. *Natural and synthetic flavors.*

13. *Flavor potentiators*—Monosodium glutamate (as in Accent ®) is the most common example.
14. *Nonnutritive sweeteners*—Controversy has existed for over a decade about the use of saccharin and cyclamates in diet soft drinks and other foods and beverages.

The original intents of food additives, as the list above indicates, included safety from harmful bacteria, long shelf-life for packaged products, replacement of nutrients lost in processing, or augmentation of nutrients missing and flavor and texture enhancement. None of these intents is controversial. The problems arise over the safety of individual additives, problems not easily solved to the satisfaction of all well-meaning people. Definitive tests are very difficult to conduct inasmuch as direct experimentation on humans is generally unacceptable. Interpreting test data obtained from experiments with animals is always difficult. What relation exists, for example, between a higher incidence of bladder tumors in mice when cyclamate is directly implanted in the mice and the possibility of bladder tumors in humans when cyclamates are taken orally and in vastly lower daily amounts? We do not know for certain. In fact, certain knowledge is impossible. How do we get reliable data concerning humans without exposing them to the risk of cancer? We cannot, unless throngs volunteer (as occurred, in a way, in the cigarette versus lung cancer controversy). In the study of the molecular basis of life we have just concluded we have found ample cause to be conservative in our use of alien and untested substances.

Brief Summary

Nutrition. Good nutrition entails the ingestion of all those substances needed for health: water, oxygen, food energy, essential amino acids and fatty acids, vitamins, and minerals. Foods that best supply the organic substances and minerals have been identified. Recommended daily allowances issued by the Food and Nutrition Board of the National Academy of Sciences—which allow for variations in individual needs and in the digestibility of foods—represent intakes adequate to meet known nutritional needs of practically all healthy people. Some need more; most people need less, but neither more nor less of anything is necessarily better. The RDAs have to be interpreted statistically. Moreover, they should be obtained through a varied diet; some trace substances may not yet be identified but yet supplied sufficiently in a varied diet. Food processors often use food additives to improve the looks, shelf-lives, and textures of foods and to inhibit disease microorganisms.

Water Needs. If you are thirsty, you need water. You cannot train to go long without water. You need more when it is hot and you are working hard, conditions in which you also may need to replace electrolytes.

Energy. The chemical energy we need is in food and oxygen combined. Both carbohydrates and lipids should be used as sources of food energy. Some lipid in the diet tends to insure ingestion of essential fatty acids. Some carbohydrate insures against too heavy a use of gluconeogenesis.

Protein. We need both a certain total nitrogen as well as several essential amino acids. Proteins supplying these needs are in most foods, but meat and dairy products are generally superior both in coefficients of digestibility and in biological value. The most important factor in biological value is the limiting essential amino acid, with human milk protein or whole egg protein serving as standards of excellence. For several reasons—low proportions of some essential amino acids, relatively low digestibility coefficients, low concentrations of proteins—several foods cannot be exclusive or major components of a healthy diet. These include corn (maize), rice, potatoes, cassava, or manioc. Youngsters on such diets can be expected to develop marasmus or kwashiorkor and experience retarded physical and mental development.

Vitamins. Besides carbohydrates, lipids, and proteins, several organic compounds—the vitamins—must be in the diet because they cannot be made and are vital to health. If any is absent, a specific disease or a set of unhealthy responses result. Each corresponds to one vitamin but can be relieved often by the vitamin or one of a few substances with vitamin activity. Excesses of the fat-soluble vitamins (A, D, E, and K), especially A and D, should be avoided. A small group of scientists led by Linus Pauling urges the use of large quantities of vitamin C (10 times the RDA) to prevent or cure the common cold, and they advocate a similar megavitamin usage of water-soluble vitamins to help alleviate mental illness. Most psychiatrists disagree with Pauling.

Minerals. The trace elements are needed in enzyme systems. The other minerals are important in bones and teeth (calcium, phosphorus especially) or in regulating metabolic activities, body fluids, and osmotic pressure balances.

Selected References

Books

1. National Academy of Sciences. *Recommended Dietary Allowances,* 8th ed., 1974.
2. F. E. Deatherage. *Food for Life.* 1975. Plenum Publishing Corporation, New York. A survey of the interacting principles of physical, biological, and social sciences that affect how food is produced, processed, distributed, and consumed.
3. S. R. Williams. *Nutrition and Diet Therapy,* 2nd ed., 1973. The C. V. Mosby Company, St. Louis. A textbook in nutrition directed primarily to professional nursing students as well as those in other allied health fields.
4. J. L. Kee. *Fluids and Electrolytes with Clinical Applications.* 1971. John Wiley & Sons, Inc., New York. A programmed learning book (soft cover) for students in the allied health sciences.
5. T. E. Furia, ed. *Handbook of Food Additives.* 1968. The Chemical Rubber Co., Cleveland. A discussion of the properties, antimicrobial activities, safety, regulatory status, and applications of food additives.
6. *Nutritive Value of Foods,* revised 1971. Home and Garden Bulletin No. 72, U.S. Department of Agriculture. U.S. Government Printing Office, Washington, D.C. An extensive tabulation of nutritive values for household measures of commonly used foods.

Articles

1. E. Frieden. "The Chemical Elements of Life." *Scientific American,* July 1972, page 52.
2. L. Pauling. "Orthomolecular Psychiatry." *Science,* 19 April 1968, page 265.

3. G. O. Kermode. "Food Additives." *Scientific American,* March 1972, page 15.
4. D. A. Levitsky. "Ill-Nourished Brains." *Natural History,* October 1976, page 6. Will the world's 400 million malnourished children suffer permanent intellectual retardation? The author opens with this question and the answer is "Not necessarily." These children do have smaller brain weights and they do learn slower than better-fed youngsters, but is the latter caused by the former? Is slow learning the result of the smaller brain or the result of the mental apathy of malnutrition, an apathy that means disinterest in learning? Definitive answers are not known yet, but some evidence suggests that the malnutrition is the cause and that on improved diets learning will improve.

Questions and Exercises

1. What is the difference between a nutrient and a food?
2. What are three areas of study that concern the research activities of nutritionists?
3. Why does the National Academy of Sciences set its recommended daily allowances higher than what, on strictly experimental grounds, are minimum daily requirements?
4. What are seven situations requiring special therapeutic nutrition?
5. Why ought our daily diet be drawn from a variety of foods?
6. What can be expected eventually to happen on each diet?
   (a) A carbohydrate-free diet
   (b) A lipid-free diet
   (c) An all brown rice diet
7. What can be expected to happen eventually on a diet free of each?
   (a) Vitamin C (b) Niacin (c) Selenium
   (d) Potassium (e) Vitamin A (f) Iodide ion
8. If we are able to make all the nonessential amino acids, why must our daily protein intake include much more protein than that represented by the essential amino acids?
9. Which probably has a higher coefficient of digestibility and why: keratin or albumin?
10. What does the milling of grains do, nutritionally?
11. Why does the protein in corn have a lower biological value than the protein in whole eggs?
12. Which proteins generally have higher biological values, those of meats or of fruits?
13. How do kwashiorkor and marasmas differ?
14. Why is the intake of adequate protein particularly important to the mental growth of an infant?
15. On a strict vegetarian diet—no meats, eggs, or dairy products of any sort—what one vitamin is the hardest to obtain?
16. Why ought strict vegetarians use two or more different sources of proteins?
17. What problem arises if a vegetarian eats rice in the morning and beans in the evening and ingests protein from no other source?

18. Why are the essential amino acids not classified as vitamins?
19. Make a table showing each vitamin, at least one good source of each and a serious consequence of a deficiency of each (according to information in this book). Use these headings:

Vitamin	Source(s)	Problem(s) If Deficient

20. What are the major minerals in each?
    (a) Bone (b) Fluids inside cells (c) Fluids outside cells
21. Make a table of the 10 trace elements and at least one particular function of each.
22. What are the four general intents for using food additives?

appendix 1

Important Concepts of General Chemistry

This appendix provides a brief review of basic concepts from general chemistry that are particularly vital to the study of organic chemistry. We assume that general chemistry has been studied at the college level before and that all that now is needed is an identification and a review of what is important. The terms in boldface must be learned.

The fundamental building units of all matter are **atoms,** and nature has about 90 chemically different kinds, each comprising a **chemical element.** (Counting man-made elements there are about 105.) All atoms are alike in consisting of an atomic **nucleus** and **electrons.** They differ in three important ways, physically: their masses, the positive charges in their nuclei (**atomic numbers**), and their electronic configurations.

Atomic nuclei consist principally of **protons** (positively charged) and **neutrons** (neutral). The sum of the neutrons and protons is the **atomic mass** of the atom in atomic mass units. Atoms of the same element that differ in their numbers of neutrons (but not protons) are called **isotopes.**

The atomic number also equals the number of electrons in an atom. The arrangement of these electrons about a nucleus is the **electronic configuration** of the atom. The electrons reside in regions of space near the nucleus called **atomic orbitals,** no more than two electrons per orbital. The four kinds of orbitals are the *s*-, *p*-, *d*-, and *f*-orbitals. Only the *s*- and *p*-orbitals are important to our study of organic chemistry. The region defining an **s-orbital** has the general shape of a sphere and that of a **p-orbital,** the general shape of two spheres that barely touch, as seen in Figure A.1. The axes of *p*-orbitals, which always occur in sets of three, point in mutually perpendicular directions.

Orbitals generally cluster in **sublevels** that make up broader regions called **principal energy levels.** The first level, which corresponds to lowest energy, has but one sublevel consisting of only one orbital, called the 1*s*-orbital. At the second principal energy level are two sublevels with two kinds of orbitals—one of the *s*-type, called the 2*s*-orbital, and three of the *p*-type, called the $2p_x$, the $2p_y$ and the $2p_z$. (The *x*, *y*, and *z* correspond to the three perpendicular axes.) The details about level three and higher are not needed for our study.

The electronic configurations of elements 1 through 20 are given in Table A.1. Of these, only carbon, hydrogen, oxygen, nitrogen, sulfur, phosphorus, and chlorine are important to our study. A knowledge of their electronic configurations is important because so much of their chemistry can be

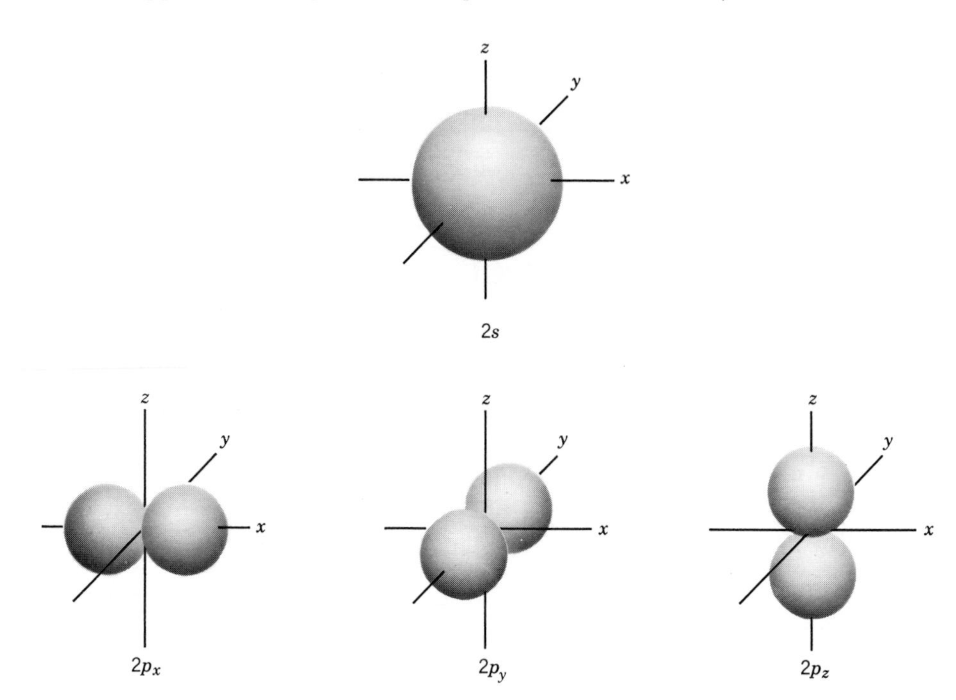

Figure A.1 Principal energy level number 2 contains both the *s*- and the *p*-types of orbitals.

correlated in those terms. The electronic configuration of the atoms of an element determines the chemistry and the chemical-bonding ability of an element, and the configuration in the outside level is the most significant of all.

Elements that have the same outside level configuration generally have very similar chemical properties and are members of the same **chemical family.** These families are best displayed in a **periodic table** (see inside cover), where they make up the vertical columns, also called **groups.** For example, the Group 7 elements, the Halogens (flourine, chlorine, bromine and iodine), consist of atoms all of which have seven electrons in their outside levels. The Alkali Metals, Group 1, atoms have one electron in their outside levels. The Noble Gas atoms (Group 0), the outside levels hold what are called **outer octets**—eight electrons—except that helium, the smallest in atomic mass and number, has level number 1 as its outside level, and it can hold only two electrons. The Noble Gases are the most chemically inert of all elements, and their unreactivity is associated with their outer level configurations.

Metals generally have 1, 2, or 3 outside level electrons, sometimes more, particularly when they have high atomic weights. **Nonmetals** generally have 4, 5, 6, 7, or 8 outside level electrons. Virtually all elements important structurally in organic chemistry are nonmetals.

In addition to the elements, there are two other important classes of matter—**compounds** and **mixtures.** Elements consist of atoms of identical atomic numbers, a different atomic number for each element. Compounds consist of at least two elements chemically combined and present in definite

TABLE A.1
Electronic Configurations of Elements 1 to 20 Showing Distributions Among the Orbitals[a]

Atomic Number	Element	1s	2s	$2p_x$	$2p_y$	$2p_z$	3s	$3p_x$	$3p_y$	$3p_z$	4s
1	H	↑									
2	He	⇅									
3	Li	⇅	↑								
4	Be	⇅	⇅								
5	B	⇅	⇅	↑							
6	C	⇅	⇅	↑	↑						
7	N	⇅	⇅	↑	↑	↑					
8	O	⇅	⇅	⇅	↑	↑					
9	F	⇅	⇅	⇅	⇅	↑					
10	Ne	⇅	⇅	⇅	⇅	⇅					
11	Na	⇅	⇅	⇅	⇅	⇅	↑				
12	Mg	⇅	⇅	⇅	⇅	⇅	⇅				
13	Al	⇅	⇅	⇅	⇅	⇅	⇅	↑			
14	Si	⇅	⇅	⇅	⇅	⇅	⇅	↑	↑		
15	P	⇅	⇅	⇅	⇅	⇅	⇅	↑	↑	↑	
16	S	⇅	⇅	⇅	⇅	⇅	⇅	⇅	↑	↑	
17	Cl	⇅	⇅	⇅	⇅	⇅	⇅	⇅	⇅	↑	
18	Ar	⇅	⇅	⇅	⇅	⇅	⇅	⇅	⇅	⇅	
19	K	⇅	⇅	⇅	⇅	⇅	⇅	⇅	⇅	⇅	↑
20	Ca	⇅	⇅	⇅	⇅	⇅	⇅	⇅	⇅	⇅	⇅

[a] From 21 to 30, electrons will go into the available 3*d* orbitals until they are filled and then 4*p* orbitals start to fill. Elements 21 to 30 make up the first group of transition elements.

proportions. Mixtures consist of two or more elements or compounds physically intermingled and not chemically combined; they are characterized by being able to exist in a wide range of proportions. They do not obey the **law of definite proportions** as do compounds (and, of course, elements). One very important kind of mixture is the **solution,** particularly the solution of a gas in a liquid or a solid in a liquid.

Two atoms can become strongly joined by a **chemical bond** principally in one of two ways—by changing into **ions** that attract each other or by sharing electrons between them and toward which their nuclei are mutually attracted. In the first way, one atom may give up one, two (or occasionally three) electrons to another atom. The donor becomes a **cation,** a positively charged ion. The acceptor becomes an **anion,** a negatively charged ion. These oppositely charged ions attract each other and aggregate in whatever definite ratio insures an overall electrical neutrality. The product is called an **ionic compound** and the force of attraction between the oppositely charged ions is called the **ionic bond.** Generally, these bonds form when one atom is a metal and the other is a nonmetal. Ionic compounds are typically salts and solids at room temperature. All those containing any of the following ions are soluble in water (at least, to the extent of 3-4 grams per 100 ml): Na^+, K^+, NH_4^+, NO_3^-, $C_2H_3O_2^-$ (acetate) and nearly all salts involving Cl^-.

A chemical bond formed by sharing electrons results from the interpenetration of atomic orbitals to make a larger orbital, called a **molecular orbital.** It encompasses both nuclei and contains the **shared pair** of electrons. This pair provides enough electron density between the two nuclei to attract both and hold them near each other. This bond is called a **covalent bond.** Compounds featuring covalent bonds consist of **molecules,** rather than oppositely charged ions. Molecules are tiny, electrically neutral particles in which two or more atoms are held together by covalent bonds.

The covalent bond often is a **polar bond,** because one of the two atoms may get more than a 50:50 share of the shared electrons. It will have a fractional excess of negative charged (symbolized by $\delta-$) leaving the other with a deficiency (symbolized by $\delta+$). The atom with a $\delta-$ is more **electronegative** than the one at the other end of the bond with a $\delta+$. The most electronegative of all elements is fluorine, and an abbreviated version of the order of relative electronegativities is: F > O > N > Cl > Br > C > H.

Covalent bonds have definite directions because the overlapping atomic orbitals that make them have direction. If polar bonds in a molecule project in a way that does not cancel each others polarity, the molecule as a whole will be polar. Thus the water molecule, which is angular, is quite polar whereas the carbon dioxide molecule, which is linear, is not.

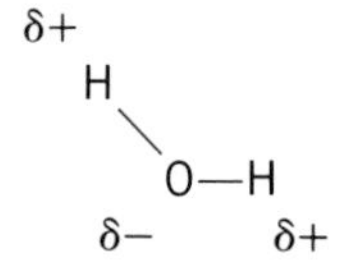

δ+
H
\
O—H
δ− δ+

Water molecule

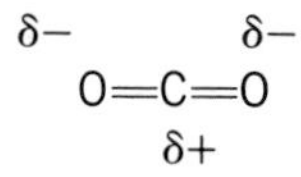

δ− δ−
O=C=O
δ+

Carbon dioxide molecule

These symbols illustrate typical **structural formulas,** the most important kind of formula in organic chemistry. Each line to or from an atom represents one shared pair of electrons. We note that more than one pair may be shared and that carbon dioxide has double bonds. (Triple bonds are also possible, as in nitrogen, N N. Double and triple bonds are common among organic substances.)

One very important polar molecule is that of water. Between water molecules exists a force of attraction which, although weak, is strong enough to be called a bond—the **hydrogen bond.** The hydrogen bond is a force of attraction between a $\delta+$ on a hydrogen atom when it is covalently bound to a strongly electronegative atom (e.g., oxygen or nitrogen) and a $\delta-$ on another electronegative atom (O or N) to which that hydrogen is not covalently bound. The hydrogen bond is usually symbolized by a dotted or dashed line, as illustrated here for water:

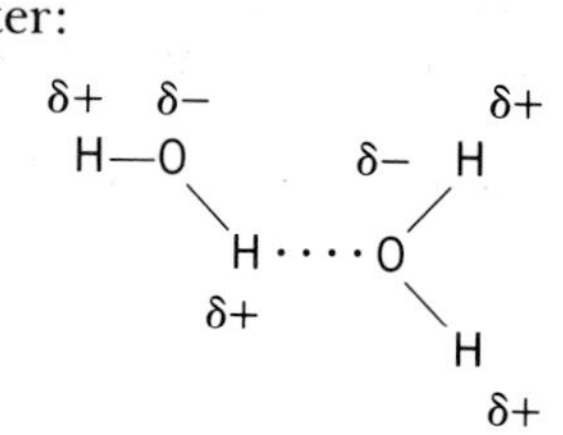

δ+ δ− δ+
H—O δ− H
\ /
H····O
δ+ \
H
δ+

This bond also occurs among substances in several organic families, e.g., the alcohols, amines, phenols and acids.

The **octet rule** governs to which extent an atom may donate or share electrons. This rule states that the tendency for donating or sharing extends to the atom's acquisition of a full outer octet or, at least, a share in such an octet. (At level 1, a filled level of two electrons constitutes the condition of stability.) The rule works best among the first 20 elements in the periodic system and among those in the main groups numbered 1-7 of the periodic table. Thus elements in Group 1, whose atoms all have but one electron in their outside levels, in their reactions generally lose just that one electron and become ions bearing charges of +1 (e.g., the sodium ion, Na^{+}). Elements of Group 2 either lose two electrons or do not react; they become ions with charges of +2 (e.g., the magnesium ion, Mg^{2+}). Elements in Group 7 may accept one electron per atom and become negatively charged ions (e.g., Cl^{-}, the chloride ion) or they may form one covalent bond by accepting a share in one more electron. Elements in Group 6 may accept two electrons (e.g., the sulfide ion, S^{2-} from a sulfur atom) or they may accept a share in two electrons and establish two convalent bonds. Elements in Groups 4 and 5, particularly the lighter ones (e.g., carbon and nitrogen), for all practical purposes never form ions. Instead those in Group 4 establish four covalent bonds and thereby acquire a share in an outer octet. Those in Group 5 can form three covalent bonds. The number of covalent bonds that an atom of an element can have is called its **covalence number.** The important covalence numbers are these: carbon, 4; nitrogen, 3; oxygen, 2; sulfur, 2; any of the halogens, 1; and hydrogen, 1.

A **chemical reaction** is an event in which substances are changed to other substances and observable changes in physical qualities take place. Heat is sometimes evolved. At the atomic-molecular-ionic level, a chemical reaction is an event in which a more or less permanent rearrangement or redistribution of electrons relative to atomic nuclei takes place. **Physical changes** do not involve such rearrangements. In chemical reactions the interacting particles always react in definite proportions by particles. For purposes of describing the quantities that react in weight units, a reacting unit of a substance called the **mole** is defined as that quantity of the substance that has as many unit particles (atoms, molecules, ions, for example) as there are atoms in 12 grams of the carbon-12 isotope. To compute the weight of one mole of any chemical, compute the **formula weight**—the sum of the atomic weights of everything in the formula—and add the word gram. Equal numbers of moles contain equal numbers of particles.

Three particularly important families of chemicals are **acids, bases** and **salts.** They are variously defined by the Arrhenius, the Brønsted, and the Lewis theories, each one broader than and including the preceding theory. For the study of organic chemistry, the **Brønsted theory** is generally the most useful, although the Lewis occasionally is needed. According to the Bronsted theory, an acid is a proton-donor and a base is a proton-acceptor, the **proton** being in this situation identical with a hydrogen ion, H^+. These definitions include the **hydronium ion** (the one Arrhenius acid), H_3O^+, and the **hydroxide ion** (the one Arrhenius base), OH^-. A **strong acid** is a powerful donor of a proton and in water is characterized by a high percent ionization. A **strong base** is a powerful binder of protons. Among the inorganic or mineral acids the strong acids are perchloric ($HClO_4$), sulfuric (H_2SO_4), nitric (HNO_3), hydrochloric (HCl), hydrobromic (HBr) and hydriodic acid (HI). All other acids are weak acids. The strong bases are sodium and potassium hydroxide (NaOH and KOH). All others are weak bases. The **conjugate base** of a weak acid, the particle resulting when the weak acid yields up a proton, is a relatively strong base. The **conjugate acid** of a weak base, the particle resulting when the weak base acquires a proton, is a relatively strong acid.

Strong mineral acids react with and are neutralized by metal hydroxides, metal carbonates, metal bicarbonates, and certain metal oxides, in each reaction forming water and a salt. (Carbon dioxide is also produced from the carbonates and bicarbonates.) A **salt** is an ionic compound in which a positive ion (other than H^+) is combined with a negative ion (other than OH^-).

Reactions are symbolized by **chemical equations** in which the chemical symbols of the reactants, separated by plus signs, are followed by an arrow pointing to the symbols the products. In a **balanced equation** identical numbers of each kind of atom must appear on both sides of the arrows and the net electrical charge on one side must equal that on the other. **Coefficients,** numbers in front of formulas in equations, are juggled to achieve the correct balance once correct formulas have all been identified and written.

The most useful kind of equation is the **ionic equation,** one that shows formulas only of the actual reactants and their products (whether they are ions or molecules or atoms), and it omits all "spectator" particles. With a knowledge of which are the strong acids and bases and which are the water-soluble salts, a complete equation may be rewritten in an ionic form by breaking up particles that are known to exist as ions anyway and crossing out those that undergo no change (the "spectators"). We illustrate this with the reaction of sodium hydroxide and sulfuric acid that gives sodium sulfate and water. (Water is not ionized.)

Complete Equation

$2NaOH$	+	H_2SO_4	$\longrightarrow$	Na_2SO_4	+	$2H_2O$
Sodium hydroxide		Sulfuric acid		Sodium sulfate		Water
(strong base)		(strong acid)		(soluble salt)		

Full Ionic Equation (an aid to obtaining the Ionic Equation)

$(2Na^+ + 2OH^-)$	+	$(2H^+ + SO_4^{2-})$	$\longrightarrow$	$(2Na^+ + SO_4^{2-})$	+	$2H_2O$
[Sodium hydroxide is fully ionized since it is a strong base]		[Sulfuric acid is fully ionized —or nearly so— since it is a strong acid]		[Sodium sulfate is fully ionized since it is a water-soluble salt]		[Water is not broken up; it is not ionized]

Ionic Equation. (We cancel all spectators and reduce remaining coefficients to their lowest whole numbers.)

$$OH^- + H^+ \longrightarrow H_2O$$

We need the complete equation when we prepare to carry out the experiment and must measure out the proper amounts of the reactants. We use the ionic equation when we want to focus our attention on the key chemical event in the reaction—in our example, simply the joining of OH^- and H^+. Some error lurks in using "fully ionized" in describing the substances above, but the problem is very minor in our study.

Concentrations of acids and bases are often expressed as **molar concentrations**—the number of moles of solute per liter of solution.

We have not by this review touched base on all the important topics of general chemistry. Those that remain and are needed in our study will be discussed where their need is encountered. If the foregoing seems a bit sketchy, consider it an indication that a more extensive review is in order. Either go back to your freshman chemistry textbook or use one of several paperback chemistry outlines on the market.

appendix 2

The R/S Families of Absolute Configuration

The *R/S* system, unlike the D/L system, is independent of reference compounds. Family membership is determined solely by a set of rules. Two sets of rules are involved.

The first set establishes a priority sequence for the four groups at a chiral center. The lowest priority is assigned number 1. The highest priority is assigned number 4. (The rules will follow.)

The second set of rules tells you first to hold up the molecular model, or draw it on paper, or imagine that the atom or group with the lowest priority—1—is behind the chiral carbon. Imagine it being behind a steering wheel and along the steering column with you seated in the driver's seat. This places the remaining three groups around the rim of the steering wheel. See Figure A-1. Next, trace a circle by moving your fingers from the group of priority 4, then to 3, and then to 2. Remember, first 4, then 3, then 2. If your finger moves clockwise, we say the groups are in a rectus or *R*-configuration. ("Rectus" means "to the right" in Latin.) But if your hand has to move counterclockwise to go from 4 to 3 to 2, then the configuration is said to be sinister, or *S*-configuration. ("Sinister" means "to the left" in Latin.)

The Priority Rules

1. Consider the atoms directly attached to the chiral center. The lower the atomic number, the lower the priority. Hence, a hydrogen atom, if present, will always have priority number 1. ("Low priority—low number.")

2. If two or more first atoms are the same, look at the atoms attached directly to them; "the lower their atomic numbers, the lower the priority".

For example, the whole ethyl group at a chiral carbon will have a higher priority than a methyl group.

$$—CH_2—H \qquad —CH_2—CH_3$$

In each case the first atom is carbon. The "second atom" (actually, of course, a set of three second-atoms) are—H, —H, —H for methyl and —H, —H, and —CH_3 for ethyl. Its set includes an atom of higher atomic number than any in the CH_3 set. Hence the ethyl group has a higher priority than the methyl group. Thus in 2-chlorobutane, $CH_3\underset{\displaystyle Cl}{\underset{|}{C}}HCH_2CH_3$, the groups have these priority numbers: 1, H; 2, CH_3; 3, CH_2CH_3; and 4, Cl. If you cannot make a decision at the second atom, go to the third, and so forth. Thus, in

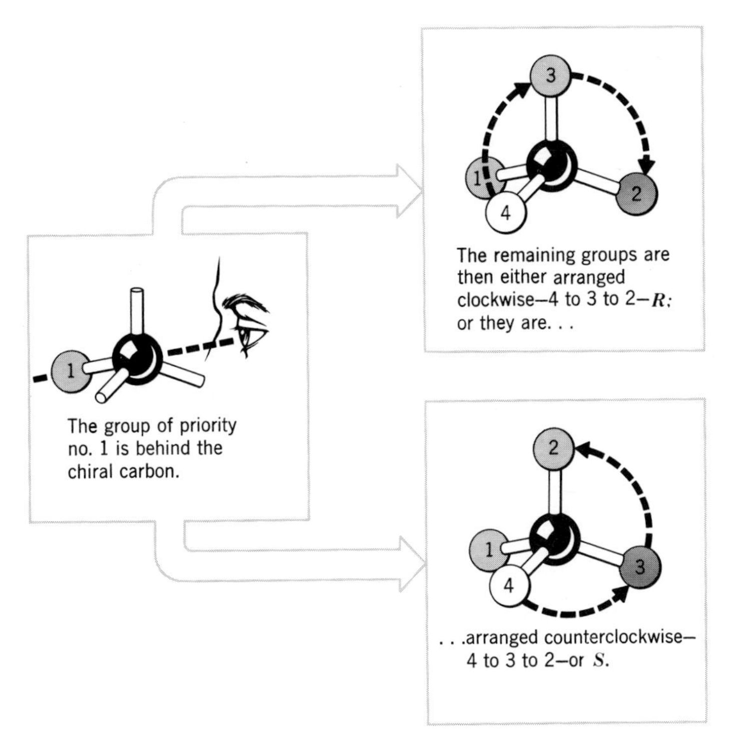

Figure A-1 The *R/S* system of configurational families.

3-methylhexane, $CH_3CH_2CH_2—\underset{\displaystyle CH_3}{\underset{|}{CH}}—CH_2CH_3$, groups at the chiral carbon have these priority numbers: 1, H; 2, CH_3; 3, CH_2CH_3; and 4, $CH_3CH_2CH_2$.

3. If you come to a second (or higher) atom with a double bond or a triple bond, rewrite the bond at that point to "double" or "triple" the atom at the other end. For example, imagine that $—CH{=}CH_2$ is, instead, $—CH\begin{matrix} \diagup CH_2—\overset{|}{\underset{|}{C}}— \\ | \\ \diagdown \overset{|}{\underset{|}{C}}— \end{matrix}$. The ketone group, $\rangle C{=}O$, is imagined to be $—\overset{|}{\underset{|}{C}}—O—\overset{|}{\underset{|}{C}}—$ (with $O—C\langle$ bonded below). Thus in glyceraldehyde, $CH_3—\underset{\displaystyle OH}{\underset{|}{CH}}—\overset{\displaystyle O}{\overset{\|}{C}}—H$, the priority numbers for the groups at the chiral carbon are: 1, H; 2, $CH_3—$; 3, $—\overset{\displaystyle O}{\overset{\|}{C}}H$; and 4, —OH.

In a molecule having two or more chiral centers, each center is handled individually and separately. Each chiral center is either in the *R* or the *S* family.

TABLE A-1
Sequence Rule Priorities—The *R/S* System

Priority No.	Group	Priority No.	Group	Priority No.	Group	Priority No.	Group
1	Hydrogen	11	Benzyl	21	Methylamino	31	Phenoxy
2	Methyl	12	Isopropyl	22	Ethylamino	32	Acetoxy
3	Ethyl	13	Vinyl	23	Phenylamino	33	Benzoyloxy
4	*n*-Propyl	14	*sec*-Butyl	24	Acetylamino	34	Fluoro
5	*n*-Butyl	15	*t*-Butyl	25	Dimethylamino	35	Sulfhydryl (HS—)
6	*n*-Pentyl	16	Phenyl	26	Nitro	36	Sulfo (HO_3S—)
7	Isopentyl	17	*p*-Tolyl	27	Hydroxy	37	Chloro
8	Isobutyl	18	Acetyl	28	Methoxy	38	Bromo
9	Allyl	19	Benzoyl	29	Ethoxy	39	Iodo
10	Neopentyl	20	Amino	30	Benzyloxy		

Source: From *J. Org. Chem.* Vol. 35, 1970; page 2866. This is a partial list of the groups listed in this paper.)

These are the principal rules. They will not handle every problem, but those that the rules do not cover are few. (The complete set of rules is in the original article: *Experientia,* Vol. 12, 1956, page 86.) Table A-1 is a list of several groups arranged in their correct priority sequence.

If you have trouble visualizing perspective drawings, and you are shown one with the atom or group of lowest priority number not in the rearward location where you would like it to be, there is a very simple solution. If you interchange any two groups at a chiral carbon, you change its configuration from *R* to *S* or from *S* to *R*. Each time you change any two groups, you make this simple change from one family to the other (see Figure A-2).

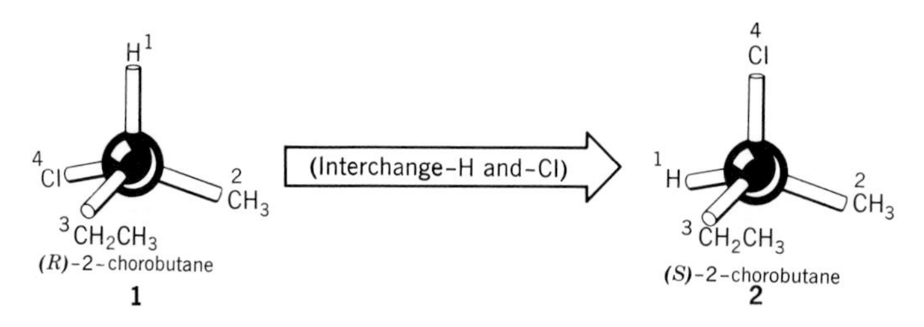

Figure A-2

In **2** the group of priority number 1 is behind the chiral carbon, and it is easy to see that the configuration of **2** is *S*. Because we developed **2** from **1** by exchange of just two groups, **1** must be *R*.

Examples of structures that have been correctly assigned to configurational families are shown in Figure A-3.

The *E/Z* System for Geometric Isomers

The priority rules for *E/S* configurational families are also used for describing *cis/trans* relations. The advantage is the complete lack of ambiguity. The procedure is first to consider separately the two sets of groups, one at one

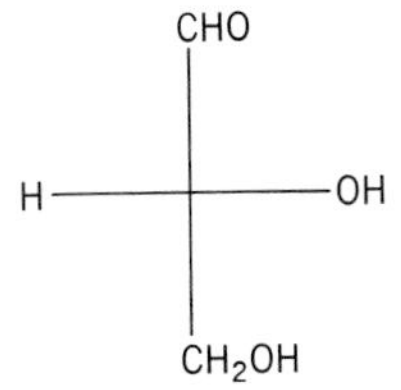

CHO

H — OH

CH_2OH

D-(+)-Glyceraldehyde
(plane projection)

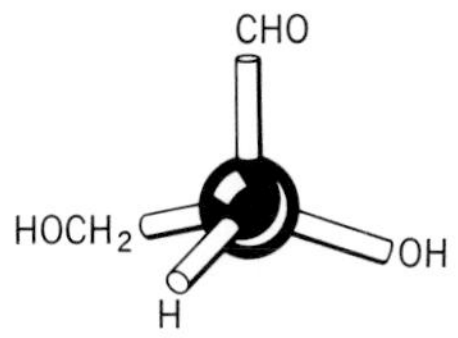

≡

CHO

$HOCH_2$ OH

H

R-(+)-Glyceraldehyde
(perspective)

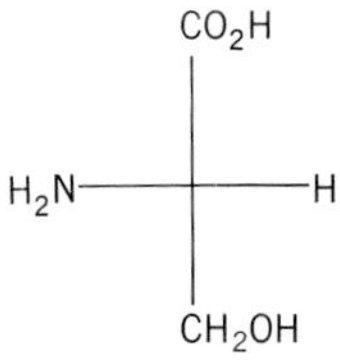

CO_2H

H_2N — H

CH_2OH

L-(−)-Serine
(plane-projection)

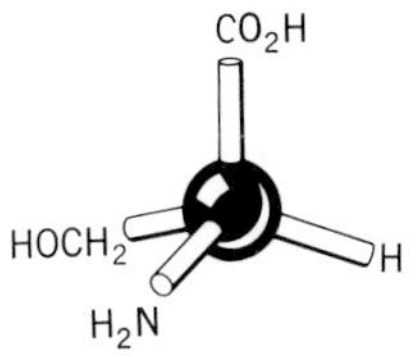

≡

CO_2H

$HOCH_2$ H

H_2N

S-(−)-Serine
(perspective)

Figure A-3

end of the double bond, the other at the opposite end. Each set has two groups.

The isomer whose higher priority groups are on the same side of the double bond (formerly, in a *cis* relation) is called the *Z*-isomer. The isomer in which the higher priority groups are on the opposite side of the double bond (formerly, the *trans* relation) is called the *E*-isomer. ("*Z*" is from the German, *zusammen,* or "together"; "*E*" is from the German, *entgegen,* or "opposite".) These examples illustrate how the system works. (The higher priority groups of each set are marked by an asterisk.) In none of these examples could you

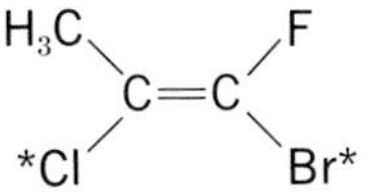

H_3C, F, *Cl, Br* — C=C

(Z)-1-Fluoro-1-bromo-2-chloropropene

H_3C, Br*, *Cl, F — C=C

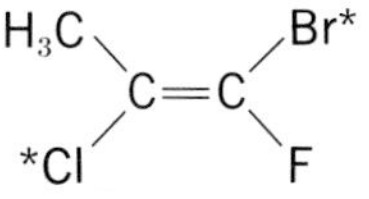

(E)-1-Fluoro-1-bromo-2-chloropropene

H_3C, $*CH_3CH_2$ — C=C — $CH(*CH_3)CH_2CH_3$, $CH_2CH(CH_3)CH_3$

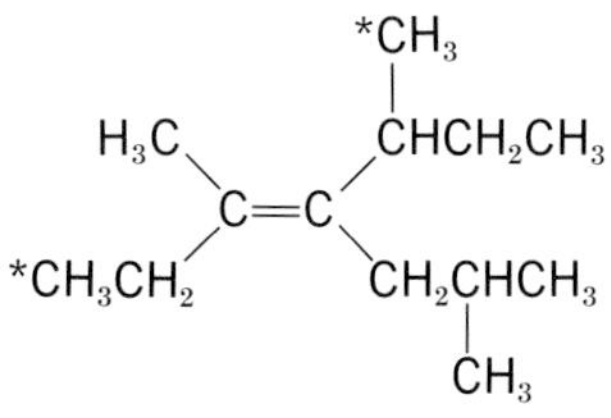

(E)-3,5-Dimethyl-4-isobutyl-3-heptene

use either *cis* or *trans* in the name and have it mean anything without adding a statement that specified which groups you personally have picked to be *cis* or *trans* to each other.

Answers to In-Chapter Questions and Exercises

Chapter 1

1.1 (a) $CH_3—CH_3$ or CH_3CH_3

(b) $CH_3—\underset{CH_3}{\underset{|}{CH}}—CH_3$ or $CH_3\underset{CH_3}{\underset{|}{CH}}CH_3$ or $(CH_3)_2CHCH_3$

(c) $CH_3—\underset{CH_3}{\underset{|}{\overset{CH_3}{\overset{|}{C}}}}—\overset{CH_3}{\overset{|}{CH}}—\overset{CH_3}{\overset{|}{CH}}—CH_3$ or $(CH_3)_3C\overset{CH_3}{\overset{|}{CH}}CH(CH_3)_2$

1.2 (a)

```
    H   H   H
    |   |   |
H — C — C — C — H
    |   |   |
    H   H   H
```

(b)

```
    H   H   H   H   H
    |   |   |   |   |
H — C — C — C = C — C — H
    |   |           |
    H   |           H
    H — C — H
        |
        H
```

(c)

```
              H
              |
    H H   —   C — H H     H     H     H
    |         |     |     |     |     |
H — C ——————— C ——— C ——— C ——— C ——— C — H
    |         |     |     |     |     |
    H H   —   C — H H     H H — C — H H
              |                 |
              H                 H
```

Chapter 2

2.1 (a) $Br—CH_2\underset{NO_2}{\underset{|}{CH}}CH_2CH_2CH_3$

(b) $CH_3\underset{CH_3}{\underset{|}{\overset{CH_3}{\overset{|}{C}}}}—\underset{CH_3}{\underset{|}{\overset{CH_3}{\overset{|}{C}}}}—\underset{CH_3}{\underset{|}{\overset{CH_3}{\overset{|}{C}}}}—\underset{CH(CH_3)_2}{\underset{|}{CH}}CH_2CH_2CH_3$

(c) $CH_3\underset{I}{\underset{|}{\overset{I}{\overset{|}{C}}}}—\underset{CH_3}{\underset{|}{CH}}—\underset{CH(CH_3)_2}{\underset{|}{CH}}—\underset{C(CH_3)_3}{\underset{|}{\overset{CH_3CHCH_2CH_3}{\overset{|}{CH}}}}—CHCH_2CH_2CH_3$

(d) $Br—\underset{Cl}{\underset{|}{CH}}—\overset{CH_3}{\overset{|}{CH}}—CH_3$

(e) 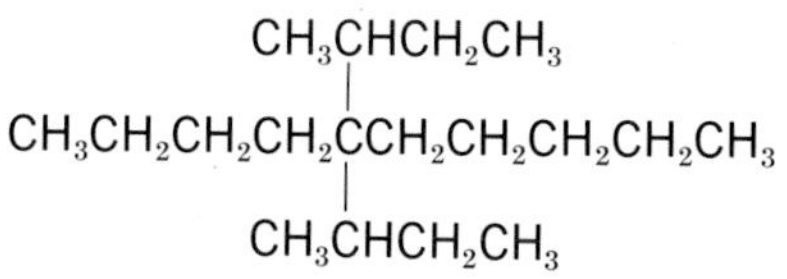
$$
\begin{array}{c}
CH_3CHCH_2CH_3 \\
| \\
CH_3CH_2CH_2CH_2CCH_2CH_2CH_2CH_2CH_3 \\
| \\
CH_3CHCH_2CH_3
\end{array}
$$

2.2 (a) 3-Methylhexane
(b) 2,3-Dimethyl-4-*t*-butylheptane
(c) 2,4-Dimethyl-5-*sec*-butylnonane
(d) 1-Bromo-3-chloro-2-iodopropane (This follows the alphabetical order; other orders and the other sequence of numbers should be accepted.)
(e) 4-*t*-Butyl-5-isopropyloctane or 4-isopropyl-5-*t*-butyloctane
(f) 1,3-Dinitro-2,2-dimethylpropane
(g) 1-Chloro-2,3-dimethylpentane
(h) 2,6-Dimethyl-3-ethylheptane
(i) 2-Methylpentane
(j) 5-Methyl-4-ethylnonane

2.3
$$
\begin{array}{l}
\qquad\qquad\quad \underline{CH_3} \quad\; \underline{CH_3} \qquad\quad \underline{CH_3} \\
\underline{CH_3}-CH_2-(CH)-(CH)-(CH)-CH_2-(CH)-\underline{CH_3} \\
\qquad\qquad\qquad\quad CH_2-CH_2-CH_2-\underline{CH_3}
\end{array}
$$

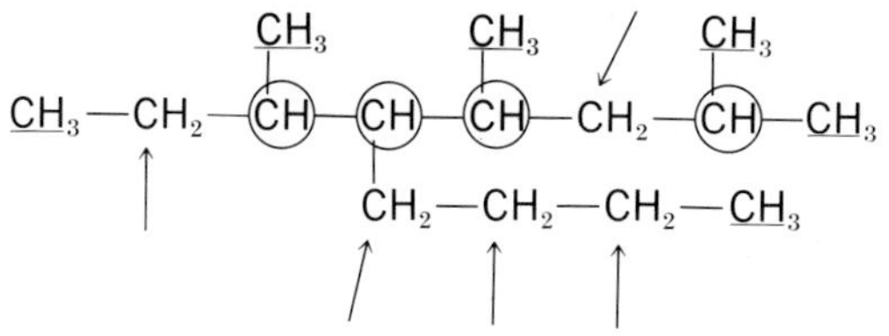

2.4 (a) $C_5H_{12} + 8O_2 \rightarrow 5CO_2 + 6H_2O$
(b) $2H{-}C{\equiv}C{-}H + 5O_2 \rightarrow 4CO_2 + 2H_2O$

2.5 (a) Two: 1-chlorobutane (*n*-butyl chloride)
2-chlorobutane (*sec*-butyl chloride)
(b) Two: 1-chloro-2-methylpropane (isobutyl chloride)
2-chloro-2-methylpropane (*t*-butyl chloride)

Chapter 3

3.1 (At this stage, only common names are expected.)
(a) *trans*-β-butylene
(b) α-butylene
(c) isobutylene
(d) *cis*-β-butylene

3.2 (a) Not possible
(b) Possible
(c) Not possible
(d) Possible
(e) Not possible

3.3 (a) $CH_3CH{=}CHCH(CH_3)_2$

(b) 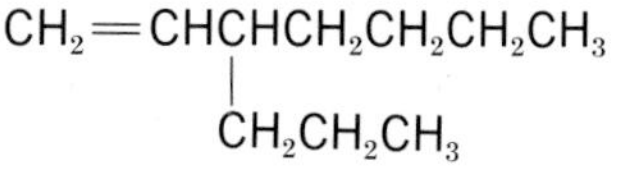
$$
\begin{array}{l}
CH_2{=}CHCHCH_2CH_2CH_2CH_3 \\
\qquad\quad\;\; | \\
\qquad\quad CH_2CH_2CH_3
\end{array}
$$

(c) 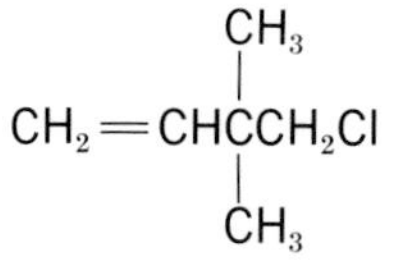
$$
\begin{array}{l}
\qquad\quad\;\; CH_3 \\
\qquad\qquad | \\
CH_2{=}CHCCH_2Cl \\
\qquad\qquad | \\
\qquad\quad\;\; CH_3
\end{array}
$$

(d) $(CH_3)_2C{=}C(CH_3)_2$

3.4 (a) 2-Methylpropene (isobutylene)
(b) 3,6-Dimethyl-4-isobutyl-3-heptene
(c) 1-Chloro-1-propene
(d) 3-Bromo-1-propene (allyl bromide)
(e) 4-Methyl-1-hexene

(f) [cyclohexene structure: H_2C–CH_2–CH=CH–CH_2–H_2C ring] [cyclopentene structure]

(g) 3,4-Dimethylcyclohexene

3.5 (a) $CH_3CH(Cl)CH_2CH_3$ (b) $(CH_3)_3CBr$

(c) $CH_3CH_2C(CH_3)(OH)$–cyclohexyl (d) 3-methylcyclohexanol + 4-methylcyclohexanol (H_3C–C_6H_{10}–OH + H_3C–C_6H_{10}–OH)

3.6 (a) $CH_3CH{=}CHCH_3$ or $CH_2{=}CHCH_2CH_3$

(b) Cannot be made this way.

(c) [cyclopentene structure]

(d) $CH_3C(CH_3){=}CHCH_3$ or $CH_2{=}C(CH_3)CH_2CH_3$

(e) Cannot be made this way.

3.7 (a) (circled) $CH_3CH_2\overset{+}{C}HCH_3$ $CH_3CH_2CH_2CH_2^+$ $CH_3CH_2CH(Cl)CH_3$

(b) (circled) $(CH_3)_3C^+$ $(CH_3)_2CHCH_2^+$ $(CH_3)_3CCl$

(c) (circled) 1-methylcyclohexyl cation (CH_3, + on ring carbon bearing CH_3); 2-methylcyclohexyl cation (CH_3, + on adjacent carbon); 1-chloro-1-methylcyclohexane (H_3C, Cl)

(d) $CH_3CH_2\overset{+}{C}HCH_3$ (only possible carbonium ion) $CH_3CH_2CH(Cl)CH_3$

(e) $CH_3\overset{+}{C}HCH_2CH_2CH_3$ $CH_3CH_2\overset{+}{C}HCH_2CH_3$ (both form)

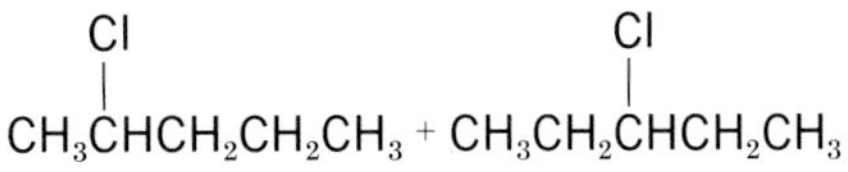

$$CH_3\underset{Cl}{C}HCH_2CH_2CH_3 + CH_3CH_2\underset{Cl}{C}HCH_2CH_3$$

(f) 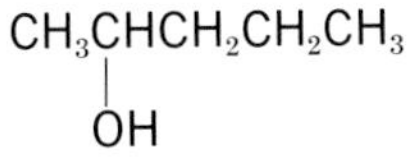 $CH_3\underset{OH}{C}HCH_2CH_2CH_3$ (2-pentanol, from $CH_3\overset{+}{C}HCH_2CH_2CH_3$)

and

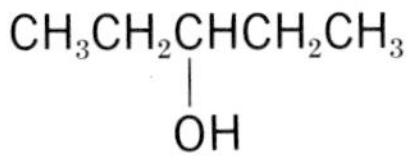

$CH_3CH_2\underset{OH}{C}HCH_2CH_3$ (3-pentanol, from $CH_3CH_2\overset{+}{C}HCH_2CH_3$)

From 2-pentene the two possible carbonium ions are both secondary and they are therefore of comparable stabilities. Hence both may be expected to form. From propylene the two possible carbonium ions are in different classes; the secondary carbonium ion, being more stable than the primary, will be the one that forms.

Chapter 4

4.1 (a) Ether (b) Alcohol, monohydric, 2° (Not a phenol; the —OH is attached to a saturated carbon.) (c) None of these (Not an alcohol; the —OH is not attached to a saturated carbon.) (d) Alcohol, monohydric, 3° (e) Alcohol, dihydric (f) None of these (Not an ether; one carbon attached to oxygen also has a carbon-oxygen double bond. The structure is of an ester.) (g) Phenol (h) Alcohol, monohydric, 1° (Not a phenol; the —OH is not attached directly to the benzene ring but to a saturated carbon.)

4.2 (a) 4-Methyl-1-pentanol (b) 2-Methyl-2-propanol
(c) 2-Methyl-2-ethyl-1-pentanol (or, 2-ethyl-2-methyl-1-pentanol)
(d) 3-Bromo-1-propanol

4.3 1,2,3-Propanetriol

4.4 4-Methyl-1,3-cyclohexanediol. (6-Methyl-1,3-cyclohexanediol would be incorrect because after giving precedence to the locations of the hydroxyl groups, numbering around the ring must be in that direction giving the methyl group the lower number.)

4.5 (a) Methyl ethyl ether (or ethyl methyl ether)
(b) di-*n*-Propyl ether
(Parts of each name are written as separate words. "Di-*n*-propyl" is hyphenated to make that part one word as in "dimethyl" or "diethyl.")

4.6 Hydrogen bonding is stronger in propylene glycol because its molecules each have two —OH groups, whereas the molecules of *n*-butyl alcohol have only one. The boiling point of the glycol is 189 °C, whereas that of the alcohol is only 117 °C.)

4.7 (a) No reaction

(b) 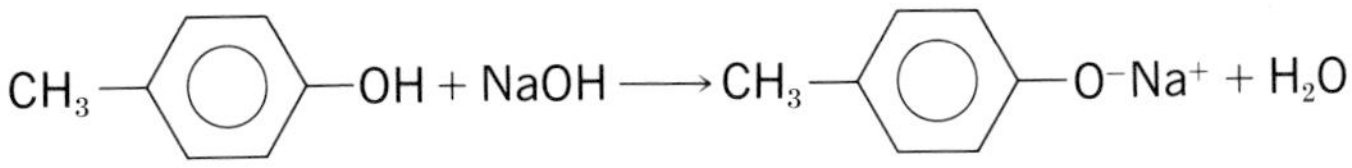 $CH_3-C_6H_4-OH + NaOH \longrightarrow CH_3-C_6H_4-O^-Na^+ + H_2O$

4.8 (a) 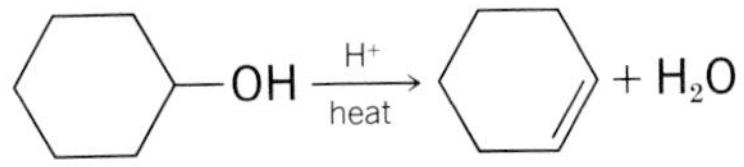 $C_6H_{11}-OH \xrightarrow[heat]{H^+}$ cyclohexene $+ H_2O$

(b) 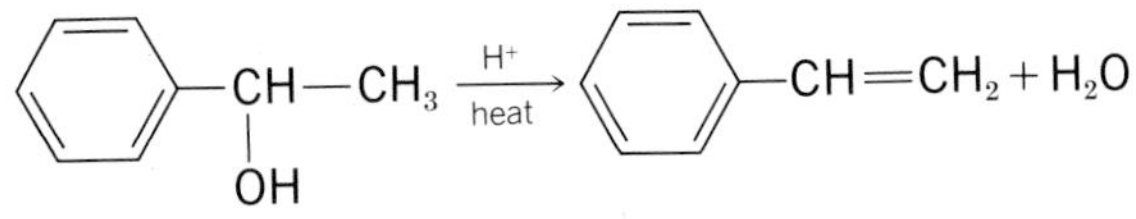 $C_6H_5-\underset{OH}{CH}-CH_3 \xrightarrow[heat]{H^+} C_6H_5-CH{=}CH_2 + H_2O$

4.9 (a) $CH_3CH_2CH_2OH$ (b) C_6H_5—CH_2—OH

4.10 (a) H—C(=O)—H (b) $CH_3C(=O)$—H

(c) cyclopentanone (ring C=O) (d) No reaction (a 3° alcohol)

Chapter 5

5.1 (a) CH_3CH_2OH ethyl alcohol
(b) CH_3CH_2SH ethyl mercaptan
(c) $CH_3CH_2NH_3^+$ ethylammonium ion, before loss of proton
$CH_3CH_2NH_2$ ethylamine, after loss of proton

5.2 Cl—C_6H_4—C(=CCl_2)—C_6H_4—Cl
DDE

5.3 (a) CH_3CH_2—S—S—CH_2CH_3

(b) C_6H_{11}—S—S—C_6H_{11}

5.4 (a) $2CH_3CH_2CH_2SH$ (b) $CH_3SH + CH_3CH_2SH$

5.5 (a) $(CH_3CH_2)_3N$ (b) $CH_3CH_2N(CH_3)_2$

(c) $(CH_3)_3C$—N($CH_2CH_2CH_2CH_3$)—$CH(CH_3)CH_2CH_3$ (d) NO_2—C_6H_4—NH_2

(e) NH_2—C_6H_4—OH

(f) C_6H_{11}—NH_2

5.6 (a) di-*n*-Butylethylamine
(b) 1,2-Diaminoethane
(c) *N,N*-Diethylaniline
(d) Cyclopentylamine

5.7 (a) $CH_3NH_2 + HCl_{aq} \rightarrow CH_3NH_3^+Cl^-$
(b) $CH_3NHCH_2CH_3 + HCl_{aq} \rightarrow CH_3N^+H_2CH_2CH_3Cl^-$
(c) $(CH_3)_3N + HCl_{aq} \rightarrow (CH_3)_3NH^+Cl^-$

5.8 (a) $CH_3CH_2NH_3^+Cl^- + NaOH_{aq} \rightarrow CH_3CH_2NH_2 + H_2O + NaCl$
(b) $(CH_3)_2NH_2^+Br^- + NaOH_{aq} \rightarrow (CH_3)_2NH + H_2O + NaBr$

Chapter 6

6.1 (a) $CH_3-CH(Cl)-C(=O)-H$ (b) $CH_3-CH(CH_3)-C(=O)-CH(CH_3)-CH_3$

(c) $HO-CH_2-CH(CH_3)-C(=O)-H$ (d) $C_6H_5-C(=O)-CH_2-CH_3$

6.2 (a) Isobutyraldehyde. ("α-Methylpropionaldehyde" fulfills the rules, too, but the shorter, simpler name is preferred.)
(b) Methyl isopropyl ketone. ("Isopropyl methyl ketone," of course, is also acceptable. Note, however, that the parts of the common names of ketones are not run together but are kept apart as separate words.)
(c) γ-Chlorobutyraldehyde
(d) α-Hydroxypropionic acid
(e) β-Bromo-α-methylbutyraldehyde. ("α-Methyl-β-bromobutyraldehyde would also fit the rules, but parts of a name that relate to alkyl groups or the carbon chain are preferably grouped close together. This compound, in other words, is a bromine derivative of α-methylbutyraldehyde and the preferred name reflects that better than the alternate name.)
(f) di-*t*-Butyl ketone

6.3 (a) 2-Methylpropanal (b) 3-Methyl-2-butanone
(c) 4-Chlorobutanal (d) 2-hydroxypropanoic acid
(e) 3-Bromo-2-methylbutanal (f) 2,2,4,4-Tetramethyl-3-pentanone

6.4 (a) 2-methyl-3-pentanone is the correct name for

$CH_3CH_2C(=O)CH(CH_3)CH_3$

In this case the carbonyl group falls at position-3, no matter from which end the chain is numbered. Hence one picks the direction of numbering to give the substituent the lower number.
(b) A substituent in the 1-position of an aldehyde is impossible. Otherwise, the compound could not be an aldehyde.
(c) This name mixes the two systems of nomenclature. The IUPAC system requires the use of numbers, not Greek letters.
(d) This name also mixes the two systems. A common name for the "parent" system in the structure ("acetaldehyde") requires the use of Greek letters to designate positions.

6.5 (a) $CH_3CH(OH)OCH_2CH_3$ (b) $CH_3CH_2CH(OH)OCH_3$ (c) $(CH_3)_2CHCH(OH)OCH_3$

6.6 (a) $CH_3CH(OCH_2CH_3)_2$ — CH_3CH bearing two OCH_2CH_3 groups (b) $CH_3CH_2CH(OCH_3)_2$ — CH_3CH_2CH bearing two OCH_3 groups (c) $(CH_3)_2CHCH(OCH_3)_2$ — $(CH_3)_2CHCH$ bearing two OCH_3 groups

6.7 (a) $CH_2O + 2CH_3OH$ (b) $CH_3CHO + 2CH_3CH_2OH$

(c) $CH_3CHO + 2CH_3OH$ (d) Not an acetal

6.8 (a) $CH_3CH_2CH(OH)CH(CH_3)CH{=}O$ (b) $CH_3CH_2CH_2CH(OH)CH(CH_2CH_3)CH{=}O$ (c) $C_6H_5CH_2CH(OH)CH(C_6H_5)CH{=}O$

Chapter 7

7.1 (a) $CH_3CH_2CO_2^-Na^+$ (b) $CH_3{-}O{-}C_6H_4{-}CO_2^-Na^+$ (para-substituted benzene ring)

(c) $CH_3CH{=}CHCO_2^-Na^+$ (Neither ether groups, as in (b), nor alkene double bonds, as in (c) react with aqueous sodium hydroxide at room temperature.)

7.2

Carboxylic acids	methanoic acid	ethanoic acid	propanoic acid	butanoic acid
Carboxylic acid salts	sodium methanoate	sodium ethanoate	potassium propanoate	ammonium butanoate
Acid chlorides	—	ethanoyl chloride	propanoyl chloride	butanoyl chloride
Anhydrides	—	ethanoic anhydride	propanoic anhydride	butanoic anhydride
Esters	ethyl methanoate	isopropyl ethanoate	ethyl propanoate	methyl butanoate
Amides	methanamide	ethanamide	propanamide	butanamide

Note: The trickiest names are the common and IUPAC names for derivatives of the C_3 acid—propionic acid (common) and propanoic acid (IUPAC).

7.3 (a) $CH_3{-}O{-}C_6H_4{-}CO_2H$ (para-substituted benzene ring) (b) $CH_3CH_2CO_2H$ (c) $2CH_3CH{=}CHCO_2H$

7.4 (a) $CH_3CO_2CH_3$, methyl acetate
(b) $C_6H_5CO_2CH(CH_3)_2$, isopropyl benzoate
(c) $(CH_3)_2CHCO_2CH_2CH_2CH_3$, propyl isobutyrate (or *n*-propyl isobutyrate)
(d) $CH_3CH_2CO_2CH_2CH_2CH_2CH_3$, butyl propionate (or *n*-butyl propionate)

7.5 (a) $CH_3CON(CH_3)_2$ *N,N*-dimethylacetamide
(b) $CH_3CH_2CH_2CON(CH_2CH_3)_2$, *N,N*-diethylbutyramide
(c) $C_6H_5CONHC_5H_6$, *N*-phenylbenzamide (actually, more commonly it is called benzanilide)
(d) No amide can form

7.6 (a) $CH_3CH_2CO_2H + CH_3OH$
(b) $CH_3CH_2OH + CH_3CO_2H$
(c) $C_6H_5CO_2H + (CH_3)_2CHOH$

(d) $CH_3(CH_2)_{14}CO_2H + CH_3(CH_2)_{12}CO_2H + CH_3(CH_2)_{16}CO_2H +$
$HOCH_2CHCH_2OH$ (glycerol) (with OH on the central CH)

(e) $CH_3CH_2CH_2OH + (CH_3)_2CHCO_2H$

7.7 (a) $CH_3CH_2CO_2^-Na^+ + CH_3OH$
(b) $CH_3CH_2OH + CH_3CO_2^-Na^+$
(c) $C_6H_5CO_2^-Na^+ + (CH_3)_2CHOH$
(d) $CH_3(CH_2)_{14}CO_2^-Na^+ + CH_3(CH_2)_{12}CO_2^-Na^+ + CH_3(CH_2)_{16}CO_2^-Na^+$
$+ HOCH_2CHCH_2OH$ (with OH on the central CH)

(e) $CH_3CH_2CH_2OH + (CH_3)_2CHCO_2^-Na^+$

7.8 (a) $CH_3CO_2H + NH_3$
(b) $CH_3NH_2 + HO_2CCH_2CH_3$
(c) $C_6H_5NH_2 + HO_2CCH_3$
(d) $CH_3CH_2CO_2H + (CH_3)_2NH$
(e) $NH_2CH_2CO_2H + NH_2CHCO_2H$ (with CH_3 on the CH)

7.9 (a) Cannot decarboxylate; not a β-keto acid
(b) $CH_3CH_2CO_2H$ (from a β-dicarboxylic acid)
(c) cyclohexanone (ring=O) (from a β-keto acid)

7.10 (a) $CH_3-C(=O)-CH_2-CO_2C_2H_5$ (b) $CH_3CH_2CH_2-C(=O)-CH(CH_2CH_3)-CO_2C_2H_5$

Chapter 8

8.1 $-CH_2-CH(CO_2H)-CH_2-CH(CO_2H)-CH_2-CH(CO_2H)-CH_2-CH(CO_2H)-$

Polyacrylic acid

8.2 Since ethylene glycol is a diol and polyacrylic acid has several carboxyl groups per molecule, and since acids and alcohols can form esters, we might envision cross-linking as follows (wherein not all possible sites for cross-linking are used):

$-CH_2-CH-CH_2-CH-CH_2-CH-CH_2-CH-$ (top chain: first CH bears C=O, second bears CO_2H, third bears C=O, fourth bears CO_2H)

Each C=O of the top chain is linked through $-O-CH_2-CH_2-O-$ (Ethylene glycol units) to a C=O on the bottom chain.

$-CH_2-CH-CH_2-CH-CH_2-CH-CH_2-CH-$ (bottom chain: first CH bears C=O, second bears CO_2H, third bears C=O, fourth bears CO_2H)

Chapter 9

9.1 (a) a and c are chain isomers
(b) b and d are position isomers
(c) a and b; a and d; a and e; b and c; b and e; c and d; c and e; and d and e are functional group isomers.

9.2 (a) CH_3CHCH_3 (with OH on C-2) (None) (b) $CH_3\overset{*}{C}HCOOH$ (with NH_2 on the starred C) (c) $HOOCCH_2CH_2\overset{*}{C}HCOOH$ (with NH_2 on the starred C)

(d) CH_2OCH_3—$\overset{*}{C}H$(—OH)—CH_2OH (e) $CH_3CH\overset{*}{C}HCH_3$ (with CH_3 on C-2 and OH on the starred C)

9.3 (a) Four. The carbons marked by the asterisks are chiral:

$$HO{-}CH_2{-}\overset{*}{C}H(OH){-}\overset{*}{C}H(OH){-}\overset{*}{C}H(OH){-}\overset{*}{C}H(OH){-}C({=}O){-}H$$

(b) All four are different. Each has four different groups; in comparing each set of four groups, at least one difference can be found for each set.
(c) 16. $n = 4$; $2^4 = 16$ (The equation applies *only* when all chiral carbons are different.)

9.4 (+)-mandelic acid. When $\alpha = +7.77°$ (observed rotation); $c = 2.50$ g/100 ml; $l =$ 2 dm (20 cm), then

$$[\alpha]^{20} = \frac{100 \times (+7.77)}{2.5 \times 2} = +155.4°, \text{ the value for (+) mandelic acid.}$$

9.5 77 g/100 cc. $[\alpha]^{20} = -5.41$ for (−)-asparagine, from Table 19.2.

$$-5.41 = \frac{100 \times (-8.3)}{c \times 2};\ c = \frac{100 \times (-8.3)}{(-5.41) \times 2} = 76.7 \text{ g/100 ml which}$$

rounds to 77 g/100 ml.

9.6 The compounds represent two sets of optical isomers. In the first set are structures 1, 3, 4 and 6; in the second: 2, 5, and 7. None in the first set can be a stereoisomer in general or an optical isomer in particular to any in the second set. Those in the first set are various stereoisomers of 1,2-cyclohexanediol; those in the second are stereoisomers of 1,3-cyclohexanediol. Any one of the first set is therefore only a position isomer of any in the second set.
(1) Only 3 and 6 are related as enantiomers—as object and mirror image that cannot be superimposed.
(2) 1 is identical with 4
2 is identical with 5
(3) 1 [or 4] is a meso compound; 2 [or 5] is a meso compound. The planes of symmetry are at right angles to the dotted lines in their structures:

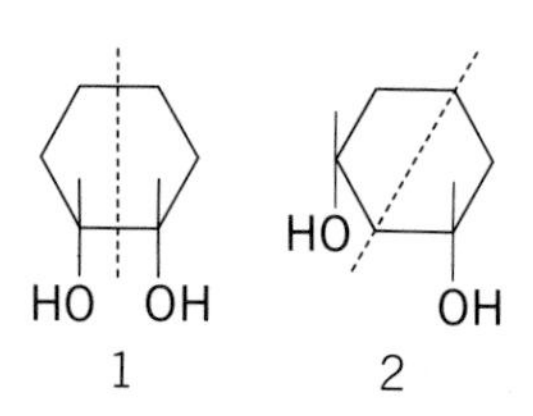

(4) 1 [or 4] and 3; 1 [or 4] and 6; 2 [or 5] and 7 are related as diastereomers. Members of each pair belong to the same set of optical isomers—we may not mix the first set with the second—but are not related as object to mirror image. A meso compound may be a diastereomer.

(5) A 50:50 mixture of 3 and 6 (compare part 1) would constitute a racemic mixture.

Chapter 10

10.1 Structures (a) and (b) are identical. Structure (a) can be slid over on top of (b) and they superimpose. Structures (c) and (e) are likewise identical. Turn one in the plane of the paper and slide it over on top of the other and they superimpose.

Structure (a) [or (b)] is an enantiomer of (d). The dashed line below represents a mirror between the two and suggests how the two are related as object to mirror image. Since the carboxyl group is not a chiral center, we may write it in

$$\begin{array}{ccc|ccc}
 & CO_2H & & & CO_2H & \\
H & — & OH & HO & — & H \\
HO & — & H & H & — & OH \\
 & CO_2H & & & CO_2H & \\
 & \text{(a)} & & & \text{(d)} &
\end{array}
\qquad
\begin{array}{ccc}
 & CO_2H & \\
H & — & OH \\
\hline
H & — & OH \\
 & CO_2H & \\
 & \text{(c)} &
\end{array}$$

any direction when constructing plane projection structures. We need not, in other words, write them as: CO_2H, HO_2C.

Structure (c) [or (e)] is a meso compound. The horizontal line in the structure above shows where the plane of symmetry occurs. Moreover, we saw above that (e), the mirror image of (c) is still superimposable on (c), and that is the ultimate test of being identical.

10.2

$$\begin{array}{ccc}
 & CO_2H & \\
H & — & OH \\
 & CH_2OH & \\
\end{array}
\qquad
\begin{array}{ccc}
 & CO_2H & \\
HO & — & H \\
 & CH_2OH & \\
\end{array}$$

D-Glyceric acid L-Glyceric acid

Note again that the nonchiral CO_2H and CH_2OH groups may be written in any direction we choose.

Chapter 11

11.1 $CH_3(CH_2)_{24}CH_2—O—\overset{O}{\overset{\|}{C}}(CH_2)_{26}CH_3$

11.2 $CH_3(CH_2)_4(H)C{=}C(H)—CH_2—(H)C{=}C(H)(CH_2)_7CO_2H$ (both double bonds cis)

11.3

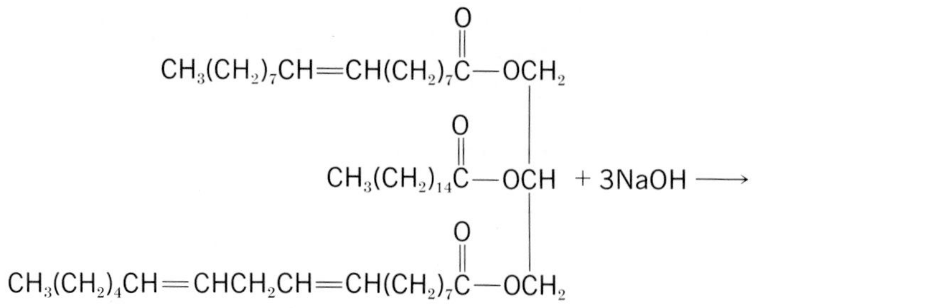

$$\begin{array}{l} CH_3(CH_2)_7CH{=}CH(CH_2)_7\overset{O}{\overset{\|}{C}}{-}OCH_2 \\ CH_3(CH_2)_{14}\overset{O}{\overset{\|}{C}}{-}OCH \quad + 3NaOH \longrightarrow \\ CH_3(CH_2)_4CH{=}CHCH_2CH{=}CH(CH_2)_7\overset{O}{\overset{\|}{C}}{-}OCH_2 \end{array}$$

1

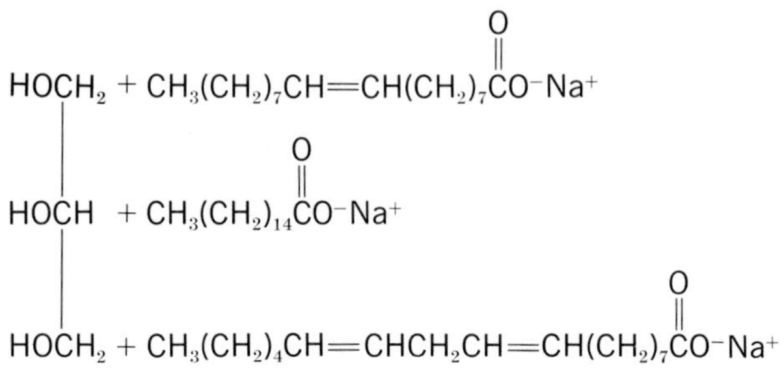

$$\begin{array}{l} HOCH_2 + CH_3(CH_2)_7CH{=}CH(CH_2)_7\overset{O}{\overset{\|}{C}}O^-Na^+ \\ HOCH \quad + CH_3(CH_2)_{14}\overset{O}{\overset{\|}{C}}O^-Na^+ \\ HOCH_2 + CH_3(CH_2)_4CH{=}CHCH_2CH{=}CH(CH_2)_7\overset{O}{\overset{\|}{C}}O^-Na^+ \end{array}$$

11.4

$$1 + 3H_2 \xrightarrow[\text{heat, pressure}]{\text{catalyst}} \begin{array}{l} CH_3(CH_2)_{17}\overset{O}{\overset{\|}{C}}{-}O{-}CH_2 \\ CH_3(CH_2)_{14}\overset{O}{\overset{\|}{C}}{-}O{-}CH \\ CH_3(CH_2)_{17}\overset{O}{\overset{\|}{C}}{-}O{-}CH_2 \end{array}$$

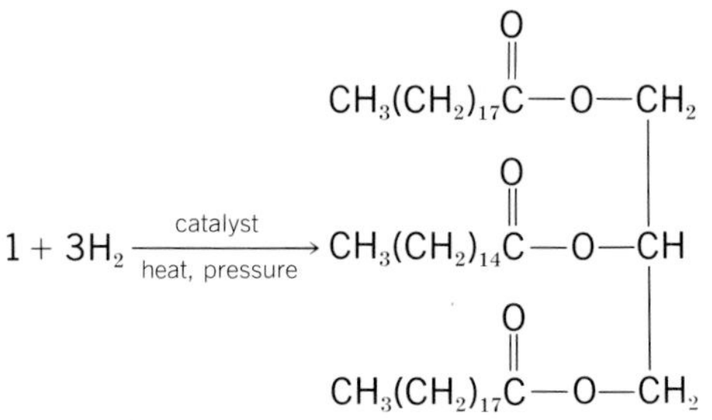

Chapter 12

12.1 Alanine exists as the dipolar ion in the solid state, a form creating strong forces of attraction between neighboring particles. The ethyl ester is an ordinary covalent substance, polar to be sure, but not one bearing full electrical charges.

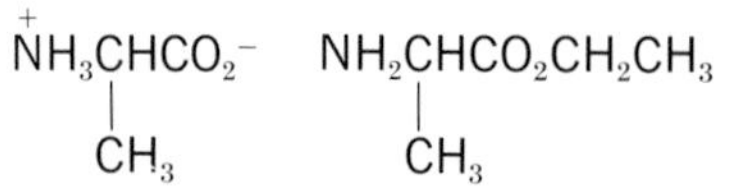

$\overset{+}{N}H_3CHCO_2^-$ (with CH_3 on the CH) — Alanine

$NH_2CHCO_2CH_2CH_3$ (with CH_3 on the CH) — Alanine, ethyl ester

12.2

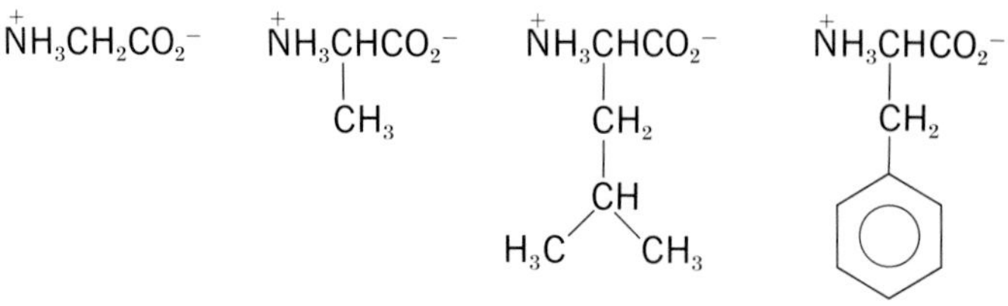

$\overset{+}{N}H_3CH_2CO_2^-$ — Glycine

$\overset{+}{N}H_3CHCO_2^-$ (with CH_3 on the CH) — Alanine

$\overset{+}{N}H_3CHCO_2^-$ (with $CH_2CH(CH_3)_2$: CH_2–CH, H_3C and CH_3) — Leucine

$\overset{+}{N}H_3CHCO_2^-$ (with CH_2–phenyl ring) — Phenylalanine

12.3 (a)

$$\overset{+}{N}H_3CHCO_2^- \quad | \quad CH_2 \quad | \quad CO_2^-$$

Aspartic acid

$$\overset{+}{N}H_3CHCO_2^- \quad | \quad CH_2 \quad | \quad CH_2 \quad | \quad CO_2^-$$

Glutamic acid

(b)

$$\overset{+}{N}H_3CHCO_2^- \quad | \quad CH_2 \quad | \quad CONH_2$$

Asparagine

$$\overset{+}{N}H_3CHCO_2^- \quad | \quad CH_2 \quad | \quad CH_2 \quad | \quad CONH_2$$

Glutamine

12.4

$$\overset{+}{N}H_3CHCO_2^- \quad | \quad CH_2CH_2CH_2CH_2NH_3^+$$

Lysine

12.5

$$CO_2^- \text{ (top)}, \quad \overset{+}{N}H_3 \text{ (left)}, \quad H \text{ (right)}, \quad CH_3 \text{ (bottom)}$$

12.6 (Using simplified structures, not dipolar structures),

$$NH_2CH(CH_3)\overset{O}{\overset{\|}{C}}—NHCH(CH_2C_6H_5)\overset{O}{\overset{\|}{C}}OH$$

Alanylphenylalanine

$$NH_2CH(CH_2C_6H_5)\overset{O}{\overset{\|}{C}}—NHCH(CH_3)\overset{O}{\overset{\|}{C}}OH$$

Phenylalanylalanine

12.7 In the same order: Ala · Phe and Phe · Ala

12.8 (a)

$$NH_2CH(CH_2CH_2CH_2CH_2NH_2)\overset{O}{\overset{\|}{C}}—NHCH(CH_2SH)\overset{O}{\overset{\|}{C}}OH$$

Lys · Cys

(b)

$$NH_2CH(CH_2CH_2CO_2H)\overset{O}{\overset{\|}{C}}—NHCH(CH_3)\overset{O}{\overset{\|}{C}}OH$$

Glu · Ala

(c)

$$NH_2CH(H)\overset{O}{\overset{\|}{C}}—NHCH(H)\overset{O}{\overset{\|}{C}}OH$$

$$\text{or,}\quad NH_2CH_2\overset{\overset{\displaystyle O}{\|}}{C}-NHCH_2\overset{\overset{\displaystyle O}{\|}}{C}OH$$

Gly · Gly

Chapter 13

13.1 (a) Sucrose (b) Glucose (c) A protein (d) A phosphate ester (e) An ester in general, but one probably between a carboxylic acid and an alcohol. Although the names of the enzymes given do not reveal the kind of reaction they catalyze on these substrates, usually the reaction is hydrolysis.

13.2 Because enzymes are mostly or wholly proteins, they become denatured and permanently inactivated by higher temperatures.

Chapter 14

14.1 (a) Proline (b) Arginine (c) Glutamic acid (d) Lysine

14.2 (a) Proline. (The codon is CCU.)
(b) Serine. (The codon is AGU.)
(c) Lysine. (The codon is AAG.)
(d) Leucine. (The codon is CUA.)
(Remember that *T* appears only in anticodons. It acts like *U*, which is paired to *A* in a codon.)

Chapter 15

15.1

Enzymes	Digestive Juice
(a) Pepsin	Gastric juice
Trypsin	Pancreatic juice
Chymotrypsin	
Elastin	
Carboxypeptidase	
Aminopeptidase	
(b) Intestinal lipase	Intestinal juice
Pancreatic lipase	Pancreatic juice
(Bile salts are needed to assist these enzymes.)	
(c) α-Amylase	Saliva and pancreatic juice
(d) Sucrase	Intestinal juice

15.2 Oxygenation of HHb releases H^+ that helps release CO_2 from $HbCO_2^-$ and HCO_3^-.

15.3 Waste CO_2 releases H^+ when it combines with HHb or with H_2O. The H^+ reduces the oxygen affinity of hemoglobin.

15.4 Extra H^+ as well as that released (as in the answer for 15.3) helps reverse the following equilibrium, sending to the left.

$$O_2 + HHb \rightleftharpoons HbO_2^- + H^+$$

15.5 (a) Metabolic acidosis. Hyperventilation helps drive out CO_2 that otherwise is making H_2CO_3 whose ionization releases H^+.
(b) Metabolic alkalosis. Hypoventilation helps retain CO_2, which can make a neutralizer of OH^-, namely H_2CO_3.

15.6 CO_2 is retained producing H_2CO_3, which ionizes to give H^+ and acidosis. The patient cannot help it. The respiratory center is functioning poorly and the alveoli are damaged.

15.7 CO_2 is driven out, reducing the level of both H_2CO_3 and H^+.

Chapter 16

16.1 3.74 kcal/gram glucose
9.38 kcal/gram palmitic acid

Solution: Formula weights: glucose ($C_6H_{12}O_6$) = 180
palmitic acid ($C_{16}H_{32}O_2$) = 256

For glucose: 180 g ≡ 673 kcal
therefore: 1 g ≡ $\frac{673}{180}$ kcal = 3.74 kcal

For palmitic acid: 256 g ≡ 2400 kcal
therefore: 1 g ≡ $\frac{2400}{256}$ kcal = 9.38 kcal

16.2 (a) Cannot go; ATP's phosphate-group transfer potential is too low to phosphorylate and make 1,3-diphosphoglycerate with a higher potential.
(b) Can go.

16.3 Pyruvate ion. Since it cannot be converted to other substances in the absence of the enzyme assembly, it will accumulate.

16.4 ADP. The reaction makes ATP from ADP and P_i; hence, if ADP levels are high the system will respond by lowering them and making ATP.

Chapter 17

17.1 (a) Alanine (b) Aspartic acid

17.2 36 ATP from the chemical energy of two lactates, using the citric acid cycle and the respiratory chain. At 6ATPs per glucose via gluconeogenesis, a maximum of six glucose molecules could be made from the energy in two lactates or three glucose molecules for one lactate.

Chapter 18

18.1 11 kg. $40 \text{ days} \times 2500 \frac{\text{kcal}}{\text{day}} = 100\ 000$ kcal of energy needed

$$100\ 000 \text{ kcal} \times 0.11 \frac{\text{g fat}}{\text{kcal}} = 11\ 000 \text{ g fat}$$
$$= 11 \text{ kg fat}$$

18.2 60 kg. $11 \text{ kg fat} \times \frac{0.6 \text{ g wet glycogen/kcal}}{0.11 \text{ g fat/kcal}} = 60 \text{ kg wet glycogen}$

18.3 39% $7.3 \frac{\text{kcal}}{\text{ATP}} \times 129 \frac{\text{ATP}}{\text{palmitic acid}} =$ 942 kcal required to make 129 ATP from palmitic acid

$$\frac{942 \text{ kcal needed}}{2400 \text{ kcal available}} \times 100 = 39\% \text{ efficiency}$$

Chapter 19

19.1 (a) $C_6H_5CH_2C(=O)CO_2H$ (b) $(CH_3)_2CHCH_2C(=O)CO_2H$ (c) $(CH_3)_2CHC(=O)CO_2H$

The answers are written as the free acids; they could just as correctly been written as their corresponding anions.

19.2 $CH_3-CH(OH)-CH(NH_3^+)CO_2^- \longrightarrow CH_3-CH{=}C(NH_2^+)-CO_2^- \longrightarrow CH_3-CH_2-C({=}NH)-CO_2^- \longrightarrow$

$CH_3CH_2C(=O)-CO_2H$ (or anion)

Index

Pages given in italics are to tables.

Families of Organic Compounds Important to the Study of Biochemistry

	Alkane	Alkene	Aromatic System	Alcohol	Ether	Mercaptan	Disulfide	Amine	Aldehyde	Ketone	Carboxylic Acid	Ester	Amide
Example	CH_3CH_3	$CH_2{=}CH_2$	(benzene ring)	CH_3CH_2OH	CH_3OCH_3	CH_3CH_2SH	$CH_3S{-}SCH_3$	CH_3NH_2	$CH_3\overset{O}{\overset{\|}{C}}H$	$CH_3\overset{O}{\overset{\|}{C}}CH_3$	$CH_3\overset{O}{\overset{\|}{C}}OH$	$CH_3\overset{O}{\overset{\|}{C}}OCH_3$	$CH_3\overset{O}{\overset{\|}{C}}NH_2$
Common Name	Ethane	Ethylene	Benzene	Ethyl alcohol	Dimethyl ether	Ethyl mercaptan	Dimethyl disulfide	Methyl amine	Acetal-dehyde	Acetone	Acetic acid	Methyl acetate	Acetamide
IUPAC Name	Ethane	Ethene	Benzene	Ethanol	Methoxy-methane	Ethanethiol	Dimethyl disulfide	Amino-methane	Ethanal	Propanone	Ethanoic acid	Methyl ethanoate	Methane-amide
General Formula	$R{-}H$	$RCH{=}CH_2$ $RCH{=}CHR$ $R_2C{=}CHR$ $R_2C{=}CR_2$	ArH	ROH	ROR	RSH	RSSR	RNH_2 R_2NH R_3N	$R\overset{O}{\overset{\|}{C}}H$	$R\overset{O}{\overset{\|}{C}}R$	$R\overset{O}{\overset{\|}{C}}OH$	$R\overset{O}{\overset{\|}{C}}OR$	$R\overset{O}{\overset{\|}{C}}NH_2$ $R\overset{O}{\overset{\|}{C}}NHR$ $R\overset{O}{\overset{\|}{C}}NR_2$
Functional Group	C—C and C—H bonds	$\rangle C{=}C\langle$	aromatic ring	$-\overset{\vert}{\underset{\vert}{C}}-\ddot{\underset{..}{O}}H$	$-\overset{\vert}{\underset{\vert}{C}}-\ddot{\underset{..}{O}}-\overset{\vert}{\underset{\vert}{C}}-$	$-\overset{\vert}{\underset{\vert}{C}}-\ddot{\underset{..}{S}}H$	$-\overset{\vert}{\underset{\vert}{C}}-\ddot{\underset{..}{S}}-\ddot{\underset{..}{S}}-\overset{\vert}{\underset{\vert}{C}}-$	$-\overset{\vert}{\underset{\vert}{C}}-\underset{\vert}{\ddot{N}}-$	$-\overset{:O:}{\overset{\|}{C}}-H$	$-\overset{\vert}{\underset{\vert}{C}}-\overset{:O:}{\overset{\|}{C}}-\overset{\vert}{\underset{\vert}{C}}-$	$-\overset{:O:}{\overset{\|}{C}}-\ddot{\underset{..}{O}}H$	$-\overset{:O:}{\overset{\|}{C}}-\ddot{\underset{..}{O}}-\overset{\vert}{\underset{\vert}{C}}-$	$-\overset{:O:}{\overset{\|}{C}}-\underset{\vert}{\ddot{N}}-$